Pond Treatment Technology

Integrated Environmental Technology Series

The *Integrated Environmental Technology Series* addresses key themes and issues in the field of environmental technology from a multidisciplinary and integrated perspective.

An integrated approach is potentially the most viable solution to the major pollution issues that face the globe in the 21st century.

World experts are brought together to contribute to each volume, presenting a comprehensive blend of fundamental principles and applied technologies for each topic. Current practices and the state-of-the-art are reviewed, new developments in analytics, science and biotechnology are presented and, crucially, the theme of each volume is presented in relation to adjacent scientific, social and economic fields to provide solutions from a truly integrated perspective.

The *Integrated Environmental Technology Series* will form an invaluable and definitive resource in this rapidly evolving discipline.

Series Editor

Dr Ir Piet Lens, The University of Wageningen, The Netherlands (piet.lens@wur.nl).

Published titles

Biofilms in Medicine, Industry and Environmental Biotechnology: *Characteristics, analysis and control*

Biofuels for Fuel Cells: *Renewable energy from biomass fermentation*

Decentralised Sanitation and Reuse: *Concepts, systems and implementation*

Environmental Technologies to Treat Sulfur Pollution: *Principles and engineering*

Pond Treatment Technology

Phosphorus in Environmental Technology: *Principles and applications*

Water Recycling and Resource Recovery in Industries: *Analysis, technologies and implementation*

Forthcoming title

Advanced Biological Treatment Processes for Industrial Wastewaters: *Principles and application*

www.iwapublishing.com

Pond Treatment Technology

Edited by Andy Shilton

Publishing
LONDON • SEATTLE

Published by IWA Publishing, Alliance House, 12 Caxton Street, London SW1H 0QS, UK
Telephone: +44 (0) 20 7654 5500; Fax: +44 (0) 20 7654 5555; Email: publications@iwap.co.uk
Web: www.iwapublishing.com

First published 2005
Reprinted 2006, 2008, 2012

British Library Cataloguing in Publication Data
A CIP catalogue record for this book is available from the British Library

Library of Congress Cataloging- in-Publication Data
A catalog record for this book is available from the Library of Congress

ISBN 1843390205
ISBN13: 9781843390206

Contents

Foreword

Pond treatment technology is used in tens of thousands of applications serving many millions of people across the globe – why? Simply because it is efficient and effective.

While pond treatment technology offers relative simplicity in its application, it incorporates a host of complex and diverse mechanisms that work to treat and cleanse polluted waters before their return to our environment.

This book offers a comprehensive review of the pond technology field including the newest ideas and latest findings. Topics covered include:

- The physical, chemical and biological characteristics of the pond environment;
- A detailed review of pond treatment mechanisms and performance;
- Comprehensive guidance on pond design, operation and upgrade options;
- A range of chapters summarising new and emerging pond technologies;
- The integration of ponds with wetlands and aquaculture systems and their use as storage reservoirs;
- Special applications of pond technology in cold climates, for agricultural wastes and for treatment of stormwater.

The objective of this book is to get this wealth of knowledge 'out there' to the users to ensure the continuous improvement and ongoing success of this crucial technology.

Andy Shilton

Dedication

Many people have worked hard to produce this text. Many more have laboured to produce the hundreds and hundreds of research publications upon which the chapters are based.

While we all rush to live our busy lives, I am sure my colleagues will agree that what is most important is what we leave behind. Working to improve a sustainable technology that acts to safeguard our waterways seems a particularly worthwhile contribution to have made.

"....es ist alles für die Kinder,....für die Kinder...."

("....it's all for the children,....for the children....")

Andy Shilton

Contributors

V. Dean Adams
Associate Dean
College of Engineering/256
University of Nevada, Reno
Reno, NV 89557, USA
Email: vdadams@unr.nevada.edu

Charles Banks
University of Southampton
School of Civil Engineering &
Environment
Southampton SO17 1BJ, UK
Email: cjb@soton.ac.uk

Stuart Bilby
Bruce Wallace Partners
PO Box 9123
Newmarket
Auckland, New Zealand
Email: stu@bwpl.co.nz

Rupert Craggs
National Institute of Water and
Atmospheric Research
PO Box 11-115
Hamilton, New Zealand
Email: r.craggs@niwa.co.nz

Tom Curtis
School of Civil Engineering &
Geosciences
University of Newcastle upon Tyne
Newcastle upon Tyne NE1 7RU, UK
Email: tom.curtis@ncl.ac.uk

Rob Davies-Colley
National Institute of Water and
Atmospheric Research
P.O. Box 11-115
Hamilton, New Zealand
Email: r.davies-colley@niwa.co.nz

Sonia Heaven
University of Southampton
School of Civil Engineering &
Environment
Southampton SO17 1BJ, UK
Email: sh7@soton.ac.uk

Marcelo Juanicó
Juanicó-Environmental Consultants Ltd
2 Aliah St
18392 Afula, Israel
Email: juanico@juanico.co.il

Thammarat Koottatep
Asian Institute of Technology
School of Sustainable Development
P.O. Box 4
Klong Luang
Pathumthani 12120, Thailand
Email: thamarat@ait.ac.th

Ian Lawrence
eWater CRC
PO Box 1, Belconnen
ACT 2616, Australia
Email:
Ian_Lawrence@lake.canberra.edu.au

Barry Lloyd
School of Engineering
University of Surrey
Guildford GU2 7XH, UK
Email: B.Lloyd@surrey.ac.uk

Duncan Mara
School of Civil Engineering
University of Leeds
Leeds LS2 9JT, UK
Email: d.d.mara@leeds.ac.uk

Jiri Marsalek
National Water Research Institute
867 Lakeshore Road, Burlington
Ontario, Canada L7R 4A6
Email: jiri.marsalek@ec.gc.ca

Piet Meiring
Meiring, Turner & Hoffmann
PO Box 36693
0102 Menlo Park, South Africa
Email: meiring5@mweb.co.za

E. Joe Middlebrooks
2128 Imperial Lane
Superior
CO 80027, USA
Email: Joemiddle@aol.com

Charlotte Paterson
School of Civil Engineering &
Geosciences
University of Newcastle upon Tyne
Newcastle upon Tyne NE1 7RU, UK
Email:
Charlotte.Paterson@newcastle.ac.uk

Howard Pearson
Sitio Araticum 4
Zona Rural
CEP 58117-000
Lagoa Seca
Paraíba, Brazil
Email: howard_william@uol.com.br

Chongrak Polprasert
Asian Institute of Technology
School of Sustainable Development
P.O. Box 4
Klong Luang
Pathumthani 12120, Thailand
Email: chongrak@ait.ac.th

Andy Shilton
Centre for Environmental Technology and Engineering
Institute of Technology and Engineering
Massey University
Private Bag 11222
Palmerston North, New Zealand
Email: A.N.Shilton@massey.ac.nz

Oleg Shipin
Asian Institute of Technology
School of Sustainable Development
P.O. Box 4
Klong Luang
Pathumthani 12120, Thailand
Email: oshipin@ait.ac.th

James Sukias
National Institute of Water and Atmospheric Research
P.O. Box 11-115
Hamilton, New Zealand
Email: j.sukias@niwa.co.nz

David Sweeney
United Water International
G.P.O Box 1875
Adelaide SA 5001, Australia
Email: david.sweeney@uwi.com.au

Chris Tanner
National Institute of Water and Atmospheric Research
P.O. Box 11-115
Hamilton, New Zealand
Email: c.tanner@niwa.co.nz

Ben Urbonas
Urban Drainage and Flood Control District
2480 W 26th Ave., Suite 156-B
Denver, CO 80211, USA
Email: burbonas@udfcd.org

Nick Walmsley
GHD
Level 11 Guardian Trust House
15 Willeston Street
Wellington, New Zealand
Email: nick.walmsley@ghd.co.nz

Corrections

End of page 91

For temperatures below 20°C:

$$C_e = \frac{C_o}{1 + \left[\left(\frac{A}{Q}\right)(0.0038 + 0.000134T)e^{((1.041 + .044T)(pH - 6.6))}\right]}$$

For temperatures above 20 °C:

$$C_e = \frac{C_o}{1 + \left[\left(0.005035\left(\frac{A}{Q}\right)\right)e^{(1.540(pH - 6.6))}\right]}$$

Page 92 line 4

Q is the wastewater flow ($m^3\ d^{-1}$)

Page 92 line 6

pH $=7.3e^{(0.0005(Alk))}$

Page 92 equation in middle of page

$$C_e = \frac{C_o}{1 + d(0.000576T - 0.00028)e^{(1.080 - 0.042T)(pH - 6.6)}}$$

Page 92 4th line from end

pH = pH near pond surface $pH=7.3e^{(0.0005(Alk))}$

Page 92 last equation on page

$$C_e = C_o\ e^{-[0.0064(1.039)^{(T-20)}]\ [t + 60.6(pH - 6.6)]}$$

Page 177 equation 9.18 should read:

$Ce = Ci / \{1 + [(A/Q)(0.0038 + 0.000134T) \exp((1.041 + 0.044T)(pH - 6.6))]\}$

Page 177 equation 9.20 should read:

pH = 7.3exp(0.0005**A**)

1

Introduction to pond treatment technology

Andy Shilton and Nick Walmsley

Pond treatment technology serves the wastewater treatment needs of agriculture, industry, cities and towns around the world and is one of the most common treatment technologies in use today. Indeed, for thousands of communities with many millions of people, from developing countries to modern industrialised nations, the only thing standing between raw wastewater and a local waterway is often a pond treatment system.

1.1 THE POND ENVIRONMENT

The main advantage of these systems is their simplicity to build and operate. Although these systems are often termed 'low tech', the mechanisms involved in the way they treat and stabilise pollution are as numerous and involved as those in conventional 'concrete and steel' technologies.

ISBN: 1843390205. Published by IWA Publishing, London, UK.

Thirumurthi (1991, pg. 231) noted, "the biology and biochemistry involved are the most complex of all the engineered biodegradation systems known to man."

To a large degree, the success of these systems can be attributed to the diverse environment that is established within the pond. Chapter 2 explains the microbiology and Chapter 3, the physical and chemical environment of the pond system.

Pond systems incorporate all aspects of conventional treatment including settlement of solids and BOD removal (see Chapter 4), disinfection (see Chapter 6), as well as offering some capability in terms of removal of nutrients and heavy metals (see Chapters 5 and 7).

1.2 THE 'STANDARD' POND SYSTEM

As pond technology has developed over the decades numerous names, for example sewage lagoons or oxidation ponds, have been used to describe the same thing. Thankfully in the last decade the work of various leading researchers, for example the design manual produced by Mara and Pearson (1998), has brought clarity to this confusion. Today there are reasonably well-established terminology and design procedures for what might be called the 'standard pond system'.

Figure 1.1 illustrates two variations of the common 'standard pond systems'. In the first of these the wastewater enters a facultative pond and then enters a series of maturation ponds. Because there is no prior treatment before the facultative pond (with the exception of screening and in some cases, grit removal) the term *primary* facultative pond is used. In the second illustration the pond system firstly, incorporates an anaerobic pond. The inclusion of an anaerobic pond can substantially decrease the size of the following ponds because, in this case, the wastewater is pre-treated by the anaerobic pond and the term *secondary* facultative pond is used. At the end of both these systems is a series of maturation ponds.

The main function of the maturation ponds is to provide for pathogen removal. In Figure 1.1 three maturation ponds are shown but in reality the number required is determined by design (see Chapters 8 and 9).

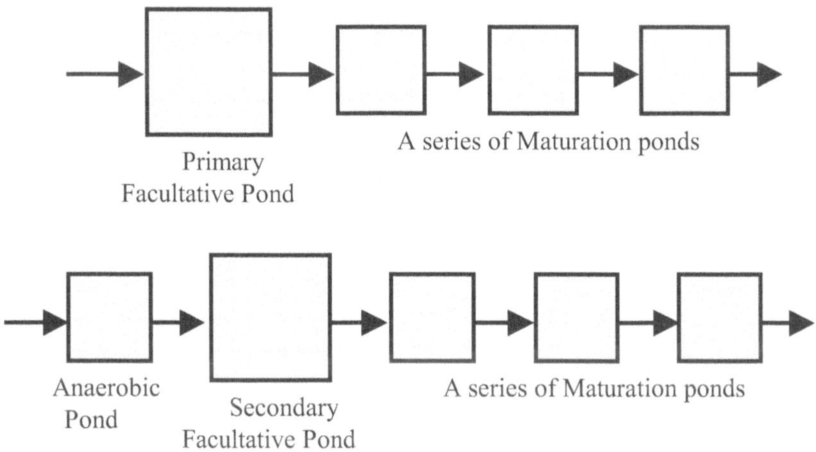

Figure 1.1 Illustration of the standard pond systems (not to scale)

1.2.1 Anaerobic ponds

Designed to receive high organic loading, their treatment function is to undertake bulk removal of the organic load. They are normally absent of dissolved oxygen and contain no significant algal population. They are particularly effective in warmer climates but even in cold conditions they provide at least primary settling. With relatively short retention times of just a few days they can reduce the organic load by 40 to 70%. This significantly reduces the size requirements of subsequent ponds resulting in substantial land and cost savings. Many practical applications have shown that odour is *not* a problem if the recommended design loadings and sulphate concentrations are not exceeded.

1.2.2 Facultative ponds

Undoubtedly this is the most common type of pond in use throughout the world. The term facultative refers to the fact that these ponds operate with both aerobic and anaerobic zones. The lower layer functions with similar characteristics to an anaerobic pond. It consists of an anaerobic sludge layer overlaid with an anoxic zone in the water column. At higher levels in the water column the water becomes oxygenated due to the presence of high concentrations of oxygen producing algae. As will be discussed repeatedly throughout this book there is a classic relationship between the bacteria (an animal) and the algae (a plant). The algae produces oxygen which is then utilised by bacteria in the water column as they oxidise organic waste for energy and in turn produce carbon dioxide (and release nutrients from the waste material) which is used by the algae. In contrast to anaerobic ponds facultative ponds are relatively shallow (typically 1.5 metres) with retention times measured in weeks. Because these ponds depend on algae, which are 'driven' by sunlight, they are designed on an area basis as opposed to anaerobic ponds that are designed on a volume basis.

1.2.3 Maturation ponds

Maturation ponds typically follow facultative ponds in series. They have also been used for 'polishing' following conventional treatment. Their primary function is to remove pathogens, but they can also achieve significant nutrient removal (Mara *et al.*, 1992). Although similar in appearance to facultative ponds, they have a low organic loading and as a result are well oxygenated. Typically, a series of smaller maturation ponds are used rather than a single large pond in order to ensure good hydraulic efficiency, which is particularly important when good pathogen removal is being sought.

1.3 POND DESIGN AND OPERATION

Numerous design equations have been proposed for ponds creating a large degree of confusion. Chapter 8 of this book reviews the various design methodologies that have been developed and in Chapter 9 a recommended design method is clearly outlined that will yield the sizing required for a standard pond system.

In addition to calculating the required pond size, the designer should also be aware of how the pond hydraulics (fluid flowpaths) are affected by design of inlets, outlets, baffles, wind and so forth and this is covered in Chapter 10.

In Chapter 11 the application and design of a range of techniques for further upgrading pond effluent quality, such as sand filters, rock filters, dissolved air floatation and so forth, are reviewed.

A well-designed pond system can still perform poorly if not operated effectively. Indeed, many of the problems encountered with pond systems simply result from operational issues such as the lack of regular sludge removal. Pond operation and maintenance is reviewed in Chapter 12.

1.4 OTHER POND TYPES AND SPECIAL APPLICATIONS

1.4.1 Fermentation/digestion pits

A recent innovation has been the concept of fermentation pits, as discussed by Oswald *et al.*, (1994). Built within a facultative pond, this is a semi-enclosed pit operating under anaerobic conditions like a low-rate digester. The pit receives the raw influent and has a retention time of around one day. This design is reported to remove solids and organic waste more effectively than conventional anaerobic ponds. Because the oxygenated facultative pond overlies this anaerobic pit it is noted that these systems have less potential for odour release. The term 'advanced facultative ponds' is commonly used to describe this integrated pond/pit system. Further information on this technique can be found in Chapters 13 and 14.

1.4.2 Hi-rate algal ponds

Originally developed by Oswald at the University of California in the sixties (Shelef and Azov, 1987), these systems are shallower than a facultative pond (0.2 to 0.8 metres) and operate at shorter hydraulic retention times of around a week or less. A paddlewheel is incorporated to drive the water around a 'race-track' shaped pond. The oxygen production is reported to be significantly higher than typical facultative pond designs. The algae produced in these systems are also reported to have good settling properties (Green *et al.*, 1996). Further information on this technique can be found in Chapter 13.

1.4.3 Advanced pond systems

There is increasing interest in the use of an integrated pond system which integrates an advanced facultative pond (with a built in fermentation pit) followed by a high rate algal pond (with recycle back to the facultative pond) followed by a series of maturation ponds. While still relatively limited in terms of the number of installations compared to the more standard pond systems previously detailed in Section 1.2, this system is one of the most popular areas of current research in the pond technology area. These integrated systems are reviewed in Chapter 13.

1.4.4 The PETRO process

The term PETRO stands for 'Pond Enhanced Treatment and Operation'. The PETRO concept basically involves using a waste stabilisation pond as a first stage to tackle the bulk of the organic load and then using a second stage process, such as a trickling filter (horizontal and vertical) or an activated sludge system for 'polishing' to improve the final effluent for removal of solids in nutrients. This technique has been particularly useful for upgrading overloaded trickling filter and activated sludge treatment systems. Refer to Chapter 14 for full details on ponds integrated with trickling filters and activated sludge processes and the PETRO process development.

1.4.5 Integrated ponds and wetland systems

Like ponds, wetland wastewater treatment systems are another type of 'natural' treatment technology. Wetland treatment technology developed after it was found that natural wetlands receiving wastewater discharges were actually able to provide significant treatment. Today artificial wetlands are constructed either as a 'surface flow system' (like a planted pond) or as a subsurface flow system (essentially a planted filter operated with either horizontal or vertical flowpaths). While not as widespread in application as pond systems (approximately 1/10) wetlands have a high deal of public appeal, in part due to the bird life they attract, and are rapidly growing in number. Because ponds and wetlands have the similar advantages of offering simple operation they are often used together to provide an integrated wastewater treatment solution. The application of wetland treatment systems is discussed in Chapter 15.

1.4.6 Aquaculture ponds

Throughout Africa and Asia it is not uncommon to add a fish or 'aquaculture' pond to the end of pond wastewater treatment system. The basic idea is that the fish will graze the algae reducing solids and subsequent harvesting of the fish then provides a source of protein and a method of recovering nutrients. Chapter 16 discusses the integration of aquaculture with pond treatment technology.

1.4.7 Storage ponds/reservoirs

There can be advantages in storing effluent within a pond as opposed to allowing to it continuously discharge. For example, effluent may be stored during winter periods when treatment is less effective due to colder temperatures. Other applications include avoiding discharge to a sensitive waterway such as a small stream at times when the stream flow is too low or during periods of algal blooms in the ponds. Apart from storing pond effluent for environmental reasons, storage is also used when the nutrient rich effluent is valued as a resource for irrigation during dry periods. Because these ponds are deep, to provide adequate storage volume, they are often referred to as 'reservoirs'.

Chapter 11, which details an operation known as controlled discharge, and Chapter18, which explores the use of ponds in cold climates, both discuss this technique. Chapter 17, however, is specifically focused on the design and application of storage reservoirs.

1.4.8 Cold climate ponds

Ponds are strongly influenced by climatic conditions because they are large water bodies that are exposed to the environment. This applies in a number of places. Because higher temperatures improve most treatment mechanisms the application of ponds are very effective in tropic and temperate regions. However, because of the advantages that ponds offer, particularly in terms of cost in regions where land is relatively inexpensive, they have still been widely applied in cold climates even when freezing conditions exist. A large amount of experience has been built up for the design and operation of ponds in cold climates and this is reviewed in Chapter 18.

1.4.9 Agricultural wastewater ponds

Perhaps the most common application of pond treatment technology is the numerous small pond systems that treat wastewater from dairy milking sheds, piggeries and other farming activities. As for domestic wastewater treatment, these applications typically utilise the standard anaerobic/facultative/maturation type pond system. However, as is discussed in Chapter 19, agricultural waste is very different to domestic wastewater requiring special consideration and design.

1.4.10 Stormwater ponds

There is increasing awareness that stormwater flushed off an urbanised catchment is not simply clean rainwater but contains a range of contaminants such as solids and heavy metals. Ponds, often supplemented with wetland plantings, are increasingly being installed to treat stormwater. Stormwater ponds have short retention times, typically of just a few days. They provide buffer storage to reduce runoff peaks, and also provide enhancement of stormwater quality by various treatment processes such as settlement of solids. As part of their design strong emphasis is also placed on the creation of recreational and habitat amenities. Chapter 20 details the function and design of stormwater ponds.

1.5 WATER QUALITY AND REGULATORY ISSUES

While algae are of critical importance to the effectiveness of pond systems, its growth in the pond and subsequent discharge does contribute to elevated unfiltered BOD and solids concentrations in the final effluent if not removed prior to discharge (see Chapter 11 for algae removal techniques). Where regulators implement strict effluent standards for BOD and/or solids this can create an issue. However, a number of researchers have questioned the appropriateness of applying strict standards to pond effluent containing algae given that the algae is a plant rather than a sewage solid. In many areas regulators recognise this difference and set standards to allow for the algae in the effluent. Chapter 9 discusses this issue further.

In addition to BOD and suspended solids, pathogens are also often monitored. Pond systems can be very effective at disinfection – see Chapter 6. Discharge standards for nutrients have been less common, but this is changing and is certainly an important issue for future consideration.

Chapter 9 provides a process design methodology for sizing ponds in relation to achieving water quality standards.

1.6 EVALUATION OF THE TECHNOLOGIES

Perhaps the two most critical factors that influence selection of any particular wastewater treatment option are performance and cost.

The level of performance of a pond system is obviously flexible depending on the system design. A simple anaerobic pond in a cold climate can give primary level treatment whereas more sophisticated pond systems can yield high removal of organics and effective disinfection. There are also possibilities of adding on additional units such as filters to further enhance the effluent quality. In Chapter 11 a review is presented that compares ponds systems with other treatment technologies and shows how a pond system can produce effluent qualities as good as or better than other conventional treatment options such as activated sludge.

The cost advantages of ponds were analysed by Arthur (1983), in an often-referenced World Bank Technical Paper, and shown to be most cost effective provided that land costs were not high. This was reconfirmed in more recent times by UNEP (1999) as summarised in Table 1.1. It is important to note that in addition to the BOD and nitrogen removal cited in this table, ponds are also capable of providing a high level of pathogen removal as part of their standard design.

Table 1.1 Generally applied wastewater treatment methods for reduction in organic matter and nutrients (UNEP, 1999)

Method	Goal	Efficiency with good practice (%)	Costs (year 2000) ($/100m^3)
Mechanical treatment	BOD_5 reduction	20-35	3-8
Biological treatment	BOD_5 reduction	70-90	25-40
Flocculation	Phosphorus removal BOD_5 reduction	30-60 40-60	6-9
Chemical precipitation $Al_2(SO_4)_3$ or $FeCl_3$	Phosphorus removal BOD_5 reduction	65-95 50-65	10-18
Chemical precipitation $Ca(OH)_2$	Phosphorus removal BOD5 reduction	85-95 50-70	12-18
Ammonia stripping	Ammonia removal	70-95	25-40
Nitrification	Ammonium ⇒ nitrate	80-95	20-30
Denitrification	Nitrogen removal	70-90	15-25
Ion exchange	Phosphorus removal Nitrogen removal	80-95 80-95	70-100 45-60
Waste stabilisation ponds	Reduction of BOD5 Nitrogen removal	70-90 50-70	2-8
Constructed wetland	Reduction of BOD5 Nitrogen removal Phosphorus removal	20-50* 70-90 0-80**	5-15
Activated carbon adsorption	Reduction of organic toxic compounds, BOD_5	40-95	60-90

* *Presumes a pretreatment (BOD5< about 75 mg/l)*

** *The removal is dependent on the adsorption capacity of the soil applied and whether harvest of the plants is foreseen*

1.6.2 An appropriate technology

While performance and cost are obviously key bottom line requirements, the importance of selecting a technology that is 'appropriate' to the needs and constraints of the local situation where it is installed is essential to achieving long term reliability and success. Throughout the world there have been many examples where wastewater process technologies have been successfully installed and commissioned only to fail soon afterwards due to lack of access to technical support when operational or maintenance problems arise.

The inherent simplicity of pond systems is often cited as one of their primary advantages over conventional treatment technologies. As discussed in previous sections there are, however, a range of advancements/modifications that can further enhance pond performance.

The application of ponds as an 'appropriate technology' has been traditionally considered in the context of developing countries. In these situations local conditions are likely to require the most 'passive' type of pond design for it to be truly appropriate. However, Bhamidimarri and Shilton (1996) note that appropriate technology doesn't necessary imply 'low technology'. This means that the use of more sophisticated pond designs can be quite appropriate if suited to the local conditions.

1.6.3 A sustainable energy technology

In recent decades environmental engineers and scientists have been very focused on protection of our waterways. However, as we look to the future it is clear that, in addition to managing our water resources, much greater consideration of our energy resources and the associated issues of carbon management must also be a key concern for the environmental engineer/scientist.

Ponds technology offers some important advantages and interesting possibilities when we view it in the light of sustainable energy, for example, it offers:

i) low cost biogas generation from anaerobic ponds;
ii) solar powered aeration via algal respiration;
iii) solar powered pH increase, and resultant disinfection and improved nutrient removal, via algal respiration;
iv) significantly lower energy consumption compared to other energy intense wastewater treatment technologies (see Table 9.23).

With regard to controlling greenhouse gases, algal growth essentially 'scrubs' carbon dioxide from the atmosphere. If this biomass is then removed and sequestered then the mechanism represents a carbon sink. While building ponds for this purpose alone would seem expensive, consider that these ponds are already in widespread existence for wastewater treatment (in some cases with subsequent algal removal). When this opportunity is compared to other alternatives being proposed for carbon dioxide removal, this pond-based approach appears to deserve further investigation.

It is not suggested that pond technology is necessarily implemented as a sustainable energy or carbon sink option in its own right. However, the fact that these issues are of rapidly growing importance to our societies clearly presents a new angle to pond technology that has perhaps previously been overlooked when assessing the feasibility of various treatment options.

1.7 SUMMARY

The practical proof of the advantages of using pond technology is simply evident in the fact it is one of the most widely applied technologies for sewered communities. However, pond technology, like any particular treatment option, cannot offer the ideal solution to all situations. It has a number of disadvantages including (developed from Mara *et al.*, 1992 and Shelef and Kanarek, 1995):

i) large land area requirements;
ii) growth of algae in ponds which increases unfiltered effluent BOD and SS concentrations;
iii) performance is influenced by variable climatic conditions and algal blooms;
iv) inconsistent nutrient removal.

Its biggest disadvantage is with regard to land area required. However, it is interesting to note that, as researchers grow to understand this technology more, physical and process designs are repeatedly being refined to reduce this land requirement from rather conservative previous guidelines. For example, in the case of New Zealand, a historical figure of 84kg BOD/ha.day has been routinely used for facultative pond design regardless of the marked differences in environmental conditions throughout the country.

Using a more modern design approach, such as that detailed in Chapter 9, which incorporates an anaerobic pond and higher organic loading on a subsequent facultative pond, sees this area reduced by around two-thirds. Furthermore, a number of researchers are now proposing further advances that could see land area requirements slashed further still, in some cases by a further 50%. Like any modern technology, as our understanding of the system mechanisms improve, so can its design be further optimised.

Odour and insect breeding are often cited as disadvantages of pond technology, but haven't been included in the above listing as these are a function of poor design and/or operation rather than an unavoidable disadvantage. Good design and operational protocol is detailed in Chapters 9 and 12.

If the presence of algae is deemed a problem by a regulatory authority then methods do exist to remove it as outlined in Chapter 11. If variable performance is due to algal blooms, cold climates or other reasons then provision of some storage and control over the effluent release is a solution that is discussed in Chapters 11, 16 and 17.

Variable performance with regards to nutrient removal still remains an issue for ponds just as it does for other natural treatment technologies like wetlands. Of course this also used to be the case for technologies such as activated sludge, which more typically serve the needs of large cities. However, once regulators started to impose stricter standards on cities, research followed that delivered the required nutrient removal solutions. Methods do exist today for upgrading ponds systems for nutrient removal, for example, chemical dosing for phosphorus removal, but these are costly and not in keeping with the natural simplicity of pond technology. However, as a result of regulators now turning their attention to smaller communities served by ponds, researchers are increasingly focusing on this issue and several relatively lost cost 'appropriate' nutrient upgrade options are emerging to meet this technology gap.

Around the world many billions of dollars have been invested in installing pond treatment technology. In a country like New Zealand the number of communities served by pond systems outnumber traditional mechanical treatment plants (for example trickling filters or activated sludge) by around 5 to 1. Clearly there are good reasons for the widespread application of pond treatment technology and this becomes evident in the following summary of the relative advantages of pond systems (developed from Mara *et al.*, 1992 and Shelef and Kanarek, 1995):

i) simple construction with relatively low capital costs;
ii) removal of solids and organic pollutants;
iii) high level of disinfection achievable;
iv) integrated sludge digestion within system;
v) biogas generation by anaerobic ponds;
vi) solar powered aeration/disinfection;
vii) minimal (or nil) mechanical/electrical equipment reducing breakdown risk;
viii) low maintenance costs and low (or nil) energy requirements;
ix) relatively minimal labour requirement without requirement for sophisticated technical training;
x) simple process operation - an appropriate technology;
xi) long retention periods gives ability to buffer flow/load fluctuations;

xii) robust process with ability to maintain reasonable levels of treatment even when subjected to prolonged overloading;

xiii) well suited to coping with summer tourist and/or food processing loads as reaction rates increase with temperature thereby tending to balance extra loading;

xiv) potential for resource recovery via algae harvesting as a source of protein or fertiliser.

For thousands of communities and many millions of people, pond technology is the most appropriate solution to their treatment needs, but effective use of any technology requires sound understanding of how it works, how best to use it and what new developments are available. This is the focus of the following chapters.

REFERENCES

Arthur, J.P. (1983). Notes on the Design and Operation of Waste Stabilization Ponds in Warm Climates of Developing Countries. Technical Paper No. 7. Washington, DC:The World Bank.

Bhamidimarri, R. and Shilton, A. (1996). How appropriate are appropriate technologies – defining the future challenge. *Water Science and Technology* **34**(11), 173-176.

Green, F., Bernstone, L., Lundquist, T. and Oswald, W. (1996). Advanced integrated wastewater pond systems for nitrogen removal. *Water Science and Technology* **33**(7), 207-217.

Mara, D. and Pearson, H. (1998). Design Manual for Waste Stabilization Ponds in Mediterranean Countries. Lagoon Technology International; Leeds, England, UK.

Mara, D., Mills, S., Pearson, H. and Alabaster, G. (1992). Waste stabilization ponds: A viable alternative for small community treatment systems. *Journal of the IWEM* **6**, 72-79.

Oswald, W., Green, F., and Lundquist, T. (1994). Performance of methane fermentation pits in advanced integrated wastewater pond systems. *Water Science and Technology* **30**(12), 287-295.

Shelef, G. and Kanarek, A. (1995). Stabilization ponds with recirculation. *Water Science and Technology* **31**(12), 389-397.

Shelef, G. and Azov, Y. (1987). High-rate oxidation ponds: the Israeli experience. *Water Science and Technology* **19**(12), 249-255.

Thirumurthi, D. (1991). Biodegradation in waste stabilization ponds (facultative lagoons). Biological Degradation of Wastes. Elsevier; London, England; 231-246.

UNEP. (1999). Planning and Management of Lakes and Reservoirs: An Integrated Approach to Eutrophication. IETC Technical Publication Series No 11. International Environmental Technology Centre, United Nations Environment Programme, Japan.

2

Microbiology of waste stabilisation ponds

Howard Pearson

2.1 INTRODUCTION

Biological wastewater treatment is a combination of aerobic and anaerobic processes involving a broad range of microorganisms. These microorganisms and processes are basically the same as those that are responsible for the self-purification of rivers and lakes polluted by urban wastewaters. However, wastewater treatment technologies are designed to optimize the conditions for microbial growth and thus optimise the treatment processes, which lead to the removal of organic carbon, nutrients and pathogenic microorganisms and thus the production of an effluent suitable for discharge to the environment.

What sets waste stabilisation ponds (WSP) apart from all other treatment technologies is the involvement of micro-algae in the process and in this respect their microbiology more closely mimics that of a polluted lake system than do

ISBN: 1843390205. Published by IWA Publishing, London, UK.

other treatment technologies. In a waste stabilisation pond there is less control over the environmental conditions that determine the rate of microbial growth and thus the efficiency of the treatment process is relatively slow compared to conventional "electro-mechanical" treatment plants (EMWT). This is why WSP require longer treatment times and thus larger land areas to provide adequate treatment. It also follows that since the rate of microbial metabolic processes essentially doubles for every 10°C rise in temperature WSP systems are more efficient and thus have a smaller land area requirement in tropical climates than in colder ones. In compensation for the slower treatment rate, WSP systems provide conditions for better pathogen removal than EMWT. Aspects of the microbiology of wastewater treatment in WSP will now be discussed in more detail.

2.2 ANAEROBIC PROCESSES AND CARBON REMOVAL IN PONDS

2.2.1 Anaerobic digestion and methanogenesis in ponds

Sedimentation followed by anaerobic digestion is the principal mechanism for the stabilisation of organic carbon (as measured by BOD or COD) in anaerobic ponds and is also a major mechanism in facultative ponds. Picot *et al.* (2002) studying the mass balance of carbon in an anaerobic pond concluded that 74% of the organic carbon removed was converted to methane, 13% into dissolved inorganic carbon and 15% stored as sludge thus emphasising the importance of the methanogenesis process in terms of organic carbon removal in WSP. The gases released during these processes also help by mixing sediments with their bound microbial flora back-up into suspension, ensuring better contact with the wastewater and thereby improving treatment efficiency.

The first step in anaerobic digestion involves the hydrolysis and solubilisation of the constituent proteins, fats and polysaccharides by fermentative bacterial genera (e.g. *Pseudomonas*, *Flavobacteria*, *Alcaligenes*, *Escherichia* and *Aerobacter*). These bacteria possess hydrolytic exo-enzymes, which are exported by the cells outside the periplasmic membrane and may even be released into the medium. These enzymes facilitate hydrolysis of the organic matter. The resultant soluble molecules of amino acids, long chain fatty acids and mono and disaccharides produced are then assimilated by the same bacteria for their metabolism and also by other fermentative species incapable of hydrolysing the original polymeric material. This is followed by the acetogenic phase (production of organic acid anions) in which the soluble products of fermentation are converted into a mixture of short chain fatty acids; ethanol and other alcohols; other organic acids (e.g. lactate); H_2 and CO_2. The fatty acids, ethanol and lactate are then further fermented by various groups of obligate hydrogen-producing acetogenic bacteria to acetate, and CO_2 and H_2 that are the key substrates for methanogenesis. In the final stage of methanogenesis various methanogenic bacteria generate methane gas by either one of two processes. The acetoclastic reaction in which the methanogens e.g. *Methanosarcina barkeri* convert acetic acid to methane is shown in equation 1 below. The CO_2 reducing reaction in which methanogens such as *Methanosarcina hungatei* produce methane from hydrogen and carbon dioxide is shown in equation 2.

$$CH_3COOH \longrightarrow CH_4 + CO_2 \quad (1)$$

$$4H_2 + CO_2^- \longrightarrow CH_4 + 2H_2O \quad (2)$$

Despite the fact that more than 10 substrates for methanogenesis have been identified it has been estimated that over 70% of the methane produced during the anaerobic digestion of sewage is obtained via acetate cleavage. Methanogenesis is the rate-limiting step in the overall anaerobic digestion process because methanogens have cell growth 'doubling times' of a few days compared with a few hours in the case of the acetogenic bacteria. Methanogens are strict anaerobes and require very precise environmental conditions e.g. they have a pH optimum between 7 and 8, and require a negative redox (< -0.24 E_0,V). Thus organic loadings must be such that volatile acid concentrations do not exceed 3000mg/L and alkalinity is greater than 2000mg/L (Parker, 1979).

The general belief that little or no methanogenesis occurs below 13°C is not supported by studies on high altitude ponds that have pointed to lower but significant levels of methanogenesis in facultative and anaerobic WSP at water temperatures of between 7-9°C (Pearson *et al.* 1987; Juanico *et al.* 2000). This is presumably due to the presence of pyschrophilic species of methanogens. Juanico *et al.* (2000) suggest that where water temperatures are above 13°C for most of the year methanogens developing at these temperatures show very decreased activity during the colder months whereas methanogens consistently exposed to low temperatures maintain their activity.

2.2.2 Sulphate reduction and the risk of odour production

The production of H_2S and thus the risk of bad odour in ponds is the result of the activity of sulphate reducing bacteria (SRB) e.g. the genera *Desulfovibrio and Desulfobacter*. These are obligate anaerobic bacteria and are present in the anaerobic layer and sediments of facultative and anaerobic ponds. They require organic material (e.g. organic acids) or hydrogen as a source of reductant and sulphate (but also sulphur or sulphite) as the terminal electron acceptor to re-oxidise their electron transport chains under anoxic conditions during the production of energy (ATP) required for their growth. Examples of equations for H_2S production by SRB are as follows:

$$CH_3COO^- + SO_4^{2-} + 3H^+ \longrightarrow 2CO_2 + \boldsymbol{H_2S} + 2H_2O$$
$$4H_2 + SO_4^{2-} + H^+ \longrightarrow \boldsymbol{HS^-} + 4H_2O$$

Both excess sulphate and excess organic material (organic overloading) will stimulate the growth and activity of SRB resulting in odour production. Acidic conditions (<pH 6) or alkaline conditions (>pH 8) will favour SRB proliferation over pH sensitive methane-producing bacteria, in the anaerobic sediments. SRB compete with methanogens for the same organic substrates particularly acetate and hydrogen. The proliferation of SRB results in more H_2S production and less (or even complete) inhibition of methane production with the consequent risk of odour production.

2.2.3 Importance of the surface crust and the bottom sludge layer

Anaerobic ponds treating predominantly dairy waste at Kitale in Kenya produced a better effluent quality when the surface crust was left in place than when the crust was removed. The impact of wind mixing of the surface pond layers was prevented by the presence of the thick surface crust of solids congealed by the high level of fat in the dairy waste and thus the surface crust helped in the maintenance of anaerobic conditions. This seems to have been very important since the ponds were seriously under-loaded receiving only about one tenth of the permissible volumetric loading for the ambient conditions and sulphate concentrations in the influent wastewaters were up to 500 mg/L. Even so there was lower than normal concentrations of sulphide (< 1.65 mg/L) in the effluents of the anaerobic ponds and no odour release despite the absence of photosynthetic bacteria (Mara *et al.* 1997).

It was established some time ago that the bottom sludge layer has an important influence on the microbiological activity of anaerobic ponds. Parker (1950) and Parker and Skerry (1968) showed that recently de-sludged anaerobic ponds performed less well compared to ones with an active sludge layer, this is believed to be because methanogens are biologically more active and multiply more rapidly when associated with solid surfaces. A recent study by Paing *et al.* (2000) has demonstrated some spatial separation of the processes of anaerobic degradation in the sludge layer of an anaerobic pond. These workers measured greater rates of acidogenesis and higher volatile fatty acid (VFA) concentrations, with pH values <6.6 occurring near the inlet. In contrast higher levels of potential methanogenesis were measured near the outlet where pH values for methanogenesis were more favourable. This supports the findings of Parker and Skerry (1968), who also found high concentrations of VFA in the sludge layer near the inlet of an anaerobic pond. Paing *et al.* (2000) suggested that this sequential distribution of microbial activity in the sludge layer was probably responsible for the increase in efficiency of anaerobic digestion in anaerobic ponds compared to septic tanks.

2.3 AEROBIC PROCESSES AND CARBON REMOVAL IN PONDS

2.3.1 Aerobic bacteria

A wide range of aerobic chemo-organotrophic (heterotrophic) bacterial genera were found in ponds by Gann *et al.* (1968), including *Pseudomonas*, *Achromobacter*, *Flavobacterium* and *Bacillus*, although much less is known of their activities than the photosynthetic organisms. The bacteria present in the aerobic layers of facultative ponds are deemed to be essentially those saprophytes present in the incoming sewage and also include *Beggiatoa*, *Sphaerotilus*, *Alcaligenes* species in addition to those already mentioned (EPA 1983). In other words most aquatic bacterial groups are present and it is assumed that the microbial degradation of organic matter in ponds is essentially similar to that in other biological sewage treatment systems although the biomass concentrations are obviously much lower. Lower active biomass concentration compared to, say, activated sludge, requires greater reactor (pond) volume and hence the longer hydraulic retention times needed for effective treatment.

2.4 PHOTOSYNTHETIC PROCESSES IN PONDS

2.4.1 Photosynthetic oxygen production by pond algae

This relationship between the phototrophic micro-algae and the aerobic chemo-organotrophic bacteria is often classically illustrated as a mutual relationship between the aerobic bacteria that utilize the oxygen produced by algal photosynthesis to oxidize organic material for growth and energy production. The algae benefit by utilizing the carbon dioxide produced by bacterial respiration along with the released nutrients (N and P) to derive energy and fix carbon for growth via photosynthesis (Figure 2.1). This view of the role of the micro-algae is, however, somewhat narrow since the algae are now known: (a) to be important in facilitating in the disinfection processes (see Chapter 6), (b) to be involved directly and indirectly in nutrient removal (see Chapter 5) and (c) that certain species are also capable of assimilating organic carbon for growth (see later in this chapter).

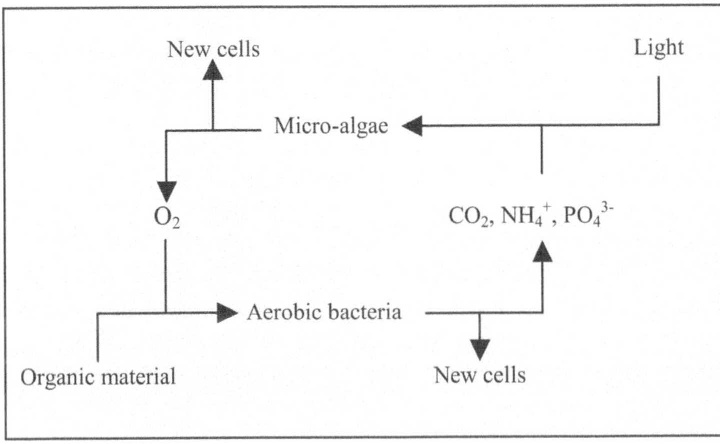

Figure 2.1 The mutualistic relationship between micro-algae and the aerobic chemo-organotrophic bacteria in WSP

It has been estimated that at least 80% of the dissolved oxygen in waste stabilisation ponds results from the photosynthetic activity of the phytoplankton population and thus the aeration of ponds depends heavily on algal activity rather than through surface mass transfer. To give some idea of the sorts of rates of oxygen production that can occur in ponds maximum net photosynthetic oxygen production rates in tropical northeast Brazil using the light and dark bottle technique gave values of $1.7gO_2/m^2/h$ in a primary facultative pond with a surface loading of $250kgBOD_5/ha/d$ and $1.2gO_2/m^2/h$ in a maturation pond with a surface loading of $50kgBOD_5/ha/d$. Numerous studies have indicated that the ratio of molecular oxygen released to algal material synthesized varies with algal species, the age of the algal cells and the availability of nutrients notably nitrogen. It has been calculated that for algae having an average age of 3 to 6 days and using ammonia as the nitrogen source the oxygen: algae quotient is between 1.5 and 1.6. That is for every g of algae synthesized (ash-free dry weight) between 1.5 and 1.6g of oxygen is released from the water (see Oswald 1988). Therefore the maintenance of a healthy algal population is fundamental to the efficient oxidation of organic material by the bacterial population.

There is an inverse relationship between surface organic loading, algal biomass concentration and oxygen production per m^2 of pond surface in facultative ponds (Konig 1984). This relationship suggests that if the algal biomass concentration falls too low, in this case $< 300\mu g$ chl *a*/L there is a risk of the facultative pond turning anoxic since net oxygen production only just matches oxygen demand. Thus at a water temperature of 24°C the maximum permissible BOD_5 surface loading should be approximately $400kg\ BOD_5/ha/day$, which is in line with the values suggested by various design equations for facultative ponds in tropical regions (Yanez 1984; Mara 1987; Mara *et al.* 1992) and is within the limits of the "envelope of failure equation" proposed by McGarry and Pescod (1970).

The depth profile of algal photosynthesis and thus dissolved oxygen concentration also varies with pond type and organic loading. Oxygen can reach super saturation levels in the surface layers of ponds during the hours of maximum photosynthesis. In the cleaner, less turbid maturation ponds photosynthetic activity can extend down to 60cm or more and the complete water column may be aerobic during daylight hours, if not for the whole 24 hours. In contrast the photosynthesis activity in facultative ponds may only extend down to a depth of 20-30cm from the surface with dissolved oxygen only measurable in the top 20cm during daylight hours and with the entire water column turning anoxic at night. Photosynthetic activity also varies with the time of day usually increasing in response to increasing levels of solar radiation incident upon the pond surface. Solar radiation at the pond surface can reach levels that inhibit photosynthesis but the micro-algae can adjust their position in the water column by using flagella movements or altering their buoyancy so as to optimize light levels for photosynthesis. Thus maximum photosynthetic activity may occur some 20cm below the water surface during periods of high light intensities.

2.4.2 Factors controlling algal photosynthesis

Light is a key factor affecting algal photosynthesis and this is discussed in detail in Chapter 3. However other factors also influence algal photosynthesis in ponds including nutrient availability (CO_2, N and P). At high light intensities when there are high concentrations of dissolved O_2, and low availability of CO_2 (HCO_3^-) in the water column the process known as photorespiration occurs. This accounts for why under such conditions the oxygen consumption by the algae in the light is higher than that normally associated with light insensitive (dark) respiration. This happens because under such conditions the enzyme ribulose bisphosphate carboxylase (RubisCO) that is normally responsible for carbon fixation (carboxylase activity) instead functions as an oxygenase consuming oxygen at the cost of carbon fixation and leading to the extra-cellular release of short chain organic acids notably glycollate into the water column. This process is more probable under the conditions prevailing in maturation ponds than in facultative ponds and can reduce algal growth rates. Certainly such algae are stressed and photorespiration may be implicated in the phenomenon of spontaneous auto-flocculation of algae that can sometimes be observed in maturation ponds at high pH. However the significance of photorespiration if at all in terms of overall pond performance is still unclear.

Ammonia and hydrogen sulphide in sufficient concentrations can inhibit algal activity and thus oxygen production in ponds. Inhibition is however reversible in the short term. This topic is discussed in more detail later in relation to algal speciation in ponds.

Another factor that can be useful in determining the "health" of the algal population in WSP is the ratio of chlorophyll *a* to phaeophytin. Phaeophytin is the degradation product of chlorophyll *a* but has the same absorption spectrum however it absorbs more weakly and is photosynthetically non-active. In healthy algal cells the ratio of chlorophyll *a* to phaeophytin is about 1.6 but this ratio drops to 1.0 in dead algal cells where all the chlorophyll has been degraded to phaeophytin (Marker *et al.* 1980; Pearson and Konig, 1986). Thus a falling ratio of chlorophyll to phaeophytin can act as an early warning signal of pond malfunction.

2.4.3 Impact of algal photosynthesis on pond pH

The high pH that is often seen in ponds is a direct result of rapid algal photosynthesis removing dissolved CO_2 from the pond water more rapidly than it can be replaced by either bacterial respiration or across the air-water interface. This results in a shift in the carbonate-bicarbonate equilibrium to produce CO_2 and hydroxyl ions. The CO_2 is fixed by the micro-algae and hydroxyl ions accumulate with a consequent rise in pH:

$$2HCO_3^- \leftrightarrow 2CO_3^{2-} + H_2O + CO_2$$
$$CO_3^{2-} + H_2O \leftrightarrow 2OH^- + CO_2$$

This process can lead to pH values >10 notably in maturation ponds where CO_2 production via bacterial respiration is less than in facultative ponds. This raise in pH has implications not only for the natural disinfection process but also for nutrient removal as discussed further in Chapters 5 and 6.

2.4.4 Algal dynamics and stratification

Algal stratification in the surface layers of ponds can be quite pronounced particularly in facultative ponds where the turbid conditions limit light penetration. At certain times of the day the bulk of the algal biomass may occupy a 15-20 cm band in the water column whereas at other times the band may be broader and appears multi-layered in terms of chlorophyll concentration. Flagellate species usually comprise the bulk of the algal biomass in facultative ponds (see the next section) and the algal layer moves up and down in the surface layers of the water column quite rapidly in response to changes in surface light intensity although it is mostly confined to the top 40cm. In maturation ponds where less turbid conditions prevail the algal layer extends deeper into the water column and stratification is usually less pronounced although the bulk of the algae are still confined to the top 50-60cm layer of the water column, again as a consequence of limited light penetration.

Algal stratification, particularly in facultative ponds, can lead to big differences in effluent quality at different times of the day depending on the position of the algal layer in relation to the pond effluent take-off level (Pearson and Konig 1986). In the case of facultative ponds there is a case for locating the effluent take-off level just below a depth of 50cm from the pond surface since the dense algal band rarely reaches such a depth and in this way the carry-over of algal solids in the effluent to the first maturation can be minimized.

Algal concentrations are more evenly distributed throughout the water column at night in response to the mixing caused by thermal convection. However it has been observed that species of flagellate algae such as *Euglena* moved down in the water column at night and may actually enter the surface sediments of facultative lagoons. This seems to be associated with their chemo-organotrophic metabolism discussed later in this chapter (Section 2.5.1).

2.5 ALGAL DIVERSITY AND FACTORS CONTROLLING ALGAL DOMINANCE

There is a wide diversity of algae in WSP and the algal genera and species that predominate in a pond appears to be a function of the surface organic loading the pond receives. In general, at higher organic loadings species diversity decreases (Caldwell, 1946; Fitzgerald and Rohlich, 1958; Shillinglaw and Piertese, 1977; Konig 1984; Konig *et al.* 1987; Athayde *et al.* 2000) and so facultative ponds have fewer algal genera than maturation ponds and flagellate genera tend to predominate. In maturation ponds diversity is greater and non-flagellate genera frequently predominate. These facts serve as a quick on-site means of determining whether the ponds in a series are under or overloaded as the dominant algal species can be quickly identified with a simple field microscope. Typical algal genera found in facultative and maturation ponds are listed in Table 2.1.

Studies by Mills (1987) and Pearson *et al.* (1987a) suggested that if the non-motile species were able to maintain their position at a favourable level in the water column they would be able to out-compete the motile species such as *Euglena*. Presumably this is not normally the case under turbid facultative pond conditions where their relatively inefficient buoyancy mechanisms cannot compete with the rapid motility of the flagellate species in terms of re-adjusting their position in the water column in response to changes in light intensity and to physical mixing of the photic zone.

Table 2.1 Key algal genera present in facultative and maturation ponds

Algal genus	Facultative ponds	Maturation ponds
Euglenophyta		
Euglena	+	+
Phacus	+	+
Chlorophyta		
Chlamydomonas	+	+
Chlorogonium	+	+
Eudorina	+	+
Pandorina	+	+
Pyrobotrys	+	+
Ankistrodesmus	-	+
Chlorella	+	+
Micractinium	-	+
Scenedesmus	-	+
Selenastrum	-	+
Carteria	+	+
Coelastrum	-	+
Dictyosphaerium	-	+
Oocystis	-	+
Rhodomonas	-	+
Volvox	+	-
Chrysophyta		
Navicula	+	+
Cyclotella	-	+
Cyanobacteria*		
Oscillatoria	+	+
Arthrospira	+	+
*Spirulina***	-	+

+ = present; - = absent ; * photosynthetic prokaryotes; present at high conductivity

Total algal biomass decreases with increased BOD surface loadings in facultative ponds (Figure 2.2) and the impact appears greater in shallow ponds than in deeper ones. At high organic loads the algal population tends towards monoculture of flagellate genera with *Chlamydomonas* proving to be the most tolerant to high loadings. In fact this genus is sometimes found forming a thin surface film on anaerobic ponds that are not sealed by a surface crust but it is unlikely that they contribute anything to the treatment process in these ponds.

High concentrations of ammonia and sulphide are associated with high organic loadings and it may be these rather than the organic load itself which controls algal biomass concentration and algal species dominance (Abelovich and Azov 1976; Konig 1984; Konig *et al.* 1987; Mills 1987; Pearson *et al.* 1987a; Athayde 2001). It is the non-ionic species of both ammonia (which is the increasingly dominant form at pH values above 7) and sulphide (which predominates at neutral to acid conditions) that rapidly enter algal cells leading to toxicity. Therefore, both the pH of the pond water (which is influenced by algal photosynthesis), and the concentrations of ammonia and sulphide are important.

Almasi and Pescod (1996) coined the term anoxic ponds for ponds that operate in the "grey" area between anaerobic and strictly facultative conditions. They noted that motile flagellate algae *Euglena* and *Chlamydomonas* were the only species found to exist under these conditions and whilst *Chlamydomonas* exhibits high tolerance to sulphide the same cannot be said of *Euglena*.

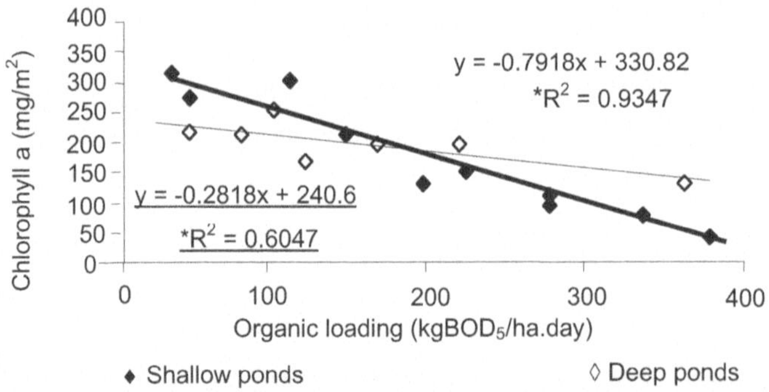

Figure 2.2 The effect of surface organic loading on chlorophyll a values expressed on an area basis in shallow (1.0m) and deep (2.20) pond series in Northeast Brazil (Athayde, 2001)

A toxic bloom of the cyanobacterial *Synechocystis* occurred in experimental ponds in Marrakech, Morocco (Oudra *et al.* 2000) during the hot, arid summer period although details of the conditions such as organic loading were not given. This unicellular coccoid cyanobacterial is very small (0.6-1.2µm in diameter) and such planktonic species are referred to as picoplankton species. There is very little information on this group because they can easily be missed in studies on the micro-algae in WSP because of their size, however given their toxicity further research on the prevalence of these species in WSP would be informative although it is doubtful they are widespread.

The genus *Chlorella* is an enigma in that it is probably the most persistent genera, occurring across a very wide range of organic loadings in WSP, however in terms of maximum biomass concentration it exhibits a preference for low loadings in the region of 28kgBOD$_5$/ha.d as normally found in maturation ponds.

Athayde (2001) concluded on the basis of changes in cell numbers, that *Chlamydomonas* was the most pollution tolerant WSP algal genera and *Scenedesmus*, which favours low organic loadings, were the best indicator algae to use when monitoring the impact of changes in organic loading in pond systems.

Total algal biomass as determined by chlorophyll *a* concentration is usually (but not always) higher in facultative ponds than in the subsequent maturation ponds of the same series. This probably reflects the reduction in available nutrients and also the increased grazing pressures by the zooplankton population that occurs in the more aerobic conditions prevailing in maturation ponds.

2.5.1 Algal photo-organotrophy and chemo-organotrophy

Certain WSP algae are capable of chemo-organotrophic growth (i.e. growth in the dark on organic substrates) and photo-organotrophic growth (i.e. growth on organic substrates but only in the light). What is significant in terms of pond algae is their ability to grow on organic acids released into the sediments and water column as a result of anaerobic degradation of organic material in the pond sludge layer. *Chlamydomonas*, *Chlorella* and *Euglena* could all grow in the dark on acetate under aerobic conditions and *Euglena* could also utilize butyrate. Furthermore only the growth of *Euglena* was not inhibited by propionate (Mills, 1987; Pearson *et al.* 1987a). In contrast none of these algae were capable of growing on the organic acids anaerobically in the dark in contrast to reports for the growth of certain WSP algae on glucose and mannose under such conditions (Weidman and Bold, 1965; Weidman, 1969). However these latter substrates are unlikely to have any significance in WSP in terms of algal metabolism, because they would be more rapidly consumed by the chemo-organotrophic bacteria.

Photo-organotrophy may play a role in the growth and dominance of algal species in ponds, particularly facultative ponds, where light penetration is very limited since it occurs at light intensities well below the photosynthetic compensation point. Purified WSP isolates of *Chlorella*, *Euglena*, *Chlamydomonas* and *Scenedesmus* all grew better in the light with acetate present than when growing photosynthetically. The algae *Scenedesmus* and *Chlamydomonas* were obligate photo-organotrophs in terms of acetate utilization and *Chlorella* grew better on acetate in the light than in the dark. In contrast *Euglena* grew almost equally as well in the dark as in the light on both acetate and butyrate. *Euglena* species have been observed migrating to the lower anaerobic organic-rich sediments in facultative ponds at night returning to the surface layers well before dawn (Konig, 1984). The significance of this in terms of overall organic carbon removal in ponds, however, is not clear and the role of algal chemo-organotrophy and photo-organotrophy merits further investigation.

2.5.2 Algal predation by pond fauna and the impact on pond efficiency

The degree of grazing of algae by the invertebrate fauna in ponds can affect effluent quality in two ways, one is beneficial the other detrimental. A reduction in the algal standing crop in the final maturation pond of a series by grazing will improve effluent quality in terms of BOD since the algal concentration leaving the pond will be diminished. In fact theoretically, zooplankton grazing can reduce algal BOD_5 in the effluent by 80-90% since the energy transfer from algae to the next trophic level, the grazers (e.g. *Daphnia*), will be between 10 and 20% with 80% energy loss via heat and detritus sedimentation.

In contrast excessive zooplankton grazing can reduce the effectiveness of the natural disinfection process that is reliant upon an active algal population to produce the high oxygen concentrations and raised pH levels via photosynthesis, which are fundamental to the overall process (see chapter 6 on *Pathogen Disinfection* for further details).

The impact of grazing was noted in pond systems in Portugal (Mills, 1987; Pearson *et al.* 1987b). A large *Daphnia* population developed in only one of two parallel maturation ponds receiving equal influent flows from the same facultative pond but the reasons for this were not entirely clear although the *Daphnia*-pond also contained significant amounts of submerged vegetation not present in the other. The impact of grazing by the *Daphnia* was to reduce the standing crop of micro-algae and so reduce effluent COD concentrations, reduce maximum pH values in the surface water and to reduce FC die-off rates (Table 2.2). Depth profiles also showed a tendency for reduced dissolved oxygen levels throughout the water column of the *Daphnia* pond (anoxic below ~50cm) compared to the non-*Daphnia* pond that was aerobic throughout its depth.

Kawai *et al.* (1987) worked at laboratory and pilot scale with a series of ponds treating domestic sewage that included anaerobic pre-treatment ponds, followed by algal ponds (containing predominantly *Scenedesmus*) and finally a zooplankton pond (containing *Daphnia*). They showed that such a system could reduce BOD_5 by 95-97% and remove total N and P by 42-59% and 37-48% respectively. Algal concentrations in the final zooplankton pond were reduced from an influent concentration of 10^6cells/ml to 10^2 cells/ml that corresponds to an algal removal efficiency of 99.99% and the water transparency improved considerably. However, the *Daphnia* population reduced during the winter months and only recovered slowly as temperatures increased again.

Table 2.2 Differences in various parameters between a maturation pond containing a large *Daphnia* population and one without. Both ponds received equal flows from the same facultative pond. (Compiled from Mills, 1987; Pearson *et al.* 1987b)

Maturation Pond	Maximum pH below surface	Chl *a* μg/L mean for water col	Light $Qs^{-1}cm^{-2} \times 10^{15}$ 10cm depth	Effluent unfiltered COD mg/L	Effluent filtered COD mg/L	Effluent FC's cfu/100ml
With *Daphnia*	8.40	1622	20	138	111	4.65×10^3
Without *Daphnia*	9.35	2534	10	360	155	2.85×10^2

This phenomenon of extensive zooplankton grazing is in fact quite widespread and there is an argument for including *Daphnia* ponds as a final stage in the treatment process once BOD and pathogen removal have been achieved particularly if the final effluent is to enter a sensitive water body. However routinely maintaining a good standing crop of *Daphnia* appears to be more complicated than might be anticipated and more work needs to be done in this area.

2.5.3 Anoxic photosynthesis and the photosynthetic bacteria

The photosynthetic purple and green sulphur bacteria utilise hydrogen sulphide (H_2S), generated by sulphur reducing bacteria as an electron donor for CO_2 reduction in photosynthesis. The sulphide is oxidised to elemental sulphur (S^o) and in the case of the 'purple' bacteria, is stored within the cells (or more accurately in the periplasm, which is the space between the outer membrane and cytoplasmic membrane) of the cell. In the case of the 'green' bacteria it is deposited outside the cells. If no H_2S is available the purple species utilise the stored elemental sulphur oxidising it to sulphate. They can also utilise other reduced sulphur compounds such as thiosulphate and sulphite. Unlike most of the Cyanobacteria and the eukaryotic microalgae, the photosynthetic bacteria cannot utilise water as an electron donor and thus do not liberate oxygen as a by-product of photosynthesis. This non oxygen-evolving photosynthetic process is termed anoxygenic photosynthesis as opposed to oxygen-evolving photosynthesis that is termed oxygenic photosynthesis.

The purple sulphur bacteria (e.g. *Thiopedia*) are common members of the microbial flora of properly functioning facultative lagoons and are normally found in the water column at a depth of approximately 50cm i.e. in the anoxic but illuminated zone below the surface micro-algae layer. They do not compete for light with the algae as they utilise wavelengths longer than 800nm, which largely pass unabsorbed through the algal zone. They obtain H_2S by diffusion from the lagoon sediments where it is produced by sulphate reducing bacteria.

Under normal conditions in ponds, purple and green sulphur bacteria are important components of the natural 'odour filtration system' as they oxidise a proportion of the H_2S before it reaches the higher aerobic layers where the oxygen produced during algal photosynthesis completes the process. They also protect the algae from photosynthetic inhibition by sulphide (Houghton and Mara 1992).

On occasions facultative (and indeed maturation ponds) turn purple as a result of an increase in the purple bacterial population to the exclusion of the algae. This is usually triggered by the presence of high H_2S concentrations generated by sulphur reducing bacteria activity in the sediments sufficient enough to support the photosynthetic requirements of a large population of photosynthetic bacteria. This can lead to the removal of all the dissolved oxygen from the pond. A larger than normal anoxic illuminated (photic) zone is formed.

Concentrations of H_2S greater than ~8mg/L are known to inhibit oxygenic photosynthesis leading to the death of many pond micro-algae (Mills 1987; Pearson *et al.* 1987a), and so further exacerbating the formation of anaerobic conditions. However it is interesting to note that in Morocco WSP systems close to the sea with purple ponds continued to function well in terms of BOD removal although information on faecal coliform removal was sparse. These ponds were dominated by *Thiopedia* species, as this genus out-competes other photosynthetic purple sulphur bacteria in illuminated, anoxic environments where relatively high salinity/conductivity conditions persist.

The anoxygenic, photo-organotrophic purple non-sulphur bacteria (genus *Rhodopseudomonas*) were found to dominate facultative ponds in Yemen (Veenstra *et al.* 1995) where high ammonia (150-200mg/L) and high sulphide concentrations were responsible for suppressing algal growth. The high sulphide levels were generated in the anaerobic pond feeding the facultative pond as a result of high sulphate levels in the incoming sewage. It is known that certain species of *Rhodopseudomonas* can oxidise sulphide to sulphate in the light under anoxic conditions.

Thus purple lagoons are often an indicator not only of organic overloading or high sulphate concentrations in the sewage but also of saline intrusion into the sewerage system notably in coastal locations.

2.5.4 Impact of algal biomass on effluent quality

The final effluent from a pond series often contains significant concentrations of algae (>500μg chl *a*/L) and these algae can on occasions account for a significant proportion of the total (unfiltered) BOD_5 and suspended solids in the effluent. A linear relationship exists between the concentrations of algal chlorophyll *a* and COD as was demonstrated by Pearson and Konig (1986). Approximately 1mg of algal chlorophyll *a* is equivalent to 300mg of COD in a pond effluent, although this value can vary with algal genera.

The presence of algae can affect compliance with effluent standards and in certain countries provision is made for this by treating algal BOD and algal solids differently to "conventional" or non-algal BOD. The reasons for "relaxing" standards in terms of algal solids and BOD are based on the realization that fish and invertebrate grazing rapidly remove the algae. Furthermore the micro-algae will continue to photosynthesise and be net producers of oxygen in the water body during daylight hours. Algal buoyancy and motility also ensure their rapid dispersion throughout the water body rather than sedimentation near the effluent entry point into the lake or river as tends to occur with non-algal solids. Thus the majority of organic carbon and nutrients in the algae enter the grazing food chain rather than the detritus food chain and are not immediately available for chemo-organotrophic bacterial growth. This is in contrast to the bulk of the non-algal BOD and solids. The detrimental impact on a water body of effluent BOD is therefore minimised when that BOD is in the form of algae. This said the fewer algae that are released into "sensitive" water bodies the better. If however, the final effluent is to be re-used for aquaculture or crop irrigation the algae are a bonus and should not be removed from the effluent since the fish eat them and they essentially represent slow release fertilizers. The algae are also soil conditioners that improve the microbial activity in the soil and aid soil water retention. Contrary to popular belief pond micro-algae do not clog drip irrigation equipment. This is usually caused by non-algal solids (e.g. sand grains) building up inside the emitters and simple filtration technologies which remove such solids but allow the smaller algal cells to pass are usually all that is necessary to avoid such blocking (Taylor *et al.* 1995).

2.6 MICROBIAL PROCESSES AND NUTRIENT REMOVAL IN PONDS

While a more detailed examination of nutrient removal in pond systems can be found in Chapter 5, the microbiology of the key processes is discussed in the following section.

2.6.1 Nitrogen transformations and removal

The removal of nitrogen in WSP systems has been largely attributed to nitrogen accumulation in pond sediments as part of the non-biodegradable microbial biomass and high rates of ammonia volatilisation through the pond surface as a result of high pH conditions caused by the photosynthetic activity of the phytoplankton population (Pano and Middlebrooks 1982; Ferrara and Avci 1982; Reed 1985; Shilton 1996). In properly designed and well-operated WSP systems comprising several maturation ponds in series ammonia removal can reach 90% (but is frequently less) and TKN removal 78.5% (Pano and Middlebrooks, 1982; Silva *et al.* 1995). Ammonia removal was highest in shallow maturation ponds where organic loadings were lowest and pH values highest and in shallow stirred high rate algal ponds that in fact were more efficient than standard WSP (Pearson *et al*, 1996; Gómez *et al.* 1995; Silva *et al.* 1995). In all these studies the implication is that ammonia volatilisation is the key nitrogen removal mechanism and furthermore the high pH conditions that favour this process are compatible with efficient faecal coliform die-off (Silva *et al.* 1995).

In general the nitrification/denitrification route for nitrogen removal in ponds has gained little favour on the grounds that the nitrifying first stage is inefficient because of the apparently low concentrations of nitrifying bacteria isolated from the water column of ponds using conventional culture media techniques and because of the low concentrations of nitrite and nitrate found within them (Reed, 1985). The presence of only low concentrations of nitrate and nitrite might however be a result of denitrification in bottom sediments and is not an argument for excluding the presence of nitrifying activity.

In contrast, several studies have suggested that in fact nitrification in the water column followed by denitrification in the pond sediments is, indeed, a significant nitrogen removal mechanism in the warm summer months in ponds. The aerobic conditions in the water column of maturation ponds have been shown to support a good nitrifier population (up to 10^7 organisms /ml) and sufficient numbers of denitrifiers were present in the sediments to support good denitrification activity (Morrison, 1984 cited by McLean *et al.* 2000).

Detailed studies have been carried out at the Werribee WSP complex (comprising 1650ha of ponds), Melbourne, Australia that included nitrogen balance calculations, diurnal, weekly and seasonal variations in nitrogen removal pathways (Constable, *et al.* 1989; Gross, *et al.* 1994). It was concluded from this work that nitrification and denitrification mechanisms do play an important role in nitrogen removal from maturation ponds. The nitrifier population (and thus the nitrification rate) was however unstable and was subject to washout under various hydraulic regimes. It was also established that it was dependent on the levels of pH, dissolved oxygen, total inorganic carbon, chlorophyll *a* and ammonia. For example at pH levels above 8.5 inhibition of nitrification appeared to occur, as was also the case when dissolved oxygen levels dropped below 6mg/L (but see the findings of Azov and Tregubova, 1995 in Section

2.7.4). Cool winter temperatures below 15°C and rapid changes in temperature also reduced nitrification rates.

A more detailed understanding of the *in situ* microbiology and activity of the potentially diverse nitrifying and denitrifying populations in ponds is vital if a true evaluation of the importance of nitrification/denitrification processes is to be achieved and optimised in terms of nitrogen removal in WSP.

2.6.2 Phosphorus transformation and removal

Since phosphorus does not have a gaseous form the efficiency of total phosphorus removal in WSP depends on how much leaves the water column and enters the sludge layer through sedimentation (mostly as organic P in the microbial biomass) and precipitation (as insoluble phosphates) compared to the amount that returns to the water column via mineralization and re-solubilisation.

Although the bacterial population will assimilate phosphorus the algae constitute the largest component of the organic phosphorus fraction in the water column as they incorporate large amounts of orthophosphate from the inorganic pool. An accurate value for the amount of phosphorus within the algal biomass of a WSP can however be difficult to estimate since the amount of cellular phosphorus varies with algal species and with growth conditions and so there is not a simple relationship between the concentration of chlorophyll a and cellular phosphorus. Some algae are able to utilize phosphorus from phosphorus-based detergents while other algae and the Cyanobacteria are able to store large quantities of phosphorus as polyphosphate granules within their cells. This "luxury" uptake of phosphorus by the micro-algae has analogies with the excessive assimilation of phosphorus by *Acinetobacter ssp.* in the activated sludge process, although the underlying processes are different.

The algae are also indirectly responsible for the chemical precipitation of phosphorus in WSP since it is the high pH values and aerobic conditions resulting from their photosynthetic activity particularly in maturation ponds that leads to the formation of insoluble hydroxyapatite at pH levels above 9.5. Whether the phosphorus enters the pond sediments as microbial material or as a chemical complex the maintenance of aerobic conditions in the surface sediments of maturation ponds is essential if phosphorus is not to be re-solubilized and returned to the water column. The amounts of phosphorus immobilised in pond sediments has been reported to be between 21 and 48% of the influent load (McKinney, 1976; Houng and Gloyna, 1984). Houng and Gloyna in pilot-scale studies on a series of ponds also showed that re-release of P from the sediments was 25 to 50 times faster from the sediments of anaerobic and facultative ponds than from maturation ponds. They produced a model for phosphorus removal and recycling on the assumption of first order kinetics and plug flow and calculated that if BOD_5 removal efficiency is 90% then the total phosphorus efficiency will be close to 45%.

2.7 MICROBIOLOGICAL ASPECTS OF SPECIAL POND SYSTEMS

2.7.1 Macrophyte ponds

The use of water plants or aquatic macrophytes in wastewater treatment is well established and only the relevant aspects of their biology in comparison with WSP will be considered here. Two types of macrophyte ponds exist, one type in which the plants float on the water surface with their roots extending down into the water column and are thus referred to as floating macrophyte ponds and the other type the rooted macrophyte ponds in which the plants are planted into a soil and gravel base of a shallow pond with a water depth of about 50cm. The roots of the plants are therefore in the sediments and their stems in contact with the water column. Rooted macrophyte systems are frequently referred to as constructed wetlands. These two types of macrophyte ponds differ from subsurface flow macrophyte systems where there is no free surface water but instead the wastewater passes through a horizontal filter media comprising soil, sand and gravel layers in which the rooted macrophytes are planted.

Commonly used plants in floating macrophyte ponds are species of *Lemna* (Duckweed), *Pistia* (Water Lettuce), *Eichhornia* (Water Hyacinth), *Azolla* (Water Velvet) and *Salvinia* (Water Fern) and in rooted macrophytes systems are *Phragmites* (Cane Grass), *Scirpus* (Bulrush), *Typha* (Cattail) and *Juncus* (Rush) although other species are being tested (Browning and Greenway, 2001). Both types of macrophyte ponds differ from WSP in that the macrophytes form a leaf canopy at or above the water surface so drastically reducing light penetration of the water column. Consequently the micro-algal population cannot develop to a significant extent in the water column. This leads to anoxic conditions in the water column because photosynthetic oxygen production in the leaves of the macrophytes, unlike that from the algae, is predominantly lost to the atmosphere above the pond and so does not aerate the pond water during daylight hours. Lack of algal photosynthesis also means that the pH never increases much above neutrality in the water column. Thus the near neutral pH, poor light penetration and anoxic conditions in the water column contrast with the conditions in maturation ponds and because of this bacteria and virus removal is reduced as is ammonia removal by volatilisation and phosphorus precipitation.

Duckweed ponds have been reported to contain populations of aerobic sulphide oxidizing bacteria of the genus *Beggiatoa* and photosynthetic purple sulphur bacteria of the genus *Chromatium* associated with the surface plant layer. These bacteria may act as a filter preventing odorous H_2S release since, despite a concentration of 9.7mg S^{2-} in the pond water, volatilisation of H_2S and thus odour was negligible (van der Steen *et al.* 2002).

It seems that processes other than volatilisation and precipitation afford nutrient removal in macrophyte ponds. For example, the productivity of the macrophytes is such that they assimilate significant amounts of nitrogen and phosphorus into their tissues. In terms of carbon removal, the surfaces of the extensive root systems either in the water column or in the sediments are important as they provide a large surface area for chemo-organotrophic and chemo-lithotrophic bacterial biofilms to develop. In the case of rooted macrophytes, the stems below the water surface are also covered in biofilms and close to the water surface these biofilms also contain algae. These biofilms with their consortia of different bacteria are important in terms of the removal of organic material in macrophyte ponds and contrast with the situation in WSP where practically no surfaces exist for biofilm development except for the pond walls and baffles. Since the water column is anoxic, the roots of the macrophytes rely on oxygen from the air diffusing down from the leaves via the stems in a system of specialised lacunae for their aerobic metabolic processes. A proportion of this air leaks out to the stem and root surfaces providing aerobic conditions at their surfaces and possibly out into the root rhizosphere (Brix, 1997). Thus, these biofilms provide both anaerobic and aerobic microhabitats for bacterial activity. Populations of bacteria (e.g. Pseudomonas) exhibiting antibiotic activity have been reported and it is argued that this may be a mechanism for coliform die-off (Broadbent *et al.* 1971). The biofilms may also enable significant rates of nitrification and denitrification to occur with more aerobic conditions prevailing close to the biofilm/root surface interface and anaerobic conditions at the biofilm/water interface. Certainly Ottová *et al.* (1997) found quite large numbers of nitrifying bacteria on the roots and rhizomes of *Glyceria* although none were present in the wastewater. In fact, it is generally believed that these biofilms may be responsible for the majority of the microbial processes occurring in macrophyte ponds (Gumbricht, 1993; Chappell and Goulder, 1994; Brix, 1997).

The lack of algal biomass in the water column also means that the amount of suspended solids and BOD attributable to algae leaving in the pond effluents are reduced. However it must be remembered that the aquatic plants must be routinely harvested or otherwise they decay generating considerable amounts of soluble BOD and suspended solids that will leave in the pond effluents. In temperate and cold climates some species of macrophytes dieback considerably in winter and this can have negative implications for the overall efficiency of the wetland treatment system.

A point of particular importance in the tropics in terms of public health is that the macrophytes in surface flow wetlands provide areas of shade and relatively static water at the pond surface and these conditions are conducive to mosquito breeding of both dirty and clean water breeding types. Putting larvae-eating fish such as *Gumbusia* (mosquito fish) and *Peocelia* into the wetland can help reduce this problem, but they cannot reach the larvae that develop above the water surface in the clean water that accumulates in the shaded leaf axils as a result of plant guttation and rain. This is a very significant problem in the case of the floating species *Eichhornia* but *Pistia* and *Lemna* species are less of a problem since their plant habit does tend to provide such niches above the pond surface. Rooted macrophyte systems often provide breeding grounds and refuges for water birds and this could lead to re-infection of pond effluents by pathogenic bacteria such as *Salmonella* species, particularly when they are being used to polish the final effluent. The above problems are, of course, not relevant to subsurface flow macrophyte systems, as they do not have an open water surface.

For further detail on the use of integrated pond and wetland systems refer to Chapter 15.

2.7.2 High rate algal pond systems (HRAP)

HRAP systems were originally pioneered and developed by Oswald and his co-workers in California (Oswald, 1963; 1988a; 1988b) and are discussed in detail elsewhere (Chapter 13) but certain aspects of their microbiology warrant comparison here with the more standard WSP designs. The key differences in these systems compared to WSP are that they are shallower (a maximum of 60cm deep), have a much longer length to width ratio since they are baffled to form a so-called "race-track" configuration and most importantly they are mixed by paddle wheels. The mixing and the shallow depth ensure that the entire micro-algal population makes best use of the incident light and receives a good supply of nutrients. Their permissible surface organic loading rates compare with those for facultative ponds. The standing crop of algal biomass can exceed 5000 μg chl *a* /L which is considerably higher than is found in WSP and yields in excess of 100,000kg dry weight of algae/ha/yr have been estimated (Shelef *et al.* 1980). The microbial population is homogeneous over the depth of the pond rather than layered as in the case of the deeper WSP. The algae found in HRAP are similar to those frequently found in WSP and include the genera *Euglena*, *Chlorella*, *Scenedesmus*, *Chloromonas Microactinium* and *Pediastrum*. Non-flagellate species tend to dominate over flagellate species presumably since the continuous mixing in HRAP reduces the advantage that flagellate species have in facultative ponds. In temperate climates HRAP tend to exhibit seasonal variation in algal species dominance in relation to light intensity, temperature and grazing pressures by zooplankton (Azov *et al.* 1980; Lincoln *et al.* 1983; Canovas *et al.* 1996). Mixing seems to induce the formation of an obvious biofloc comprising algae, bacteria and zooplankton the so-called ALBAZOD (Soeder, 1984; Cromar and Fallowfield, 1992) not noticeable in WSP and this readily settles if mixing is stopped. Cromar and Fallowfield (2002) have used image analysis to assist the determination of relative algal and bacterial biomass concentrations in the biofloc. The photo-organotrophic uptake of organic carbon by the micro-algae present in the biofloc is an important mechanism of carbon removal in HRAP and it has been suggested that the algae consume more organic carbon than the chemo-organotrophic bacterial population (Abeliovich and Weisman, 1978).

In temperate climates nitrification appears to occur seasonally in HRAP notably during the winter when there is less ammoniacal-N assimilation by the algae and reduced ammonia volatilisation from the pond surface because of lower pH values in the water column (Nurdogan and Oswald, 1995). During the rest of the year nitrification is probably inhibited by a combination of high pH and lack of substrate (Craggs *et al.* 2002). However higher light intensities incident upon the surface of a continuously mixed HRAP system in summer may also be significant since ammonia oxidisers are sensitive to sunlight.

HRAP are not usually stand-alone reactors but are integrated into a pond series. Oswald and his co-workers (Oswald, 1990, 1991; Oswald *et al.* 1994; Green *et al.* 1995b) have developed the *Advanced Integrated Pond System* (AIPS), in which the HRAP is preceded by a specially designed facultative pond (with anaerobic fermentation pits built into its base), followed by a series of maturation ponds. The first maturation pond (or settling pond) allows for settling of the biofloc from the HRAP while the subsequent maturation ponds provide for better pathogen removal. The facultative pond allows the system to accept higher organic loadings than a stand-alone HRAP and the ability to re-circulate effluent from the HRAP to the facultative pond assists nitrogen removal by denitrification once nitrification has occurred in the HRAP.

2.7.3 Attached-growth ponds

Recent studies on attached-growth WSP found that the biomass growing on attached-growth media, for example, fine strings of polyvinylidene installed in the pond water column, improved removal of nitrogen and organic matter (Shin and Polprasert, 1987; 1988; Valentis and Lesavre, 1990; Baskaran *et al.* 1992). Kilani and Ogunrombi (1984) in laboratory-scale studies showed that baffled ponds gave a better performance than un-baffled ponds and they attributed this to the improved hydraulic efficiency. However Polprasert and Agarwalla (1995) re-interpreted this improved performance as being due to the increased biofilm formation on the baffles and suggested that the biofilm could significantly assist the suspended biomass in biodegrading the incoming substrate in WSP. They produced a model of a facultative pond that took account of the biomass growing on the walls and sediment surface of the ponds. The model involved substrate mass balances in the bulk liquid flow and in the biofilm, with substrate transport through the liquid sub-layer acting as a link between the two. This model was tested using ponds in Bangkok and New Mexico and they concluded that biofilm biomass was responsible for 46 and 49% of the BOD_5 removal emphasising the importance of biofilm bacteria in organic matter degradation in facultative ponds. Pearson *et al.* (1995) also noted that better pathogen removal and organic material removal was obtained in a highly baffled pond where the baffles had significant biofilm growth. Muttamara and Puetpaiboon (1996) studied nitrogen removal in laboratory scale baffled waste stabilisation ponds in which the aim of the baffles was to increase surface area for algal/bacterial biofilm development and to create more 'plug-flow' conditions. Their results showed that the baffled ponds gave better removal efficiencies for total nitrogen, NH_3-N, COD and BOD_5 compared to unbaffled WSP. They also claimed that the biofilms on the baffles showed increase potential for nitrification. McLean *et al.* (2000) using pilot-scale attached-growth ponds for treating combined domestic and industrial sewage showed that the ammonia removal rate via nitrification was 50% better in ponds containing vertical panels of polypropylene geo-textile for added biofilm support when compared to the control ponds without geotextile panels. Good ammonia removal rates were only obtained when the geotextile was suspended in the photic zone (the top 50cm of pond depth) when algae formed part of the biofilm. COD and suspended solids in the effluents of the biofilm containing ponds were also lower than in the control pond effluents.

Nitrification was studied in dairy farm waste stabilisation ponds by Craggs *et al.* (2000) in which both HDPE sheet and polypropylene mesh was arranged on frames at different depths in a facultative pond to increase the surface area available for biofilm production. Their results again showed increased nitrification potential (measured as the rate of ammoniacal-N removal in bioassays under controlled conditions), in the facultative ponds with mechanical aeration and biofilm attachment surfaces. These authors suggested in their conclusions that if biofilms were suspended in the photic zone where algal photosynthesis was sufficient to keep the water column aerobic during the day aeration need only be applied at night to give good nitrification.

2.7.4 Wastewater storage and treatment reservoirs (WSTR)

WSTR were pioneered in Israel where they are fundamental to the effective use of limited water resources. They are discussed in detail elsewhere in this book (Chapter 17) but it is pertinent here to consider various aspects of their microbiology. They differ from WSP in that they are normally operated as batch-fed reactors and are much deeper. They can be filled with raw sewage or effluent from anaerobic ponds. Their surface loading rates are the same as those used for facultative ponds (Mara and Pearson 1992; Mara *et al.* 1996). Theoretically the sorts of changes that occur through a series of WSP should occur with time in a WSTR. This appears true for some but not all parameters and Abeliovich (1982) who studied the Ram WSTR in Israel over an eight month period found it biologically stable in terms of both algal species and chlorophyll *a* concentration in spite of considerable diurnal and seasonal variability in water quality. Dor *et al.* (1987) who made a very detailed study of another WSTR in Israel, the Na'an reservoir, obtained similar results and concluded that the reduced, specialised community seemed to be more resistant to disturbance, e.g. changes in light, temperature, pH and dissolved oxygen than more diversified communities. They put forward the hypothesis that a hyper-eutrophic ecosystem (i.e. one rich in mineral nutrients), which also receives an input of organic matter, increases bacterial activity and becomes selective towards algae. These few, specialised algal species remain because they are both facultative organotrophs and also resistant to bacterial toxins. The bacterial community is also selected on the basis of resistance to high DO, pH and algal excretions.

The micro-algae that dominated these Israeli reservoirs irrespective of season and organic loading regime were non-motile genera including *Chlorella* and *Microactinium* with *Scenedesmus*, *Selenastrum* and *Tetraedron* making small contributions depending on the season (Abeliovich, 1982; Dor *et al.* 1987). This compared with the dominance of the flagellate *Euglena* and *Chlamydomonas* species in the preceding oxidation ponds.

Studies on algal diversity in the tropical WSTR at EXTRABES in NE Brazil (Athayde, 2001) identified 29 different genera and of these 20 were non-flagellate species (Table 2.3). However the most dominant genera were *Chlamydomonas*, *Chlorella*, *Euglena*, *Pyrobotrys*, *Scenedesmus* and the cyanobacterium *Oscillatoria*. In neither the filling nor the resting phases could any clear pattern be seen between algal frequency and changes in surface organic loading on the WSTR, and filling with either raw sewage or anaerobic pond effluent made no difference.

Total algal biomass in terms of chlorophyll *a* concentration is generally lower than that recorded for facultative ponds. Values usually ranged between 180 and 400 μg chl *a* /L in the top 50cm of the water column in the Na'an reservoir (Dor *et al.* 1987) and similarly in the pilot-scale WSTR in NE Brazil (Mara *et al.* 1996), but surprisingly reached mean values around 1,500 μg chl *a* /L for the whole water column in the Ram reservoir (Abeliovich, 1982). However the operating regime in this latter case included periods where the inflow of wastewater into the WSTR was in equilibrium with the outflow mimicking conditions in a deep facultative WSP.

Table 2.3 Algae identified in WSTR in NE Brazil (genera are in italics) (modified from Athayde 2001)

Cyanobacteria	Chlorococcales
Nostocales	Oocystaceae
Nostocaceae	*Chlorella*
Nodularia	*Ankistrodesmus*
Anabaena	Oocystis
Rivulariacea	Scenedesmaceae
Rivularia	*Coelastrum*
Chroococcales	*Scenedesmus*
Chroococcaceae	Palmellaceae
Gomphosphaeria	*Sphaerocystis*
Anacystis	Zygnematales
Oscillatoriales	Desmidiaceae
Oscillatoriaceae	*Closterium*
Oscillatoria	Zygnemataceae
Arthrospira	*Zygonema*
Euglenophyta	Chaetophorales
Euglenales	Chaetophoraceae
Euglenaceae	*Phytoconis*
Euglena	Ulotrichales
Phacus	Ulotrichaceae
Lepocinclis	*Ulothrix*
Chlorophyta	Chrysophyta
Volvocales	Centrales
Chlamydomonadaceae	Coscinodiscaceae
Chlamydomonas	*Cyclotella*
Chlorogonium	Pennales
Spondylomoraceae	Naviculaceae
Pyrobotrys	*Navicula*
Volvocaceae	Fragilariaceae
Pandorina	*Fragilaria*
Eudorina	Cryptophyta
	Cryptochrysidaceae
	Rhodomonas

A detailed study of algal biomass concentrations in tropical pilot WSTR during the filling and resting phases by Athayde (2001) showed that the algal population increased steadily during the filling phase and reached a maximum only after 20 to 40 days into the resting phase. Subsequently the algal biomass concentration in the WSTR decreased with time from 3000 to 500μg chl *a* /m^2

although the reasons for this were not clear as there was no clear correlation between chlorophyll *a* and either ammonia or sulphide toxicity or with nutrient concentrations. The most probable answer was predation by zooplankton although no studies on the fauna were carried out. Dor *et al.* (1987), however noted various populations of zooplankton in Israeli WSTR that varied with season and included the rotifer genera *Epiphanes*, *Brachionus* and *Hexarthra*. The crustacean fauna were represented by the copepod genus *Microcyclops* and the cladocerans by *Moina* and *Daphnia*.

Since they are relatively deep, WSTR have a much larger dark, anoxic depth of water above the sediments than WSP. Dor *et al.* (1987) showed that net photosynthesis activity and thus oxygen production, occurred only in the top 30cm in winter extending down to 110 cm in spring and summer. Thus the effective photic zone in terms of photosynthesis never exceeded about 1m. Below this level there was a net consumption of oxygen by microbial respiration. This affected the oxygen and pH profiles in the water column, for example the water column was anoxic at a depth of 20cm at night in November but was in excess of 10mgO_2/L at 60cm in June. In the case of pH, values in the range 8.5-9.2 were often recorded down to depths of about 1m in summer but rarely exceeded 8.0 in the surface waters in winter. Athayde (2001) noted in tropical WSTR in diurnal profile studies that O_2 decreased from supersaturated levels at the surface to zero by a depth of 50-70cm during the day and the entire water column was anoxic between 02.00h and 06.00h. Interestingly, the pH never rose above 8.1 even at the surface and was always between 7.0-7.3 below 1m. Thus in WSTR the biological disinfection zone (the top 1m or less) is relatively narrow compared to their total depth. This said, microbiological purification of the stored water was rapid and in the Brazilian WSTR where water temperatures were around 25°C the concentration of faecal coliforms dropped to <1000cfu/100ml throughout the water column within 28 days during the resting phase (Athayde, 1999; Athayde *et al.* 2000).

Nitrogen removal rates are low in WSTR compared to WSP and consequently ammonia concentrations remain high with recorded values reaching between 15-25mg/L (Abeliovich, 1982; Dor *et al.* 1987; Mara *et al.* 1996). This is probably due to reduced rates of NH_3 volatilisation, although this is still considered to be the main route of nitrogen removal (Azov and Tregubova, 1995). Lack of consistent nitrification rates in WSTR in Israel were attributed to the reduced microbial activity at low temperatures (15°C), which notably inhibited the nitrite oxidising bacteria so leading to the accumulation of nitrite. It was also linked to inhibition by solar radiation, which stops all nitrification processes, and to mixing which inhibited the nitrite to nitrate step (Abeliovich, 1987; Abeliovich and Vonshak, 1993). Azov and Tregubova, (1995) found the optimum pH for nitrification in WSTR was highly, around 9.0.

This occurs in the upper layer where algal photosynthesis increases pH levels but this is also where the nitrifying bacteria would be vulnerable to inhibition by light. Populations of nitrifying bacteria are nevertheless commonly found in the water column of lakes particularly eutrophic ones (Head *et al.* 1993) and this area warrants further research (see later).

Araújo *et al.* (2000) found total phosphorus removal in a tropical WSTR was generally lower (9-33%) than in a WSP operating under similar environmental conditions. Soluble orthophosphate concentrations (2.5-4.2mg/L) remained virtually unchanged. They concluded that environmental conditions, favourable to mechanisms for phosphorus removal, such as high pH and high dissolved oxygen which favour removal by precipitation, particularly with calcium ions to form hydroxyapatite, were unlikely to occur, except for short periods in the afternoons in the top 1m layer where the pH was above 8.2 and the water aerobic. Thus microbial uptake particularly by the micro-algae would appear to be an important route of phosphate removal from the water column near the surface. Conversely, anaerobic conditions in the lower part of the WSTR would favour the release of phosphorus back into the water column.

High sulphide levels occurred in tropical WSTR as a result of *in situ* sulphate reduction. Mara *et al.* (1996) recorded the highest concentrations of sulphide (23mg/L), during the first few days of the resting phase when the surface organic loading during the filling phase was 267kgBOD$_5$/ha/d at a temperature of 24-27°C. Three to four months were required for the sulphide levels to fall to < 1.0mg/L throughout the water column but sulphide concentrations were virtually zero in the 1m layer after 2 months as a result of oxidation in the oxygen-rich surface water layers (Araújo *et al.* 2000). As with ammonia, the high levels sulphide did not appear to limit algal biomass concentrations, emphasising the tolerance of these algae to existing conditions in these reactors (Athayde 2001).

2.8 THE NEED FOR FUTURE MICROBIOLOGICAL RESEARCH IN PONDS

WSP are extremely complex ecosystems and our knowledge of their biology and their underlying treatment processes is improving but there is still much to do and understand if biologists are to aid the engineer in the quest for improving WSP design. The following sections give some examples of areas of further research and applications that may aid that quest.

2.8.1 The nitrogen removal process

Conventional isolation techniques for nitrifiers, notably ammonia-oxidising bacteria, are difficult and time-consuming using sequential enrichment in an ammonium salts medium. It can also result in a collection of isolates that are not representative of the species diversity that exists in WSP systems. Techniques are now available for the detection of 16S rRNA genes and the ammonia mono-oxygenase gene (*amo* A) of *Nm. europaea* to determine the presence, distribution and activity of the two main groups of nitrifiers in the environment and have been successfully applied to lake water samples, sediments and activated sludge (Head *et al.* 1993). Their application to WSP systems would greatly facilitate our understanding of the "conventional" nitrification process in the various types of ponds.

The presence of the planctomycete-like anamox bacteria *Brocadia anammoxidans*, capable of converting ammonia and nitrite under anoxic conditions (Strous *et al.* 1999) and the flexibility of the metabolism of 'aerobic' nitrifiers (including aerobic denitrification by N. *europaea*), warrant further studies in WSP. These two groups might even be natural partners in ecosystems with limited oxygen supply (Schmidt *et al.* 2001) and this might well include ponds.

2.8.2 The algae

The identification of pond algae in most studies (but not all) has been limited to the genus level. This is not surprising since, for example, there are allegedly over 200 species of *Euglena* and more than 400 species of *Chlamydomonas*. Only by the use of molecular techniques will it be possible to determine whether a species is miniaturised or modified in form due to prevailing conditions or represents a different species of the same genus that is proliferating under the changing conditions. Our knowledge of the occurrence and concentration of picoplankton in WSP is also virtually non-existent yet these tiny algae play a key role in primary production in lakes and oceans. The significance of organotrophy in WSP algae particularly under anoxic conditions is also not well documented. Such studies on WSP algae will help considerably in understanding the complexity of the waste stabilisation pond environment and the roles played by algae in the treatment process.

2.8.3 Biofilms and bacterial consortia

There is a growing awareness of the significance of biofilms and bioflocs in WSP treatment but our knowledge of their structure in WSP is poor. The application of fluorescent *in situ* hybridisation (FISH) techniques in combination with a confocal laser microscope for observing through layers of biofilms and consortia of microbes as has already been applied to sewage sludges would thus be very informative. These techniques also allow one to phylogenetically characterise an entire habitat and would greatly aid our investigations into the microbial populations and their activities in all types of ponds.

2.8.4 Zooplankton

Our knowledge of the zooplankton of WSP is woefully limited and compares unfavourably with the information available and the time and effort that has been spent on the micro-algae. Despite this, it is known that the zooplankton and insect larvae contribute to the treatment process in activated sludge and biological filters and must be of fundamental importance to treatment efficiency in ponds. *Daphnia* ponds if they can be made into a sustainable and controllable technology may prove to be an effective way of polishing algal-rich effluents and could have applications for aquaculture. Protozoa exist whose cells are packed full of methanogens but to date little is known of their existence or role in ponds.

2.9 CONCLUDING REMARKS

Our knowledge of the microbiology of pond systems has improved greatly over the past twenty years notably in terms of the dynamics of the algal populations in relation to the efficiency of carbon removal and pond disinfection processes. This information has already helped to improve the physical design of pond systems. However there is still much to understand about the relationship between physico-chemical parameters and microbial activity. Only with an even more detailed understanding of the microbial processes in WSP will it be possible to:

(a) Improve efficiency and produce more cost-effective WSP designs with important savings in land area.
(b) Understand more precisely the causes for pond failure and provide quick and effective remedies should it occur.
(c) Make reliable predictions on how much additional organic load may be placed on an existing pond complex without risk of its failure.
(d) Recognise early symptoms of potential problems before actual pond failure occurs.
(e) Expand the use of ponds for the treatment of industrial effluents.

The application of molecular biological techniques to help resolve engineering and environmental problems is still in its infancy but is developing fast and future studies on ponds could significantly benefit from these techniques.

REFERENCES

Abeliovich, A. (1982). Biological equilibrium in a wastewater reservoir. *Water Research* **16**, 1135-1138.

Abeliovich, A. (1987). Nitrifying bacteria in wastewater reservoirs. *Applied Environmental Microbiology* **53**, 754-760.

Abeliovich, A. and Vonshak, A. (1993). Factors inhibiting nitrification of ammonia in deep wastewater reservoirs. *Water Research* **27**(10), 1585-1590.

Abeliovich, A. and Weisman, D. (1978). Role of heterotrophic nutrition in the growth of the alga *Scenedesmus obliquus* in high rate oxidation ponds. *Applied and Environmemtal Microbiology* **35**, 32-37.

Abeliovich, A. and Azov, Y. (1976). Toxicity of ammonia to algae in sewage oxidation ponds. *Applied and Environmental Microbiology* **31**, 801-806.

Almasi, A. and Pescod M.B. (1996). Wastewater treatment mechanisms in anoxic stabilisation ponds. *Water Science and Technology* **33**(7), 125-132.

Araújo, A.L.C., de Oliveira, R., Mara, D.D., Pearson, H.W. and Silva, S.A. (2000). Sulphur and phosphorus transformations in wastewater storage and treatment reservoirs in northeast Brazil. *Water Science and Technology* **42**(10-11), 203-210.

Athayde, G.B. (1999). *On the design and operation of wastewater treatment and storage reservoirs in Northeast Brazil.* PhD Thesis University of Leeds UK.

Athayde, S.T.S. (2001). *Algal and Bacterial Dynamics in Waste* Stabilisation *Ponds and Wastewater Storage and Treatment Reservoirs*. PhD Thesis University of Liverpool UK.

Athayde, S.T.S., Pearson, H.W., Silva, S.A., Mara, D.D., Athayde Jr. G.B., and de Oliveira, R. (2000). Algological studies in waste stabilisation ponds. In: I Conferencia Latinoamericana en Lagunas de Estabilizacion y Reuso. Edit. M.P.Varón. pp132-139. Cali, Colombia.

Azov, Y. and Tregubova (1995). Nitrification processes in stabilisation ponds. *Water Science and Technology* **31**(12), 313-319.

Azov, Y., Shelef, G., Moraine, R. and Levy, A. (1980). Controlling algal genera in high rate wastewater oxidation ponds. In: *Algal Biomass: Production and Use*. Ed. G. Shelef and C.J. Soeder. pp 245- 255. Elsevier, Amsterdam.

Baskaran, K., Scott, P.H. and Connor, M.A. (1992) Biofilms as an aid to nitrogen removal in sewage treatment lagoons. *Water Science and Technology* **26**(7-8), 1707-1716.

Brissaud, F., V.Lazarova, C. Ducoup, C.Joseph, B. Levine and Tournoud., M.G. (2000). Hydrodynamic behaviour and faecal coliform removal in a maturation pond. *Water Science and Technology* **42**(10-11), 119-126.

Brix, H. (1997). Do macrophytes play a role in constructed wetlands? *Water Science and Technology* **35**(5), 11-17.

Broadbent, P., Baker, K.F. and Waterworth, Y. (1971). Bacteria and actinomycetes antagonistic to fungal roots pathogens in Australian soils. *Australian Journal of Biological Sciences* **24**, 925-944.

Browning, K. and Greenway, M. (2002). Suitability of four macrophyte species in subsurface flow constructed wetlands. In: *Conference Papers of 5th International IWA Specialist Group Conference on Waste* Stabilisation *Ponds: Pond Technology for the New Millennium*. pp. 229-256. NZ Water and Wastes Association, Auckland.

Caldwell, D.H. (1946). Sewage oxidation ponds, performance, operation and design. *Sewage Works*, **18**, 433-458.

Canovas, S., Picot, B., Casellas, C., Zulkifi, H., Dubois, A. and Bontoux, J. (1996). Seasonal development of phytoplankton and zooplankton in a high-rate algal pond. *Water Science and Technology* **33** (7), 199-206.

Chappell, K.R. and Goulder, R. (1994). Seasonal variation of epiphytic extracellular enzyme activity on 2 freshwater plants, Phragmites australis and Elodea canadensis. *Archives in Hydrobiology* **132**, 273-253.

Constable, J.D., Conner, M.A. and Scott, P.H. (1989). The comparative importance of different nitrogen removal mechanisms in 5 west lagoon, Werribee treatment complex. *13th Australian Water and Wastewater Association Conference*, Canberra.

Craggs, R. J., Tanner, C.C., Sukias, J.P.S. and Davies-Colley, R.J. (2002). Dairy farm wastewater treatment by an advanced pond system. In: *Conference Papers of 5th International IWA Specialist Group Conference on Waste* Stabilisation *Ponds: Pond Technology for the New Millennium.* pp.105-111. NZ Water and Wastes Association, Auckland.

Craggs, R.J., Tanner, C.C., Sukias, J.P.S. and Davies-Colley, R.J. (2000). Nitrification potential of attached biofilms in dairy farm waste stabilisation ponds. *Water Science and Technology* **42**(10-11), 195-202.

Cromar, N.J. and Fallowfield, H.J. (2002). Use of image analysis to determine algal and bacterial biomass in a high rate algal pond following Percoll fractionation. In: *Conference Papers of 5th International IWA Specialist Group Conference on Waste* Stabilisation *Ponds: Pond Technology for the New Millennium.* pp.105-111. NZ Water and Wastes Association, Auckland.

Cromar, N.J. and Fallowfield, H.J., (1992). Separation of components of the biomass from high rate algal ponds using Percoll density gradient centrifugation. *Journal of Applied Phycology* **4**, 157-163.

Dor, I., Schechter, H. and Bromley, H. J. (1987). Limnology of a hypertrophic reservoir storing wastewater effluent for agriculture at Kibbutz Na'an, Israel. *Hydrobiologia*, **150**, 225-241.

Environmental Protection Agency (1983). *Design Manual for Municipal Wastewater* Stabilisation *Ponds*. EPA 625/1-83-015.

Ferrara, R.A. and Avci, C.B. (1982). Nitrogen dynamics in waste stabilisation ponds. *Journal of the Water Pollution Control Federation* **54**(4), 361-369.

Fitzgerald, G.P. and Rohlich, G.A., (1958). An evaluation of stabilisation pond literature. *Sewage and Industrial Wastes* **30**, 1213-1224.

Gann, J.D., Collier, R.E. and Lawrence, C.H. (1968). Aerobic bacteriology of waste stabilisation ponds. *Journal of the Water Pollution Control Federation* **40**, 185-191.

Gómez E., Casellas, C., Picot, B. and Bontoux, J. (1995). Ammonia elimination processes in stabilisation ponds and high-rate algal pond systems. *Water Science and Technology* **31**(12), 303-312.

Green, F.B., Lundquist, T.J. and Oswald, W.J. (1995a). Energetics of advanced integrated wastewater pond systems. *Water Science and Technology* **31**(12), 9-20.

Green, F.B., Bernstone, L., Lundquist, T.J., Muir, J., Tresan, R.B. and Oswald, W.J. (1995b). Methane fermentation, submerged gas collection, and the fate of carbon in advanced integrated wastewater pond systems. *Water Science and Technology* **31**(12), 55-65.

Gross, P.M., Scott, P.H. and Connor, M.A. (1994). Development of a management procedure for maintaining nitrification in sewage treatment lagoons. Nutrient Removal from Wastewaters. Ed. N.J.Horan, P.Lowe, and E.I.Stentiford. Technomic Publishing Co., Lancaster, pp.47-54.

Gumbricht, T. (1993). Nutrient removal processes in freshwater submersed macrophyte systems. *Ecological Engineering* **2**, 1-30.

Head, I.M., Hiorns, W.D., Embley, T.M. and McCarthy, A.J. (1993). The phylogeny of autotrophic ammonia-oxidising bacteria as determined by analysis of 16S ribosomal RNA gene sequences. Journal of General Microbiology, 139, 1147-1153.

Houghton, S.R. and Mara, D.D. (1992). The effects of sulphide generation in waste stabilisation ponds on photosynthetic populations and effluent quality. *Water Science and Technology* **26**, 1759-1768.

Houng, H. and Gloyna, E. (1984). Phosphorus models for waste stabilisation ponds. *Journal of the Environmental Engineering Division of the American Society of Civil Engineers.* **110**, 550-561.

Juanico, M. and Shelef, G. (1991). The performance of stabilisation reservoirs as a function of design and operation parameters. *Water Science and Technology* **23**(7-9), 1509-1516.

Juanico, M. and Shelef, G. (1994). Design, operation and performance of stabilisation reservoirs for waste water irrigation in Israel. *Water Research* **28**(1), 175-186.

Juanico, M., Weinberg, H. and Soto, N. (2000) Process design of waste stabilisation ponds at high altitude in Bolivia. *Water Science and Technology* **31**(12), 9-20.

Kawai, H., Jureidini, P., da Conceição Neto, J., Motter, O.F. and Rossetto, R. (1987). The use of an algal-microcrustacean polyculture system for domestic wastewater treatment. *Water Science and Technology* **19**(12), 65-70.

Kilani, J.S. and Ogunrombi, J.A. (1984) Effects of baffles on the performance of model waste stabilisation ponds. *Water Research* **18**, 941-944.

Konig, A. (1984). Ecophysiological studies on some algae and bacteria of waste stabilisation ponds. PhD Thesis, University of Liverpool, UK.

Konig, A., Pearson, H.W. and Silva, S.A. (1987). Ammonia toxicity to algal growths in waste stabilisation ponds. *Water Science and Technology* **19**(12), 115-122.

Lincoln, E.P., Hall, T.W. and Koopman, B.L. (1983). Zooplankton control in mass algal cultures. *Aquaculture*, **32**, 321- 337.

Mara, D.D. (1987). Waste stabilisation ponds: problems and controversies. *Water Quality International* **1**, 20-22.

Mara, D.D. and Pearson, H.W. (1992). Sequential batch-fed effluent storage reservoirs: a new concept of wastewater treatment prior to unrestricted crop irrigation. *Water Science and Technology* **26**(7-8), 1449-1458.

Mara, D.D., Alabaster, G.P., Pearson, H.W. and Mills, S.W. (1992) Waste stabilisation ponds: A design manual for eastern Africa. Lagoon Technology International, Leeds, England.

Mara, D.D., Pearson, H.W., Alabaster, G. and Mills, S. (1997). An Evaluation of Waste Stabilisation Ponds in Kenya. Research Monographs in Tropical Health Engineering (11). Ed., D.D. Mara. University of Leeds, England.

Mara, D.D., Pearson, H.W., Silva, S.A., Araújo, A., de Oliveira, R.E. and Soares, J. (1996). Process performance of an experimental deep effluent storage reservoir under different organic loadings in northeast Brazil. *Water Science and Technology* **33**(7), 243-249.

Marker, A.F.H., Nusch, E.A., Rai, H. and Reiman, B. (1980). The measurement of photosynthetic pigments in freshwaters and standardization of methods: Conclusions and recommendations. *Archives in Hydrobiology* **14**, 91-106.

McGarry, M.G. and Pescod, M.B. (1970). Stabilisation pond design criteria for tropical Asia. In: *Proceedings of the Second International Symposium on Waste Treatment Lagoons*. Ed. R.E. McKinney, 114-132. Laurence: University of Kansas.

McKinney, R.E. (1976). In: *Ponds as a Wastewater Treatment Alternative*. Ed. E.F.Gloyna, J.F. Malina and E.M. Davis. pp 317-325. University of Texas, USA.

McLean, B.M., Baskaran, K. and Connor, M.A. (2000) The use of biofilms to enhance nitrification rates in lagoons: experience under laboratory and pilot-scale conditions. *Water Science and Technology* **42**(10-11), 187-194.

Mills, S.W. (1987). *Sewage Treatment in Waste Stabilisation Ponds: Physiological Studies on the Microalgal and Faecal Coliform Populations*. PhD Thesis, University of Liverpool.

Muttamara, S. and Puetpaiboon, U. (1996) Nitrogen removal in baffled waste stabilisation ponds. *Water Science and Technology* **33**(7), 173-181.

Nurdogan, Y. and Oswald, W.J. (1995). Enhanced nutrient removal in high rate ponds. *Water Science and Technology* **31**(12), 33-45.

Oswald, W.J. (1963). The high-rate pond in waste disposal. *Developments in Industrial Microbiology* **4**,112-119.

Oswald W.J. (1988a). Micro-algae and wastewater treatment. In: *Micro-algal biotechnology* Ed. M.A Borowitzka and L.J. Borowitzka. pp 305-328. Cambridge University Press, Cambridge.

Oswald W.J. (1988b). Large-scale algal culture systems (engineering aspects). In: *Micro-algal biotechnology* Ed. M.A Borowitzka and L.J. Borowitzka. pp 357-394. Cambridge University Press, Cambridge.

Oswald, W.J. (1990). Advanced integrated wastewater pond systems. In: *Supplying Water and Saving the Environment for Six Billion People. Proceedings of the 1900 ASCE Convention, Environmental Engineering Division*. pp73-81. New York.

Oswald, W.J. (1991). Introduction to advanced integrated wastewater ponding systems. *Water Science and Technology* **24**(5), 1-7.

Oswald, W.J., Green, F.B. and Lundquist, T.J. (1994). Performance of methane fermentation pits in advanced integrated wastewater pond systems. *Water Science and Technology* **30**(12), 287-295.

Ottová, V., Balcarová, J. and Vymazal, J. (1997) Microbial characteristics of constructed wetlands. *Water Science and Technology* **35**(5), 117-123.

Oudra, B., El Andaloussi, M., Franca, S., Barros, P., Martins, R., Oufdou, K., Sbiyyaa, B., Loudiki, M., Mezrioui, N. and Vasconcelos, V. (2000). Harmful cyanobacterial toxic blooms in waste stabilisation ponds. *Water Science and Technology* **42** (10-11), 179-186.

Paing, J., Picot, B., Sambuco, J.P. and Rambaud, A. (2000). Sludge accumulation and methanogenic activity in an anaerobic pond. *Water Science and Technology* **42** (10-11), 247-255.

Pano, A.E. and Middlebrooks, J. (1982). Ammonia nitrogen removal in facultative wastewater stabilisation ponds. *Journal of the Water Pollution Control Federation* **54**(4), 344-351.

Parker, C.D. (1950). Purification of sewage in lagoons. *Sewage and Industrial Wastes* **22**, 760-775.

Parker, C.D. (1979). Biological mechanisms in lagoons. *Progress in Water Technology* **11**, 71-85.

Parker,C.D. and Skerry, G. (1968). Functions of solids in anaerobic lagoon treatment of wastewater. *Journal of the Water Pollution Control Federation* **40**, 192-204.

Pearson, H.W. and Konig, A. (1986). The biology of waste stabilisation ponds. *Seminario Regional de Investigacion Sobre Lagunas de Estabilizacion*. 26-39. Lima.

Pearson, H.W., Mara, D.D., Thompson, W. and Maber, S. (1987). Studies on high altitude waste stabilisation ponds in Peru. *Water Science and Technology* **19**(12), 349-353.

Pearson, H.W., Mara, D.D. and Arridge, H.A. (1995). The influence of pond geometry and configuration on facultative and maturation pond performance and efficiency. *Water Science and Technology* **31**(12), 129-139.

Pearson, H.W., Mara, D.D., Cawley, L.R., Arridge, H.M. and Silva, S.A. (1996). The performance of an innovative tropical experimental waste stabilisation pond system operating at high organic loadings. *Water Science and Technology* **33**(7), 63-73.

Pearson, H.W., Mara, D.D., Mills, S.W. and Smallman, D.J. (1987a). Factors determining algal populations in waste stabilisation ponds and the influence of algae on pond performance. *Water Science and Technology* **19**, 131-140.

Pearson, H.W., Mara, D.D., Mills, S.W. and Smallman, D.J. (1987b). Physiochemical parameters influencing faecal bacterial survival in waste stabilisation ponds. *Water Science and Technology* **19**, 131-140.

Picot, B., Paing, J., Sambuco, J.P., Costa, R.H.R. and Rambaud, A. (2002) Biogas production, sludge accumulation and mass balance of carbon in anaerobic ponds. In: *Conference Papers of 5th International IWA Specialist Group Conference on Waste* Stabilisation *Ponds: Pond Technology for the New Millennium*. pp.381-388. NZ Water and Wastes Association, Auckland.

Polprasert C. and Agarwalla, B.K. (1995) Significance of biofilm activity in facultative pond design and performance. *Water Science and Technology* **31**(12), 119-128.

Reed, S.C. (1985). Nitrogen removal in stabilisation ponds. *Journal of the Water Pollution Control Federation* **57**(1), 39-45.

Schmidt, M., Schmitz-Esser, S., Jetten, M. and Wagner, M. (2001). 16S-23S rDNA of anaerobic ammonium oxidising bacteria: implications for phylogeny and *in situ* detection. Environmental Microbiology, 3, 450-459.

Shelef, G., Oron, G., Moraine, R. and Azov, Y. (1980). Algal biomass production as an integrated part of wastewater treatment and reclamation. In: *Algal Biomass*. Ed. G. Shelef and C.J. Soeder. Elsevier Biomedical Press.

Shillinglaw, S.N. and Piertese, A.J.H. (1977). Observations on algal populations in an experimental maturation pond system. *Water S.A.* (Pretoria), **3**, 183-192.

Shilton, A. (1996). Ammonia volatilisation from a piggery pond. *Water Science and Technology* **33**(7), 183-189.

Shin, H.K. and Polprasert, C. (1987). Attached-growth waste stabilisation pond treatment evaluation. *Water Science and Technology* 19(12), 229-235.

Shin, H.K. and Polprasert, C. (1988). Ammonium nitrogen removal in attached-growth ponds. *J. Env. Eng.Div., ASCE* **114**, 846-863.

Silva, S.A., de Oliveira, R., Soares, J., Mara, D.D. and Pearson, H.W. (1995). Nitrogen removal in pond systems with different configurations and geometries. *Water Science and Technology* **31**(12), 321-330.

Soeder, C.J. (1984). Aquatic bioconversions of effluents in ponds. In: *Animals as Waste Converters*. Ed. E.H. Ketelaars and S.B. Iwewa. pp 130-136. Pudoc, Wageningen, Netherlands.

Strous, M., Fuerst, J., Kramer, E., Logemann, S., Muyzer, G., van de Pas, K., Webb, R., Kuenen, J.G. and Jetten, M.S.M. (1999). Missing lithotroph identified as new planctomycete. *Nature* **400**, 446-449.

Taylor, H.D., Bastos, R.K.X., Pearson, H.W. and Mara, D.D. (1995). Drip irrigation with waste stabilisation pond effluents: solving the problems of emitter fouling. *Water Science and Technology* **31**(12), 417-424.

Valentis, G. and Lesavre, J. (1990). Wastewater treatment by attached-growth microorganisms on a geotextile support. *Water Science and Technology* **22**(1-2), 43-51.

van der Steen, N.P., Nakiboneka, P., Mangalika, L., Ferrer, A.V.M. and Gijzen, H.J. (2002). Effect of duckweed cover on greenhouse gas emissions and odour release from waste stabilisation ponds. In: *Conference Papers of 5th International IWA Specialist Group Conference on Waste* Stabilisation *Ponds: Pond Technology for the New Millennium*. pp. 257-266. NZ Water and Wastes Association, Auckland.

Veenstra, A., Al-Nozaily, F.A. and Alaerts, G.J. (1995). Purple non-sulphur bacteria and their influence on waste stabilisation pond performance in the Yemen Republic. *Water Science and Technology* **31**(12), 141-149.

Weidman, V. (1969). Heterotrophic nutrition of waste stabilisation pond algae. In: *Properties and Products of Algae*. Ed. J.E Zajic. Pp.107-114. Plenum Press, New York.

Weidman, V. and Bold, H. (1965). Heterotrophic growth of selected waste stabilisation pond algae. *Journal of Phycology* **1**, 66-69.

Yanez, F.A. (1984). Reduccíon de organismos patogenos y eliseno de lagunas de estabilizacíon en paises en desarrollo. XIX *Congresso Interamericano de Ingeneria Sanitaria y Ambiental*, Santiago, Chile.

3

Physical and chemical environments

Charlotte Paterson and Tom Curtis

3.1 THE DYNAMIC ENVIRONMENT

The design and management of a waste stabilisation pond is in effect the design and management of a small lake. Though waste stabilisation ponds (and lakes) may appear to be calm and placid bodies of water they are in fact in a constant state of flux as virtually every conceivable physico-chemical parameter changes in response to the seasonal and diurnal changes in sunlight, wind and temperature as well as the changes in the influent quantity and quality.

Thus successful design and management of a waste stabilisation pond cannot, does not and should not require the creation of certain fixed conditions. Rather it requires the creation of a range of conditions compatible with success. This means by implication, avoiding the range of conditions associated with failure which is consistent with a traditional engineering maxim (Petroski, 1985) of success as the avoidance of failure. Those who are monitoring pond performance and conditions must also be aware of the exceptionally dynamic nature of WSP

ISBN: 1843390205. Published by IWA Publishing, London, UK.

to ensure that they obtain representative and repeatable measurements of the WSPs characteristics.

In practice, one achieves a satisfactory physicochemical environment by adhering to one of several, typically empirical, design procedures. These procedures dictate the organic load and the depth of the WSP which, in turn, in conjunction with the prevailing climatic and meteorological conditions dictate the biological and physicochemical characteristics of the WSP.

The pond's physical and chemical environment is not only dynamic but it is also complex. Light promotes the growth of algae, but algae impede light penetration and produce oxygen. However, oxygen will affect the concentration of sulphide. Sulphide is toxic to the algae and the algae also affect pH. pH affects toxicity and so on.

Fortunately, for most practicable purposes the behaviour of a pond can be related to three important variables: light, dissolved oxygen and pH. The behaviour of other physicochemical factors can largely be related to these variables.

3.2 LIGHT

3.2.1 Introduction

Light has two very important roles: it drives photosynthesis by algae, which ultimately drives the production of oxygen and the pH, and it kills pathogens. Thus, much of the climatic variation in the performance of WSP is ultimately related to variation in the amount of light. Light can be thought of as varying as a function of time, varying with the season (from month to month), the weather, the time of day, and spatially (i.e. through the pond).

Algae produces oxygen, which in turn ensures efficient bacterial degradation of organic waste to minimise odour emission, and provide conditions for enhanced pathogen reduction and ammonia nitrogen removal.

The functional relationship between light intensity and photosynthesis forms the basis of most models of algal production (Marra, 1978 cites Platt *et al.* 1977) and is fundamental to algal ecology (Neale and Marra, 1985).

Visible and UV light kills pathogens in WSP via a mechanism known as photo-oxidation. In photo-oxidation the light energy is absorbed by a sensitizer that reacts with oxygen to generate a 'singlet' of oxygen, which can kill pathogens (Curtis *et al.*, 1992). The sensitizer can be inside or outside the cell. It has been established that humic substances (which are outside the cell) can act as sensitizers. Humic substances can absorb all visible and UV wavelengths. Even red light has enough energy to generate singlet oxygen and kill pathogens, but only at high pH and oxygen concentrations. Refer to Chapter 6 for a detailed redraw of the disinfection mechanisms.

3.2.2 The nature of light

Light intensity, also known as irradiance, can be measured in terms of the photon flux per unit area (µmol m^{-2} s^{-1}). Light in the 400 to 700 nm waveband is generally accepted as the range for photosynthetically active radiation, and is commonly abbreviated to PAR (Kirk, 1994). Photosynthetically active radiation usually constitutes 45% to 50% of the total light energy received at the earth's surface (Talling, 1957; Kirk, 1994 cites Monteith, 1973). Although it is recognised that photosynthetic rates (Hall and Rao, 1999) and attenuation (Curtis *et al.*, 1994) vary with the spectral composition within this range, the broad band total is considered an appropriate level of accuracy for most pond research related to photosynthetic activity.

Light below 400nm is known as the ultra violet (UV) and is commonly divided into UVB (from 300-320 nm) UVA (from 320-400nm). These wavelengths are commonly associated with disinfection but all the wavelengths > 700nm can kill bacteria in the presence of oxygen. Though UVB is theoretically capable of killing in the absence of oxygen, it is functionally irrelevant because the intensity at the wavelengths is very low and these wavelengths are the most rapidly absorbed in water (Curtis *et al.*, 1994).

This chapter provides guidance on improving pond design by better consideration of these effects. However, it must be recognised that the current understanding of some of these 'inputs' and 'influences' is still very limited.

3.2.3 Light attenuation

The unidirectional nature of light gives rise to a vertical gradient of light intensity as a function of depth (Huisman *et al.*, 1999). Light penetration is of fundamental importance to the functioning of facultative and maturation ponds (Curtis *et al.*, 1994). The light intensity decays rapidly in clean water and in waste stabilisation ponds, and where the light attenuation is greater, this gradient is even steeper. The high algal concentrations have significant influence on the total light attenuation, often limiting algal growth itself. This phenomenon is known as self-shading.

Light intensity diminishes approximately exponentially with depth, as described by Beer's Law. It is illustrated in Figure 3.1 where $\mathbf{I_1}$ represents the light intensity at a given depth, $\mathbf{I_2}$ represents the light intensity at a greater depth, and $\mathbf{\Delta y}$ represents the distance between the two observations.

The equation given in Figure 3.1 can be rearranged to the equation shown in Figure 3.2 to calculate the overall vertical light attenuation coefficient (**b**). If the units of **Δy** are metres (m), the units of **b** are m^{-1}.

In many cases, $\mathbf{I_1}$ is taken as the light intensity just beneath the water surface, and a series of $\mathbf{I_2}$ observations are made at several depths. The data are then transformed and plotted as in Figure 3.2, which typically results in a straight line of points from which the overall light attenuation coefficient can be estimated by the slope of the linear regression. Most data sets plotted in this way are well represented by a straight line, and have high coefficients of determination.

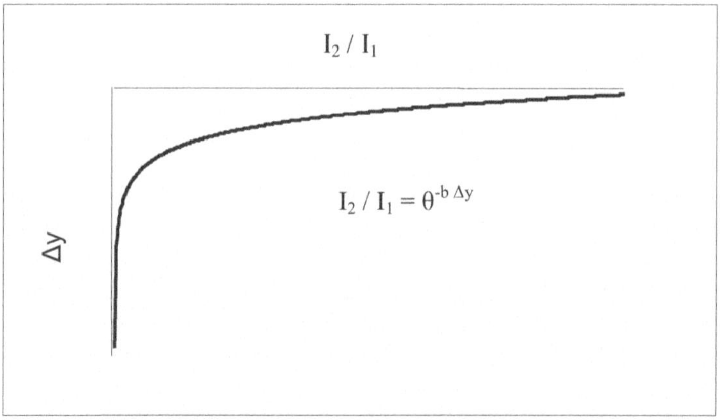

Figure 3.1 Illustration of Beer's Law

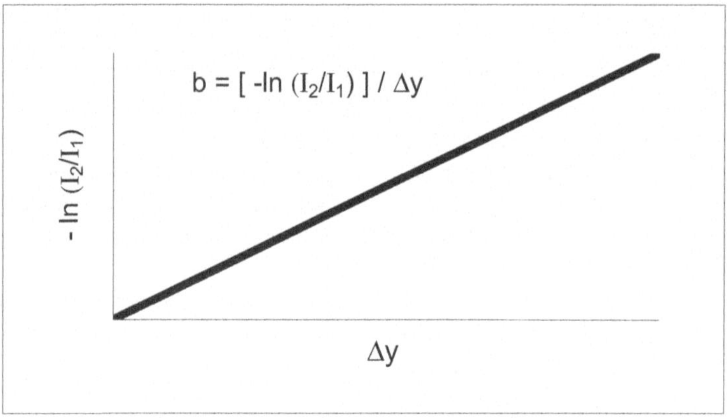

Figure 3.2 Calculation of overall light attenuation coefficient

3.2.4 Algal and non-algal light attenuation

Four factors contribute to light attenuation: water, gilvin (dissolved yellow humic matter), tripton (inanimate particulate matter) and algae (Curtis *et al.*, 1994). In practice only algal concentration varies. It is therefore convenient to express total light attenuation in terms of algal and non-algal attenuation, where the latter is the sum of attenuation due to water, gilvin and tripton and is approximately fixed.

Algal and non-algal vertical light attenuation coefficients (**p** and **q** respectively) can be estimated by plotting overall vertical light attenuation coefficients (**b**) against column average algal concentration, usually measured as m^{-1} and μg/l chlorophyll *a*, respectively. The data will typically produce a straight line of points, as illustrated in Figure 3.3, from which **p** can be estimated from the gradient of the regression line, and **q** can be estimated from the intercept with the y-axis. If chlorophyll *a* concentration is measured in μg/l, and the units of **b** are m^{-1}, then the units of **p** are $(\mu g \text{ chl } a \text{ } l^{-1})^{-1} \text{ } m^{-1}$ and the units of **q** are m^{-1}.

The equation representing Beer's Law can thus be rewritten in the following form:

$$I_2 / I_1 = e^{-(p \text{ chla} + q) \Delta y}$$

3.2.5 Measurement of light

Underwater light intensity can be measured by a spherical quantum sensor (e.g. Li-193SA, Li-Cor Inc., Lincoln, USA). The sensor shape and size resembles a standard light bulb and can be supported in an upright position by a weighted frame. The sensor consists of a silicon photodiode and spherical acrylic diffuser (Li-Cor, 1991). The diffuser is spherical in order to collect light with equal efficiency regardless of direction. This feature is particularly applicable to waste stabilisation ponds since algal cells are randomly oriented in the water and make equal use of scattered light from any direction (Kirk, 1994). The photodiode is designed to measure the total PAR. The sensor is connected to a light meter (e.g. Li-250, Li-Cor Inc., Lincoln, USA), which is calibrated for the sensor type and medium (water or air), and gives a resolution of at least 1 $\mu mol \text{ } m^{-2} \text{ } s^{-1}$.

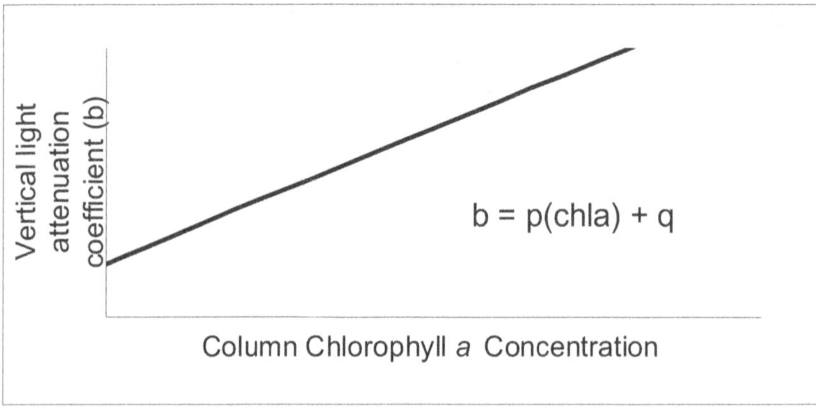

Figure 3.3 Estimation of algal and non-algal vertical light attenuation coefficients

Secchi depth is a simple alternative measure of light penetration. A white disc, around 30 cm in diameter, is attached to a graduated stick or tube and lowered into the pond until invisible. The depth at which this occurs is known as the Secchi depth.

The vertical light attenuation coefficient, **b**, is an inverse measure while Secchi depth is a positive measure of relative light availability. Correlation analysis of observations in waste stabilisation ponds have confirmed significant positive linear correlation of **b** with column average chlorophyll *a* concentration and the significant negative correlation with both of these with Secchi depth (Curtis *et al.*, 1994; Weatherell, 2001).

3.3 DISSOLVED OXYGEN

3.3.1 Significance of oxygen

Two critical roles for oxygen are the control of odour and disinfection. While oxygen is obviously needed for aerobic degradation of organic matter, it is apparent that satisfactory organic matter removal will occur in ponds that have relatively low oxygen concentrations, for example, in winter in temperate climates (Abis and Mara, 2002) or in heavily loaded ponds in the tropics (Silva, 1982).

Oxygen assists in reducing odour by facilitating the oxidation of sulphides and other smelly chemicals produced in the sediments at the bottom of the pond. Heavily loaded ponds often have flecks of white material on their surface which could well be elemental sulphur. The oxidation may occur directly (though this has not been intensively studied) or be biologically mediated, for example by photosynthetic sulphur bacteria or sulphur oxidising bacteria.

Curtis *et al.* (1992) and later Davies-Colley *et al.* (1999) demonstrated the significant influence of dissolved oxygen (DO) and pH on sunlight inactivation of faecal micro-organisms. Their experiments indicate that sunlight disinfection is directly dependent on DO concentration, which strongly suggests the involvement of a photo-oxidative process. Algae are inextricably linked to these three factors, depending on light for photosynthetic activity, promoting high DO and pH (Mara *et al.*, 1992), and influencing light penetration through the water column (Curtis *et al.*, 1994). Pathogen removal is an important consideration, especially in water scarce regions where the pond effluent is used for irrigation of crops. The treatment of pathogens in pond systems is discussed more fully in Chapter 6.

3.3.2 Sources of Oxygen

In nearly all waste stabilisation ponds photosynthesis is the most important source of oxygen, although modest aeration may also occur from the surface (Ellis, 1983). The relative importance of surface aeration depends on the extent of the photosynthesis. Thus, in winter, in temperate climates surface re-aeration may have a more significant role.

Photosynthesis is the process by which the energy of sunlight is used by the algae to synthesise carbohydrates, or 'fix' carbon in algal cells from carbon dioxide and water. Although involving many complex mechanisms, the overall process can simply be represented by the following overall equation (Kirk, 1994):

$$\underset{\text{(carbon dioxide)}}{CO_2} + \underset{\text{(water)}}{2H_2O} \rightarrow \underset{\text{(carbohydrate)}}{CH_2O} + \underset{\text{(water)}}{H_2O} + \underset{\text{(oxygen)}}{O_2}$$

Of course algae also consume oxygen in a process known as respiration. Respiration is the consumption of fixed carbon in maintenance (Reynolds and Irish, 1997), and results in the uptake of oxygen and the evolution of carbon dioxide. Measurement of dissolved oxygen changes over time can be used to directly determine both gross and net photosynthetic rates, and the respiration rate.

3.3.3 Oxygen dynamics

Oxygen dynamics are driven by photosynthesis, which is a function of light, light attenuation and organic load. Consequently, oxygen varies within a pond diurnally, spatially, with depth, and between ponds with different organic loads. Maturation ponds are typically aerobic all the way through the water column, whilst in facultative ponds dissolved oxygen drops to zero at some depth (this point is called the oxypause).

For any given location and organic load, oxygen concentrations are lowest, and the oxypause is highest towards the end of the night. This is because at night algae continue to respire and thus become net consumers of oxygen. Then as dawn breaks and the sunrises photosynthesis will commence, the oxygen concentration will gradually increase and the oxypause, if present, will sink.

It is quite common for the water in the upper levels of ponds to become supersaturated with sometimes greater than 20mg/l of oxygen. In the winter months in temperate climates the nights are longer and the days shorter and thus less oxygen is produced.

Increasing organic load also decreases the amount of oxygen available. Not only by using up more oxygen, but also by decreasing the productivity and variety of the algae (Konig, 1982). Thus heavily loaded ponds have lower dissolved oxygen concentrations. The relationship between algae microbiology and organic loading is discussed in more detail in Chapter 2.

3.3.4 Measurement of oxygen

Since oxygen concentration is very dynamic, it must be measured at a particular position and hour if being monitored over a period of time. Oxygen concentration was originally measured by the iodometric or Winkler method (APHA, 1995), which is a titrimetric procedure based on the oxidising property of dissolved oxygen. This method has been largely superseded by the use of membrane electrodes, also known as Clark-type electrodes (Dubinsky *et al.*, 1987; van der Heever and Grobbelaar, 1997) which measure the rate of diffusion of molecular oxygen across a membrane. Portable DO meters with convenient electronic readouts are now available, which are useful both for field sampling and laboratory experiments. Rapidly advancing technology has developed membrane electrodes with high sensitivity and fast response (Ludyanskiy and Pasichny, 1992).

Most DO meters cannot measure concentrations above 20 mg/l. Although such a high value is unlikely to occur in most oceans and lakes, supersaturation is common in high rate algal ponds (HRAP) and can also occur in conventional waste stabilisation ponds during periods of high photosynthetic activity. For example, Mara *et al.* (1997) listed observations for pond systems in Kenya; of which two pond systems regularly produced DO readings off the scale (maximum 20 mg/l). However, DO meters with the capacity to measure DO concentrations of up to 50 mg/l are now available, such as the YSI model 95 DO meter (YSI Inc., Dayton, USA).

3.3.5 Redox and redox potential and its relationship with oxygen

Redox is short for reduction and oxidation (oxidation is the loss of electrons and reduction is the gain of electrons). Carbonaceous wastes will be the most important source of electrons in a WSP and, oxygen, where present, would be the most important electron acceptor. A redox measurement represents the potential of a system to donate electrons and is expressed as Volts or as the negative logarithm of the electron capacity, pE. In practice the measurement of redox potentials is widely regarded as difficult and unreliable and therefore rarely reported in the WSP literature. As a rule of thumb, at any given pH, the redox potential is determined by the availability and type of electron acceptor. Thus where oxygen is plentiful redox potentials are high and decline as successively less thermodynamically favourable acceptors are consumed in the sequence: oxygen, nitrate, iron, sulphate, and carbon dioxide. Thus methanogenesis, which employs carbon dioxide as an electron acceptor is unlikely to proceed at elevated redox potentials (as this would imply the presence of more energetically favourable acceptors). In practice it is easier to examine for the redox sensitive chemicals (and infer the redox potential) than to measure the redox itself. Since redox reactions consume or generate protons, pH will also affect redox potential.

3.4 pH

3.4.1 The nature and significance of pH

pH is a measure of the hydrogen concentration. Specifically it is the negative logarithm of the hydrogen concentration. This means that a change in one pH unit is equivalent to an order of magnitude change in hydrogen ion concentration.

pH, like oxygen, is important in pathogen removal, nutrient removal and odour control. High pH values are generally regarded as beneficial. pH seems to affect the survival of pathogens both directly and indirectly. The destruction of bacterial pathogens happens when the pH exceeds a certain threshold of around 9 (Parhad and Rao, 1974; Pearson *et al.*, 1987). Elevated pH may also promote pathogen removal by enhancing photo-oxidation. Refer to Chapter 6 for further detail on pathogen disinfection. Volatisation of ammonia and precipitation of phosphorus are also improved at higher pH, and is discussed further in Chapter 5.

pH is also very important in odour control because pH affects the disassociation of H_2S. The sulphide ion may exists in three forms: H_2S, HS^-, S^{2-}. Only H_2S smells, and this form predominates at pH values below about 7.5.

$$H_2S \Leftrightarrow HS^- \Leftrightarrow S^2$$

3.4.2 The carbonate/bicarbonate buffering system

The pH of waste stabilisation ponds (and all bodies of water) is controlled by the carbonate bicarbonate buffering system (for an excellent overview, see Snoeyink and Jenkins, 1980), in which the balance of hydrogen ions and hydroxyl ions is moderated via their interaction with the carbon dioxide present in the water:

$$CO_2 + H_2O \Leftrightarrow H_2CO_3 \Leftrightarrow H^+ + HCO_3^- \Leftrightarrow H^+ + CO_3^{2-}$$

pH changes may be caused by increases or decreases in acids and the addition, or more importantly, the removal of carbon dioxide. Volatile fatty acids (VFAs) are an obvious and important source of the former. VFAs are monobasic organic acids that have the formula R.COOH, where R is generally relatively short (e.g. butyric acid: C_3H_7COOH). VFAs are formed when anaerobic bacteria break down organic matter in the incoming waste. Thus heavy organic loads tend to decrease the pH of waste stabilisation ponds. This affects odour both because it promotes the reassociation of H_2S from HS^- and H^+ and because many VFAs are odourous in their own right.

pH is also affected by the loss (or gain) of CO_2 from the water column. Algal photosynthesis in ponds therefore acts to consume and remove CO_2 from the water column. The loss of this carbon dioxide leads to a shift to the left in the carbonate-bicarbonate buffering system, thus the loss of hydrogen ions, and thus an increase in pH. While this may be offset, to some extent, by the formation of CO_2 through algal and bacterial respiration the pH in ponds can become relatively high as discussed previously.

3.4.3 Temporal and spatial variation in pH

The variation of pH depends on the variation in photosynthesis and organic load. Thus pH and oxygen have similar dynamics. In any given pond, pH is lowest at night and increases during the day, and may reach values in excess of 9.0 in a moderately loaded WSP in warm weather. However, high pH values are generally confined to the upper reaches of the pond (where photosynthesis occurs) and decrease as light penetration declines through the water column. It follows that pH values tend to be lower in winter in temperate climates. Higher organic loads invariably lead to a reduction of pH but the effect of any given loading at any given site is difficult to predict. However, the following equation has been proposed (Pano and Middlebrooks, 1982):

$$pH = 7.3\exp(0.0005A)$$

where A is the alkalinity (as $CaCO_3$) in mg/l, (Pano and Middlebrooks, 1982).

3.4.4 Measurement of pH

The pH values of ponds must be measured *in situ* and thus some form of portable pH measuring device is required. The measurement of pH is relatively straightforward using standard electrodes. However, submersible pH meters must be used if the variation of pH with depth is to be assessed. If the pH value is to be compared between days or ponds, the time of sampling should be the same.

3.5 TEMPERATURE

3.5.1 Why temperature is important

Temperature cannot obviously be controlled in a pond, and so it is something that we must understand and allow for in the design process. The pond temperature itself is largely a function of the location of the pond and may vary from less than 0°C to over 30°C.

Temperature has two distinct roles in waste stabilisation ponds. Firstly, the temperature of the water dramatically affects the rate of the biological processes and so ponds generally operated at lower temperatures need to be larger for equivalent performance compared to those in warmer regions. It is, therefore, not surprising that temperature features prominently in a number of design equations. However, these relationships are empirical and do not imply a mechanistic relationship. Temperature may be a good guide to performance partly because it is positively correlated with the amount of sunshine. Secondly, temperature is important because it affects the hydraulic properties of the water as waste stabilisation ponds stratify and de-stratify under the influence of the sun and wind (Marais, 1966; Llorens *et al.*, 1992).

3.5.2 Stratification

Stratification occurs when the sun shining on the surface of the pond causes the uppers layers to warm up and thus become less dense than the cooler waters below. Stratification is observed as a distinct change in temperature at a given depth in the pond, this inflexion is known as the thermocline. Stratification is typically lost at night when the pond surface cools. It is possible that if surface cooling is rapid, the cooler (denser) surface layer could sink and cause the lower layers to rise, a process known as turnover. This effect has however been poorly studied. Further discussion on temperative effects such as this can be found in Chapter 10.

3.5.3 Thermal short-circuiting

Temperature is important for hydraulics because the influent water may fall or rise depending on whether it is warmer or colder than the water in the pond. This may, in certain circumstances, lead to significant short-circuiting. For example, where the influent is warmer than the bulk of the pond water the influent may rise to a narrow volume of water at the top of the pond and thus move relatively quickly to the take-off point. This can lead to a reduction in the effective volume of the waste stabilisation pond (Kellner and Pires, 2002; Torres *et al.*, 2000).

3.5.4 Implications for hydraulics

Chapter 10, which deals with pond hydraulics, presents a further discussion on these topics. It draws focus to the relative magnitude and impact of mixing energy from various sources such as the inlet and wind. Whether stratification will develop and/or if thermal short-circuiting will occur is dependent on the degree of mixing energy as, obviously, a highly mixed system will not suffer these effects. Research in this area is still extremely limited but there is no doubt that an examination of mixing energy input is the key to assessing and potentially remedying these types of problems.

3.6 SALINITY

The salinity of the water is a vital consideration in wastewater reuse and detailed guidelines on appropriate salinity are available (Pettygrove and Asano, 1985). The salinity of a particular wastewater is essentially governed by the salinity of the influent. Typical domestic wastewater may have a salinity of about 0.5-2.0 dS/m, regarded as suitable for irrigation. Some industrial wastewaters and all wastewaters subject to saline intrusion or mixed with seawater will have a higher salinity. Treatment in waste stabilisation ponds will not decrease the salinity. However, excessive evaporations may increase it.

3.7 ELEMENTAL CYCLING

3.7.1 Nitrogen

It is sometimes desirable to remove ammonia (to decrease oxygen demand or decrease toxicity to fish) or all nitrogen (to decrease the risk of eutrophication) from wastewaters prior to disposal or reuse. Although ammonia nitrogen removal is usually associated with high technology processes, high removal (over 90%) can be achieved in waste stabilisation ponds, by volatilisation, assimilation in algal biomass and possibly biological nitrification (Pano and Middlebrooks, 1982; Gomez *et al.*, 1995). Silva *et al.* (1995) also reported ammonia removals of up to 90% by waste stabilisation pond systems, and found that optimum conditions for high ammonia removal were compatible with optimal BOD removal and faecal coliform die-off. Most of the nitrogen in ponds is present as ammonia, which is either present in the influent or derived from the breakdown of proteins in the sludge layers. Nitrification (the conversion of ammonia to nitrite and nitrite to nitrate) has been reported in waste stabilisation ponds (Lai and Lam, 1997). However, nitrification is probably the exception, rather than the rule. Nitrite and nitrate are rarely observed at significant concentrations in ponds and sensitive molecular tools have so far failed to detect the presence of ammonia oxidising bacteria in ponds known to be removing ammonia (Kartal, 2002).

Nitrogen removal in pond systems is discussed in more detail in Chapter 5.

3.7.2 Sulphur

The reduced form of sulphur, H_2S, is of particular significance both because it smells and because it is toxic. Sulphur will typically enter the ponds as either reduced forms of sulphur present in the amino acids that make up proteins (which is one reason that wastes high in protein can smell so awful) or from sulphate present in the influent water. Sulphides in ponds are mostly produced from sulphate by sulphate-reducing bacteria in anaerobic zones of the pond. These bacteria use carbon as an electron donor and sulphate as an electron acceptor. Sulphide may also be released when sulphur-containing amino acids are broken (Toprak, 1997). Oxygen production by algae prevents anaerobic conditions through most of the pond depth, and oxidise dissolved sulphide thereby preventing the release of hydrogen sulphide gas, to which unpleasant pond odours are usually attributed. The increase in pH caused by algal photosynthetic activity also influences the form of the dissolved sulphide and the rate of oxidation. Sulphide may also be oxidised by anaerobic photosynthetic bacteria which split H_2S to provide reducing power for photosynthesis (from the hydrogen) releasing sulphate and elemental sulphur which can sometimes be observed on the surface of heavily loaded ponds (Houghton and Mara, 1992; Veenstra *et al.*, 1995). They are often called "pink ponds" because of the distinctive colour associated with the unusual photosynthetic pigments in these organisms. If the oxidised form of sulphur finds itself in the reducing zone again then it may well be reduced again. Indeed there is no reason to doubt that there is extensive internal cycling of sulphur in waste stabilization ponds. However, sulphide is not all bad. It will combine with certain heavy metals to form insoluble sulphides that will precipitate out and remove this particularly pernicious form of pollution.

3.7.3 Phosphorus

Phosphorus is primarily important because of its role in eutrophication. Total phosphorus levels in domestic influent are typically 4-15 mg/l. Two removal mechanisms have been identified: algal uptake and precipitation by combination with iron or calcium. The phosphorus in the sediment is mostly inorganic with reports of between 60-70% (Ortuno *et al.*, 2000) and 95% (Gomez *et al.*, 2000), of the phosphorus being combined with either iron or calcium. These findings suggest that chemical precipitation is the major removal mechanism. The phosphorus in the sediment will feed back into the water column in both aerobic and anaerobic conditions though the latter is faster (Ortuno *et al.*, 2000). Phosphorus removal is discussed in more detail in Chapter 5.

3.7.4 Carbon

Though there is no question that carbon is removed in waste stabilisation ponds, quite where it goes is a matter of debate. Traditionally, it has been assumed that organic carbon present in the influent was broken down to CO_2 by heterotrophic bacteria and then taken up by algae which are themselves subsequently broken down, and so on (Mara, 1976). There is, of course, no net loss of carbon in this cycle and this explanation persists despite longstanding and contemporary critiques (McKinney, 1962; El Ouarghi *et al.*, 2002). Moreover it is significant that the loss of organic carbon occurs in even the most heavily loaded waste stabilization ponds (Silva, 1982; Abis and Mara, 2002). It is evident that anaerobic processes are also very important. Of course some carbon accumulates in the sludge and is removed during desludging and some is lost as biomass in the effluent. However, a considerable, but as yet undocumented proportion of the carbon is converted into methane and volatile fatty acids. This methane may then be converted to carbon dioxide by methane oxidising bacteria which are almost certainly present in aerobic waste stabilization ponds. See Chapter 4 for further information on the removal of organic carbon.

3.8 SUMMARY

The physical and chemical environment is, if nothing else, dynamic. The basic roots of change are twofold: the climate and the design of the pond. Moreover, the most powerful effect of both design and climate is its ability to effect photosynthesis since the algae (and to some extent the bacteria) are driven by light and temperature.

3.9 RESEARCH RECOMMENDATIONS

There is still much to understand about the physico-chemical conditions in WSP. The basic concepts are well established in the WSP and the limnological literature. However, we still do not have a sufficiently detailed or quantitative understanding of even the carbon cycle and we cannot predict *a priori* pH, dissolved oxygen concentrations or sulphur concentrations and speciation: key parameters governing WSP performance. This kind of quantitative understanding is essential. WSP are large and their inner workings are complex. Consequently, a purely empirical approach in which depth, climate, load and so on are varied is not only expensive to undertake but is also remarkably uninformative. Very expensive studies often yield relatively situation bound results of low predictive value. Detailed quantitative studies are challenging and a significant number of cooperative studies are required to yield usable quantitative models. Coordination of research effort is required. This is difficult, but worthwhile. The beguiling physical simplicity of WSP should not allow us to forget that the ability to quantitatively understand and thus truly engineer WSP requires engineering science of the very highest calibre.

REFERENCES

Abis, K.L. and Mara, D.D. (2002) Research on waste stabilisation ponds in the United Kingdom-I Initial results from pilot-scale facultative ponds. *Proceedings 5th International IWA specialist group conference on Waste Stabilisation Ponds* 1-10.

APHA (1995) *Standard Methods for the Examination of Water and Wastewater*, 19th Edition. American Public Health Association, Washington DC.

Curtis, T.P., Mara, D.D. and Silva, S.A. (1992) The influence of humic substances oxygen and pH on the effect of light on faecal coliforms in waste stabilization ponds, *Applied and Environmental Microbiology* **58**, 1335-1343.

Curtis, T.P., Mara, D.D., Dixo, N.G.H. and Silva, S.A. (1994) Light penetration in waste stabilization ponds. *Water Research* **28**(5), 1031-1038.

Davies-Colley, R.J., Donnison, A.M., Speed, D.J., Ross, C.M. and Nagels, J.W. (1999) Inactivation of faecal indicator micro-organisms in waste stabilisation ponds: interactions of environmental factors with sunlight. *Water Research* **33**(5), 1220-1230.

Dubinsky, Z., Falkowski, P.G., Post, A.F. and van Hes, U.M. (1987) A system for measuring phytoplankton photosynthesis in a defined light-field with an oxygen-electrode. *Journal of Plankton Research,* **9**(4), 607-612.

Ellis, K.V. (1983). Stabilization Ponds: Design and Operation. *Critical Reviews in Environmental Control* **13**, 69-102.

El Ouarghi, H., Praet, E., Jupsin, H. and Vasel, J-L. (2002). About the contribution of algae to the WSP process. *Proceedings 5th International IWA specialist group conference on Waste Stabilisation Ponds* 817-214.

Gomez, E., Casellas, C., Picot, B. and Bontoux, B. (1995) Ammonia elimination processes in stabilization and high rate algal pond systems. *Water Science and Technology* **31**, 303-312.

Gomez, E., Paing, J., Casellas, C. and Picot, B. (2000) Characterisation of phosphorus in sediments from waste stabilization ponds. *Water Science and Technology* **42**(10), 257-264.

Hall, D.O. and Rao, K.K. (1999). *Photosynthesis*, 6th Edition. University Press, Cambridge.

Huang, H.J.S and Gloyna, E.F. (1984) Phosphorous models for waste stabilization ponds. *J. Environmental Engineering Division ASCE* **110**, 550-561.

Houghton, S.R. and Mara, D.D. (1992) The effects of sulphide generation on photosynthetic populations and effluent quality. *Water Science and Technology* **26,** 1759-1768.

Huisman, J., Jonker, R.R., Zonneveld, C. and Weissing, F.J. (1999) Competition for light between phytoplankton species: experimental tests of mechanistic theory. *Ecology* **80**(1), 211-222.

Kartal, B (2002), MSc Thesis, University of Newcastle, UK.

Kayombo, S., Mbwette, T.S.A., Mayo, A.W., Katima, J.H.Y. and Jorgensen, S.E. (2000) Modelling diurnal variation of dissolved oxygen in waste stabilization ponds. *Ecological Modelling* **127**, 21-31.

Kellner, E., and Pires, E.C. (2002) The influence of thermal stratification on the hydraulic behaviour of waste stabilization ponds, *Wat. Sci. Tech.*, **45**, 41-48.

Kirk, J.T.O. (1994) *Light and Photosynthesis in Aquatic Ecosystems,* 2nd Edition. Cambridge University Press, Cambridge.

Konig, A (1982) PhD Thesis, University of Liverpool, UK.

Lai, P.C.C. and Lam, P.K.S. (1997) Major pathways for nitrogen removal in wastewater stabilization ponds. *Water air and soil pollution* **94**, 125-136.

Lawlor, D.W. (1993) *Photosynthesis: Molecular, Physiological and Environmental Processes,* 2nd Edition. Longman Scientific and Technical, Harlow.

Li-Cor (1991) *Li-Cor Underwater Radiation Sensors, Type SA: Instruction Manual,* 2nd Edition. Li-Cor, Lincoln.

Llorens, M., Saez, J. and Soler, A. (1992) Influence of thermal stratification on the behaviour of a deep wastewater stabilization pond. *Water Research* **26**, 569-577.

Ludyanskiy, M.L. and Pasichny, A.P. (1992) A system for water toxicity estimation. *Water Research* **26**(5), 689-694.

Mara, D.D. (1976) *Sewage Treatment in Hot Climates.* John Wiley and Sons, Chichester, UK.

Mara, D.D., Alabaster, G.P., Pearson, H.W. and Mills, S.W. (1992) *Waste Stabilisation Ponds: A design manual for eastern Africa.* Lagoon Technology International, Leeds, UK.

Mara, D.D., Pearson, H.W., Alabaster, G.P. and Mills, S.W. (1997) *An Evaluation of Waste Stabilization Ponds in Kenya*. Research Monographs in Tropical Public Health Engineering, No. 11, University of Leeds, UK.

Marais, G.V.R. (1966) New factors in the design and operation and performance of waste stabilization ponds. *Bulletin of the World Health Organisation* **34**, 737-763.

Marra, J. (1978) Effect of short-term variations in light intensity on photosynthesis of a marine phytoplankter: a laboratory simulation study. *Mar. Biol.* **46**, 191-202.

McKinney, R.E. (1962) *Microbiology for Sanitary Engineers*. McGraw-Hill, New York,

Monteith, J.L. (1973) *Principles of Environmental Physics*. Edward Arnold, London, UK.

Neale, P.J. and Marra, J. (1985) Short-term variation of P_{max} under natural irradiance conditions: a model and its implications. *Mar. Ecol. Prog. Ser.* **26**, 113-124.

Ortuno, J.F., Saez, J., Llorens, M. and Soler, A. (2000) Phosphorus release from sediments of a deep wastewater stabilization pond. *Water Science and Technology* **42**(10), 265-272.

Pano, A. and Middlebrooks, E.J. (1982) Ammonia nitrogen removal in facultative wastewater stabilization ponds. *J. Wat. Pollut. Control Fed.* **54**(4), 344-351.

Parhad, N.M. and Rao, N.U. (1974) Effect of pH on survival of *Escherichia coli*. *J. Wat. Poll. Cont. Fed.* **46**(5), 980-986.

Pearson, H.W., Mara, D.D., Mills, S.W. and Smallman, D.J. (1987) Physicochemical parameters influencing the fecal bacterial survival in waste stabilization ponds, *Water Science and Technology* **19**, 145-152.

Petroski, H. (1985) *To engineer is human: the role of failure in successful design.* St Martins Press, New York, USA.

Pettygrove, T. and Asano, T. (1985) *Irrigation with reclaimed municipal wastewater: a guidance manual*, Lewis, Boca Raton.

Platt, T., Denman, K.L. and Jassby, A.D. (1977) Modeling the productivity of phytoplankton. In *The Sea: ideas and observations on progress in the study of the seas* (Edited by Goldberg, E. D.), pp. 807-856. John Wiley, New York, USA.

Polprasert, C., Dissanayake M.G. and Thanh, N.C. (1983) Bacterial die off kinetics in waste stabilization ponds. *J. Wat. Pollut. Cont. Fed.* **55**(3), 285-296.

Reynolds, C.S. and Irish, A.E. (1997) Modelling phytoplankton dynamics in lakes and reservoirs: the problem of in-situ growth rates. *Hydrobiology* **349**, 5-17.

Snoeyink, V.L. and Jenkins, D. (1980) *Water Chemistry*, Wiley, New York, USA.

Schulze, E.-D. and Caldwell, M.M. (Ed.) (1995) *Ecophysiology of Photosynthesis*. Springer-Verlag, Berlin, Germany.

Silva, S.A. (1982) PhD Thesis, University of Dundee, UK.

Silva, S.A., de Oliveira, R., Soares, J., Mara, D.D. and Pearson, H. W. (1995) Nitrogen removal in pond systems with different configurations and geometries. *Wat. Sci. Tech.* **31**(12), 321-330.

Talling, J.F. (1957) Photosynthetic characteristics of some freshwater plankton diatoms in relation to underwater radiation. *New Phytol.* **56**, 29-50.

Toprak, H. (1997) Hydrogen sulphide emission rates originating from anaerobic waste stabilization ponds. *Environmental Technology* **18**, 795-805.

Torres, J.J., Soler, A., Saez, J. and Llorens, M. (2000) Hydraulic performance of a deep stabilisation pond fed at 3.5 m depth. *Water Research* **34**, 1042-1049.

van der Heever, J.A. and Grobbelaar, J.U. (1997) The use of oxygen evolution to assess the short-term effects of toxicants on algal photosynthetic rates. *Water SA* **23**(3), 233-237.

Veenstra, S.A., Alnozailyh, F.A. and Alaerts, G.J. (1995) Purple sulphur bacteria and their influence on waste stabilization pond performance in the Yemen. *Water Science and Technology* **31**(12), 141-149.

Weatherell, C.A. (2001) *Predicting Algal Concentration in Waste Stabilisation Ponds*. PhD Thesis, University of Newcastle, UK.

4

Solids and organics

Nick Walmsley and Andy Shilton

4.1 WASTEWATER CHARACTERISTICS

4.1.1 General overview

The solid and organic constituents of wastewater, which are the main focus of this chapter, are examined and discussed in more detail in the following sections, but prior to these and the subsequent chapters on nutrients (Chapter 5), pathogens (Chapter 6) and heavy metals (Chapter 7), this section provides a general overview of wastewater characteristics. Typical waste constituents in municipal sewage are summarised in Table 4.1, (for information on agricultural wastes refer to Chapter 19).

It is worth noting that the term 'municipal' allows for discharges from the numerous small trade and commercial premises in most communities as distinct from 'domestic' which relates solely to discharges from people's homes. Few community sewers discharge only domestic wastes to a pond treatment system. In addition to the parameters listed in Table 4.1, there are a range of minor

ISBN: 1843390205. Published by IWA Publishing, London, UK.

constituents, including heavy metals (see Chapter 7), sulphur compounds and synthetic organics. Some of these constituents are actually micronutrients and necessary for full biological activity but can still be toxic at high concentrations.

Concentrations of wastewater vary widely depending on water usage and stormwater infiltration/inflows to the sewer. Considering concentrations alone can therefore be misleading and it is important to instead quantify the mass load (concentration x flow) when undertaking a design (see Chapter 9) or when generally assessing the strength of a particular wastewater (or effluent).

Table 4.1 Typical constituents of municipal sewage (Imhoff *et al.*, 1972; Rössle and Pretorius, 2001; Metcalf and Eddy, 2003; WEF, 1998; Fair, 1968)

Parameter	**Typical Values** Per capita load	Concentration
Dry Weather Flow	150-250 l/p/d	
pH	6.8-7.6	
Grease and Oil	12-25 g/p/d	48-167 mg/L
BOD_5	50-80 g/p/d	200-350 mg/L
COD	100-200 g/p/d	400-800 mg/L
SS	50-80 g/p/d	200-350 mg/L
Amm-N	6-8 g/p/d	24-53 mg/L
Org-N	3-4 g/p/d	12-27 mg/L
TKN	9-11 g/p/d	35-75 mg/L
TP	2-4 g/p/d	8-25 mg/L
Anionic Detergents (MBAS)	1.2-2.5 g/p/d	5-13 mg/L

The nitrogenous compounds in raw sewage exist primarily in the form of organic (proteinaceous) nitrogen and ammonia nitrogen. The sources are primarily human excretion of which some 75% is excreted as urea and the rest as organic nitrogen. Oxidised nitrogen compounds such as nitrate exist only in very low quantities. Proteins and urea undergo deamination, releasing ammonia as the wastewater flows to the treatment plant. The longer the sewage is held in the sewer the greater will be the release of ammonia.

Phosphorous compounds in raw sewage are predominantly phosphates with typically some 50% deriving from detergents. Low phosphorus detergents, which are becoming more common in developed communities, may be the most efficient and cheapest way of reducing phosphorus discharges to the environment. A full review of nutrients in relation to pond treatment can be found in Chapter 5.

In domestic sewage chloride, nitrogen, phosphate and potash occur mainly in the urine (70-90%) compared to faeces (Fair, 1968; Metcalf and Eddy, 2003).

Alkalinity and specific conductance vary according to the dissolved mineral content of the water supply. For municipal sewage i.e. no major industry waste content, alkalinity values are often the water supply values plus 50-100 mg/L as $CaCO_3$.

Saltwater must be prevented from entering ponds and sewerage systems serving ponds. Salt water may be contained in some industrial discharges such as from ion exchange water treatment systems or can occur in wastewater reticulated in coastal areas due to infiltration or where brackish water is used as a supply. Combinations of sewage and saltwater which contain more than 5% saltwater can result in sulphates being reduced to sulphides and subsequent release of odorous hydrogen sulphide. This issue is discussed further in Chapter 12.

4.1.2 Organic constituents

Organic carbon is present in water in naturally occurring organic matter such as plant and animal detritus. These substances tend to be readily biodegradable. Organic carbon can also occur in traces of lubricants, liquid fuels, fertilisers and pesticides. These man-made substances are not all readily biodegradable.

When readily biodegradable substances are introduced into the environment they quickly decompose through the action of natural microbial populations. Some of the organic matter is oxidised to carbon dioxide and water while the rest is used for the synthesis of new microbial cells. In due course, these organisms will also die and become food for other decomposers.

When a biodegradable organic wastewater is discharged into an aquatic ecosystem such as a stream, estuary or lake, oxygen dissolved in the water is consumed due to the respiration of micro-organisms that oxidise this organic matter. The more easily biodegradable a waste, the more rapid is the rate of its oxidation and the corresponding consumption of oxygen. Because of this relationship and its significance to water quality (dissolved oxygen levels in the water), the organic content of wastewaters are usually measured in terms of the amount of oxygen consumed during their oxidation. This is termed the Biochemical Oxygen Demand (BOD) or the Chemical Oxygen Demand (COD), depending on the method of analysis used.

BOD is defined as the amount of oxygen required by bacteria to decompose a defined quantity of organic matter for a specified time (usually 5 days) under aerobic conditions. Provided that nitrification has been inhibited, the amount of oxygen reported with this method represents only the carbonaceous oxygen demand (CBOD) and is essentially the 'easily decomposed' organic matter.

COD is defined as the oxygen equivalent of the organic portion of the sample that is susceptible to oxidation by a strong chemical oxidant. COD measures both the biodegradable and the non-biodegradable or "inert" organic matter. As a consequence the COD test values are always greater than the BOD_5 test values for the same sample.

For untreated domestic sewage the ratio of COD:BOD_5 is typically about 2.2 but can vary between about 1.5 and 3 (Metcalf and Eddy, 2003). The ratio is an indication of biodegradability and if it is greater than 3 then it is likely that the wastewater contains some toxic components or that there is a nutrient or micronutrient deficiency that restricts biological activity.

As the wastewater is treated in a pond process the COD:BOD_5 ratio increases as the readily biodegradable fraction is consumed. A well treated and stabilised secondary effluent with low soluble BOD_5 can have a COD:BOD_5 ratio >5.

A discussion on the biodegradability of agricultural wastes is found in Chapter 19.

4.1.3 Solid constituents

The solids concentrations in fresh municipal sewage are split between coarse suspended matter, colloidal dispersed matter and molecular solution (dissolved) matter. The physical characterisation of sewage solids can be explained (Rössle and Pretorius, 2001) as:

- **Coarse suspended matter** – particles readily settleable of colloidal and non-colloidal nature (particle size > 1 μm); microscopically visible and filterable.
- **Colloidal dispersed matter** – fine particles not readily settleable (particle size 1 μm to 1 nm); ultra-microscopically visible, non-filterable, chemically flocculable.
- **Molecular solution (dissolved) matter** – Constituents are dissolved in true solution (particle size < 1 nm); not visible by any instrumental method, not removed by physical treatment processes.

Approximately 50-70% of the solids in municipal wastewater are readily settleable. These solids typically contain 25-40% of the BOD_5 load (Metcalf and Eddy, 2003). This is, therefore, the treatment efficiency that can be achieved by a primary settlement tank and it follows that this is what a pond could achieve in the absence of any biological activity e.g. very cold conditions (see Chapter 18 on cold climate ponds).

Of the remaining 60-75% of the BOD_5, approximately half is filterable and half is non-filterable.

The nutrients typically associated with settleable solids are:

- 10-20% of the organic nitrogen; and,
- 15-25% of the phosphorus.

4.1.4 Decay prior to treatment

If the sewage is allowed to go anaerobic through long retention in a sewerage system and/or due to warm temperatures, then decay of dissolved and solid organics will occur. This will shift the balance between the solid and the liquid phases with concentrations of fermented organic acids increasing in the liquid phase as solids are hydrolysed. This decay also releases nutrients bound up in the solid fraction.

4.2 GROWTH OF SOLIDS AND ORGANICS WITHIN A POND

When considering a treatment system such as a pond it is obvious to focus on the decay of solids and organics, but within pond systems there is also growth of biomass to consider including:

- bacterial growth;
- algal growth;
- growth of higher-level microorganisms.

4.2.1 Bacterial growth

Bacterial growth is essentially a transformation of the organic carbon (as commonly estimated by BOD or COD) that was contained in the waste. Bacteria use this organic matter for two purposes – energy and growth. As a result of the bacteria consuming the organic carbon it is either converted to new cells (growth) or ends up being converted to CO_2 if oxidised by aerobic bacteria or CH_4 and CO_2 if fermented by anaerobes.

4.2.2 Algal growth

Algal growth takes its carbon source from the CO_2 expelled by the bacteria or from CO_2 that diffuses into the pond water from the atmosphere. The interesting point to recognise here is that algal growth will fix organic carbon thereby <u>increasing</u> both the solids and the BOD concentrations, which is the opposite to what we want a pond to achieve! While it is true that this does occur, the overall effect of the bacteria breaking down the wastewater solids/organics while the algal growth supplies the oxygen and elevates the pH results in a net beneficial effect in terms of the pond's overall treatment performance.

The algae and bacterial biomass is either washed out with the pond effluent, is grazed or dies and sinks to the bottom.

It is important to prevent the development of surface mats of filamentous algae, or thin scum of encysted unicellular or blue-green algae (Hartley and Weiss, 1970).

Such growths seriously reduce the light penetration and interfere with the mass transfer of oxygen into the pond from the atmosphere. Infestations of surface-dwelling macrophytes such as *Lemna* have the same adverse effects (Hawkes, 1983).

4.2.3 Growth of higher-level organisms

Higher-level organisms grow by grazing of bacteria by protozoa/rotifers (see Chapter 2); and by grazing of algae by zooplankton (see Chapter 2) or by fish (see Chapter 16). While this grazing consumes a lot of solids it does, in itself, create some new biomass. Again, some of the consumed carbon is oxidised for energy and released as carbon dioxide and so this results in a further net loss of organic carbon from the pond system.

4.3 DECAY OF SOLIDS AND ORGANICS WITHIN A POND

4.3.1 Overview of decay in ponds

Anaerobic ponds act in several capacities. Firstly, they provide more than adequate detention time for 'primary' settlement of solids. Secondly, these settled solids are anaerobically digested in a sludge layer at the bottom of the pond – this might be compared to an unmixed, unheated sludge digester. Thirdly, in addition to sludge digestion, these ponds may also provide some anaerobic biological treatment of the fine solids that remain suspended and the dissolved organic waste in the liquid layer.

As organic matter enters a facultative pond the settleable and flocculated colloidal material settles to the bottom to form a sludge layer where organic matter is decomposed anaerobically. The remainder of the organic matter, which is either soluble or suspended, remains in the body of the water where decomposition is predominantly undertaken by aerobic or facultative bacteria.

Facultative waste stabilisation ponds are characterised by having an upper aerobic zone, a variable anoxic zone below this and a bottom anaerobic sludge zone with active decay of organics occurring in all three zones. Primary facultative ponds that receive raw wastewater (as opposed to secondary facultative ponds which receive pre-treated wastewater) obviously tend to have a greater sludge layer with more digestion activity.

Maturation ponds, which often follow facultative ponds in series, provide more pathogen treatment but less percentage reduction in organics and solids. The organic load on maturation ponds is very low so that they remain aerobic throughout their depth, except possibly in the thin sludge layer at the bottom. BOD_5 and suspended solids may increase across a maturation lagoon due to algal growth, which converts carbon dioxide to biomass.

4.3.2 Aerobic decay

In a classic relation carbon dioxide is released by bacteria (an animal) as they oxidise organic carbon from the wastewater as an energy source. This is then consumed by the algae, (a plant) which in turn releases oxygen that is used by the bacteria. An overview of the activity and transformations of solids and organics in a primary facultative pond is presented in Figure 4.1.

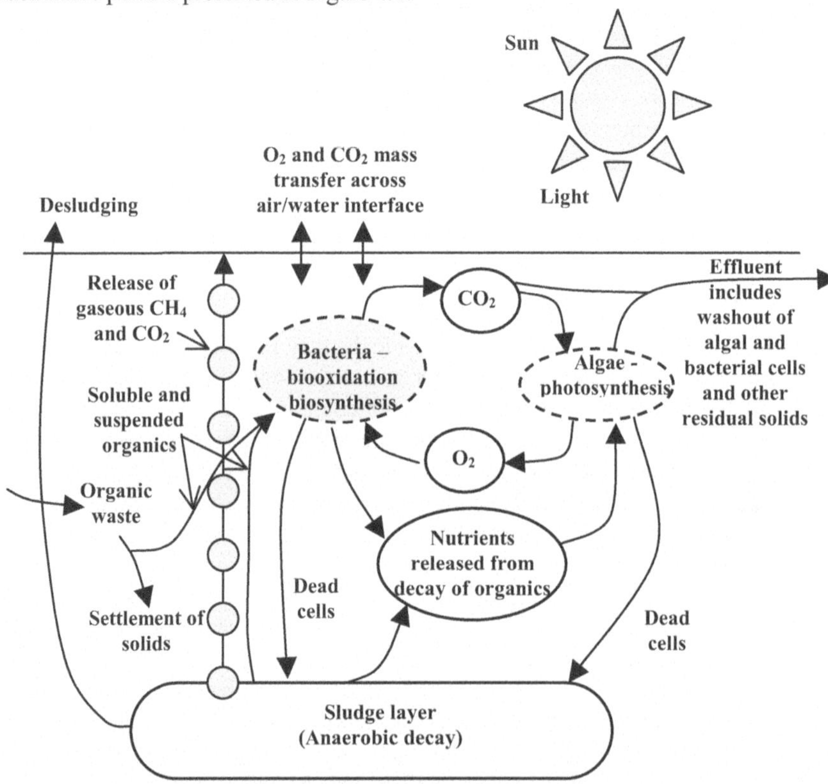

Figure 4.1 Basic biological interactions in a facultative pond with emphasis on solids and organics transformations (modified from Hawkes, 1983)

Dissolved organic matter in solution is absorbed and consumed principally by aerobic and facultative bacteria. Fine suspended organic solids are also colonised, hydrolysed and consumed by such bacteria.

The rate of BOD removal is a function of the bacterial density and activity. The organic carbon (measured as carbonaceous BOD) ultimately leaves the pond as either carbon dioxide (CO_2) or methane (CH_4) gas; is accumulated and removed as sludge;

or is washed out with the effluent as a dissolved organic compound, dissolved CO_2 or in solid form such as algae, bacteria or higher level animals.

The key contribution of algae to this treatment process is as phototrophs, producing oxygen to maintain the aerobic conditions of the pond. As the bacteria breakdown organic matter they release nitrogen and phosphorous compounds and carbon dioxide which the algae rely on for growth.

Apart from nutrient availability, temperature and solar radiation are the major factors effecting algal photosynthetic activity. Most algae grow over a wide temperature range of 4-40°C with optimum growth for the dominant green algae about 20°C. Photosynthetic activity is determined by the quantity of light entering the pond, which in turn is a function of the surface area of the pond. Organic loading rates used for design are therefore expressed as Kg BOD_5/ha/day i.e. areal loading.

4.3.3 Grazing by higher-level aerobes

Many higher life forms (animals) can develop in lagoons. These include protozoans and microinvertebrates such as rotifers, daphnia and annelids (often termed the zooplankton). Many of these organisms play a role in waste purification by feeding on bacteria and algae and promoting flocculation and settling of particulate material. See Chapter 2 for more detail.

4.3.4 Anaerobic decay in the liquid zone

Within an anaerobic pond and to a lesser extent in the lower reaches of a facultative pond, anaerobic decay can occur in the liquid zone, particularly at warmer temperatures. This decay process is aided by activity in the anaerobic sludge layer that results in gaseous build-up, which then lifts the anaerobic biomass as it erupts up to the surface. These gaseous eruptions thereby provide a mechanism for the mixing of the bacterial biomass up into the liquid layer.

If overlaid by an aerobic zone, as is the case for facultative ponds, hydrogen sulphide generated and released from the sludge layer is oxidised, thus preventing odours from being released.

4.3.5 The benthic/sludge zone

In the sludge layer the settled solids are anaerobically broken down with methane and carbon dioxide being released together with a variety of nutrients and other soluble degradation products. The gases escape to the atmosphere – in a facultative pond up to 30% of the BOD load can be dissipated via gas. Anaerobic degradation is temperature-dependent and no significant activity occurs in the sludge below a water temperature of about 15°C. A 4-fold increase in activity occurs over the temperature range of 4°-22°C (Gloyna, 1971).

The accumulated solids settle, digest and thicken on the floor of the pond. Little data about the maturation of accumulated sludges exist but one study (Carré, 1987) found the dry solids content higher near the inlet and increased with depth. The sludge profile indicated a ten-fold increase in concentration over the 55 cm depth profile that had accumulated over 10 years. This ranged from 4.7-48% dry solids at the inlet to 3.2-30.1% dry solids at the outlet. Nelson (2000) reported a similar trend in concentration for a facultative pond sludge up to 8 years old with 4.1-15.7% across the depth profile. Both studies showed similar volatile solids levels from 53% for the freshest to 33% for the oldest solids.

In practice, in situ measurement of solids below the water column is difficult to undertake accurately and most full-scale sludge surveys show lower concentrations, possibly due to mixing during the sampling. Walmsley (1995) reported the desludging of three 10-hectare facultative ponds after 10 years operation with an in situ sludge profile surveyed at an average of 6.5% dry solids and measurements up to only twice this concentration. The mass of solids measured during the full-scale desludging confirmed that the in situ survey had under-estimated the sludge quantities.

4.3.6 Benthic feedback

Soluble degradation products, such as ammonia, organic acids and inorganic nutrients are released from the benthic sludge layer during anaerobic decay and may be subsequently oxidised aerobically in the over-laying pond liquid.

This 'benthic feedback' from the sludge into the liquid can create a large extra oxygen demand on the liquid zone at particular times of the year. Overloading problems can result, particularly for an already heavily loaded pond, when over a cold winter settled solids are accumulated and then in spring, as temperatures rise, this accumulation degrades and imposes a large extra oxygen demand on the pond.

Iwema *et al.,* (1987) studied this on a 2152 m^2 facultative pond but failed to establish a short-term sedimentation-digestion balance due to difficulties with bottom sediment being re-suspended due to gas mixing. Lumbers (1988) showed the importance of benthic feedback can be an important factor regarding the overall loading into the aerobic layer but it is not yet possible to incorporate this within pond design methods.

4.4 TREATMENT PERFORMANCE

Facultative ponds normally achieve 75-85% BOD_5 (unfiltered) removal with effluent suspended solids consisting predominantly of algae and zooplankton, commonly present at concentrations between 50-70 mg/l. Anaerobic plus facultative plus maturation ponds can achieve 85-95% BOD_5 removals.

A properly designed and run facultative pond will usually have an effluent filtered BOD_5 less than 20 mg/L, and often less than 10 mg/L even from the first pond. This is important in the European Community where the requirements for discharge into surface or coastal waters are based on a filtered sample. However, this is not universal and many countries have discharge limits that are based on whole (unfiltered) sample values. Table 4.2 summarises typical effluent results from a primary (facultative) pond treating municipal sewage.

Table 4.2 Typical effluent results from a primary facultative pond treating municipal sewage (Archer, 2003; Pearson, 2004)

Contaminant	Median	90th percentile
Unfiltered BOD5 (g/m3)	25-30	35-75
Suspended solids (g/m3)	45-90	75-175
Ammoniacal-N (g/m3)	17-25	27-30

When higher BOD_5 and suspended solids levels occur in a pond effluent it is almost always due to the presence of algae. Therefore, if algae can be removed or reduced in the effluent, BOD and suspended solids will invariably be reduced. There are various techniques for upgrading pond systems to improve the removal of organics and solids and these are detailed in Chapter 11.

4.5 SUMMARY AND FUTURE RESEARCH NEEDS

Substantial knowledge now exists on wastewater characteristics and their relationships and influence on the physical, biological and biochemical reactions occurring in ponds. However, we are still unable to utilise this knowledge with certainty as part of design criteria for prediction of:

- Nitrogen removal, particularly ammonia transformations;
- Benthic feedback effect on oxygen demand and odour potential;
- Sludge accumulation and degradation over time.

REFERENCES

APHA. Standard Methods for the Examination of Water and Wastewater, 20th edition. American Public Health Association, Washington, DC, 1998.

Archer H. and O'Brien B. Waste Stabilisation Ponds – Improved Performance that is Cost-effective, NZWWA Conference, 2003.

Carré J. and Baron D. (1987) Effects of Maturation on the Characteristics of Wastewater Stabilisation Pond Sludges. *Water Science and Technology* **19**(12).

Fair G.M., Geyer J.C. and Okun D.A. (1968) Water and Wastewater Engineering, Volume 2. *Water Purification and Wastewater Treatment and Disposal.*

Gloyna E.F. Waste Stabilisation Ponds, *WHO Monograph Series No 60*, 1972.

Gray N.F. Biology of Wastewater Treatment, Oxford University Press, 1989.

Hartley W.R. and Weiss C.M. Light Intensity and vertical distribution of algae in tertiary oxidation ponds, *Water Research* **4**, 1970.

Hawkes H.A. (1983) The applied significance of ecological studies of aerobic processes. In *Ecological aspects of used water treatment, Vol. 3, The processes and their ecology*, (Ed C.R. Curds and H.A. Hawkes), Academic Press.

Imhoff K., Muller W.J. and Thistlethwayte D.K.B. Disposal of Sewage and other Water-borne Wastes, 2nd Ed., 1972.

Iwema A., Carré J. and Minot D. (1987) Sedimentation and Digestion on Pond Bottoms – An Attempt to Establish a Short-term Material Balance. *Water Science and Technology* **19**(12).

Lumbers J.P. and Andoh R.Y.G. (1998) The Identification of Benthic Feed-back in Facultative ponds. *Water Science and Technology* **19(**12).

Metcalf and Eddy Inc. Wastewater Engineering Treatment and Reuse, 4th Ed., 2003.

Nelson K.L. and Jimenez B.C. (2000) Sludge Accumulation, Properties and Degradation in a Waste Stabilisation Pond in Mexico. *Water Science and Technology* **42(**10-11), 231-236.

Pearson H.W., Silva Athayde, S.T., Athayde, G.B. and Anselmo Silva, S. The Ideal Pond Design for all Eventualities: Do We know enough to set clear unequivocal physical design guidelines? *IWA 6th International Conference on WSPs,* 2004.

Rössle W.H. and Pretorius W.A. (2001) A review of Characterisation Requirements for In-line Fermenters, Paper 1: Wastewater Characterisation. *Water S.A.* **27**(3).

Walmsley N.A. and Dougherty A.P. Desludging of Large Facultative Ponds With Controlled Sludge Disposal to Land, NZWWA Conference 1995.

WEF Manual of Practice No 8, Design of Municipal Wastewater Treatment Plants, 4th Ed., 1998.

5

Nutrients

Rupert Craggs

The effectiveness of nutrient removal by a pond treatment system is linked to the characteristics of the pond environment. For information on the microbiological, physical and chemical environment within a pond and its implications for nutrient removal refer to Chapters 2 and 3.

5.1 INTRODUCTION

Discharge of wastewater stabilization pond (WSP) effluent containing high concentrations of nutrients can have several adverse impacts on receiving water quality:

- Elevated nitrogen (N) and phosphorus (P) concentrations in receiving waters may cause eutrophication and proliferation of nuisance plants. They exert a high oxygen demand at night, when they respire rather than photosynthesis oxygen, reducing dissolved oxygen concentration (DO), which may negatively affect other aquatic life.

- Free ammonia (NH_3-N) is potentially toxic to fish and other aquatic life particularly in receiving waters containing aquatic plants, where daytime photosynthesis temporarily elevates temperature and pH (Davies-Colley *et al.*, 1995).

Regulations governing the discharge of treated wastewater increasingly require significant removal of nutrients. Therefore improving the nutrient removal capability of WSPs is becoming ever more necessary. A more thorough understanding of the nutrient removal processes that occur in WSPs and the factors affecting them is critical to developing design improvements to enhance WSP nutrient removal performance.

5.1.1 Typical wastewater nutrient concentrations

Nutrients are present in wastewaters as particulate and colloidal organic solids (mainly insoluble proteins, nucleic acids and polysaccharides), dissolved organic matter (e.g. monosaccharides, polyphosphates), and dissolved inorganic compounds (e.g. NH_4-N, NO_2-N, NO_3-N and phosphate (PO_4-P)). Dissolved inorganic nutrient concentrations depend upon the degree of decomposition of the wastewater before it reaches the WSP. Nutrients are released from the organic material by both anaerobic and aerobic mineralisation processes, for example, the aerobic process of ammonification:

$$C_{18}H_{19}O_9N + 17.5\ O_2 + H^+ \rightarrow 18\ CO_2 + 8H_2O + NH_4^+$$

Much of the nitrogen in domestic wastewaters comes from urine, and is initially in the form of urea (70-90% of total nitrogen, TN). However, urea is hydrolysed to ammoniacal-N by urease enzymes that naturally occur in urine (Silva *et al.,* 1995). Thus, typically 60% of wastewater TN is present as ammoniacal-N (Barnes and Bliss 1983).

Phosphates can comprise up to 50-70% of the total phosphorus in domestic wastewater and are released by decomposition of organic phosphorus compounds and hydrolysis of polyphosphates by phosphatase enzymes. Typical concentrations of nutrients and other parameters in domestic wastewater are given in Table 5.1.

Both ammoniacal-N and phosphate react with water to form an equilibrium with their ionic form. The balance of these equilibriums depends on the nutrient concentration, temperature and pH. Ammoniacal-N exists in an equilibrium of free (or 'free') ammonia (NH_3) and the ammonium ion (NH_4^+):

$$NH_3 + H_2O \leftrightarrow NH_4^+ + OH^-$$

Table 5.1 Ranges of median nutrient concentrations (g m^{-3}) found in domestic wastewater (Middlebrooks *et al.*, 1982; Oswald 1988; Bitton 1994; Mara 1997)

Parameter	Domestic sewage
Biochemical Oxygen Demand (BOD_5)	70-275
Chemical Oxygen Demand (COD)	200-700
Suspended Solids (SS)	100-350
Total Nitrogen (TN)	15-60
Ammoniacal-N (NH_4-N)	8-35
Nitrate-N (NO_3-N)	0.3
Total Phosphorus (TP)	4-15
Phosphate (PO_4-P)	4-10
pH	7-7.8
N:P ratio	4:1

The ammoniacal-N equilibrium is displaced to the left as the pond water pH increases above pH 7.0. As pH has a log scale, this means that there is a tenfold increase in free ammonia (NH_3) concentration between pH 7 and 8. At pH 9, 20-40% of ammoniacal-N is present as NH_3 and above pH 10 more than 80-90% is present as NH_3. Complete conversion of ammoniacal-N to NH_3 gas occurs at around pH 11.

Phosphate exists in an equilibrium of three forms (PO_4^{3-}, HPO_4^{2-} and $H_2PO_4^-$):

$$PO_4^{3-} + H_2O \leftrightarrow HPO_4^{2-} + H_2O + OH^- \leftrightarrow H_2PO_4^- + OH^-$$

At the usual pH of domestic wastewater, NH_4^+ and HPO_4^{2-} are the predominant forms (Nurdogan and Oswald, 1995).

5.2 NUTRIENT REMOVAL PROCESSES

Nutrients may be removed in WSPs by several different processes. Refer to Figure 5.1 and 5.2. The main nutrient removal processes include:

- Sedimentation of wastewater solids
- Assimilation into algal/bacterial biomass (and subsequent biomass removal)
- Volatilisation of ammonia to the atmosphere
- Precipitation of phosphate (and some ammoniacal-N)
- Adsorption to bottom sludge or pond walls
- Nitrification / Denitrification

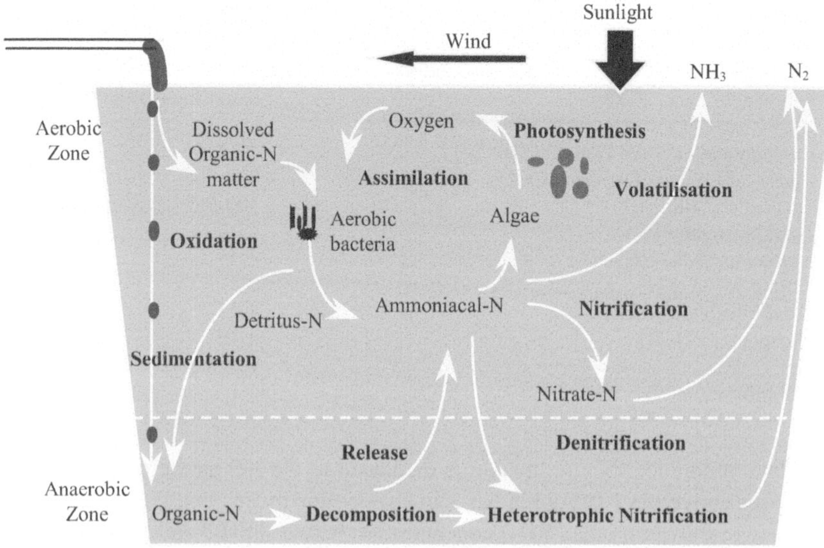

Figure 5.1 Nitrogen removal processes in WSPs

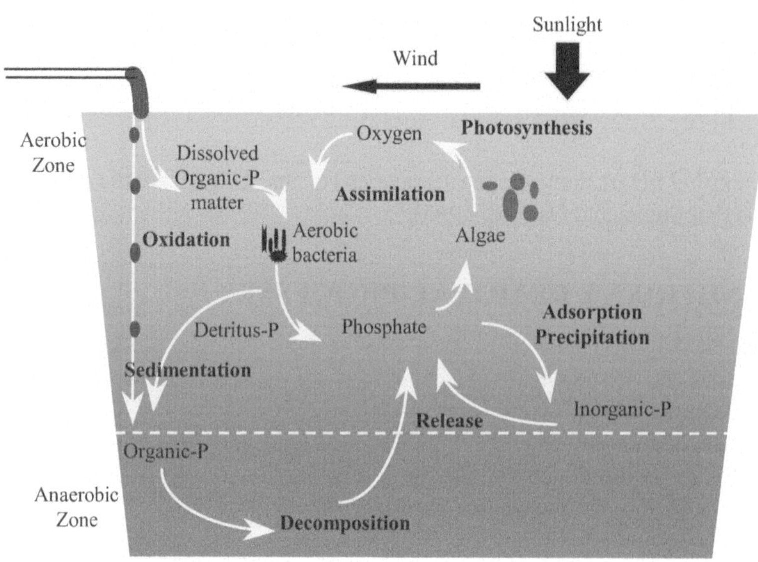

Figure 5.2 Phosphorus removal processes in WSPs

5.2.1 Sedimentation of wastewater solids

Organic N and P are removed from the incoming wastewater through simple sedimentation of wastewater solids. Sedimentation is encouraged by provision of quiescent conditions and can be enhanced by providing deep zones within ponds into which the wastewater is added with a minimum of mixing from the inlet. Alternatively, as discussed further in Chapter 10 (Hydraulics), it may be beneficial to encourage an adequate inlet momentum that distributes the solids out into the pond to avoid localised build up near the inlet.

Many authors have concluded that assimilation/sedimentation is the main process of nitrogen removal in WSPs (e.g. Ferrara and Avci, 1982; Somiya and Fujii, 1984; Reed, 1985; Wrigley and Toerien, 1990; Lai and Lam, 1997). Ferrara and Avci, (1982) estimated that 96% of total N removed was assimilated into algal and bacterial cells. Although Reed, (1985) using the same data concluded that sedimentation only accounted for 25% of total N removal. Lai and Lam, (1997) found that assimilation accounted for up to 25% of ammoniacal-N removal. Assimilation/sedimentation is also an important process for P removal in WSPs (Somiya and Fujii, 1984; Wrigley and Toerien, 1990; Mara, 1997).

5.2.2 Algal / bacterial assimilation

Algae and bacteria in WSPs exist in a classic symbiotic relationship. Bacteria metabolise organic waste for growth and energy, producing new bacterial biomass and releasing carbon dioxide (CO_2) and inorganic nutrients (NH_4-N, NO_x-N and PO_4-P). Algae then utilise the CO_2 through photosynthesis, assimilating the nutrients into algal biomass and releasing O_2. The elevated O_2 concentration, in turn, supports the aerobic bacterial activity.

Assimilation of nutrients into algal and bacterial biomass depends on the cell density, growth rate and composition, and is affected by organic load, nutrient concentration, detention time, and wastewater physical characteristics (temperature, pH and hardness) (Middlebrooks *et al.*, 1999). The nutrient composition of the algae and bacteria cells differs between species and with culture age, (for example, older algal cells contain less nitrogen). The nutrient composition of microalgae grown on domestic wastewater varies between 0.6-16% (average 8%) for nitrogen and 0.16-5.0% (average 2%) for phosphorus (Hemens and Mason, 1968).

Stoichiometric formulae for algal and bacterial biomass can be obtained by dividing the % composition of each element by the atomic weight and rationalizing so that the coefficient for phosphorus is 1. For example:

Bacteria:	$C_{118}H_{170}O_{51}N_{17}P$	(Somiya and Fujii, 1984)
Algae:	$C_{106}H_{181}O_{45}N_{16}P$	(Green *et al.*, 1996)

Both algae and bacteria have N:P atomic ratios of approximately 15:1 (Redfield, 1934). Therefore much less phosphorus than nitrogen is removed by assimilation. Moreover, given that domestic wastewater has an N:P ratio of only about 4:1 (Table 5.1), it contains insufficient nitrogen to enable complete removal of phosphorus by assimilation (de la Noüe and Basseres, 1989; Nurdogan and Oswald, 1995).

Provided light and temperature are not limiting, assimilation of nutrients into algal biomass in WSPs is limited by nutrient availability and the toxicity of ammonia or sulphide (Goldman *et al.*, 1982a and b; Wrigley and Toerien, 1990).

Nitrogen source

Algal growth rate is unaffected by inorganic nitrogen source (NH_4-N, NO_3-N, NO_2-N) (South and Whittick, 1987). However, nitrate and nitrite must be reduced to ammoniacal-N before assimilation, (Oh-Hama and Miyachi, 1988) as only free ammonia (NH_3) can be assimilated by the algae (Abeliovich and Azov, 1976; Chevalier and de la Noüe, 1985).

The reason that algae assimilate ammoniacal-N and other reduced forms of nitrogen (e.g. urea) in preference to oxidized forms of nitrogen (nitrite and nitrate) is due to the greater energy requirement for nitrate reduction (Oswald *et al.*, 1953; Fogg, 1975; Oh-Hama and Miyachi, 1988; Raven *et al.*, 1992). In addition, the presence of ammoniacal-N prevents nitrate assimilation by inhibiting the production of nitrate reductase (Thompson *et al.*, 1989).

Phosphorus source

Phosphate is the only form of phosphorus assimilated by algae but at low phosphate concentrations organic phosphorus may be hydrolysed by phosphatase enzymes produced at the cell surface (Fogg, 1975). Once assimilated, algae convert phosphate into polyphosphates, which serve as reservoirs of high-energy phosphate for ATP synthesis.

Algal synthesis

If it is assumed that ammonium is the source of nitrogen, carbon dioxide the source of carbon, phosphate the source of phosphorus and water the source of hydrogen and oxygen, the synthesis of algal biomass can be approximated by the following equation:

$$106CO_2+16NH_4^++HPO_4^{2-}+236H_2O \xrightarrow{Light+Algae} C_{106}H_{181}O_{45}N_{16}P+118O_2+171H_2O+14H^+$$

The effect of pH

Algal nutrient assimilation and photosynthesis influence the pH of pond water. Nitrate assimilation raises pH (Oh-Hama and Miyachi, 1988), while ammonia assimilation decreases pH (Azov and Goldman, 1982). By depleting the concentration of CO_2 or HCO_3^- in pond water, algal photosynthesis can raise the pH to as high as 11 or more. Pond water pH, in turn, affects many processes associated with algal growth and metabolism, as well as the availability and uptake of nutrient ions (Richmond, 1986; de la Noüe and de Pauw, 1988).

Carbon availability: Elevated pond water pH can inhibit algal growth by making free carbon dioxide unavailable through conversion to carbonate and bicarbonate (Fogg, 1975; de la Noüe and De Pauw, 1988; Azov and Goldman, 1982; Richmond, 1986).

Ammonia toxicity: High concentrations of free ammonia inhibit photosynthesis by disrupting algal cell chloroplast function (Azov and Goldman, 1982; Abeliovich and Vonshak, 1993). The inhibitory concentration of ammoniacal-N is related to pond water pH and temperature which both affect the concentration of free ammonia. An ammoniacal-N concentration of 36 g m^{-3} may reduce algal growth if the pond water pH rises above 8, and may reduce algal photosynthesis by 50% at pH 9.5 (20-25°C), while an ammoniacal-N concentration of 54 g m^{-3} may reduce algal photosynthesis by 90% at pH 9.5 (20-25°C) (Azov and Goldman, 1982; Veenstra *et al.*, 1995). Thus in ponds with high levels of ammoniacal-N, algal photosynthesis will proceed until it elevates the NH_3 concentration to inhibitory levels as a result of increasing pond water pH and temperature. However, the inhibition is naturally brought back into balance, as microbial respiration will increase the pond water CO_2 concentration, which in turn lowers the pH and returns the NH_3 concentration below inhibitory levels.

Hydrogen sulphide toxicity: Algae are even more sensitive to high levels of dissolved hydrogen sulphide (H_2S) than to free ammonia. The concentration of hydrogen sulphide is influenced by total sulphide concentration and pond water pH and temperature (Pearson *et al.*, 1987; Gomez *et al.*, 1995; Nurdogan and Oswald, 1995). At pH <6 almost all sulphide is present as H_2S gas, while at pH >9 most of the sulphide is in ionised forms (HS^- and S^{2-}) (Pearson *et al.*, 1987; Veenstra *et al.*,

1995). H_2S concentrations above 1 g S m^{-3} in pond surface waters can significantly reduce algal growth (Pearson *et al.,* 1987).

Luxury consumption

Several species or algae are capable of accumulating much more phosphate than that required for growth. This process, called luxury consumption, stores phosphate within the algal cells as polyphosphate (volutin) granules (Robinson *et al.,* 1989).

Sedimentation of algal/bacterial biomass

The algae that grow in WSPs tend not to settle easily due to their small size (*Chlorella* sp.), motility (e.g. *Euglena* sp.), or flotation (e.g. *Oscillatoria* sp.). However at high pond water temperature and pH, sedimentation may be improved by autoflocculation of the algae/bacterial biomass (Nurdogan and Oswald, 1995).

5.2.3 Ammonia volatilisation

Ammoniacal-N may be lost through the WSP surface through volatilisation of ammonia gas (Nurdogan and Oswald, 1995; Pearson *et al.,* 1996). The rate of ammonia volatilisation depends on the free ammonia concentration and factors such as the pond temperature and mixing conditions.

Ammonia volatilisation can be a dominant process of nitrogen removal in WSPs accounting for 75-98% of total N removal in domestic WSPs with pH ranges at 7 to 9, and temperature ranges from 22 to 28°C (Pano and Middlebrooks, 1982; Somiya and Fujii, 1984; Reed, 1985; Pearson *et al.,* 1996). Reed, (1985) suggested that even when the pH of the pond water is comparatively low, microsites (within algal floccs) with elevated pH could promote ammonia volatilisation.

5.2.4 Phosphate precipitation

Phosphates (PO_4^{3-}, HPO_4^{2-}, $H_2PO_4^-$) may be removed from the pond water through precipitation of insoluble complexes with cations (e.g. Ca^{2+}, Mg^{2+}, Al^{3+}, Fe^{3+}). Precipitation is dependent upon the pond pH, temperature, phosphate concentration and cation concentration (Goldman *et al.,* 1982a; Moutin *et al.,* 1992). Provided sufficient cations are present (> 50-100 g m^{-3}), for every rise of one pH unit above pH 8.2, the concentration of phosphate remaining in the pond water decreases by a factor of ten, whereas at low cation concentrations (< 50 g m^{-3}) there is only significant phosphate precipitation at pH >10 (Diaz *et al.,* 1994). Precipitation is strongly dependent on temperature and different stages of the precipitation process may be either aided or hindered by high or low temperatures (Maurer *et al.,* 1999).

Several types of phosphate complex may be formed, some of which are shown along with their solubility products in Table 5.2.

Table 5.2 Solubility products for some phosphate complexes (Moutin *et al.*, 1992; Nurdogan and Oswald, 1995; Hartley *et al.*, 1997)

Compound	Chemical formula	pK_s at 20°C	pK_s at 25°C
Magnesium ammonium phosphate	$MgNH_4PO_4$		115.0
(Struvite)			
Hydroxyapatite	$Ca_5(PO_4)3OH$	58.5	52.8-57
Octacalcium phosphate	$Ca_4H(PO_4)_3 . 3H_2O$	50.6	46.9
(Amorphous) Tricalcium	$Ca_3(PO_4)_2$		25.2-28.1
phosphate	$Ca_2(OH)_2HPO_4$		27.3
Hydroxydicalcium phosphate	$Mg(OH)_2$		10.5
Magnesium hydroxide	$CaCO_3$		8.3
Calcium carbonate (Calcite)	$MgCO_3$		7.6
Magnesium carbonate (Magnesite)	$CaHPO_4 . 2H_2O$	6.6	6.6-7.0
Calcium phosphate (Brushite)	$CaSO_4$		4.6
Calcium sulphate			
Magnesium phosphate	$Fe_3(PO_4)_2 . 8H_2O$		
Vivianite	$AlPO_4 . 2H_2O$		
Variscite			
Potassium ammonium phosphate	$Ca1_8(MgFe)H_2(PO_4)$		
Whitlockite	CaOH	1.4	
Calcium hydroxide			

Although struvite and hydroxyapatite are more thermodynamically stable than the other complexes, several other factors affect their formation. Struvite ($MgNH_4PO_4$) is unlikely to form in domestic wastewater, which typically has low Mg concentrations (Maurer *et al.*, 1999). Struvite formation is promoted by adjusting the wastewater Mg:N:P ratio to 1:1:1 at supersaturation levels and pH > 8 (Momberg and Oellermann, 1992).

At pH >8.2 hydroxyapatite precipitation can account for up to 80% of P removal in WSPs (Ellis, 1983), particularly if the wastewater Ca:P ratio is close to 2:1 (Maurer *et al.*, 1999). However, hydroxyapatite formation may be inhibited by the presence of magnesium ions and organic acids (e.g. fulvic, humic and tannic) in the wastewater (Moutin *et al.*, 1992).

Inhibition of hydroxyapatite formation results in the precipitation of other phosphate compounds including octacalcium phosphate and tricalcium phosphate (Moutin *et al.*, 1992; Diaz *et al.*, 1994). Thus, at pH values as low as 9.0, a large fraction of phosphate can be precipitated as hydroxyapatite, octacalcium phosphate and tricalcium phosphate (Nurdogan and Oswald, 1995).

Phosphate precipitated at high pH during the day may be subsequently released at night when pH declines to <8 (Diaz *et al.*, 1994). The amount of dissolution varies for each phosphate complex. If pond water pH is reduced to 7, 75-90% of octacalcium phosphate may dissolve, while nearly 50% of more stable phosphate compounds (e.g. hydroxyapatite, tricalcium phosphate, and whitlockite) remain insoluble (Diaz *et al.,* 1994).

Somiya and Fujii (1984) found that phosphorus precipitation was a less important P removal process than assimilation/sedimentation. However, in ponds with sufficient cation concentrations and appropriate pH, precipitation can be a major P removal process (Wrigley and Toerien, 1990; Moutin *et al.,* 1992; Gomez *et al.,* 2000).

5.2.5 Adsorption

Inorganic phosphates and ammoniacal-N may be removed by adsorption to pond sludge or, at high pH, to ferric (Fe^{3+}) oxyhydroxide [Fe(OOH)], aluminium hydroxide [$Al(OH)_3$] and calcium carbonate [$CaCO_3$] crystals, which are formed in significant amounts on the surface of aerobic sludge in ponds with high Fe^{2+}, Al^{3+}, and Ca^{2+} concentrations (Moutin *et al.,* 1993; Diaz *et al.,* 1994; Nurdogan and Oswald, 1995; Hartley *et al.,* 1997). Gomez *et al.,* (2000) found significant adsorption of phosphorus to pond sludge containing high concentrations of Fe and Al.

5.2.6 Nitrification / Denitrification

Nitrification

Under aerobic conditions ammoniacal-N may be oxidised by nitrifying bacteria to nitrite and then nitrate (Ferrara and Avci, 1982; Pano and Middlebrooks, 1982; Abeliovich and Vonshak, 1993; Lai and Lam, 1997).

Ammonia is oxidised to nitrite via hydoxylamine (e.g. by *Nitrosomonas* sp and *Nitrosospira* sp):

$$NH_4^+ + \frac{1}{2}O_2 \rightarrow NH_2OH + H^+$$
(enzyme: ammonia monooxygenase)

$$NH_2OH + O_2 \rightarrow NO_2^- + H_2O + H^+$$
(hydroxylamine oxidoreductase)

Nitrite is then oxidised to nitrate (e.g. by *Nitrobacter sp* and *Nitrospira sp*):

$$NO_2^- + \frac{1}{2}O_2 \rightarrow HNO_3^-$$
(nitrite dehydrogenase)

Nitrification requires oxygen (approximately 4.6 g O_2 / g NH_4-N) and reduces alkalinity by 7.14 g $CaCO_3$ / g NH_4-N oxidized (USEPA 1975). Optimal conditions for nitrification in WSPs are: DO >1 g m^{-3}, temperature >8 °C and pH 6.0-9.0 (Baskaran *et al.,* 1992; Azov and Tregubova, 1995; Villaverde *et al.,* 1997). Nitrifying bacteria grow better when attached to aerobic surfaces than when suspended in the water column and are often found at the surface of aerobic pond sludge (Baskaran *et al.,* 1992; Craggs *et al.,* 2000).

Nitrification is also inhibited at the pond surface by high levels of solar-UV light (Abeliovich and Vonshak, 1993; Azov and Tregubova, 1995). It appears that sunlight penetration may select for slow nitrifying stationary phase cells that are more resistant to UV light than actively nitrifying cells (Abeliovich and Vonshak, 1993).

Denitrification

Denitrification is an anaerobic respiration process in which denitrifying bacteria (e.g. *Pseudomonas* sp and *Bacillus* sp) oxidise organic matter by reducing nitrate to nitrous oxide (N_2O) and nitrogen (N_2) gases (Abeliovich and Vonshak, 1993).

$$2NO_3^- \xrightarrow{\text{Nitrate reductase}} NO_2^- \xrightarrow{\text{Nitrite reductase}} NO \xrightarrow{\text{Nitric oxide reductase}} N_2O \xrightarrow{\text{Nitrous oxide reductase}} N_2$$

The optimum conditions for denitrification in ponds are DO <1 g m^{-3}, pH of 7.0-8.5, temperature >10°C and sufficient organic carbon in the wastewater as an electron donor (Bitton, 1994). Denitrification increases pond water alkalinity by 3.6 g $CaCO_3$ per g NO_3-N reduced. Incomplete denitrification can occur at low COD:NO_3 ratios, low pH and in the presence of oxygen (Hanaki *et al.,* 1992) and results in the formation of nitrite and nitrous oxide.

In most WSPs, nitrification occurs intermittently and for unpredictable lengths of time and probably does not play a major role in nitrogen removal (Ferrara and Avci, 1982; Pano and Middlebrooks, 1982; Reed, 1985; Mara, 1997). The inconsistency of nitrification may be attributed to the variable DO, temperature and pH and lack of aerobic attachment surfaces in facultative pond surface waters resulting in a low and fluctuating population of nitrifying bacteria. Slow growing nitrifying bacteria are usually restricted to the aerobic surface waters in facultative ponds where they may easily be washed out of the pond or out competed by faster growing heterotrophic bacteria (Ferrara and Avci, 1982; Lai and Lam, 1997).

However, Lai and Lam, (1997) demonstrated that nitrification-denitrification could be a major N removal pathway in facultative ponds during periods of high algal abundance. High photosynthetic activity probably creates favourable conditions (pond water DO and pH) for daytime nitrification throughout the facultative pond depth (Hurse and Connor, 1999). Zimmo *et al.,* (2004) found that 16-25% of TN was removed by nitrification-denitrification depending upon temperature in pilot-scale 0.9m deep WSPs.

Denitrification mainly occurs in anoxic sludge and does not appear to be a limiting process for nitrogen removal in WSPs (Baskaran *et al.,* 1992; Zimmo *et al.,* 2004). In maturation ponds receiving nitrified wastewater, denitrification in the sludge accounted for 66% of total N removal (Somiya and Fujii, 1984).

5.2.7 Heterotrophic nitrification / denitrification

Heterotrophic nitrification is a process by which organic and inorganic nitrogenous substances are metabolised by bacteria (e.g. *Arthrobacter* sp) to nitrite and nitrate through hydroxamic acids, oximes, nitrose compounds and nitro compounds (Tate, 1980; Castignetti and Hollocher, 1984; Verstraete *et al.,* 1997). The most probable pathway for the heterotrophic nitrification of ammoniacal-N is through hydroxylamine, 1-nitrosoethanol to nitrite and nitrate. The nitrate is subsequently denitrified to N_2 gas (van de Graaf *et al.,* 1995). Diab *et al.,* (1993) suggest that heterotrophic nitrification may be particularly significant under acid conditions (pH <4.5).

Other nitrification processes (e.g. chemical nitrification and methylotrophic nitrification) may also occur in WSPs (Diab *et al.,* 1993). However, there has been no specific research on the occurrence of these processes in WSPs.

Heterotrophic nitrification followed by denitrification has been found to account for 30-50% of TN removal from deep facultative ponds with anaerobic sludge (Brockett, 1976; Oswald *et al.,* 1994; Van de Graaf *et al.,* 1995; Green *et al.,* 1996). But no information is available for most WSP systems.

5.2.8 Phosphine production

Under highly anaerobic conditions small amounts of phosphine (PH_3), an odorous, toxic and combustible gas maybe produced from decomposing wastewater sludge (Glindemann *et al.,* 1996). Phosphine may be produced directly by microbial metabolism or released from phosphine bound within the organic matter (Devai *et al.,* 1988; Roels and Verstraete, 2001).

5.3 RELATIVE IMPORTANCE OF PROCESSES

Several authors have tried to identify the key nutrient removal process occurring in WSPs. However, different processes have been suggested, even when evaluating the same pond systems (e.g. Pano and Middlebrooks, 1982; Ferrara and Avci, 1982 and Reed, 1985).

Nutrient removal in WSPs results from any combination of all of the removal processes discussed above and others, which are as yet poorly understood. The processes that occur in a pond at any one time depend upon many factors including pond design, retention time and organic loading rate, wastewater composition and seasonally variable parameters such as algal production, mixing/stratification and environmental conditions (pH, temperature, DO, sunlight).

5.4 RELEASE OF NUTRIENTS FROM POND SLUDGE

The wastewater solids and algal/bacterial biomass deposited in the pond sludge are largely biodegradable and can eventually become a source of nutrients that are released back to the pond water by decomposition, mineralisation and resolubilisation (Reed *et al.,* 1995). Therefore nutrient release from accumulated pond sludge may reduce the overall efficiency of nutrient removal by the pond system (Somiya and Fujii, 1984; Moutin *et al.,* 1993). Release of N to maturation pond waters from decomposing bottom sludge has been shown to reduce TN removal to less than 45% (Somiya and Fujii, 1984) and Picot *et al.* (1992) found that 24% of the influent TP was mineralised within a WSP system.

5.4.1 Pond sludge phosphorus content

Phosphorus may be present in several forms in WSP sludge, and sludge composition varies widely between different pond systems and pond types (Table 5.3). Aerobic pond sludges (e.g. from maturation ponds) tend to have highest P concentrations and release less P back to the pond water (Gomez *et al.,* 2000).

Table 5.3 Composition of pond sludge as a % of TP

P Fraction	WSP (8 m deep)	WSP (1.45 m)	HRP (0.35 m)
$CaCO_3$-P	3	} 33-37	} 92.7
Hydroxyapatite-P	24		
Fe/Al-P	35	57-59	3.9
Inorganic P	66	92-94	96.6
Organic P	26.1	6-8	3.1
Inert-P	8.1	-	-
TP (mg g^{-1} d. wt)	3.5	14.3-22.1	60
Reference	Ortuno *et al.*, 2000	Gomez *et al.*, 2000	Moutin *et al.*, 1992

Biological activity may affect phosphorus release directly through decomposition of organic matter and mineralisation of organic phosphorus compounds (Ortuno *et al.,* 2000), or indirectly through changes in the pond pH and redox potential (Bostrum 1984).

Release of P, from the Fe/Al-P fraction, increases under anaerobic conditions, at low pH and when the sludge has a higher P concentration than the water (Reed *et al.,* 1995; Ortuno *et al.,* 2000). Under anaerobic conditions, insoluble ferric (3^+) oxyhydroxide is reduced to soluble ferrous (2^+), therefore adsorbed phosphate is released into solution (Bostrum 1984; Ortuno *et al.,* 2000). At low pH, cations (e.g. Ca^{2+}, Fe^{3+} or Al^{3+}) may be sequestered by chealators to release adsorbed phosphorus. Phosphorus release from inorganic particulate P may also occur at high pH due to substitution of phosphate with hydroxyl ions at the adsorption sites on iron and aluminium compounds (Ortuno *et al.,* 2000).

However, not all mobilised phosphorus is released to the pond water. For example, Moutin *et al.,* (1993) found that some of the P released by dissolution of Fe(OOH)~P was readsorbed by more stable $CaCO_3$.

5.5 NUTRIENT REMOVAL EFFICIENCY

WSP nutrient removal varies greatly, both with seasons and also between pond systems of similar design (Davies-Colley *et al.,* 1995). Nitrogen removal efficiency is usually less than 70% although values range from 9% to 95% and ammoniacal-N removal efficiency from 0% to 95% (Ferrara and Avci, 1982; Mara, 1996; Pearson *et al.,* 1996; Mara, 1997; Middlebrooks *et al.,* 1999). Phosphorus removal efficiency is generally lower than that of nitrogen removal and typically ranges from 40 to 50% (Mara, 1996; Pearson *et al.,* 1996). Nitrogen and phosphorus removal efficiencies found by various researchers are shown in Table 5.4.

Table 5.4 Nitrogen and phosphorus removal efficiency (% of influent concentration removed) of WSP systems

TN		TP		Reference
Facultative Ponds				
21%	(20% NH_3-N	20%		Picot *et al.*, 1992
31%	(29% NH_3-N	40%		Racault *et al.*, 1995
40%		51%		Garcia *et al.*, 2000
21-73%	(60-93% NH_3-N	20-48%	(1-23% PO_4-P	Li *et al.*, 1991
	(52% NH_3-N			Santos and Oliveira 1987
40-80%				USEPA 1985
<79% TKN	(<92% NH_3-N			Silva *et al.*, 1995
	(90% NH_3-N			Pano and Middlebrooks 1982
	(95% NH_3-N			Middlebrooks *et al.*, 1982
Maturation Ponds				
45%		43%		Somiya and Fujii 1984
	(82% NH_3-N)		(48% PO_4-P)	Wrigley and Toerien 1990

5.5.1 Predicting Nutrient Removal Performance

Several estimates of potential nutrient removal in WSPs can be found in the literature. For example, Uhlmann (1987) derived equations for nitrogen and phosphorus removal in WSPs based on organic and nutrient load, residence time, temperature, and solar irradiation. The equations most frequently used to estimate nitrogen removal follow:

1) Pano and Middlebrooks, (1982) derived an equation for ammoniacal-N removal by volatilisation from WSPs based on pond water temperature and pH.

For temperatures below 20°C:

$$C_e = \frac{C_o}{\left(1 + [(\frac{A}{Q})(0.0038 + 0.000134\ T)^{((1.041+0.044\,T)(pH-6.6))}]\right)}$$

For temperatures above 20°C:

$$C_e = \frac{C_o}{\left(1 + [(0.005035\ (\frac{A}{Q}))^{(1.540\,(pH-6.6))}]\right)}$$

where:

C_e and C_o	are the effluent and influent concentrations (g N m^{-3})
A	is the pond area (m^2)
Q	is the wastewater flow (m^2 d^{-1})
T	is the pond temperature
pH	$= 7.3^{(0.0005\,(Alk))}$
Alk	is the expected influent alkalinity (g $CaCO_3$ m^{-3}) (see USEPA 1983).

Silva *et al.*, (1995) and Soares *et al.*, (1996) found good agreement between observed ammoniacal-N removal in WSPs and values predicted by these equations for removal through volatilisation. In Chapter 9 (Design) this approach is discussed again.

Middlebrooks also derived a more conservative equation for TKN removal (USEPA 1985). This equation is also based on ammoniacal-N volatilisation at elevated pond water temperature and pH:

$$C_e = \frac{C_o}{\left(1 + d[(0.000576T - 0.0028)^{((1.080 - 0.042T)(pH - 6.6))}]\right)}$$

where:

d	is the pond retention time (d).

2) Reed derived an equation for total nitrogen removal (g $N.m^{-3}$) in WSPs based on pond water temperature and pH (Reed 1985; Reed *et al.*, 1995):

$$C_e = C_o e^{-K_T[t + 60.6(pH - 6.6)]}$$

where:

K_T	=	temperature dependent rate constant
K_T	=	$K_{20}\,(\theta)^{(T-20)}$
K_{20}	=	rate constant at 20°C = 0.0064
θ	=	1.039
t	=	detention time in system, d
pH	=	pH near pond surface $pH = 7.3^{0.0005\,Alk}$
Alk	=	expected influent alkalinity (g $CaCO_3$ m^{-3}) (see USEPA 1983).

Thus,

$$C_e = C_o e^{-(0.0064\,(1.039)T - 20)[t + 60.6(pH - 6.6)]}$$

Pond water temperature can be estimated using the equation of Mancini and Barnhart (1976):

$$T = \frac{0.5AT_a + QT_i}{0.5A + Q}$$

where:

A = surface area of pond, m^2
T_a = ambient air temperature, ^oC
T_i = influent temperature, ^oC
Q = influent flow rate, $m^3\ d^{-1}$.

3) More recently, Middlebrooks (USEPA 1995) derived an equation for TN removal based on ammoniacal-N volatilisation incorporating pond water temperature and pH and hydraulic loading rate:

$$C_e = \frac{C_o}{1 + t(0.000576T - 0.00028)e^{(1.080 - 0.042T)(pH - 6.6)}}$$

Phosphorus removal in WSPs is usually crudely estimated using the model developed by Houng and Gloyna (1984), which predicts that 45% of P will be removed when BOD removal is 90%.

5.6 IMPROVING NUTRIENT REMOVAL

5.6.1 Aerobic ponds - maturation and high rate ponds

Both nitrogen and phosphorus removal can be improved through the addition of maturation ponds to a WSP system. Nutrient removal is enhanced in these ponds by assimilation into algal biomass and elevation of the pond surface pH, which promotes both ammoniacal-N removal by volatilisation and phosphate removal through precipitation with cations. Phosphorus will remain immobilised provided the pond bottom is aerobic (Gomez *et al.*, 2000). High Rate Ponds are specifically designed to promote the symbiosis between aerobic bacteria and algae, thus much of the nutrient removal is through assimilation into algal biomass. Intensive photosynthesis in these gently mixed ponds raises the pH of the total pond volume to ~10, thus remaining ammoniacal-N and phosphate can be reduced to very low levels through volatilisation and precipitation respectively. The colonial algal species that grow in high rate ponds are easily settled, making subsequent removal of the biomass, and its associated nutrients, a relatively affordable option.

5.6.2 Biofilm attachment surfaces

Nitrogen removal by nitrification can be improved by as much as 50% in WSPs containing biofilm (McLean *et al.,* 2000). Incorporation of biofilm attachment surfaces (baffles and geotextile supports) into the aerobic surface waters of facultative ponds promotes the establishment of nitrifying bacteria and has been shown to enhance rates of nitrification (Baskaran *et al.,* 1992; Diab *et al.,* 1993; Polprasert and Agarwalla, 1995; Craggs *et al.,* 2000; McLean *et al.,* 2000). Photosynthetic activity of algae within surface biofilms further improves rates of nitrification (Baskaran *et al.,* 1992; Craggs *et al.,* 2000; McLean *et al.,* 2000) and has also been shown to enhance nitrogen removal by assimilation and ammoniacal-N volatilisation (Rakkoed *et al.* 1999).

5.6.3 Precipitation and adsorption

Natural compounds containing alumino-ferric compounds (e.g., allophane and zeolite), calcium and magnesium (e.g., lime and dolomite) can be added to WSPs to provide cations for P precipitation and substratum for ammoniacal-N adsorption (Nurdogan and Oswald, 1995; Sakadevan and Bavor, 1998). Phosphorus removal in pond systems is often enhanced through chemical precipitation with aluminium sulphate, ferrous or ferric chloride and lime (Odegaard *et al.,* 1987; Surampalli *et al.,* 1995) and effluent concentrations of 1 g P m^{-3} can be achieved.

A new development in this area is the use of "active" filters, which are built with filter media (steel slag for example) capable of adsorbing and removing phosphorus (Shilton *et al.*, 2005).

5.7 SUMMARY

Nutrients are removed in WSPs by a combination of physical, chemical and biological processes including sedimentation, assimilation into algal and bacterial biomass, volatilisation or nitrification/denitrification of ammoniacal-N and precipitation or adsorption of phosphate. The extent to which any one process occurs within a pond is dependent upon the pond environment: light and dissolved oxygen concentration - aerobic versus anaerobic, pH - inhibition versus augmentation and temperature - stimulation. Most conventional pond designs are too simplistic to favour any individual removal process. The challenge to the pond designer of the future will be to engineer the pond environment to promote particular processes and provide consistent nutrient removal.

5.8 FURTHER RESEARCH

5.8.1 Nitrogen removal

Further research is needed to improve our current understanding of the importance of the particular nitrogen removal processes in WSPs and how these processes may be controlled. There is a need for basic research on processes such as anamox, heterotrophic nitrification, chemical nitrification and methylotrophic nitrification, all of which may occur to some degree in WSPs or could possibly be promoted if they were better understood.

5.8.2 Phosphorus removal

Poor phosphorus removal is a major barrier to the future uptake of pond technologies, particularly in the developed world. Phosphorus removal can be improved somewhat by increasing pond water hardness and high levels of removal may be achieved through the use of flocculants and polyeletrolytes. However these applications add considerably to the operation costs of the system. There is a need to develop low-cost enhanced phosphorus removal technologies for WSPs.

REFERENCES

Abeliovich, A. and Azov, Y. (1976) Toxicity of ammonia to algae in sewage oxidation ponds. *Appl. Environ. Microbiol.* **31**, 801-806.

Abeliovich, A. and Vonshak, A. (1993) Factors inhibiting nitrification of ammonia in deep wastewater reservoirs. *Water Research* **27**(10), 1585-1590.

Azov, Y. and Goldman, J.C. (1982) Free ammonia inhibition of algal photosynthesis in intensive cultures. *Applied and Environmental Microbiology* **43**(4), 735-739.

Azov, Y. and Tregubova, T. (1995) Nitrification processes in stabilisation reservoirs. *Water Science and Technology* **31**(12), 313-319.

Barnes, D. and Bliss, P.J. (1983). Theory of Nitrification. In *Biological control of nitrogen in wastewater treatment*, Spon, London. pp. 29-47.

Baskaran, K., Scott, P.H. and Conner, M.A. (1992) Biofilms as an aid to nitrogen removal in sewage treatment lagoons. *Water Science and Technology* **26**(7-8), 1707-1716.

Bitton, G. (1994) Wastewater microbiology. Ecological and applied microbiology, R. Mitchell, ed., Wiley-Liss, New York.

Bostrom, B. (1984) Potential mobility of phosphorus in different types of lake sediment. *Int. Revue ges. Hydrobiol.* **69**, 457-474.

Brockett, O.D. (1976) Microbial reactions in facultative oxidation ponds: 1. anaerobic nature of oxidation pond sediments. *Water Research* **10**, 45-49.

Castignetti, D. and Hollocher, T.C. (1984) Heterotrophic nitrification among denitrifiers. *Applied and Environmental Microbiology* **47**(4), 620-623.

Chevalier, P. and de la Noüe, J. (1985) Wastewater nutrient removal with microalgae immobilized in carrageenan. *Enz. Microb. Technol.* **7**, 621-624.

Craggs, R.J., Tanner, C.C., Sukias, J.P.S. and Davies-Colley, R.J. (2000) Nitrification potential of attached biofilms in dairy farm waste stabilization ponds. *Water Science and Technology* **42**(10-11), 195-202.

Davies-Colley, R.J., Hickey, C.W. and Quinn, J.M. (1995) Organic matter, nutrients and optical characteristics of sewage lagoon effluents. *New Zealand Journal of marine and freshwater research* **29**, 235-250.

de la Noüe, J. and Bassères, A. (1989) Biotreatment of anaerobically digested swine manure with microalgae. *Biol. Wastes* **29**, 17-31.

de la Noüe, J. and De Pauw, N. (1988) The potential of microalgal biotechnology: A review of production and uses of microalgae. *Biotech. Adv.* **6**, 725-770.

Devai, I., Felfoldy, L., Wittner, I. and Plosz, S. (1988) Detection of phosphine: new aspects of the phosphorus cycle in the hydrosphere. *Nature* **333**, 343-345.

Diab, S., Kochba, M. and Avnimelech, Y. (1993) Nitrification pattern in a fluctuating anaerobic-aerobic pond environment. *Water Research* **27**(9), 1469-1475.

Diaz, O.A., Reddy, K.R. and Moore, P.A.J. (1994) Solubility of inorganic phosphorus in stream water as influenced by pH and calcium concentration. *Water Research* **28**(8), 1755-1763.

Ellis, K.V. (1983) Stabilization Ponds: Design and Operation. Critical Reviews in Environmental Control, **13**, 69-102.

Ferrara, R.A. and Avci, C.B. (1982) Nitrogen dynamics in waste stabilization ponds. *Journal of the Water Pollution Control Federation* **54**(4), 361-369.

Fogg. (1975) *Algal Cultures and Phytoplankton Ecology.*, Univ. Wisconsin Press.

Garcia, J., Mujeriego, R., Bourrouet, A., Penuelas, G. and Freixes, A. (2000) Wastewater treatment by pond systems: experiences in Catalonia, Spain. *Water Science and Technology* **42**(10-11), 35-42.

Glindermann, D., Stottmeister, U. and Bergmann, A. (1996) Free phosphine from the anaerobic biosphere. *Environ. Sci. and Pollut. Res.* **3**, 17-19.

Goldman, J.C., Azov, Y., Riley, C.B. and Dennett, M.R. (1982a) The effect of pH in intensive microalgal cultures. I. Biomass regulation. *J. Exp. Mar. Biol. Ecol.* **57**, 1-13.

Goldman, J.C., Azov, Y., Riley, C.B. and Dennett, M.R. (1982b) The effect of pH in intensive microalgal cultures. II. Species competition. *J. Exp. Mar. Biol. Ecol.* **57**, 15-24.

Gomez, E., Casellas, C., Picot, B. and Bontoux, J. (1995) Ammonia elimination processes in stabilisation and high-rate algal pond systems. *Water Science and Technology* **31**(12), 303-312.

Gomez, E., Paing, J., Casellas, C., Picot, B. (2000) Characterisation of phosphorus in sediments from waste stabilization ponds, *Water Science and Technology* **42**(10), 257-264.

Green, F.B., Bernstone, L.S., Lundquist, T.J. and Oswald, W.J. (1996) Advanced integrated wastewater pond systems for nitrogen removal. *Water Science and Technology* **33**(7), 207-217.

Hanaki, K., Zheng, H. and Matsuo, T. (1992) Production of nitrous oxide gas during denitrification of wastewater. *Water Science and Technology.* 26(5/6) 1027-1036.

Hartley, A.M., House, W.A., Callow, M.E. and Leadbeater, B.S.C. (1997) Coprecipitation of phosphate with calcite in the presence of photosynthesizing green algae. *Water Research* **31**(9), 2261-2268.

Hemens, J. and Mason, M.H. (1968) Sewage nutrient removal by shallow algal stream. *Water Research* **2**, 277-287.

Houng, H-S. and Gloyna, E.F. (1984) Phosphorus models for waste stabilization ponds. *J. Environ. Eng.* **110**(3), 550-561.

Hurse, T.J. and Connor, M.A. (1999) Nitrogen removal from wastewater treatment lagoons. *Water Science and Technology* **39**(6), 191–198.

Lai, P.C.C. and Lam, P.K.S. (1997) Major pathways for nitrogen removal in waste water stabilization ponds. *Water, Air, and Soil Pollution* **94**(1-2), 125-136.

Li, J., Wang, J. and Zhang, J. (1991) Removal of nutrient salts in relation with algae in ponds. *Water Science and Technology* **24**(5), 75-83.

Mancini, J.L. and Barnhart, E.L. (1976). Industrial waste treatment in aerated lagoons. In *Ponds as a Wastewater Treatment Alternative*, Water Resources Symposium No. 9, University of Texas.

Mara, D.D. (1996) Waste stabilization ponds: effluent quality requirements and implications for process design. *Water Science and Technology* **33**(7), 23-31.

Mara, D.D. (1997) *Design Manual for Waste Stabilization Ponds in India*, Lagoon Technology International Ltd, Leeds, UK.

Maurer, M., Abramovich, D., Siegrist, A.H. and Gujer, W. (1999) Kinetics of biologically induced phosphorus precipitation in wastewater treatment. *Water Research* **33**(2), 484-493.

McLean, B.M., Baskaran, K. and Connor, M.A. (2000) Use of algal-bacterial biofilms to enhance nitrification rates in lagoons: Experience under laboratory and pilot-scale conditions. *Water Science and Technology* **42**(10-11), 187-194.

Middlebrooks, E.J., Middlebrooks, C.H., Reynolds, J.H., Watters, G.Z., Reed, S.C. and George, D.B. (1982) *Wastewater Stabilization Lagoon Design, Performance and Upgrading*, MacMillan Publishing Co., Inc, New York.

Middlebrooks, E.J., Reed, S.C., Pano, A. and Adams, V.D. (1999) Nitrogen removal in wastewater stabilization lagoons. In *6th National Drinking Water and Wastewater Treatment Technology Transfer Workshop. August 2-4, 1999*, Kansas City, Missouri.

Momberg, G. and Oellermann, R.A. (1992) The removal of phosphate by hydroxyapatite and struvite crystallisation in South Africa. *Water Science and Technology* **26**(5-6), 987-996.

Moutin, T., Gal, J.Y., El Halouani, H., Picot, B. and Bontoux, J. (1992) Decrease of phosphate concentration in a high rate pond by precipitation of calcium phosphate: Theoretical and experimental results. *Water Research* **26**(11), 1445-1450.

Moutin, T., Picot, B., Ximenes, M.C. and Bontoux, J. (1993) Seasonal variations of P compounds and their concentrations in two coastal lagoons (Herault, France). *Hydrobiologia* **252**, 45-59.

Nurdogan, Y. and Oswald, W.J. (1995) Enhanced nutrient removal in high-rate ponds. *Water Science and Technology* **31**(12), 33-43.

Odegaard, H., Balmer, P. and Hanaeus, J. (1987) Chemical precipitation in highly loaded stabilization ponds in cold climates: Scandinavian experiences. *Water Science and Technology* **19**(12), 71-77.

Oh-Hama, T. and Miyachi, S. (1988). Chlorella. In *Micro-algal Biotechnology*, (eds M.A. Borowitzka and L.J. Borowitzka), vol, pp. 3-26, C.U.P., Cambridge.

Ortuno, J.F., Saez, J., Liorens, M. and Soler, A. (2000) Phosphorus release from sediments of a deep wastewater stabilization pond. *Water Science and Technology* **42**(10-11), 265-272.

Oswald, W.J., Gotaas, H.B., Ludwig, H.F. and Lynch, V. (1953) Algal symbiosis in oxidation ponds III. Photosynthetic oxidation. *Sewage Industrial Wastes* **25**, 690-692.

Oswald, W.J. (1988). Micro-algae and waste-water treatment. In *Micro-algal Biotechnology*, (eds M.A. Borowitzka and L.J. Borowitzka), vol, pp. 305-328, Cambridge University Press, Cambridge.

Oswald, W.J., Green, F.B. and Lundquist, T.J. (1994) Performance of methane fermentation pits in advanced integrated wastewater pond systems. *Water Science and Technology* **30**(12), 287-295.

Pano, A. and Middlebrooks, E.J. (1982) Ammonia nitrogen removal in facultative wastewater stabilization ponds. *Journal of the Water Pollution Control Federation,* **54**(4), 344-351.

Pearson, H.W., Mara, D.D., Mills, S.W. and Smallman, D.J. (1987) Factors determining algal populations in waste stabilization ponds and the influence of algae on pond performance. *Water Science and Technology* **19**(12), 131-140.

Pearson, H.W., Mara, D.D., Cawley, L.R., Arridge, H.M. and Silva, S.A. (1996) Performance of an innovative tropical experimental waste stabilisation pond system operating at high organic loadings. *Water Science and Technology* **33**(7), 63-73.

Picot, B., Bahlaoui, A., Moersidik, S., Baleux, B. and Bontoux, J. (1992) Comparison of the purifying efficiency of high rate algal pond with stabilization pond. *Water Science and Technology* **25**(12), 197-206.

Polprasert, C. and Agarwalla, B.K. (1995) Significance of biofilm activity in facultative pond design and performance. *Water Science and Technology* **31**(12), 119-128.

Racault, Y., Boutin, C. and Seguin, A. (1995) Waste stabilization ponds in France: A report on fifteen years experience. *Water Science and Technology* 31(12), 91–101.

Rakkoed, A., Danteravanich, S. and Puetpaiboon, U. (1999) Nitrogen removal in attached growth waste stabilization ponds of wastewater from a rubber factory. *Water Science and Technology* **40**(1) 45–52.

Raven, J.A., Wollenweber, B. and Handley, L.L. (1992) A comparison of ammonium and nitrate as nitrogen sources for photolithotrophs. *New Phytol.* **121**, 19-32.

Redfield, A.C. (1934) *On the proportions of organic derivatives in sea water and their relation to the composition of Plankton.* In James Johnstone Memorial Volume (ed. Daniel, R. J.) Liverpool Univ. Press. 176-192.

Reed, S.C. (1985) Nitrogen removal in wastewater stabilisation ponds. *J. Wat. Pollut.Contr. Fed.* **57**(1), 39-45.

Reed, S.C., Crites, R.H. and Middlebrooks, E.J. (1995) *Natural Systems for Waste Management and Treatment*, McGraw-Hill Inc, NY, USA.

Richmond, A. (1986). Cell response to environmental factors. In *CRC Handbook of Microalgal Mass Culture*, (ed A. Richmond), vol, pp. 69-99, CRC Press inc., Boca Raton, Florida.

Robinson, P.K., Reeve, J.O. and Goulding, K.H. (1989). Phosphorous uptake kinetics of immobilized Chlorella in batch and continuous-flow culture. *Enzyme Microb. Technol.* **11**: 590-596.

Roels, J. and Verstraete, W. (2001) Biological formation of volatile phosphorus compounds. *Bioresource Technology* **79**, 243-250.

Sakadevan, K. and Bavor, H.J. (1998) Phosphate adsorption characteristics of soils slags and zeolite to be used as substrates in constructed wetland systems. *Water Research* **32**(2), 393-399.

Santos, M.C.R. and Oliveira, J.F.S. (1987) Nitrogen Transformations and Removal in Waste Stabilization Ponds in Portugal: Seasonal Variations. *Water Science and Technology* **19**(12), 123-130.

Shilton, A., Drizo, A., Mahmood, B., Banker, S., Billings, L., Glenny, S. and Luo, D. (2005) 'Active' filters for upgrading phosphorus removal from pond systems. *Water Science and Technology* **51**(12): 111-116.

Silva, S.A., Mara, D.D. and de Oliveira, R. (1987) Performance of a Series of Five Deep Waste Stabilization Ponds in Northeast Brazil. *Water Science and Technology* **19**(12), 61-64.

Silva, S.A., de Oliveira, R., Soares, J. and Mara, D.D. (1995) Nitrogen removal in pond systems with different configurations and geometries. *Water Science and Technology* **31**(12), 321-330.

Soares, J., Silva, S.A., de Oliveira, R., Araujo, A.L.C., Mara, D.D. and Pearson, H.W. (1996) Ammonia removal in a pilot-scale WSP complex in northeast Brazil. *Water Science and Technology* **33**(7), 165-171.

Somiya, I. and Fujii, S. (1984) Material balances of organics and nutrients in an oxidation pond. *Water Research* **18**(3), 325-333.

South, G.R. and Whittick, A. (1987) *Introduction to Phycology*, Blackwell Sci. Publ., Oxford.

Surampalli, R.Y., Banerji, S.K., Pycha, C.J. and Lopez, E.R. (1995) Phosphorus removal in ponds. *Water Science and Technology* **31**(12), 331-339.

Tate, R.L.I. (1980) Variation in heterotrophic and autotrophic nitrifier populations in relation to nitrification in organic soils. *Applied Environmental Microbiology* **40**(1), 75-79.

Thompson, P.A., Levasseur, M.E. and Harrison, P.J. (1989) Light-limited growth on ammonium vs. nitrate: What is the advantage for marine phytoplankton. *Limnol. Oceanogr.* **34**(6), 1014-1024.

Uhlmann, D. (1987) Possibilities and limits of the phosphorus and nitrogen elimination in unaerated sewage ponds. *Acta Hydrochim. Hydrobiol.* **15**(6), 623-636.

U.S.E.PA. (1975) *Process design manual for nitrogen removal.* US EPA, Centre for Environmental Research Information, Cincinnati, OH.

U.S.E.PA. (1983) *Technology transfer process design manual for municipal wastewater stabilization ponds.* EPA 625/1-83-015, US EPA, Centre for Environmental Research Information, Cincinnati, OH.

U.S.E.PA. (1985) *Wastewater Stabilization Ponds: Nitrogen Removal.* US EPA, Centre for Environmental Research Information, Cincinnati, OH.

U.S.E.PA. (1995), US EPA, Centre for Environmental Research Information, Cincinnati, OH.

van de Graaf, A.A., Mulder, A., de Bruijn, P., Jetten, M.S., Robertson, L.A. and Kuenen, J.G. (1995). Anaerobic oxidation of ammonium is a biologically mediated process. *Appl. Environ. Microbiol.* **61**(4), 1246-1251.

Veenstra, S., Al-Nozaily, F.A. and Alaerts, G.J. (1995) Purple non-sulfur bacteria and their influence on waste stabilisation pond performance in the Yemen Republic. *Water Science and Technology* **31**(12), 141–149.

Verstraete, W., Tanghe, T., DeSmul, A. and Grootaerd, H. (1997). Anaerobic biotechnology for sustainable waste treatment. In *Biotechnology in the Sustainable Environment*, (eds G.S. Sayler, J. Sanseverino and K.L. Davis), vol, pp. 343-359, Plenum Press Div Plenum Publishing Corp, 233 Spring St/New York/NY 10013.

Villaverde, S., García-Encina, P.A. and Fdz-Polanco, F. (1997) Influence of pH over nitrifying biofilm activity in submerged biofilters. *Water Research* **31**(5), 1180-1186.

Wrigley, T.J. and Toerien, D.F. (1990) Limnological Aspects of Small Sewage Ponds. *Water Research* **24**(1), 83-90.

Zimmo, O.R., van der Steen N.P. and Gijzen H. J. (2004). Quantification of Nitrification and Denitrification Rates in Algae and Duckweed Based Wastewater Treatment Systems. *Environmental Technology* 25(3) 273-283.

6

Pond disinfection

Rob Davies-Colley

6.1 INTRODUCTION

Disinfection – the removal of pathogenic micro-organisms – is a key function of any waste treatment process for sewage and other faecally-contaminated wastewaters. Waste stabilisation ponds (WSPs) are remarkably efficient and effective at removing a great variety of pathogenic (disease-causing) organisms. This is one of the main reasons these comparatively low cost, simple treatment systems are so popular and appropriate for use in the developing world (Mara, 2001; Maynard *et al.*, 1999).

This chapter discusses the types of micro-organisms that are of concern in faecally-contaminated wastewaters, overviews processes of disinfection in WSP systems and then considers removal efficiencies and processes. Disinfection in *maturation* ponds, which are usually specifically designed for removal of pathogens (Maynard *et al.*, 1999), is emphasised, but other pond types (and the influences of pond design features) are considered. Post-disinfection of pond effluents is considered because it may be demanded in developed countries to meet stringent

ISBN: 1843390205. Published by IWA Publishing, London, UK.

standards. Finally research needs for improving WSP disinfection performance are outlined. Disinfection processes are very dependent on the biological, physical and chemical environments within ponds, which are reviewed in Chapters 2 and 3.

6.2 PATHOGENS AND INDICATOR ORGANISMS

6.2.1 Pathogens

Four main categories of pathogen may occur in faecally-contaminated wastewaters (Table 6.1). Bacterial pathogens include *Salmonella*, *Shigella*, and *Vibrio cholerae* (the agent of cholera) (Table 6.1) as well as several so-called 'zoonotic' diseases such as *Campylobacter* spp for which domestic or wild animals may be reservoirs of infection. Not listed in Table 6.1 are the so-called 'opportunistic' bacterial pathogens (a range of genera, including *Pseudomonas* and *Klebsiella*) that are ubiquitous and sometimes pathogenic. Many thousands of people die every year from bacterial diseases, mainly in the developing world where public health engineering is comparatively primitive. Pond treatment, with its potential for excellent natural disinfection, has a crucial role to play in reducing the incidence of bacterial diseases and avoiding epidemic outbreaks (Mara, 2001).

The water-borne enteric diseases of perhaps greatest concern in the developed world, and still of major importance in the developing world, are those caused by enteric ('gut') viruses (Table 6.1). At least 140 types have been reported (Bitton, 1999), including Hepatitis A virus and rotavirus. Viral diseases are seldom lethal for people with healthy immune systems, but can be debilitating and unpleasant, and some (e.g. hepatitis A) have life-long effects. Because the enteric viruses occur mainly at low concentrations in waters and wastewaters, special methods of concentration such as adsorption to and elution from, laboratory columns must be used for their detection. Some viruses are rather resistant to standard disinfectants such as chlorine (often more so than indicator bacteria), but are usually rather efficiently removed in WSPs as will be discussed later in this chapter.

Of increasing concern, following some highly publicised outbreaks, are the protozoan parasites (Table 6.1) including *Giardia*, and *Cryptosporidium*, (Bitton, 1999; Robertson *et al.*, 1999). These organisms cause unpleasant symptoms of diarrhoea, abdominal pain and nausea, and, although rarely fatal, the infections can last for many months. The cysts of *Giardia* and *Entamoeba*, and oocysts of *Cryptosporidium* (the infective stages) are remarkably robust in waters and wastewaters, and are resistant to standard disinfection methods such as chlorination. Concern about these organisms appears to be one of the main drivers in developed countries towards increasing use of ultraviolet disinfection (to which they are susceptible, although more resistant than indicator bacteria – Bitton, 1999). However, WSPs in which sedimentation and exposure to ultraviolet radiation (UV) in sunlight occurs through multiple stages, are capable of efficient removal of these parasites.

Finally there are the worm parasites (Table 6.1), which are unpleasant organisms that can sometimes cause severe complications or death (Bitton, 1999). Recent estimates suggest that at least 50% of the world's population may be infected with one or more helminth species (Chan, 1997). The infectious stage of helminth worms are the eggs (ova), which infected individuals shed in their faeces and which, again, are efficiently removed in WSPs. Removal of these organisms seems to be of more concern in the developing world than in developed countries where disease incidence is lower.

Table 6.1 Categories of pathogenic micro-organism that may be present in sewage and other organic wastes of faecal origin (after Bitton, 1999)

Category of pathogen	**Example organisms**	**Illness(es) caused**
Bacteria	*Salmonella* spp.	Typhoid fever, salmonellosis
	Shigella spp.	Bacterial dysentery
	Vibrio cholerae	Cholera
	Escherichia coli (pathogenic strains)	Diarrhoea
	*Campylobacter*spp	Campylobacteriosis
	Yersinia enterocolitica	Acute gastroenteritis
	Leptospira	Leptospirosis
Viruses	Poliovirus	Paralysis, meningitis
	Enteroviruses	Meningitis, respiratory infection
	Hepatitis A virus	Hepatitis
	Norwalk types	Gastroenteritis
	Rotavirus	Gastroenteritis and dysentery
Protozoan parasites	*Giardia* spp	Giardiasis
	Cryptosporidium spp	Cryptosporidiosis
	Entamoeba spp	Amoebic dysentery
Worm parasites (helminths)	Tapeworms (e.g., *Taenia* spp)	Parasitism (range of symptoms)
	Roundworms (e.g., *Ascaris* spp, *Trichuris trichiura*)	
	Hookworms (e.g. *Necator americanus*)	

6.2.2 Indicators

The question always arises: why do we use indicators when the concern is obviously with the infectious (and dangerous) pathogens? The problem is that a great variety of pathogens can occur in organic wastes of faecal origin, and their concentration is extremely variable (Bitton, 1999; Haas, 1986). During epidemics many sick individuals in the community actively 'shed' pathogens in their faeces, and the pathogen concentrations in the sewage may then be orders of magnitude higher than normal. This makes for difficulties in monitoring treatment processes such as WSPs and receiving waters. Which pathogen of the many that might conceivably be present do you test? Under normal (non-epidemic) conditions the pathogens may be virtually absent from the wastewater, and are difficult to detect.

Indicator organisms, however, are always present and usually at fairly reliable levels (within, say, two orders of magnitude) in domestic sewage and other faecally-contaminated wastewaters (Bitton, 1999). Thus the presence of these organisms is a reliable indication of faecal contamination. Furthermore, indicators are relatively easily, and comparatively cheaply, measured. Conversely, many pathogens are difficult and expensive to measure because they require specialist microbiological expertise and sophisticated procedures and laboratory facilities. For the moment monitoring of pathogens in wastewaters is very much in the research arena.

Some exciting advances in virological and bacteriological methods such as rapid enzymatic methods (Davies and Apte, 1996; George *et al.*, 2002) and gene probes (e.g. Sobsey *et al.*, 1998) may be expected to revolutionise both monitoring of, and scientific understanding of, pathogens in WSPs. However at present monitoring of indicators such as the faecal coliform group of bacteria is far more common.

Table 6.2 lists some ideal features of indicator organisms. No real indicator meets all these criteria, but some, notably *E. coli,* are good on most of the criteria. Most microbiological water quality guidelines and standards of relevance to WSP performance are based on tests for indicator bacteria. The performance of WSPs is therefore gauged primarily in terms of the removal of indicator bacteria through the pond sequence, although it must be recognised that their behaviour may be appreciably different from that of certain pathogens of concern, particularly non-bacterial pathogens.

Table 6.3 lists some bacterial groups that are commonly used for indicating faecal contamination, disease risk, and treatment system performance (a thorough review is given in Bitton, 1999). The most used indicator is the coliform group of bacteria. More useful than the whole coliform group (total coliforms), is the 'faecal' coliform subgroup, sometimes referred to as the 'thermo-tolerant' coliforms, recognising their ability to grow at elevated incubation temperatures. A disadvantage of the 'faecal' coliforms as an indicator is the thermal tolerance of the genus *Klebsiella*, which is not exclusively faecal in origin. Currently, the favoured bacterial indicator, particularly in freshwaters, is the species *Escherichia coli,* which is an unequivocal indicator of contamination by the faeces of warm-blooded animals, at least in temperate climates. Unfortunately, in tropical climates *E. coli* can sometimes multiply in the environment, so that this organism, while still suitable for WSP effluent monitoring, is not necessarily a reliable indicator in tropical receiving waters (Hazen, 1988).

Table 6.2 Desirable features of indicator organisms

Criterion	**Rationale**
1. Ubiquitous enteric organism	Always present in faecally-contaminated wastewater
2. Always present in presence of enteric pathogens	So that presence of indicator warns of risk of disease
3. More common than the pathogen	Easier to detect/monitor
4. More resistant than the pathogen	Pathogen will not be present if no indicator is detected
5. Does not multiply in the environment	So that contamination is not falsely indicated
6. Easily detected by inexpensive methods	So on-going, routine monitoring is affordable
7. Non pathogenic	No danger to laboratory personnel

The faecal streptococci group of bacteria (Table 6.3) has long been recognised as a potential indicator of faecal contamination, but only recently has the use of this group (or the enterococci subgroup) come into favour as an indicator, particularly in seawater, following the epidemiological work of Cabelli and co-workers (e.g., Cabelli *et al.*, 1983).

Some non-bacterial indicators are useful for studying WSP disinfection. Viruses that infect bacteria, known as 'bacteriophages', provide a useful alternative indicator of faecal contamination and a convenient model of human enteric viruses in the environment (Havelaar *et al.*, 1993), including WSPs. For this reason the removal by WSPs of phages native to sewage has been the subject of several studies (e.g., Davies-Colley *et al.*, 1999). The spore-forming anaerobic bacterium *Clostridium perfringens* is sometimes useful as a faecal tracer (Bitton, 1999) because of the resistance of its spores to environmental stressors, particularly sunlight.

Table 6.3 Bacterial indicators

Bacterial genus or species	Indicator groups		
Escherichia	*E. coli*	Faecal coliforms	Total coliforms
Klebsiella			
Enterobacter			
Citrobacter			
S. faecalis	Enterococci	Faecal streptococci	
S. faecium			
S. durans			
S. avium			
S. bovis			
S. equinus			

6.3 OVERVIEW OF DISINFECTION

A large number of factors may influence 'natural' disinfection in WSPs (Maynard *et al.*, 1999), and these factors are summarised in Table 6.4 and briefly discussed in the following sections. There is, however, considerable controversy about underlying mechanisms (Maynard *et al.*, 1999) and this is an active area of current research. Sunlight is now recognised as playing an important role in disinfection, and is discussed in the next section (6.4). More recently it has been recognised that, except at extreme values (say > 9.5-10), pH operates mainly by modifying the response to sunlight, so pH is also discussed in section 6.4, Sunlight-mediated disinfection.

Table 6.4 Factors that have been proposed to cause or influence disinfection in WSPs

Factor	Likely mechanism(s)	Micro-organisms affected[1]	Ponds where active[2]
Temperature	Affects rates of removal processes	B, V, P, H	A, F, M
HRT	Affects extent of removal (time for operation)	B, V, P, H	A, F, M
Algal toxins	Algal exudates are toxic to certain bacteria	Mainly B?	F, M
Sedimentation	Settlement of infectious agent (e.g. ova, cysts)	H	A, F, M
	OR Settlement of aggregated solids including the infectious agent	P, H (B, V?)	A, F, M
'Biological disinfection'	Ingestion by higher organisms (protozoans)	B, V (P?)	F, M
Sunlight	DNA damage by solar UV-B radiation	B[3], V, P	F, M
	OR photo-oxidation (DO-sensitive) (range of wavelengths)	B (P?)	F, M

1. Micro-organism: B – bacteria, V – viruses, P – protozoan parasites, H – helminth worms.
2. Ponds: A – anaerobic, F – facultative, M – maturation.
3. Most of the DNA damage to bacteria by UV-B radiation is repaired, and the lethal effect is related mainly to overwhelming of the repair capacity.

6.3.1 Temperature

Maturation ponds are often designed for disinfection using the Marais (1974) equation, a function only of the average temperature (Section 6.4.6). However, several studies have pointed out that temperature alone is not the primary cause of disinfection (e.g. Mills *et al.*, 1992). Temperature is itself only lethal to micro-organisms (thermal shock) at high values above around 45°C so temperature should be regarded as a secondary factor influencing the *rate* of action of primary factors.

6.3.2 Hydraulic residence time (HRT)

HRT controls the time available for action of primary inactivation or removal mechanisms within WSPs, and should thus be regarded, like temperature, as a secondary factor (Maynard *et al.*, 1999). The influence of theoretical HRT on disinfection is often confounded by short-circuiting of wastewater (refer to Chapter 10 for further detail on hydraulics). The highly inconsistent, and sometimes poor removal of indicator bacteria in conventional WSPs (e.g., Hickey *et al.*, 1989; Sukias *et al.*, 2001) is often at least partly due to short-circuiting, in which influent wastewater is rapidly conveyed to the pond outlet with minimal time for operation of 'primary' removal factors such as sunlight, sedimentation, or ingestion by antagonistic microbiota.

6.3.3 Algal toxins

Some researchers have suggested a contribution to disinfection by certain algae in WSPs that produce extracellular materials toxic to faecal bacteria. For example, Oufdou *et al.* (2001) reported that cyanobacteria occurring in WSPs were toxic to *E. coli*, *Salmonella*, and a number of other bacteria. However, Maynard *et al.*, (1999) cited other studies dismissing algal toxins as a major contributor to WSP disinfection and so the relevance of this mechanism is still unclear.

6.3.4 Sedimentation

Sedimentation in the (largely quiescent) waters of WSPs is believed to be the dominant mechanism for the removal of helminth ova (Maynard *et al.*, 1999). An approach to pond design, ensuring that standards for effluent reuse (< 1 egg/litre) will be met, was proposed by Ayres *et al.* (1992). Protozoan parasites are also fairly efficiently removed within WSPs by sedimentation (Grimason *et al.*, 1993). Since the settling velocities of isolated (oo-)cysts are rather low (2.2-2.8 cm/h for *Cryptosporidiium* oocysts) their aggregation with settleable solids seems likely. Parasite (oo)cysts and eggs can survive for long periods in pond sludge so any sludge disturbance may be expected to remobilise these pathogens (Maynard *et al.*, 1999). Bacteria and viruses might also, in principle, be removed by sedimentation if sorbed onto settleable solids (either wastewater solids or algae) within WSPs. There is little information on this potential mechanism, although WSP solids were found to sorb coliphages under aerobic conditions (Ohgaki *et al.*, 1986), suggesting the potential for viral removal by sedimentation.

6.3.5 'Biological disinfection' (ingestion by antagonistic microbes)

WSPs are inhabited by a diverse range of micro-fauna that obtain nutrition by ingestion of wastewater solids including microbes. Ingestion by these 'antagonistic' organisms of the smaller microbes of sanitary concern such as bacteria, viruses, and perhaps oocysts may cause inactivation by exposure to digestion fluids. Even if excreted microorganisms are not inactivated (i.e. they are still culturable), containment within faecal pellets seems likely to reduce infectivity and promote removal by sedimentation. The phenomenon of micro-faunal disinfection has been investigated in constructed wastewater wetlands (Decamp and Warren, 1998), although apparently not yet in WSPs. However, Manage *et al.* (2002) reported the removal of virus-like particles by flagellate ingestion in a hypereutrophic urban pond, and it seems highly likely that a similar process occurs in WSPs. Indeed, antagonistic microbiotic activity may well be the major process of bacteria and virus removal at times and places of low sunlight exposure within WSPs, such as at night and deep in the water column. Experiments in which the antagonistic action of micro-fauna is attenuated by size-selective filtration or eukaryotic cell inhibitors such as cycloheximide (e.g. Manage *et al.*, 2002) seem to confirm biological disinfection as a potentially important mechanism.

6.4 SUNLIGHT-MEDIATED DISINFECTION

An increasing body of evidence indicates that sunlight is the single most important factor causing disinfection in WSPs (Leduc and Gehr, 1990; Maynard *et al.*, 1999; Mayo, 1995). For example, when WSP water is isolated in transparent bottles fixed at different depths within the water column of WSPs, the die-off of indicator bacteria is far more rapid near the (sun-exposed) water surface than at depth where the water is comparatively dark (e.g. Mayo, 1989). Several studies, notably those of Curtis *et al.* (1992b), have shown that exposure of WSP water to sunlight causes rapid (time for 90% 'kill', $T_{90} \sim 1$ hr) removal of *E. coli*, compared with minimal decline in the dark. An example of such data is given in Figure 6.1.

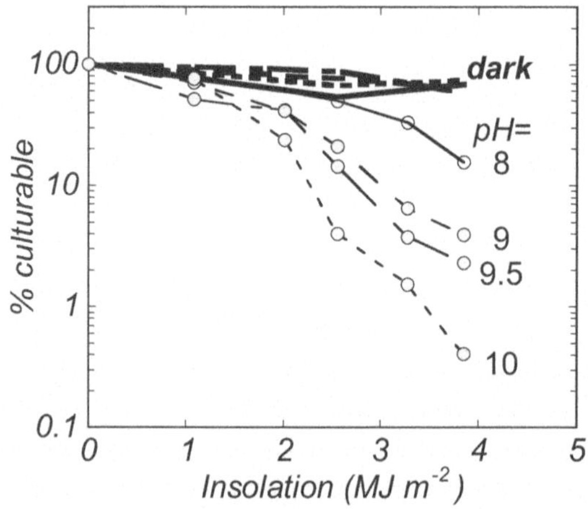

Figure 6.1 Survival curves for faecal coliforms exposed to sunlight in WSP effluent in small stirred microcosms at different pH values as indicated. The culturable count, expressed as a percentage of initial count, is plotted against sunlight exposure in insolation units. Identical microcosms were simultaneously held in the dark (wrapped in aluminium foil). The insolation scale (5 Mega-joules m^{-2}) may be compared with ~ 35 MJ of solar energy per m^2 insolation on a clear day in the tropics, or the temperate zone in summer (Davies-Colley *et al.*, 2000)

Several statistical studies of WSP influent and effluent concentrations of faecal bacteria have attempted to identify the relative importance of sunlight and other factors to their inactivation. Some of these statistical studies have concluded that sunlight is unimportant or at least masked by other factors (e.g., Pearson *et al.*, 1987; Saqqar and Pescod, 1992). Other studies have concluded that sunlight is the major factor, although perhaps interacting with other factors to modify response (Qin *et al.*, 1991; Troussellier and Legendre, 1989; Troussellier *et al.*, 1986).

A major difficulty with statistical approaches is that physico-chemical conditions in WSPs in the field are highly variable with time, particularly on a diurnal cycle owing to algal metabolism driven by sunlight, and seasonally with varying insolation. Thus temperature, pH, dissolved oxygen, and algal production are all highly correlated within WSP water (e.g., Davies-Colley *et al.*, 1999). For this reason it is very difficult to statistically separate the variables and identify those causative of microbiological removal/inactivation (Troussellier *et al.*, 1986). Experimentation using laboratory scale reactors, in which each variable can be manipulated in turn while maintaining good control of the others, is likely to be more definitive than statistical approaches.

Calkins *et al.* (1976) showed that removal of *E. coli* from a WSP system of four lagoons correlated with time-integrated solar UV-B (skin-burning UV, 300-320 nm) 'dose'. Calkins *et al.* (1976), and Moeller and Calkins (1980), inferred from this that UV-B in sunlight was the main factor causing die-off, but later work (see below) has shown that longer wavelengths (> 320 nm) in the solar spectrum may also be important for some organisms. Furthermore, it is now known that dissolved oxygen, pH, and constituents of the WSP medium, can all influence sunlight action on the indicator organisms of faecal coliforms and *E. coli* (Curtis *et al.*, 1992b; Davies-Colley *et al.*, 1999).

6.4.1 Wavelength

Shorter wavelength visible and ultra-violet radiation has long been known to be bactericidal (Whitelam and Codd, 1986). Gates (1929) may have been first to study the wavelength-dependence of bacterial inactivation in the UV range, showing that 265 nm light was most lethal and that the energy dose required for a given 'kill' was strongly correlated to wavelength, being 25 times greater at 300 nm and 250 times greater at 313 nm. This strong relationship of wavelength to inactivation efficiency reflects the absorption spectrum of DNA (Jagger, 1985), which peaks at 265 nm. It therefore appears that the disinfection mechanism involves the absorption of the UV light, by the DNA which then becomes damaged preventing successful growth by the micro-organism.

Numerous experiments have shown that the shortest wavelengths in sunlight, in the so-called UV-B range (290-320 nm, skin-burning UV) are rapidly lethal to bacteria and viruses via mechanisms (DNA damage) similar to those caused by germicidal lamps at 254 nm (Jagger, 1985). However, only a small amount of solar energy occurs in the UV-B range (for example about 0.2% in noon sunlight in the tropics). So the question arises – what is the contribution to microbial inactivation of longer wavelengths of sunlight, including ultraviolet in the so-called UV-A range (320-400 nm – containing perhaps 5% of noon sunlight, (Davies-Colley *et al.*, 1999), or visible light (400-700 nm – containing approximately 50% of solar energy)? Jagger (1985) reviewed the photo-biological literature, and asserted that the UV-A and visible radiation can certainly be damaging to bacteria and viruses, although the mechanisms seem to be generally different from that of germicidal UV radiation and UV-B. In particular, physico-chemical factors of the external medium, such as dissolved oxygen, salinity, and pH influence the lethal action of UV-A and visible radiation.

The wavelengths of sunlight that cause inactivation may be identified by experiments with optical filters (so-called long-pass filters) that extinguish all radiation in the solar spectrum with wavelengths below a certain 'cut-off'. Davies-Colley *et al.* (1997) used small microcosms with good control of DO, pH and temperature for such optical filter experiments. They reported that at saturation DO and a relatively low pH of 7.5, the UV-B, UV-A and visible range of the solar spectrum all contributed approximately equally to inactivation of enterococci and an F-specific RNA phage. However, only the UV-B caused appreciable inactivation of *E. coli* (at this pH) and an F-specific DNA phage. This divergent wavelength-dependence suggests very different mechanisms of inactivation of the different organisms by sunlight. This is discussed further in the following section.

There is strong experimental evidence that sunlight inactivation, mainly by UV wavelengths, is rapid near the surface of WSPs, but sunlight action over the whole pond water column is diminished because of strong light-attenuation in the water column (Curtis *et al.*, 1994; Davies-Colley *et al.*, 1995). Several authors (notably Mayo, 1989; Mayo, 1995; Sarikaya and Saatci, 1987) have plotted indicator bacterial removal rate in WSPs versus depth-averaged insolation (sunlight exposure), and demonstrated a good correlation, but with a relatively large intercept (at zero insolation) indicating appreciable removal by (unidentified) 'dark' processes. However, these authors concluded that sunlight remained responsible for most of the disinfection.

6.4.2 The mechanisms

Based on a critical review of the literature and their own experiments, Davies-Colley *et al.* (2000) concluded that light is probably the single most important biocidal factor in WSPs, causing inactivation by at least three main mechanisms that can operate simultaneously in WSPs (Table 6.5).

Mechanism 1 involves the absorption by solar UV-B (300-320 nm) by DNA causing direct damage to the DNA primarily by pyrimidine dimer formation (Jagger, 1985). This process is independent of oxygen and other conditions in the external medium. In bacteria and higher organisms, DNA damage is efficiently repaired by enzymatic processes (photo-reactivation, excision repair, and post-replication repair) (Jagger, 1985).

Mechanism 2 involves the absorption by short solar UV wavelengths (UV-B and some shorter-wavelength UV-A) by cell constituents (so-called 'endogeneous photosensitizers) including DNA but also other cellular constituents. The activated photosensitizers react with oxygen to form highly reactive photo-oxidising species that, in turn, damage internal targets (notably the genetic material) within the cell or viral particle. This mechanism depends on the dissolved oxygen in the external medium, which in turn controls intracellular oxygen.

Table 6.5 Features of the three main mechanisms of sunlight inactivation causing disinfection in WSPs (after Davies-Colley *et al.* 2000)

Mechanism	Contributing Wavelengths (nm)	Absorbed by	Primary target	Oxygen dependence	pH dependence	Repairable
1. Photobiological DNA damage	UV-B, 300-320	DNA	DNA	No	No	Yes (bacteria)
2. Photo-oxidative damage (primarily to DNA)	UV-B, (+UV-A?)	DNA (+ other cell constituents?)	DNA	Yes	No	Yes (bacteria)
3. Photo-oxidative damage to external structures	300-550 nm	Humics Organic solids	Cell membrane, Capsid proteins?	Yes	Some bacteria (incl. *E. coli*)	No

Mechanism 3 involves absorption by extra-cellular constituents of the medium (so-called 'exogeneous photosensitizers', notably humic material) of a wide range of ultra-violet and visible wavelengths in sunlight. The activated photosensitizers react with oxygen to form highly reactive photo-oxidising species (notably singlet oxygen and the hydroxy radical). These, in turn, damage external targets including particularly, the membrane of bacterial cells or, potentially, host-binding proteins on viral particles. This mechanism is also dependent on dissolved oxygen in the external medium. Typically UV-A wavelengths (320-400 nm) are responsible for the majority of photo-oxidation.

6.4.3 Oxygen

Mechanisms 2 and 3 (Table 6.5) are both photo-oxidation reactions that rely on the presence of dissolved oxygen. In the total absence of oxygen, mechanisms 2 and 3 are impossible and only mechanism 1 occurs. For some micro-organisms, notably DNA-viruses that lack the ability to repair DNA damage, mechanism 1 may well dominate. But bacteria and higher (eukaryotic) organisms are usually capable of repairing DNA damage such that mechanism 1, and to a lesser extent, mechanism 2 may be largely (although not totally) negated by repair, leaving mechanism 3 dominant overall.

In WSPs, dissolved oxygen is highly variable over time (with diurnal changes driven by algal metabolism) and space (with strong vertical gradients from the sunlit euphotic zone to the virtually dark hypolimnion). Conditions are most variable in the surface waters (epilimnion) of WSPs, which are typically super-saturated during the day, but reach low oxygen levels (or anoxia) at night. Several studies (e.g., Curtis *et al.*, 1992b) have shown that supersaturated oxygen concentrations are not, in their own right, toxic to pathogenic micro-organisms. It is the *interaction* of oxygen and sunlight that is toxic.

6.4.4 pH

A variety of studies, beginning with the work of Parhad and Rao (1974), have suggested that pH is a primary cause of disinfection in WSPs. However simple experiments with exposure of faecal indicator bacteria (in the dark) to elevated pH have shown that pH alone is not toxic except at extreme high values not normally encountered in WSPs (Curtis *et al.*, 1992b). Instead pH *interacts* with sunlight. Figure 6.1 shows that, at the same sunlight exposure, (MJ m^{-2}) die-off increases significantly with increasing pH. Of course, pH, like dissolved oxygen, varies in WSPs on a diurnal cycle.

6.4.5 Humic substances

Mechanism 3 in Table 6.5 involves photo-oxidation of external structures by photo-oxidising species produced in the external medium of the WSP water. This photo-oxidation is catalysed by photosensitizers – light absorbing materials – in the WSP effluent. Although much of the total light absorption by WSP effluent is by algal cells and by associated detrital material (Davies-Colley *et al.*, 1995), these particulate materials are likely to be inefficient photosensitizers of disinfection because the reactive oxygen species that they produce tend to be quenched before reaching contaminant micro-organisms. Humic-type materials dominate light absorption in WSPs (Davies-Colley *et al.*, 1995), and may be the most important photosensitizer for disinfection (Curtis *et al.*, 1992b).

6.4.6 Modelling bacterial indicator removal in WSPs

Traditionally, most maturation ponds have been designed for disinfection using the Marais (1974) equation or variants thereon – a function only of temperature:

$$k_{\mathrm{T}} = k_{20}\theta^{(T-20)}$$

where k_{T} is the first-order bacterial removal constant, T is temperature and θ is the temperature coefficient (Arrhenius formulation). Marais (1974) proposed k_{20} = 2.6 d^{-1}, and θ = 1.19, values that appear still to be widely used by pond designers, although alternative values have been proposed (see Mayo 1995 for a review).

However, sunlight much more powerfully affects disinfection than temperature, and Mayo (1995) showed that models of disinfection incorporating sunlight were much more satisfactory than those that did not. Mayo (1995) built on the earlier work of Sarikaya and Saatci (1987) to develop a semi-empirical model of the first-order disinfection rate constant (k, h^{-1}) incorporating terms for dark dieoff, sunlight inactivation, and pH:

$$k = k_{\mathrm{d}} + k_{\mathrm{S}}S/KH + k_{\mathrm{pH}}\mathrm{pH}$$

in which H is pond depth (m), K is the attenuation coefficient (m^{-1}) for (bactericidal wavelengths in) sunlight, S is solar radiation (MJ h^{-1} m^{-2}) and the factor S/KH is depth-averaged sunlight over the pond water column (depth, H). The terms in this equation were evaluated by multiple linear regression (the model explaining about 55% of the variance) using datasets for WSPs in (tropical-subtropical) Thailand, Saudi Arabia, and Tanzania.

The Mayo (1995) formulation may be the best currently available for predicting pond disinfection. This is despite the separation of pH and sunlight, which experiments by Curtis *et al* (1992b) and Davies-Colley *et al.* (1999) have shown to be strongly interactive factors, and the neglect of dissolved oxygen, which also interacts with sunlight (Curtis *et al* 1992a). The Mayo (1995) model formulation should be broadly applicable to WSPs in temperate regions. However, re-calibration would be desirable for temperate zone WSPs. Craggs *et al.* (2004) recently reported such a calibration for a high-rate pond treating dairy cattle wastes in temperate New Zealand.

6.5 BACTERIAL PATHOGEN REMOVAL

This section reviews the removal in WSPs of some important bacterial pathogens in relation to removal of faecal indicator bacteria.

6.5.1 *Salmonella*

Macdonald and Ernst (1986) found that *Salmonella* bacteria were mostly absent from the effluent of one WSP system (Camden) close to Sydney, Australia. However at a nearby WSP system (Windsor), the frequent occurrence of *Salmonella* was attributed to birds, which carry salmonellae in their digestive tracts. Pearson *et al.* (1987) added spikes of pure cultures to experimental WSPs, and found that *Salmonella typhimurium*, and other bacteria of sanitary significance, behaved rather similarly to faecal coliforms (including sensitivity to pH) but were more rapidly removed.

Arridge *et al.* (1995) found that native salmonellae were efficiently removed from sewage in an experimental WSP system. Low concentrations of the pathogen precluded detailed examination of the kinetics of removal in the different ponds within the system, but salmonellae appeared to be inactivated or removed more rapidly than *E. coli*, consistent with other studies. Castillo and Trumper (1991) reported that salmonellae were slightly more resistant than faecal coliforms. Emparanza-Knoerr and Torrella (1993) reported rather high salmonellae concentrations up to 10^3 MPN/100 ml) in the effluent of an overloaded WSP with 10^5-10^6 MPN/100 ml of faecal coliforms, suggesting little removal and perhaps even multiplication of salmonellae within the WSP system.

In general, salmonellae seem to be rather efficiently removed by WSPs, and their removal is more rapid than, although fairly well predicted by, faecal coliforms or *E. coli*..

6.5.2 *Shigella*

Shigella bacteria, the agents of bacterial dystentery, are generally recognised as less persistent in the environment than *E. coli* (e.g., Bitton, 1999). Few studies have reported on survival of *Shigella* in WSPs, but there are indications that these organisms are efficiently removed by WSPs, again more rapidly than faecal coliforms (e.g., Macdonald and Ernst, 1986; Madera *et al.*, 2002).

6.5.3 *Vibrio cholerae*

The behaviour of *Vibrio cholerae*, the infective agent causing cholera, appears to be appreciably different in WSPs from faecal coliforms (including *E. coli*) and other bacteria of sanitary significance. This divergent behaviour suggests that caution should be applied when using data on *E. coli* removal in WSPs to infer removal of *V. cholerae*.

There is evidence that *V. cholerae* sometimes grows in WSPs. Macdonald and Ernst (1986) detected *V. cholerae* strains in WSPs, with higher levels occurring in summer months. Nair *et al.* (1988) showed that summer peaks were typical of *V. cholerae* in WSPs in India. Arridge *et al.* (1995) reported experiments with *V. cholerae* spiked into WSP water contained in open-top buckets suspended near the water surface in a WSP system. Effective removal (9 orders of magnitude) occurred with approximately first order kinetics over about 13 days in all of the ponds tested. Arridge *et al.* (1995) suggested that relatively rapid removal in anaerobic ponds might be due to sulphide toxicity.

Some notable work has been done in Morocco. Lesne *et al.* (1990) reported a strong seasonal cycle of *V. cholerae* in an experimental WSP in Marrakech, with summer peaks and winter lows – a pattern opposite to that of *E. coli*. Mezrioui *et al.* (1994; 1995) suggested that algae of the genus *Chlorella* inhibited *V. cholerae*, but cyanobacteria (*Synechococcus*) promoted *V. cholerae*. The opposite seasonal patterns of *V. cholerae* and *E. coli* in the Marrakech WSP system was attributed by these authors to the seasonal pattern of algae, in which green algae, particularly *Chlorella,* dominate for most of the year, with cyanobacteria prominent in summer. Furthermore, the high pHs and high temperatures typical of WSPs in summer were interpreted as promoting *V. cholerae* whereas *E. coli* inactivation in this season is rapid. Overall, treatment did not greatly reduce *V. cholera*, despite efficient removal of *E. coli*. Oufdou *et al.* (1998) showed that *Synechocystis* exudates promoted *V. cholerae*, but inhibited *E. coli* and a *Salmonella* sp, and Oufdou *et al.* (2001) showed that compounds produced by the cyanobacteria

Pseudoanabaena spp also stimulated *V. cholerae* while inhibiting *E. coli*, and the pathogens *Salmonella* sp, *Staphylococcus aureus* and *Candida albicans*.

Apparently *V. cholerae* may occur at high levels in summer in WSPs in marked contrast to the generally excellent removal of other bacterial pathogens in this season. Furthermore, removal in anaerobic ponds may be relatively good – again in marked contrast to other bacterial pathogens, which are ineffectively removed by anaerobic ponds.

6.5.4 *Campylobacter*

Campylobacteriosis is a 'zoonotic' (harboured by animals) disease that can be important in some developed countries as well as in the developing world. A very limited amount of information suggests that *Campylobacter* spp, including the most human-pathogenic *C. jejuni*, are effectively removed in WSPs. Oragui *et al.* (1986) reported virtually complete removal of *Campylobacter* in a WSP system, and Pearson *et al.* (1987) reported that *Campylobacter jejuni* (along with *Salmonella* spp) was more rapidly removed than *E. coli* in a WSP. Given that *C. jejuni* may be the single most important cause of waterborne gastroenteritis in the developed world, quantifying the removal of this pathogen in WSPs would seem to be a research priority.

6.5.5 Other bacterial indicators and pathogens

The behaviour of a number of other pathogenic or indicator bacteria in WSPs has been studied. Several studies have shown that other indicator bacteria including faecal streptococci and enterococci, are more rapidly removed in WSPs than *E. coli*. For example Oragui *et al.* (1986), Pearson *et al.* (1987), Castillo and Trumper (1991), Arridge *et al.* (1995), and Campos *et al.* (2002) all found that faecal streptococci were rather rapidly removed within WSP systems.

Davies-Colley *et al.* (2000; 1999) showed that enterococci (part of the faecal streptococci group) were more rapidly inactivated by sunlight (by a photo-oxidative mechanism) than *E. coli* in WSP water, except at elevated pH (above about 9) under which conditions accelerated inactivation of *E. coli* makes these bacteria less persistent than enterococci. Generally it appears that *E. coli* are the better indicator except perhaps at very elevated pH (>9.5) when *E. coli* are more rapidly removed than enterococci and some pathogens (Figure 6.1).

A few studies have been done on bacteria that are opportunistic pathogens, such as *Staphylococcus* spp, *Aeromonas* spp and *Pseudomonas* spp, within WSPs. These bacteria are generally appreciably more persistent within WSP systems than *E. coli* and other standard indicators. For example, Nascimento *et al.,* (1991) found that removal of *Pseudomonas aeruginosa* in WSPs was lower than that of faecal coliforms and streptococci. Boussaid *et al.* (1991) found that *Aeromonas* spp removal (a little over 1 log unit) was much lower than that of faecal coliforms in a WSP system. Bahlaoui *et al.* (1997) reported that *P. aeroginosa* and *A. spp* had greater resilience than the standard indicator bacteria in a high-rate algal pond.

Even more persistent are the spores of *Clostridium perfringens*, which may be more useful as a tracer than an indicator in WSP systems (Campos *et al.*, 2002; Oragui *et al.*, 1995).

6.6 VIRUS REMOVAL

There have been numerous studies of the removal of viruses in WSPs, and most, but not all have reported excellent performance (reviews have been published by Chaudhuri, (1973) and Maynard *et al.*, (1999)). Few studies have investigated mechanisms of removal or inactivation of viruses in WSPs, although available data implicates sunlight, pH, and adsorption onto solids.

6.6.1 Bacteriophages as model viruses

Bacteriophages are viruses that infect bacteria. Certain bacteriophages, notably the F-specific phages, have been used as convenient tracers and faecal indicators in sewage and other faecally-contaminated wastes, and it is argued that they are reasonable models of the environmental behaviour of some enteric virus pathogens (Havelaar *et al.*, 1993; IAWPRC, 1991). Arguments favouring use of phages rather than enteric viruses include (after Campos *et al.*, 2002):

- Some are abundant in treated wastewater (much more so than enteric viruses)
- Cannot reproduce outside bacterial host
- Easy to isolate and count
- Quick to enumerate compared to enteric viruses
- Similar persistence to enteric viruses (phages are models).

At least four main types of bacteriophage have been studied for various purposes in WSPs (Table 6.6), and the cited references may be consulted for detailed information.

Table 6.6 Bacteriophages that have been used as indicator organisms and models of enteric virus behaviour in WSPs (after Sinton & Finlay, 1996)

Phage	Bacterial host	Characteristics	References
Somatic phages	Usually *E. coli* or other coliform bacteria	'Tailed' viral particles	Borrego *et al.* (1987)
F-specific phages	Usually *E. coli*	Attach to F-pili produced by some bacteria	IAWPRC (1991)
F-RNA phages		Small icosahedral particles	Havelaar *et al.* (1993)
F-DNA phages		Thin filamentous particles	IAWPRC (1991)
B. fragilis phages	*Bacteroides fragilis*	Tailed 'somatic' phages	Tartera & Jofre (1987)
Serratia marscens phage	*Serratia marscens*		Drury & Wheeler (1982)

6.6.2 Phage behaviour in WSPs

A number of studies of WSP disinfection have been reported using phages. Ohgaki *et al.* (1986) carried out experiments with coliphages in a WSP, and demonstrated the role of sunlight inactivation (strong depth-dependence and rapid inactivation in the light versus dark conditions). Adsorption of coliphage to WSP solids was also shown. Castillo and Trumper (1991) found that coliphages were more resistant than faecal coliforms or streptococci in an enclosure within a WSP. Die-off was slower in fall than in summer, but with less of a seasonal contrast than for the bacterial indicators.

Turner and Lewis (1995) reported that F-specific phage, enterococci and faecal coliform had parallel survival (about 3.5 log removal) through two maturation ponds. Donnison and Ross (1995) studied somatic and F-specific coliphages in WSPs treating sewage and meat processing wastes, and found that correlations of bacterial indicators and coliphages were not particularly strong, suggesting different behaviour within WSPs.

Frederick and Lloyd (1995) reported studies with bacteriophages infecting *Serratia marscens* in a WSP. Microcosm experiments showed rapid removal by sunlight (compared to dark conditions), but high pH (> 9.0) also appeared to play a role. Later experiments (Vorkas and Lloyd, 2000) suggested that these phages, and also phages of *Pseudomonas syringae*, *Erwinia amanas*, and *Erwinia amylovora*, would be useful tracers for evaluating hydraulic performance and disinfection in WSPs. Removal of all of these phages in microcosms was attributed to sunlight interacting with high pH (> 8.5), apparently due to a photo-oxidation process. Campos *et al.* (2002) reported 1-4.6 log removal of F-specific, somatic and *B. fragilis* phages in a two-pond WSP system that removed 0.3-4.7 log units of bacterial indicators.

Davies-Colley *et al.* (1997; 1999) studied sunlight inactivation of F-specific phages (both FRNA and FDNA phages) native to sewage spiked into WSP effluent in small stirred microcosms controlled for temperature, dissolved oxygen and pH. They showed that the F-RNA phage appeared to be inactivated (by a range of wavelengths in sunlight) by a dissolved-oxygen-dependent photo-oxidation mechanism (mechanism 3 in Table 6.5), but F-DNA phage was inactivated by direct photo-biological damage to the DNA (caused only by UV-B in sunlight, Mechanism I in Table 6.5). These experiments suggest that physico-chemical conditions (pH, DO, temperature, solids content) can strongly influence sunlight action on different phages, and, by implication, different viral pathogens too.

6.6.3 Virus removal

An early review by Chaudhiri (1973) of research on viruses in WSPs concluded that exposure to sunlight was a major removal factor for viruses, augmented by adsorption to WSP solids. A more recent review by Maynard *et al.* (1999) also implicated sunlight and adsorption/sedimentation of solids, plus elevated pH.

Sobsey and Cooper (1973) studied behaviour of poliovirus I in laboratory microcosms containing algal-bacterial mixtures and WSP water. They found that poliovirus rapidly adsorbed to WSP solids according to a Freundlich isotherm. Poliovirus was rapidly inactivated in WSP water, but not in algal cultures in sterilized sewage, suggesting a biological agent of inactivation present in the WSP water. These authors concluded that both adsorption to WSP solids and antagonistic microbial action might be involved in removing viruses from WSP systems.

Rao *et al.* (1981) reported fairly efficient (88-98%) viral removal from two WSPs with HRTs of 2.7-17 days. Macdonald and Ernst (1986) reported effective enteric virus removal from two WSPs. No enteric viruses were detected (by electron microscopy) in samples in which faecal coliforms and *E. coli* were low. Oragui *et al.* (1986) reported 3 log removal of enteric viruses (enterovirus and rotavirus) in a series of deep anaerobic, facultative and maturation ponds, which removed 4 log of faecal coliforms and faecal streptococci. Oragui *et al.* (1995) found that rotavirus was slightly more slowly removed than *E. coli* in the chain of ponds in two different experimental WSP systems.

Manage *et al.* (2002) reported removal of virus-like particles (detected by electron microscopy) in water of a hypertrophic pond (not a WSP, but presumably comparable). They attributed this removal to ingestion by nano-flagellates (0.8-5 μm size) based on the persistence of virus-like particles in 0.8 μm versus 5 μm filtrates, and also in unfiltered pond water to which cycloheximide (an inhibitor of protein synthesis in eukaryotic organisms) had been added.

6.7 REMOVAL AND VIABILITY OF HELMINTH OVA

Removal of infectious worm parasite eggs from domestic wastewater is a particularly important function of wastewater treatment in developing countries where community infection levels are often high (Mara, 2001). Multiple-pond WSP systems are capable of efficient removal of helminth eggs mainly by the process of sedimentation to the sludge. Consistent with this process, sludge concentrations are usually highest in the primary pond and thereafter decrease through WSP systems, and also tend to decrease within ponds with distance from the inlet (e.g. Bouhoum *et al.*, 2000).

Most empirical studies have reported complete helminth removal in multiple pond WSPs (e.g., Bouhoum *et al.*, 2000; Saqqar and Pescod, 1991). However 100% removal efficiency is not always guaranteed, and occasional carryover of eggs (particularly those of hookworms rather than the faster-settling eggs of *Ascaris* round worms) has been reported (e.g. Lloyd and Frederick, 2000). An empirical design equation for removal of helminth eggs from wastewater was proposed by Ayres *et al.* (1992) based on empirical data from pond systems in Kenya, Brazil and India, and has been broadly confirmed in other studies (e.g., Stott *et al.*, 2003; von Sperling *et al.*, 2003).

6.7.1 Helminths in WSP sludge

The sedimentation of helminth eggs within WSPs transfers the concern from the water to the sludge. The concentration of eggs in wastewater and therefore in the sludge is likely to reflect prevalence in the community (Lloyd and Frederick, 2000). Nelson (2003) reviewed literature on accumulation in sludge and loss of viability of helminth eggs, including experiments using dialysis chambers to determine inactivation of *Ascaris* eggs and a range of micro-organisms in sludge. *Ascaris* eggs were very persistent in the sludge of WSPs, with about 50% loss of viability in the first year. Phages were more rapidly inactivated, and indicator bacteria even more so, with approximate first order rate coefficients of 0.001, 0.01 and 0.1 d^{-1} respectively for the three categories of organism. Clearly, disturbance of the sludge has the potential to remobilise long-lived helminths, and such episodes may account for occasional detection in later ponds within WSP systems.

6.8 PROTOZOAN REMOVAL

Protozoan pathogens are persistent in the environment owing to their formation of resistant cysts or oocysts, which are the infectious stage (Robertson *et al.*, 1999). Experimental data suggests however, that, despite their environmental resistance, protozoan oocysts are effectively removed within WSPs (generally more effectively than in other types of waste treatment, with the exception of dedicated UV disinfection – Robertson *et al.*, 1999), albeit at slower rates than bacterial pathogens or indicators. Robertson *et al.* (1999) concluded that aggregation with settleable solids followed by sedimentation is probably the main removal process in WSPs (mainly in the primary pond).

Early work on protozoa in WSPs used bright-field microscopy for detection of oocysts on small (< 3 L) samples (Robertson *et al.*, 1999), but more recent work has tended to favour immuno-fluorescence techniques (e.g. Grimason *et al.*, 1993) that involve filtration of large volumes and have increased recovery and detection success.

The finding of both protozoan cysts and helminth ova in sludge suggested to Bouhoum *et al.* (2000) that sedimentation is the main removal process for both types of parasite. However, protozoan cysts and oocysts have lower settling velocities (2.2-2.8 cm/hr for Cryptosporidium oocysts – Robertson *et al.*, 1999) than helminth ova, so aggregation with wastewater solids into large, more rapidly settling flocs seems likely to contribute to removal by sedimentation.

Ultraviolet disinfection is well known to be effective in the removal of protozoan parasites (e.g. Clancy *et al.*, 2000), so a potential mechanism for their removal in WSPs could be sunlight exposure of surface water to solar UV-B *combined* with sedimentation. Support for this hypothesis comes from a study of loss of infectivity (>97%) of *Cryptosporidium* oocysts isolated in semipermeable bags suspended in surface water of a high rate algal pond (Araki *et al.*, 2001). In conventional ponds sunlight exposure of the surface mixed layer may inactivate most cysts that escape sedimentation to the sludge. The superior cyst removal in summer versus winter noted by Wiandt *et al.* (1995) might plausibly be related to seasonality in sunlight inactivation, similar to the seasonal variation in *E. coli* induced by variation in insolation (Troussellier *et al.*, 1986).

Several studies have reported virtually complete removal of oocysts from WSP effluent despite sometimes-high influent concentrations (Bouhoum *et al.* 2000, Grimason *et al.* 1993, Wiandt *et al.* 1995). Those cysts that are not inactivated by sunlight exposure appear to be concentrated in the sludge, implying a health risk with the handling of such sludges.

6.9 INFLUENCE OF PHYSICAL DESIGN

WSP disinfection primarily involves sedimentation (especially for worm and protozoan parasites), sunlight action (especially bacteria and viruses), and biological disinfection (protozoan ingestion of bacteria and viruses, and perhaps also protozoan parasites). The physical design of ponds influences disinfection by affecting the retention time available for these removal and inactivation processes, as well as their intrinsic efficiency (Chapter 10). While the hydraulic residence time is, therefore, a particularly important parameter affecting disinfection (refer to Section 6.3.2 and Chapter 10), in this section we consider the influence of other factors. Sedimentation efficiency could be influenced by the hydraulic surface-loading rate (the flow rate, m^3 d^{-1} divided by the surface area m^2). This criterion actually has the same units as velocity (m d^{-1}) and can also be thought of as the "critical" settling velocity. Clearly this rate must be lower than the settling velocity of the pathogen or aggregates of the pathogen and settleable pond solids. As a guide, anaerobic ponds should therefore have hydraulic surface loading rates of less than 2 m d^{-1}, which is approximately the settling rate of helminth ova, although in practice factors such as the inlet/outlet structures will also have a large impact.

Sunlight disinfection is affected mainly by the average exposure of pond water to biocidal wavelengths (primarily ultra-violet radiation), which in turn depends on attenuation of radiation by pond water and pond depth. Biological disinfection (ingestion by antagonistic micro-organisms) is poorly understood, and it is difficult to conceptualise effects of physical design in the current state of the science. Fortunately, several empirical studies have been conducted to investigate the influence of physical design features on overall disinfection efficiency, and they are reviewed below.

6.9.1 Pond configuration and depth

Pond configuration in plan (length:width ratio) affects disinfection primarily by affecting pond hydraulics and mixing (Chapter 10). Depth seems likely to affect disinfection by sunlight by altering the average sunlight exposure of the water column. Moreover, the depth of a pond of a given HRT determines the hydraulic surface-loading rate and therefore may affect the settling efficiency. Both of these considerations suggest that shallow ponds with high sunlight exposure and long hydraulic residence times should work best. However, pond depth cannot be reduced indefinitely (in practice 0.4 m is often considered a minimum) because of the increased cost of land purchase and pond construction, and the risk of colonisation by macrophytes and of sludge disturbance by wind.

At the other extreme, deep ponds (2m or greater) have the (theoretical) advantage of having a smaller land area for a given HRT. Moreover thermal stratification provides a hypolimnetic volume that may act to entrap sedimenting micro-organisms of concern and isolate them from the pond effluent. Deep ponds have been designed as storage reservoirs for reuse in agriculture. The microbiology of these systems has been reviewed in Chapter 2 and indeed a full chapter is devoted to these systems later in this book (Chapter 17). Disinfection in these systems is discussed in Section 6.9.3.

Oragui *et al.* (1986) reported effective removal of faecal indicator bacteria and bacterial pathogens in 3 m deep ponds, and concluded that disinfection in deep ponds is comparable to that in maturation ponds of more typical depth (around 1m).

Sarikaya *et al.* (1987) applied a simple model of coliform removal developed by Sarikaya and Saatci (1987) (including an explicit term for sunlight exposure) and showed that comparatively shallow ponds gave better coliform bacterial removal. Sarikaya and Saatci (1988) analysed pond-depth dependence theoretically, and showed that optimum depth in terms of performance versus cost of construction tended to fall in the range 0.5-1 m for a rather wide range of conditions.

Pearson *et al.* (1995) found that differences in length:breadth ratios (in the range 1:1 to 6:1) and depths (range 1-2 m) of facultative ponds had little apparent effect on faecal coliform removal. They concluded that the importance of physical design features has been over-stated. However, relatively shallow maturation ponds (0.4 m) gave better faecal coliform removal than deeper ponds of the same area in their experimental complex despite shorter residence time.

Von Sperling (1999) analysed coliform dieoff data for 33 facultative and maturation ponds in Brazil with a range of climates, physical configurations and detention times. He found that first order coliform removal rate coefficients were inversely related to pond depth (range 0.4-2.3 m) and, surprisingly, were *inversely* related to pond detention time (ranging from ca.1 to >100 days) - itself being inversely related to pond depth at constant surface area. Plausibly the greater exposure of shallower pond water columns to sunlight was responsible for the observed depth-dependence of coliform removal rate. Von Sperling (1999) concluded that shallow ponds (0.4 m) give disinfection superior to deeper ponds of the *same* land area, despite lower HRT.

In deep ponds sunlight might be expected to be relatively less important than other factors owing to lower exposure to biocidal wavelengths. However Xu *et al.* (2002) found that sunlight was still the main explanatory variable accounting for faecal coliform removal in deep (1.4-2.8 m) ponds of a WSP system.

6.9.2 Inlet and outlet structures and baffling

The most important consideration with inlet and outlet structures, regarding disinfection, is to reduce short-circuiting. Normally pond designers locate inlet and outlets at opposite ends of ponds for this reason, although tracer studies and CFD modelling show that severe short-circuiting can still occur with an opposite inlet/outlet.

CFD studies (e.g. Shilton and Harrison, 2003) suggest that baffling should markedly improve bacterial removal, consistent with empirical studies demonstrating improved microbial quality from baffled ponds. For example, Pearson *et al.* (1995) found that, at pilot scale, a baffled maturation pond (effective length:breadth of 100:1) gave better faecal coliform removal than unbaffled ponds. While baffling has generally been considered as a method of improving hydraulic, and so treatment, efficiency, to date there has been very limited practical advice with regard to the number of baffles to be used, placement of these baffles and the relative inlet/outlet placement. Shilton and Harrison (2003) have recently published guidelines on baffling as summarised in Chapter 10.

An inlet near the bottom, and an outlet taking off surface water, are typical recommendations that are important to WSP disinfection. Near-bottom inlets inject parasites close to the sludge layer so that they need settle over only a small distance. Surface off-take has the advantage that the surface wastewater has had appreciable exposure to sunlight, which inactivates bacteria and viruses of sanitary concern.

6.9.3 Wastewater treatment and storage reservoirs

Comparatively deep ponds are attractive (because of their smaller surface area for a given volume) where a long residence time is important such as wastewater storage for seasonal reuse in agricultural irrigation. Wastewater storage reservoirs have been built in Israel (depths in the range 6-15 m, Juanico and Shelef, 1994), Spain, and other semi-arid Mediterranean countries (Chapter 17). The primary purpose of these systems is wastewater storage, but useful extra 'polishing', including disinfection, has been noted by several authors (Liran *et al.*, 1994). However, the disinfection efficiency is rather low (about 90%) despite long residence times (50-180 days). Liran *et al.* (1994) found that dieoff was slow in the bottom waters (hypolimnion) of wastewater reservoirs, but much faster in the surface stratified layer (epilimnion), which they attributed to elevated pH rather than sunlight exposure. They made three suggestions to improve the microbiological quality of irrigation water:

1. vertical mixing of the storage reservoir to improve disinfection by breakdown of stratification and exposure of bottom waters to surface conditions,
2. plug-flow design to improve hydraulics, and
3. diversion of fresh wastewater, with its high burden of faecal microbial contaminants, to a separate pond so as not to contaminate disinfected pond water used for irrigation (i.e. operate the reservoir as a batch reactor during the irrigation season).

Consistent with these recommendations, Pearson *et al.*, (1996b) reported that batch-loaded pilot-scale deep wastewater storage reservoirs produced an effluent of high microbiological quality (low indicator and pathogenic bacteria, and no helminth eggs) suitable for unrestricted irrigation.

6.10 POST DISINFECTION OF WSP EFFLUENTS

Although disinfection by WSPs is generally excellent and much better than in mechanical treatment plants (George *et al.*, 2002), final effluent quality is still somewhat variable. Many pond systems, however, have never been specifically designed for pathogen removal using procedures such as those detailed in Chapter 9 and the potential exists to redesign them to improve their performance in this regard. Intermittent episodes of poor disinfection performance may be related to periods of low insolation (and consequently poor sunlight inactivation), hydraulic short-circuiting, or sludge disturbance with entrainment of sludge-concentrated microbes into the effluent. There is continuing interest in the supplementary treatment of WSP effluent to meet stringent microbiological quality standards, particularly in developed countries. A number of options for 'upgrading' pond disinfection are summarised below. A broader review of pond upgrading is given in Chapter 11.

6.10.1 Chlorine

Several studies of chlorine disinfection of WSP effluent were carried out in the 1960s and 1970s (reviewed by Polprasert and Rajput, 1984). These studies mostly concluded that chlorination was feasible despite the high chlorine demand of WSP solids (mainly algal biomass and detritus) and concerns about 'sheltering' of pathogens within algal solids. Benefits additional to disinfection were cited, including reduced suspended solids and increased water clarity. Johnson *et al.* (1978) developed a model for WSP effluent chlorination and concluded that effective disinfection could be achieved with relatively low doses (2-3 g m^{-3}, residual 0.5-1 g m^{-3}) and contact times of less than 50 minutes.

Polprasert and Rajput (1984) found that disinfection kinetics were two-phase, with rapid initial inactivation followed by a slower kill rate.

Despite its technical feasibility (and continued widespread use in potable water treatment) chlorination of WSP effluents is now not usually considered. This is probably because adequate natural disinfection is achieved by WSPs at many sites, and final microbiological quality can usually be enhanced by the addition of further maturation ponds. Furthermore, there are concerns with the environmental toxicity of chlorinated organic by-products and increasing recognition that chlorine is more toxic to bacteria than to pathogens of other types, notably viruses (Tyrrell *et al.*, 1995).

6.10.2 Ozone

Ozone seems preferable to chlorine as a disinfectant considering the toxic by-products of the latter, and, together with UV disinfection, is gaining ground versus chlorination for potable water treatment. Several studies have shown that the ozone 'kill' of viruses is better relative to indicator bacteria. Tyrrell *et al.* (1995) found that ozone was preferable to chlorine as a disinfectant of the effluent from mechanical secondary sewage treatment plants, particularly for its more potent 'virucidal' action. No studies appear to have reported the use of ozone to disinfect WSP effluent, but this is probably feasible despite the likely high ozone demand of typical WSP effluent.

6.10.3 Ultra-violet disinfection

UV disinfection of WSP effluent has been regarded as problematic because of the high solids content and related high UV attenuation of the pond water (Nelson, 2000). However, Nelson (2000) reviewed the available literature and asserted that WSP disinfection should be feasible for WSP effluent. WSP effluent solids are predominantly algae, which do not strongly adsorb bacteria, so survival curves for UV exposure of WSP effluent do not exhibit the pronounced 'tailing' typical of other effluents (Emerick *et al.*, 1999), due to shielding of solids-associated bacteria. The minimal 'tailing' means that < 10 *E. coli* cfu/100 mL can be achieved at practicable doses (say, > 1000 J/m^2). However, Nelson (2000) recognised that viruses might be less well inactivated by UV irradiation due to stronger association with WSP solids. UV treatment is increasingly becoming commonplace for wastewaters in a wide range of applications, but further research seems warranted before it can be recommended generally for WSP effluent.

6.10.4 Filters

The addition of various kinds of 'filters' as an add-on to WSPs has long held attraction for designers wanting to provide better control over final effluent quality with a smaller 'footprint' than, say, extra maturation ponds. Middlebrooks (1995) reviewed the WSP upgrading options: sand filtration, hyacinth and duckweed ponds, land application, rock filters and constructed wetlands, focussing mainly on removal of algal suspended solids in final effluent. While intermittent sand filtration is very effective at removing algal solids, and presumably also faecal microbes, the least expensive 'polishing' options for WSP effluent may be rock filters and/or wetlands (see Chapters 11 and 15).

Rock filters are an attractive 'add-on' to WSPs because they are cheap, being easily constructed out of local materials and having no need for sophisticated operational facilities. If a rock filter clogs, it can simply be dismantled and rebuilt. Rock filters rapidly acquire biofilm coatings on the media, and these presumably act to entrap microbial contaminants that may on occasion escape WSPs. The most successful rock filters are variants on a design trialled in Veneta, Oregon (Swanson and Williamson, 1980). Saidam *et al.* (1995) studied rock filters of various local materials for treating WSP effluent near Amman, Jordan, and reported 90% removal of faecal coliforms (to < 1000 MPN/100 mL).

Subsurface flow constructed wetlands may be viewed as rock filters with (wetland) plants (Kadlec *et al.*, 2000; Tanner, 2001). The presence of plants usually increases the nutrient-processing efficiency of constructed wetlands over rock filters, and also increases their removal of microbial contaminants (Tanner, 2001). A review of wetlands by Kadlec *et al.* (2001) reported fairly consistent removal of around 90-99% of faecal indicator bacteria and phages in a number of studies. There are many similarities between constructed wetlands and WSPs regarding 'natural' disinfection, the main differences being that wetlands are less sunlight-exposed and lower in pH, but in compensation may have more favourable conditions for 'dark' inactivation by sedimentation, filtration, and ingestion by antagonistic micro-fauna. Polishing WSP effluent with constructed wetlands may provide consistent natural disinfection, better than adding more maturation ponds for example, because the very different 'environment' ponds *versus* wetlands, creates the opportunity for different removal/inactivation processes to contribute to overall removal. For example, pathogens concentrated in the sludge of WSP ponds might be entrained on occasions by sludge-disturbing events, but wetlands should 'buffer' against such excursions, providing an extra level of assurance about effluent microbiological quality.

6.11 RESEARCH NEEDS

This Chapter has highlighted several important areas for future research. They are outlined below, and summarised in Table 6.7.

There is a need for improved modelling for prediction of WSP performance in terms of *E. coli* removal. Currently the best available models are semi-empirical statistical models (Curtis *et al.*, 1992a; Mayo, 1995). A truly deterministic model of *E. coli* removal in conventional WSPs is yet to be developed, even though we know enough about the role of the main factors (sunlight interacting with DO, pH, and affected by pond optics and depth) to now attempt this type of model (Davies-Colley *et al.*, 2000). Ultimately the potential exists for incorporating such mechanistic models within a CFD hydraulics model thereby truly integrating the two areas of decay kinetics and mass transfer. Such a modelling capability would probably be first used by researchers to guide further experimentation, but eventually would be useful to pond designers.

Table 6.7 Some research topics in the field of WSP disinfection

Research subtopic	Rationale (drivers)	Requirements/Notes
1. Modelling of *E. coli* removal	Needed for improved WSP design	Improved understanding of the 'dark' dieoff processes, and of WSP optics and hydraulics
2. Biological disinfection (antagonistic micro-fauna)	Needed for improved modelling	Development of suitable WSP microcosms and methods for inhibiting micro-fauna
3. Application of emerging technologies for enumeration of microbial contaminants	Improved monitoring (incl. alarm systems) More sophisticated experiments on inactivation/removal	Awaits development of fast but cheap microbiological methods.
4. Pathogen behaviour in WSPs, especially protozoans	Pathogens may not behave the same as Indicators in WSP systems	Improved enumeration techniques
5. High-rate ponds and Advanced Integrated Pond systems (AIWPS)	High microbiological quality requirements More consistent treatment than in conventional WSPs	Repeat many of the disinfection experiments done on conventional WSPs
6. Add-ons to WSP systems, e.g. wetlands, rock filters, UV disinfection	As for 5.	Experiments and monitoring to investigate buffering action (further reducing chance of contamination of receiving water)

Little is known about the 'dark' processes of disinfection in WSPs, particularly 'biological disinfection', and this is a priority for research and may also assist with modelling overall disinfection. Experiments with WSP microcosms controlled by inactivating antagonistic microfauna, should be fruitful.

New methods for enumerating microbial contaminants (e.g. Davies and Apte, 1996; Sobsey *et al.*, 1998) seem likely to revolutionise our ability to study microbiological contaminant behaviour in WSPs among other environments. This will, in turn, facilitate more sophisticated experiments on WSP disinfection, and will have a multiplier effect, giving us a better understanding of mechanisms of removal and guidance for WSP designers. Eventually these improved methods of analysis may make it possible to monitor WSP effluents continuously, and detect episodes of low microbial quality.

As we have seen, most of the information on pond disinfection (necessarily) deals with removal of indicator organisms, notably *E. coli*, and far fewer studies have been made of pathogens. More information on pathogen removal in WSPs relative to *E. coli* would be valuable to provide assurance that *E. coli* is a suitable indicator (as seems to be already established for *Salmonella*, for example) and to provide a means for predicting removal of very different micro-organisms (e.g. protozoan parasites and some viruses).

A few studies suggest that high-rate algal ponds and advanced, integrated pond systems centred on high-rate ponds (AIWPS, Oswald, 1991) are capable of faster and more consistent disinfection than conventional pond systems (Araki *et al.*, 2001; Bahlaoui *et al.*, 1997; Davies-Colley *et al.*, 2003; Fallowfield *et al.*, 1996), presumably because of the separation of pond processes into distinct modules (in order: anaerobic or facultative ponds, high rate algal ponds, settling ponds, and conventional maturation ponds). Refer to Chapter 13. High quality final effluents may be achievable with smaller footprints in advanced, integrated pond systems. More work on disinfection performance of such 'improved' pond systems seems highly desirable.

Finally, there would seem to be considerable scope for 'add-ons' to WSPs that improve consistency of microbial quality in the final effluent, and this should be a fertile ground for applied research. Supplementary treatment could include dedicated disinfection processes, such as ozonators or UV reactors (Nelson, 2000), or 'natural' means of further treatment, notably rock filters or constructed wetlands. Research on virus removal by UV irradiation of WSP effluent would seem to be a particular priority.

6.12 SUMMARY

WSPs are excellent treatment systems for disinfection of faecally-contaminated organic wastes, and are generally superior to mechanical treatment methods with the exception of dedicated disinfection processes such as ozonators or UV reactors (e.g. George *et al.*, 2002). Removal of all four of the main categories of pathogens (bacteria, viruses, protozoan parasites, and helmith parasites) is generally highly efficient in WSPs, although occasional episodes of microbial contaminant carryover can engender a demand for more consistent treatment. *E. coli* seems a reasonable indicator of disinfection by WSPs, as regards bacterial and viral pathogens, except, perhaps when pH is high, accelerating *E. coli* removal, but not the inactivation of certain pathogens. Removal of protozoan and helminth parasites in WSPs is not so well indicated by *E. coli*, because these organisms are removed mainly by sedimentation.

Combinations of complex, and relatively poorly understood processes operate in WSP systems to remove most pathogenic micro-organisms (Maynard *et al.*, 1999). These processes include particularly (1) *sedimentation* particularly for helminth ova and possibly protozoan oocysts in anaerobic and facultative ponds; (2) *action of antagonistic micro-organisms*, notably native protozoans and flagellates that can ingest bacterial and viral pathogens; and (3) *exposure to* (short-wavelengths in) *sunlight*, sometimes interacting with high dissolved oxygen and elevated pH.

Sunlight exposure (confined to WSP upper layers during daylight) seems to be the most universally important disinfection mechanism in WSPs for bacteria and viruses (Davies-Colley *et al.*, 2000), and is possibly also important for protozoan parasites. However, slow but ubiquitous and consistent 'dark' processes of disinfection, including sedimentation and ingestion by micro-fauna, can also make an important contribution to overall disinfection. The combination of sedimentation of solids and clarification of the surface water of ponds (with consequently improved sunlight penetration) may be an important, hitherto unrecognised interaction between otherwise independent disinfection mechanisms in WSPs.

WSP disinfection is sometimes poor due to short-circuiting, a common problem that can often be addressed by baffling and consideration of the inlet/outlet design and positioning. Bacterial (and possibly viral) removal is better in comparatively shallow maturation ponds, presumably due to their greater sunlight exposure. A depth of 0.4 m is recommended.

Although disinfection in WSPs is typically excellent, there will be some situations where more consistency of, and control over, final effluent microbiological quality will be demanded, particularly in developed countries with sometimes stringent standards and high public expectations. Dedicated disinfection by ozone or UV irradiation should be feasible to further treat WSP effluent. Other potentially useful 'add-ons' to WSPs to improve consistency of final effluent microbiological quality include rock filters and constructed wetlands.

Finally, there are numerous research needs on disinfection in WSPs, particularly on: (1) modelling of *E. coli* removal, (2) action of antagonistic microfauna (biological disinfection), (3) application of new microbiological methods, (4) pathogen behaviour in WSPs, (5) disinfection performance of 'improved' pond systems such as high-rate ponds, and (6) add-ons for dedicated disinfection or to provide better control of final effluent quality from WSPs.

REFERENCES

Araki, S., Martin-Gomez, S., Becares, E., De Luis-Calabuig, E., and Rojo-Vazquez, F. (2001) Effect of High-Rate Algal Ponds on Viability of Cryptosporidium parvum Oocysts. *Applied and Environmental Microbiology* **67**, 3322-3324.

Arridge, H., Oragui, J.I., Pearson, H.W., Mara, D.D., and Silva, S.A. (1995) Vibrio-Cholerae O1 and Salmonellae Removal Compared with the Die-Off of Fecal Indicator Organisms in Waste Stabilisation Ponds in Northeast Brazil. *Wat. Sci. Tech.* **31**(12), 249-256.

Ayres, R.M., Alabaster, G.P., Mara, D.D., and Lee, D.L. (1992) A Design Equation for Human Intestinal Nematode Egg Removal in Waste Stabilisation Ponds. *Water Research* **26**, 863-865.

Bahlaoui, M.A., Baleux, B., and Troussellier, M. (1997) Dynamics of pollution-indicator and pathogenic bacteria in high-rate oxidation wastewater treatment ponds. *Water Research* **31**, 630-638.

Bitton, G. (1999) Wastewater microbiology. Second edition. Wiley-Liss, New York.

Borrego, J.J., Morinigo, M.A., de Vincente, A., Cornax, R., and Romero, P. (1987) Coliphage as an indicator of faecal pollution in water. Its relationship with indicator and pathogenic micro-organisms. *Water Research* **21**, 1473-1480.

Bouhoum, K., Amahmid, O. and Asmama, S. (2000) Occurrence and removal of protozoan cysts and helminth eggs in waste stabilisation ponds in Marrakech. *Wat. Sci. & Tech.* **42**(10-11), 159-164.

Boussaid, A., Baleux, B., Hassani, L., and Lesne, J. (1991) Aeromonas species in stabilisation ponds in the arid region of Marrakesh, Morocco, and relation to faecal-pollution and climatic factors. *Microbial Ecology* **21**, 11-20.

Cabelli, V.J., Dufour, A.P., McCabe, L.J., and Levin, M.A. (1983) A marine water quality criterion consistent with indicator concepts and risk analysis. *Journal of the Water Pollution Control Federation* **55**, 1306-1314.

Calkins, J., Buckles, J.D. and Moeller, J.R. (1976) The role of solar ultraviolet radiation in 'natural' water purification. *Photochemistry and Photobiology* **24**, 49-57.

Campos, C., Guerrero, A. and Cardenas, M. (2002) Removal of bacterial and viral faecal indicator organisms in a waste stabilisation pond system in Choconta, Cundinamarca (Colombia). *Wat. Sci. Tech.* **45**(1), 61-66.

Castillo, G.C. and Trumper, B.A. (1991) Coliphages and Other Microbial Indicators in Stabilisation Ponds. *Environmental Toxicology and Water Quality* **6**, 197-207.

Chan, M S. (1997). The global burden of intestinal nematode infections - fifty years on. *Parasitology Today* **13**, 438-443.

Chaudhuri, M. (1973) Virus Removal in Waste Stabilisation Ponds. Indian Journal of Environmental Health **16**, 171-177.

Clancy, J.L., Bukhari, Z., Hargy, T.M., Bolton, J.R., Dussert, B.W. and Marshall, M.M. (2000) Using UV to inactivate Cryptosporidium. *Journal of the American Water Works Association* **92**, 97-104.

Craggs, R. J.; Zwart, A.; Nagels, J. W.; Davies-Colley, R. J. (2004) Disinfection in a pilot scale high rate pond treating dairy farm wastewater. *Ecological engineering (in press)*

Curtis, T.P., Mara, D.D., Dixo, N.G.H. and Silva, S.A. (1994) Light penetration in waste stabilisation ponds. *Water Research* **28**, 1031-1038.

Curtis, T.P., Mara, D.D. and Silva, S.A. (1992a) The Effect of Sunlight on Fecal-Coliforms in Ponds - Implications for Research and Design. *Wat. Sci. Tech.* **26**(7-8), 1729-1738.

Curtis, T.P., Mara, D.D., and Silva, S.A. (1992b) Influence of pH, Oxygen, and Humic Substances on Ability of Sunlight to Damage Fecal-Coliforms in Waste Stabilisation Pond Water. *Applied and Environmental Microbiology* **58**, 1335-1343.

Davies, C.M. and Apte, S.C. (1996) Rapid enzymatic detection of faecal pollution. *Wat. Sci. Tech.* **34**(7-8), 169-171.

Davies-Colley, R.J., Craggs, R.J. and Nagels, J.W. (2003) Disinfection in a Pilot-Scale Advanced Pond System (APS) for Domestic Sewage Treatment in New Zealand. *Wat. Sci. Tech.* **48**(2), 81-87.

Davies-Colley, R.J., Donnison, A.M., and Speed, D.J. (1997) Sunlight wavelengths inactivating faecal indicator micro organisms in waste stabilisation ponds. *Wat. Sci. Tech.* **35**(11-12), 219-225.

Davies-Colley, R.J., Donnison, A.M., and Speed, D.J. (2000) Towards a mechanistic understanding of pond disinfection. *Wat. Sci. Tech.* **42**(10-11), 149-158.

Davies-Colley, R.J., Donnison, A.M., Speed, D.J., Ross, C.M. and Nagels, J.W. (1999) Inactivation of faecal indicator micro organisms in waste stabilisation ponds: Interactions of environmental factors with sunlight. *Water Research* **33**, 1220-1230.

Davies-Colley, R.J., Hickey, C.W. and Quinn, J.M. (1995) Organic matter, nutrients, and optical characteristics of sewage lagoon effluents. *New Zealand Journal of Marine and Freshwater Research* **29**, 235-250.

Decamp, O. and Warren, A. (1998) Bacterivory in ciliates isolated from constructed wetlands (reed beds) used for wastewater treatment. *Water Research* **32**, 1989-1996.

Donnison, A.M. and Ross, C.M. (1995) Somatic and F-specific coliphages in New Zealand waste treatment lagoons. *Water Research* **29**, 1105-1110.

Drury, D.F. and Wheeler, D.C. (1982) Applications of Serratia marscens bacteriphage as a new microbial tracer of aqueous environments. *Journal of Applied Bacteriology* **53**, 137.

Emerick, R.W., Loge, F.J., Thompson, D. and Darby, J.L. (1999) Factors influencing ultraviolet disinfection performance part II: Association of coliform bacteria with wastewater particles. *Water Environment Research* **71**, 1178-1187.

Emparanza-Knoerr, A. and Torrella, F. (1993) Microbiological performance and Salmonella dynamics in a wastewater depuration pond system of southeastern Spain. *Wat. Sci. Tech.* **31**(12), 239-248.

Fallowfield, H.J., Cromar, N.J. and Evison, L.M. (1996) Coliform die-off rate constants in a high rate algal pond and the effect of operational and environmental variables. *Wat. Sci. Tech.* **34**(11), 141-147.

Frederick, G.L. and Lloyd, B.J. (1995) Evaluation of Serratia-Marcescens Bacteriophage as a Tracer and a Model for Virus Removal in Waste Stabilisation Ponds. *Wat. Sci. Tech.* **31**(12), 291-302.

Gates, F.L. (1929) A study of the bactericidal action of ultra-violet light. *Journal of general physiology* **13**, 231-260.

George, I., Crop, P. and Servais, P. (2002) Fecal coliform removal in wastewater treatment plants studied by plate counts and enzymatic methods. *Water Research* **36**, 2607-2617.

Grimason, A.M., Smith, H.V., Thitai, W.N., Smith, P.G., Jackson, M.H. and Girdwood, R.W.A. (1993) Occurrence and removal of Cryptosporidiium spp. oocysts and Giardia spp. cysts in Kenyan waste stabilisation ponds. *Wat. Sci. Tech.* **27**(3-4), 97-104.

Haas, C.N. (1986) Wastewater disinfection and infectious disease risks. *CRC Critical Reviews in Environmental Control* **17**, 1-20.

Havelaar, A.H., Olphen, M. and Drost, Y.C. (1993) F-specific RNA bacteriophages are adequate model organisms for enteric viruses in fresh water. *Applied and Environmental Microbiology* **59**, 2956-2962.

Hazen, T.C. (1988) Faecal coliforms as indicators in tropical waters: a review. *Toxicity Assessment* **3**, 461-477.

Hickey, C.W., Quinn, J.M. and Davies-Colley, R.J. (1989) Effluent characteristics of domestic sewage oxidation ponds and their potential impacts on rivers. *New Zealand Journal of Marine and Freshwater Research* **23**, 585-600.

IAWPRC (1991) Bacteriophages as Model Viruses in Water Quality Control. *Water Research* **25**, 529-545.

Jagger, J. (1985) Solar-UV Actions on Living Cells, 1st edn. Praeger Publishers, New York.

Johnson, B. A., Wight, J. L., Middlebrooks, E. J., Reynolds, J. H. and Venosa, A. D. (1978) Mathematical-Model for Disinfection of Waste Stabilisation Lagoon. *Journal of the Water Pollution Control Federation* **50**, 2002-2015.

Juanico, M. and Shelef, G. (1994) Design, operation and performance of stabilisation reservoirs for wastewater irrigation in Israel. *Wat. Sci. Tech.* **28**, 175-186.

Kadlec, R.H., Knight, R.L., Vymazal, J., Brix, H., Cooper, P. and Haberl, R. (2000) Constructed wetlands for pollution control, IWA Publishing, London.

Leduc, R. and Gehr, R. (1990) Removal of coliform bacteria from aerated stabilisation lagoons. 1. Kinetics, modelling and biotic variables. *Water Pollution Research Journal of Canada* **25**, 231-263.

Lesne, J., Baleux, B., Boussaid, A. and Hassani, L. (1990) Dynamics of non-01 Vibrio cholerae in experimental sewage stabilisation ponds under arid Mediterranean climate. *Wat. Sci. Tech.* **24**(2), 387-390.

Liran, A., Juanico, M. and Shelef, G. (1994) Coliform removal in a stabilisation reservoir for wastewater irrigation in Israel. *Water Research* **28**, 1305-1314.

Lloyd, B.J. and Frederick, G.L. (2000) Parasite removal by waste stabilisation pond systems and the relationship between concentrations in sewage and prevalence in the community. *Wat. Sci. Tech.* **42** (10-11), 375-386.

Macdonald, R.J. and Ernst, A. (1986) Disinfection Efficiency and Problems Associated with Maturation Ponds. *Wat. Sci. Tech.* **18**(10), 19-29.

Madera, C.A., Pena, M.R. and Mara, D.D. (2002) Microbiological quality of a waste stabilisation pond effluent used for restricted irrigation in Valle Del Cauca, Colombia. *Wat. Sci. Tech.* **45**(1), 139-143.

Manage, P.M., Kawabata, Z., Nakano, S. and Nishibe, Y. (2002) Effect of heterotrophic nanoflagellates on the loss of virus-like particles in pond water. *Ecological Research* **17**, 473-479.

Mara, D. (2001) Appropriate wastewater collection, treatment and reuse in developing countries. *Proceedings of the Institution of Civil Engineers-Municipal Engineer* **145**, 299-303.

Marais, G.V.R. (1974) Faecal bacteria kinetics in stabilisation ponds. *Journal of the Environmental Engineering Division, ASCE* **100**, 119-139.

Maynard, H.E., Ouki, S.K. and Williams, S.C. (1999) Tertiary lagoons: A review of removal mechanisms and performance. *Water Research* **33**, 1-13.

Mayo, A.W. (1989) Effect of Pond Depth on Bacterial Mortality Rate. *Journal of Environmental Engineering ASCE* **115**, 964-977.

Mayo, A.W. (1995) Modeling Coliform Mortality in Waste Stabilisation Ponds. *Journal of Environmental Engineering ASCE* **121**, 140-152.

Mezrioui, N., Oudra, B., Oufdou, K., Hassani, L., Loudiki, M. and Darley, J. (1994) Effect of Microalgae Growing on Waste-Water Batch Culture on Escherichia-Coli and Vibrio-Cholerae Survival. *Wat. Sci. Tech.* **30**(8), 295-302.

Mezrioui, N., Oufdou, K. and Baleux, B. (1995) Dynamics of non-O1 Vibrio cholerae and fecal coliforms in experimental stabilisation ponds in the arid region of Marrakesh, Morocco, and the effect of pH, temperature, and sunlight on their experimental survival. *Canadian Journal of Microbiology* **41**, 489-498.

Middlebrooks, E.J. (1995) Upgrading pond effluents: an overview. *Wat. Sci. Tech.* **31**(12), 353-368.

Mills, S.W., Alabaster, G.P., Mara, D.D., Pearson, H.W. and Thitai, W.N. (1992) Efficiency of Fecal Bacterial Removal in Waste Stabilisation Ponds in Kenya. *Wat. Sci. Tech.* **26**(7-8), 1739-1748.

Moeller, J.R. and Calkins, J. (1980) Bactericidal agents in wastewater lagoons and lagoon design. *Journal of the Water Pollution Control Federation* **52**, 2442-2451.

Nair, G.B., Sarkar, B.L., De, S.P. and Chakrabarti, M.K. (1988) Ecology of V. cholerae in the freshwater environs of Calcutta, India. *Microbial Ecology* **15**, 203-215.

Nascimento, M.J., Oliveira, J.S., Oliveira, L. and Mexia, J.T. (1991) Contribution for the Study of New Pathogenic Indicators Removal from Wsp in Portugal. *Wat. Sci. Tech.* **24**(2), 381-386.

Nelson, K.L. (2000) Ultraviolet light disinfection of wastewater stabilisation pond effluents. *Wat. Sci. Tech.* **42**(10-11), 165-170.

Nelson, K.L. (2003) Concentrations and inactivation of Ascaris eggs and pathogen indicator organisms in wastewater stabilisation pond sludge. *Wat. Sci. Tech.* **48**(2), 89-96.

Ohgaki, S., Ketratanakul, A. and Prasertsom, U. (1986) Effect of sunlight on coliphages in an oxidation pond. *Wat. Sci. Tech.* **18**(10), 37-46.

Oragui, J.I., Arridge, H., Mara, D.D., Pearson, H.W. and Silva, S.A. (1995) Rotavirus removal in experimental waste stabilisation pond systems with different geometries and configurations. *Wat. Sci. Tech.* **31**(12), 285-290.

Oragui, J.I., Curtis, T.P., Silva, S.A. and Mara, D.D. (1986) The removal of excreted bacteria and viruses in deep waste stabilisation ponds in northeast Brazil. *Wat. Sci. Tech.* **18**(10), 31-35.

Oswald, W.J. (1991) Introduction to advanced Integrated Wastewater Ponding Systems. *Wat. Sci. Tech.* **24**(5), 1-7.

Oufdou, K., Mezrioui, N., Oudra, B., Barakate, M. and Loudiki, M. (1998) Effect of extracellular and endocellular products from cyanobacterium, Synechocystis sp., on the growth of some sanitation system bacteria. *Archiv fur Hydrobiologie. Supplementband.* **125**, 139-148.

Oufdou, K., Mezrioui, N., Oudra, B., Loudiki, M., Barakate, M. and Sbiyya, B. (2001) Bioactive compounds from Pseudanabaena species (Cyanobacteria). *Microbios* **106**, 21-29.

Parhad, N. and Rao, N.V. (1974) Effect of pH on survival of E. coli. *Journal of the Water Pollution Control Federation* **46**, 149-161.

Pearson, H.W., Mara, D.D. and Arridge, H.A. (1995) The Influence of Pond Geometry and Configuration on Facultative and Maturation Waste Stabilisation Pond Performance and Efficiency. *Wat. Sci. Tech.* **31**(12), 129-139.

Pearson, H.W., Mara, D.D., Cawley, L.R., Arridge, H.M. and Silva, S.A. (1996a) Performance of an innovative tropical experimental waste stabilisation pond system operating at high organic loadings. *Wat. Sci. Tech.* **33**(7), 63-73.

Pearson, H.W., Mara, D.D., Cawley, L.R., Oragui, J.I. and Silva, S.A. (1996b) Pathogen removal in experimental deep effluent storage reservoirs. *Wat. Sci. Tech.* **33**(7), 251-260.

Pearson, H.W., Mara, D.D., Mills, S.W. and Smallman, D.J. (1987) Physico-chemical parameters influencing faecal bacterial survival in waste stabilisation ponds. *Wat. Sci. Tech.* **19**(12), 145- 152.

Polprasert, C. and Rajput, V.S. (1984) Study on Chlorine Disinfection of Pond Effluent. *Water Research* **18**, 513-518.

Qin, D., Bliss, P.J., Barnes, D., & Fitzgerald, P.A. (1991) Bacterial (Total Coliform) Die-Off in Maturation Ponds. Wat. *Sci. Tech.* **23**(7-9), 1525-1534.

Rao, V.C., Lakhe, S.B. and Waghmare, S.V. (1981) Virus Removal in Waste Stabilisation Ponds in India. *Water Research* **15**, 773-778.

Robertson, L.J., Smith, P.G., Grimason, A.T. and Smith, H.V. (1999) Removal and destruction of intestinal parasitic protozoans by sewage treatment processes. *International Journal of Environmental Health Research* **9**, 85-96.

Saidam, M.Y., Ramadan, S.A. and Butler, D. (1995) Upgrading waste stabilisation pond effluent by rock filters. *Wat. Sci. Tech.* **31**(12), 369-378.

Saqqar, M.M. and Pescod, M.B. (1991) Microbiological Performance of Multistage Stabilisation Ponds for Effluent Use in Agriculture. *Wat. Sci. Tech.* **23**(7-9), 1517-1524.

Saqqar, M.M. and Pescod, M.B. (1992) Modelling coliform reduction in wastewater stabilisation ponds. *Wat. Sci. Tech.* **26**(7-9), 1667-1677.

Sarikaya, H.Z. and Saatci, A.M. (1987) Bacterial Die-Off in Waste Stabilisation Ponds. *Journal of Environmental Engineering ASCE* **113**, 366-382.

Sarikaya, H.Z. and Saatci, A.M. (1988) Optimum Pond Depths for Bacterial Die-Off. *Water Research* **22**, 1047-1054.

Sarikaya, H.Z., Saatci, A.M. and Abdulfattah, A.F. (1987) Effect of Pond Depth on Bacterial Die-Off. *Journal of Environmental Engineering ASCE* **113**, 1350-1362.

Shilton, A. (2000) Potential application of computational fluid dynamics to pond design. *Wat. Sci. Tech.* **42**(10-11), 327-334.

Shilton, A. and Harrison, J. (2003) Integration of coliform decay within a CFD model of a waste stabilisation pond. *Wat. Sci. Tech.* **48**(2), 205-210.

Sinton, L.W. and Finlay, R.K. (1996) Bacteriophages as microbiological water quality indicators. Water and Wastes in New Zealand, **52-55**.

Sobsey, M.D., Battigelli, D.A., Shin, G.A. and Newland, S. (1998) RT-PCR amplification detects inactivated viruses in water and wastewater. *Wat. Sci. Tech.* **38**(12), 91-94.

Sobsey, M.D. and Cooper, R.C. (1973) Enteric virus survival in algal-bacterial wastewater treatment systems. I. Laboratory studies. *Water Research* **7**, 669-685.

Stott, R., May, E. and Mara, D.D. (2003) Parasite removal by natural wastewater treatment systems: performance of waste stabilisation ponds and constructed wetlands. *Wat. Sci. Tech.* **48**(2), 97-104.

Sukias, J.P.S., Tanner, C.C., Davies-Colley, R.J., Nagels, J.W. and Wolters, R. (2001) Algal abundance, organic matter, and physico-chemical characteristics of dairy farm facultative ponds: implications for treatment performance. *New Zealand Journal of Agricultural Research* **44**, 279-296.

Swanson, G.R. and Williamson, K.J. (1980) Upgrading lagoon effluents with rock filters. *Journal of the Sanitary Engineering Division, ASCE* **106**, 1111-1118.

Tanner, C.C. (2001) Plants as ecosystem engineers in subsurface-flow treatment wetlands. *Wat. Sci. Tech.* **44**(11-12), 9-17.

Tartera, C. and Jofre, J. (1987) Bacteriophages active against Bacteroides fragilis in sewage-polluted water. *Applied and Environmental Microbiology* **53**, 1632-1637.

Troussellier, M. and Legendre, P. (1989) Dynamics of fecal coliform and culturable heterotroph densities in an eutrophic ecosystem: Stability of models and evolution of these bacterial groups. *Microbial Ecology* **17**, 227-235.

Troussellier, M., Legendre, P. and Baleux, B. (1986) Modeling of the Evolution of Bacterial Densities in an Eutrophic Ecosystem (Sewage Lagoons). *Microbial Ecology* **12**, 355-379.

Turner, S.J. and Lewis, G. (1995) Comparison of F-specific bacteriphage, enterococci, and faecal coliform densities through a wastewater treatment process employing oxidation ponds. *Wat. Sci. Tech.* **31**(5-6), 85-89.

Tyrrell, S.A., Rippey, S.R. and Watkins, W.D. (1995) Inactivation of bacterial and viral indicators in secondary sewage effluents using chlorine and ozone. *Water Research* **29**, 2483-2490.

von Sperling, M. (1999) Performance evaluation and mathematical modelling of coliform die-off in tropical and subtropical waste stabilisation ponds. *Water Research* **33**, 1435-1448.

von Sperling, M., Chernicharo, C.A.L., Soares, A.M.E. and Zerbini, A.M. (2003) Evaluation and modelling of helminth egg removal in baffled and unbaffled ponds treating anaerobic effluent. *Wat. Sci. Tech.* **48**(2), 113-120.

Vorkas, C.A. and Lloyd, B.J. (2000) A comparative assessment of bacteriophages as tracers and models for virus removal in waste stabilisation ponds. *Wat. Sci. Tech.* **42**(10-11), 127-138.

Whitelam, G.C. and Codd, G.A. (1986) Damaging effects of light on micro organisms. *Special Publications of the Society of General Microbiology* **17**, 129-169.

Wiandt, S., Baleux, B., Casellas, C. and Bontoux, J. (1995) Occurrence of Giardia sp. cysts during wastewater treatment by a stabilisation pond in the south of France. *Wat. Sci. Tech.* **31**(12), 257-265.

Xu, P., Brissaud, F. and Fazio, A. (2002) Non-steady-state modelling of faecal coliform removal in deep tertiary lagoons. *Water Research* **36**, 3074-3082.

7

Heavy metal removal

Rupert Craggs

7.1 INTRODUCTION

Discharge of heavy metals to the environment not only results in acute toxicity to aquatic organisms, but also has longer-term effects through bioaccumulation and biomagnification in aquatic communities. It is therefore important that heavy metal removal is optimised in WSPs that receive high levels in the influent wastewater.

7.1.1 Typical wastewater heavy metal concentrations

Domestic wastewater often contains elevated concentrations of lead, cadmium, chromium, copper and zinc due to corrosion of water pipes and plumbing (Table 7.1).

ISBN: 1843390205. Published by IWA Publishing, London, UK.

Table 7.1 Ranges of heavy metal concentrations in domestic wastewater (Laxen and Harrison, 1981; Soniassy and Lemon, 1986; Kaplan *et al.*, 1987; Toumi *et al.*, 2000)

	Concentration (mg m^{-3})	
Metal	Minimum	Maximum
Cd	2	280
Cr	1	40
Cu	34	1010
Fe	250	6680
Pb	0.3	400
Mn	-	1000
Zn	210	4360
Ni	1	60
Ar	2	10
Hg	0.05	0.4
Ag	0.4	1000

7.2 HEAVY METAL REMOVAL PROCESSES

Heavy metals may be removed by a variety of processes in WSPs. The main heavy metal removal processes include (Figure 7.1):

- Sedimentation of wastewater solids
- Adsorption to algal/bacterial biomass and bottom sludge
- Bioaccumulation into algal/bacterial biomass (and subsequent sedimentation)
- Chelation
- Precipitation

7.2.1 Sedimentation of wastewater Solids

Most heavy metals in raw wastewater are associated with particulate matter (Kaplan *et al.*, 1987) and are deposited in the WSP sludge (Suffern *et al.*, 1981). Heavy metal concentrations in WSP sludge are therefore typically much higher than in the overlying pond water (Balasubramanian *et al.*, 1997).

7.2.2 Adsorption

Adsorption of heavy metals onto the surface of algae and bacteria cells is a rapid process occurring at the same rate with both living and dead cells (Khummongkol *et al.*, 1982; Gadd, 1990). Adsorption involves attraction of the positively charged metal ions to the numerous negatively charged sites on the surface of algae and bacteria cells (e.g. carboxyl groups of proteins and fatty acids, phosphate groups of nucleic acids, phospholipids and polyscaccharides and organo-sulphate groups), but can also involve physical (van der Waals forces) and chemical attraction (Rai *et al.*, 1981; Tam and Wong, 1983; Wang and Wood, 1984; Wilkinson *et al.,* 1989; Gadd, 1990). The amount of heavy metal adsorption varies with algal or bacterial species, as it is dependent on the particular cell surface composition (Gale and Wixon, 1979). Adsorption of heavy metals results in the displacement of divalent or monovalent cations (e.g. Ca^{2+}, Mg^{2+}, K^+, Na^+) that are normally associated with the cell surface (Gale and Wixon, 1979; Gadd, 1990).

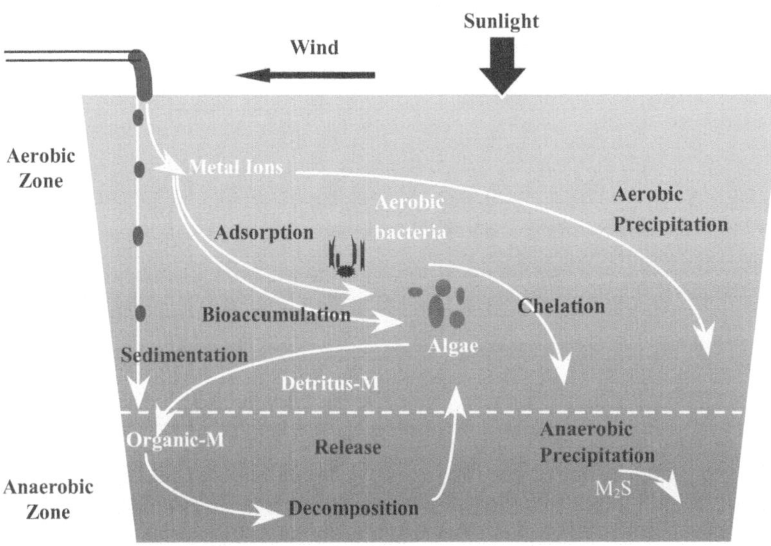

Figure 7.1 Heavy metal removal processes in WSPs

7.2.3 Bioaccumulation

Algae and bacteria are well known for their capacity to accumulate heavy metals from wastewaters since many heavy metals (e.g. Cu, Fe, M, Zn, Co and Mo) are required as essential micronutrients (Aaronson *et al.,* 1980; Tam and Wong, 1983; Wilkinson *et al.,* 1989; Gadd, 1990). Heavy metal bioaccumulation within the cells is an active process requiring energy (Jennett *et al.,* 1980), thus bioaccumulation is a much slower process than adsorption, which is a passive process (Khummongkol *et al.,* 1982).

Algal cell heavy metal concentrations tend to correlate with the concentrations in the surrounding wastewater (Tam and Wong, 1983; Wong and Tam, 1984; Wang and Wood, 1984; Maeda *et al.,* 1990) but may also be several thousand times higher (Tarifeno-Silva *et al.,* 1982; Becker, 1983; see Wilde and

Benemann, 1993 for a review). The accumulated metal ions are usually compartmentalized within the cell in vacuoles or are converted to less toxic forms by binding or precipitation (Gadd, 1990).

Heavy Metal Toxicity: At high concentrations heavy metals can inhibit the growth of algae and bacteria and may even cause death (Gale and Wixon, 1979; Tam and Wong, 1983; Gadd, 1990). Concentrations of several heavy metals that have been found to inhibit algal growth are shown in Table 7.2.

Table 7.2 Inhibitory concentrations of heavy metals to microalgae (Moshe *et al.*, 1972; Wong and Tan, 1984)

Heavy Metal	Inhibitory Concentration (g m^{-3})
Cu	0.0 - 2.75
Mn	3.0 -50.0
Cd	1
Cr	1.5
Ni	2
Zn	0.1 - 11.2

Heavy metals are most toxic to aquatic life in their free ionic form (Kaplan *et al.,* 1987), therefore, toxicity decreases as pH is high due to formation of insoluble precipitates (Rai *et al.,* 1981). Heavy metal toxicity also declines with increasing algal culture age and following previous heavy metal exposure (Neilson *et al.*, 1979). Therefore, WSPs that have comparatively long residence times and high pH are generally much less sensitive to heavy metal toxicity than conventional secondary mechanical treatment processes (WHO, 1987). Moreover, some WSP algae can tolerate high heavy metal concentrations (Wang and Wood, 1984) and Moshe (1972) found that WSPs were unaffected by heavy metal concentrations as high as 60 g m^{-3}.

7.2.4 Chelation

Many WSP algae and bacteria release extracellular secretions that contain phytochetalins (for example, polysaccharides, peptides and organic acids (humic, fulvic and nitrilotriacetic)). The amount of chelating agent released increases with cell age and with previous exposure to high heavy metal concentrations (Kaplan *et al.,* 1987). These chelating agents form complexes with free heavy metal ions, and when present at high concentrations, will reduce bioaccumulation (and hence toxicity) of heavy metals present in the wastewater (Suffern *et al.,* 1981; Wong *et al.,* 1984; Kaplan *et al.,* 1987). Pond water pH also affects the stability of heavy metal chelates with increased stability at higher pH (Gale and Wixon, 1979).

7.2.5 Precipitation

Heavy metals may precipitate under both anaerobic and aerobic conditions. Under anaerobic conditions heavy metals precipitate with sulphide (Jackson, 1978). Whereas, under aerobic conditions and at high pH, some heavy metal cations combine with anions such as hydroxide and phosphate to form insoluble precipitates (e.g. $Fe(OH)_2\ .\ 4H_2O$) (Moshe *et al.,* 1972; Rai *et al.,* 1981; Tam and Wong, 1983; Tack *et al.,* 1996).

7.3 RELEASE OF HEAVY METALS FROM POND SLUDGE

Pond water heavy metal concentrations may gradually increase over time due to release from decomposing organic sludge (Kaplan *et al.,* 1987). Heavy metals that are bound within the organic sludge may be released under acidic conditions by exchange with hydrogen ions (Tam and Wong, 1983). However, the relationship between heavy metal release/removal and pH is extremely complex and is different for chemical and biological processes and for different metals (Maynard *et al.,* 1999). Under aerobic conditions heavy metal sulphides in pond sludge will dissolve releasing the metal ions to the pond water (Tack *et al.,* 1996).

7.4 HEAVY METAL REMOVAL EFFICIENCY

Little research has been conducted on the removal of heavy metals in WSPs (Smillie and Loutit, 1982; Kaplan *et al.,* 1987; Toumi *et al.,* 2000). Most removal occurs in primary ponds (anaerobic or facultative) and is due to sedimentation of solids to which the heavy metals have adsorbed (Smillie and Loutit, 1982; Nejmeddine *et al.,* 2000; Toumi *et al.,* 2000). Heavy metal concentrations in pond sludge generally decline with distance from the inlet and with increasing pond depth (Smillie and Loutit, 1982).

Uptake (adsorption or bioaccumulation) by pond algae and bacteria has also been shown to be an important heavy metal removal process (Smillie and Loutit, 1982; Soniassy and Lemon, 1986), with WSP bacteria capable of concentrating more heavy metals than WSP algae (Smillie and Loutit, 1982).

Heavy metal removal efficiencies in WSPs vary between different metals and type of WSP system, but in general, heavy metal removal improves as the number of ponds in the WSP system increases, particularly if final ponds are aerobic maturation ponds (Table 7.3).

Table 7.3 Heavy metal removal efficiency of in WSPs.

Heavy Metal	System	% Removal	Reference
Cu	AP	21	Nejmeddine *et al.*, 2000
Pb	AP	11	
Zn	AP	28	
Cu	FPs	60	Soniassy and Lemon 1986
Cd	FPs	90	
Cr	FPs	0	
Fe	FPs	0	
Hg	FPs	70	
Ni	FPs	99	
Pb	FPs	83	
Zn	FPs	90	
Cu	2 FPs	76	Smillie and Loutit 1982
Cr	2 FPs	0	
Fe	2 FPs	84	
Mn	2 FPs	26	
Pb	2 FPs	9	
Zn	2 FPs	100	
Cu	SP/AP/FP/WSR	30	Kaplan *et al.*, 1987
Cd	SP/AP/FP/WSR	45	
Pb	SP/AP/FP/WSR	41	
Zn	SP/AP/FP/WSR	42	
Cu	WSP algae culture	70-90	Filip *et al.*, 1979
Cd	WSP algae culture	70-90	
Cr	WSP algae culture	20	
Cu	AP/ 3 FP,/2MP	92	Toumi *et al.*, (2000)
Pb	AP/ 3 FP,/2MP	71	
Zn	AP/ 3 FP,/2MP	91	
Cu	AP/FP/MPs	40-44	Chughtai and Ahmad 1988
Cd	AP/FP/MPs	100	
Co	AP/FP/MPs	100	
Cr	AP/FP/MPs	40-44	
Bo	AP/FP/MPs	100	
Ni	AP/FP/MPs	0	
Pb	AP/FP/MPs	100	
Zn	AP/FP/MPs	100	

(SP, Settling Pond; AP, Anaerobic Pond; FP, Facultative Pond; MP, Maturation Pond)

7.5 SUMMARY

Heavy metals may be removed in WSPs by a variety of processes including sedimentation of wastewater solids, adsorption to algal/bacterial biomass and bottom sludge, bioaccumulation into algal/bacterial biomass (and subsequent sedimentation), chelation and precipitation. There is very little information on heavy metal removal in WSPs but the most important processes would appear to be sedimentation of wastewater solids or adsorption and bioaccumulation followed by sedimentation of pond algae and bacteria.

7.6 FURTHER RESEARCH

Further research is required on ways to enhance heavy metal uptake by pond algae and bacteria and potentially use them to mine heavy metals from wastewaters. More research is needed on reducing the release of heavy metals from pond sludge back to the pond water.

REFERENCES

Aaronson, S., Berner, T. and Dubinsky, Z. (1980). Microalgae as a source of chemicals and natural products. In *Algae Biomass*, (eds G. Shelef and C.J. Soeder), 575-601, Elsevier/ Nth Holland Biomedical Press, Amsterdam.

Balasubramanian, S., Pappathi, R., Jayanthi Bose, A. and Raj, S.P. (1997) Bioconcentration of copper, nickel and cadmium in multicell sewage fed fish ponds. *Journal of Environmental Biology* **18**(2), 173-179.

Becker, E.W. (1983) Limitations of heavy metal removal from wastewater by means of algae. *Water Research* **17**, 459-466.

Chughtai, M.I.D. and Ahmad, K. (1988) Removal of heavy metals by a system of waste stabilization ponds. *Pakistan Journal of Biochemistry* **21**, 1-2.

Filip, D.S., Peters, T., Adams, V.D. and Middlebrooks, E.J. (1979) Residual heavy metal removal by an algae-intermittent sand filtration system. *Water Research* **13**(3), 305-313.

Gadd, G.M. (1990). Heavy metal accumulation by bacteria and other microorganisms. *Experientia* **46**, 834-840.

Gale, N.L. and Wixson, B.G. (1979) Removal of heavy metals from industrial effluents by algae. *Dev. Ind. Microbiol.* **273**(1979).

Jackson, T.A. (1978) The biogeochemistry of heavy metals in polluted lakes and streams at Flin Flon, Canada, and a proposed method for limiting heavy-metal pollution of natural waters. *Environmental Geology* **2**(3), 173-189.

Jennett, J.C., Hassett, J.M. and Smith, J.E. (1980) The use of algae to control heavy metals in the environment. *Miner. Environ.* **2**(1), 26-31.

Kaplan, D., Abeliovich, A. and Ben-Yaakov, S. (1987) Fate of Heavy Metals in Wastewater Stabilization Ponds. *Water Research* **21**(10), 1189-1194.

Khummongkol, D., Canterford, G.S. and Fryer, C. (1982) Accumulation of heavy metals in unicellular algae. *Biotech. Bioeng.* **24**(2643-2660).

Laxen, D.P.H. and Harrison, R.M. (1981) The physicochemical speciation of Cd, Pb, Cu, Fe and Mn in the final effluent of a sewage treatment works and its impact on speciation in the receiving river. *Water Research* **15**, 1053-1065.

Maeda, S., Mizoguchi, M., Ohki, A. and Takeshita, T. (1990) Bioaccumulation of zinc and cadmium in fresh-water alga, Chlorella vulgari. 1. Toxicity and accumulation. *Chemosphere* **21**(8), 953-963.

Maynard, H.E., Ouki, S.K. and Williams, S.C. (1999) Tertiary lagoons: A review of removal mechanisms and performance. *Water Research* **33**(1), 1-13.

Moshe, M., Betzer, N. and Kott, Y. (1972) Effect of Industrial Wastes on Oxidation Pond Performance. *Water Research* **6**(10), 1165-1171.

Neilson, A.H., Blankley, W.F. and Lewin, R.A. (1979). Growth with organic carbon and energy sources. In *Handbook of phycological methods*, (ed J.R. Stein), vol, pp. 276-284, C.U.P., N.Y.

Nejmeddine, A., Fars, S. and Echab, A. (2000) Removal of dissolved and particulate form of metals (Cu, Zn, Pb, Cd) by an anaerobic pond system in Marrakesh (Morocco). *Environmental Technology* **21**(2), 225-230.

Rai, L.C., Gaur, J.P. and Kumar, H.D. (1981) Phycology and Heavy metal pollution. *Biol. Rev.* **56**, 99-151.

Smillie, R.H. and Loutit, M.W. (1982) Removal of metals from sewage in an oxidation pond system. *New Zealand Journal of Science* **25**, 371-376.

Soniassy, R.N. and Lemon, R. (1986) Lagoon treatment of municipal sewage effluent in a subarctic region of Canada (Yellowknife, N.W.T.). *Water Science and Technology* **18**(2), 129-139.

Suffern, J.S., Fitzgerald, C.M. and Szluha, A.T. (1981) Trace Metal Concentrations in Oxidation Ponds. *J. Water Pollut. Contr. Fed* **53**(11), 1599-1608.

Tack, F.M., Callewaert, O. and Verloo, M.G. (1996) Metal solubility as a function of pH in a contaminated, dredged sediment affected by oxidation. *Environmental Pollution* **91**(2), 199-208.

Tam, F.Y. and Wong, M.H. (1983) Sewage sludge for cultivating freshwater algae and the fate of heavy metals at higher trophic organisms I. Different methods of extracting sewage sludge on the properties of sludge extracts. *Arch. Hydrobiol.* **96**(4), 475-485.

Tarifeño-Silva, E., Kawasaki, L.Y., Yu, D.P., Gordon, M.S. and Chapman, D.J. (1982) Aquacultural approaches to recycling of dissolved nutrients in secondarily treated domestic wastewaters. III Uptake of dissolved heavy metals by artificial food chains. *Water Research* **16**(1), 59-65.

Toumi, A., Nejmeddine, A. and El Hamouri, B. (2000) Heavy metal removal in waste stabilization ponds and high rate ponds. *Water Science & Technology* **42(**10-11), 17-21.

Wang, H.-K. and Wood, J.M. (1984) Bioaccumulation of Nickel by Algae. *Environ. Sci. Technol.* **18**, 106-109.

Wilde, E.W. and Benemann, J.R. (1993) Bioremoval of heavy metals by the use of microalgae. *Biotech. Adv.* **11**, 781-812.

Wilkinson, S.C., Goulding, K.H. and Robinson, P.K. (1989) Mercury accumulation and volatilization in immobilized algal cell systems. *Biotechnol. Lett.* **11**(12), 861-864.

Wong, M.H. and Tam, F.Y. (1984) Sewage sludge for cultivating freshwater algae and the fate of heavy metals at higher trophic organisms II. Heavy metal contents of *Chlorella pyrenoidosa* cultivated in various extracts. *Arch. Hydrobiol.* **100**(2), 207-218.

8

Pond process design - an historical review

Andy Shilton and Duncan Mara

Essentially there are four approaches to wastewater stabilisation pond design: loading rates; empirical design equations; reactor theory and mathematical modelling.

8.1 LOADING RATES

This approach involves a 'black box' type of design, where a parameter such as population, flow or BOD is used to determine the required pond volume or area. This simplified approach to the process design of pond systems has been very commonly used throughout the world. For example, in New Zealand a figure of 84kg BOD/ha day (MWD, 1974) has been routinely used for facultative pond design, regardless of the marked differences in environmental conditions throughout the length of the country.

Most of the loading rate design approaches take little or no account of pond

ISBN: 1843390205. Published by IWA Publishing, London, UK.

shape and layout, the characteristics of the wastewater, or environmental factors such as temperature. These factors can, however, have a significant effect on pond performance. For example, a study by Finney and Middlebrooks (1980) reviewed the performance of four facultative pond systems that all had similar design values for organic loading and retention time. They found that one of the systems, which consisted of a larger number of smaller ponds in series, produced a consistently superior effluent, thereby highlighting the benefits of its more efficient hydraulic design.

There have been improvements made to this design approach to take account of temperature. Gloyna (1976) presented an equation, based on his experience, that a pond at 35°C would achieve 90 percent BOD removal in 3.5 days, which incorporated the van't Hoff-Arrhenius relationship to determine the corresponding retention time at other temperatures:

$$D = 3.5.\theta^{(35-T)}$$

where:

D	=	pond retention time (days);
θ	=	temperature coefficient;
T	=	temperature (°C).

McGarry and Pescod (1970) presented a paper based on a large quantity of pond loading/performance data and proposed the following equation which gives the failure limit for maximum facultative pond loading:

$$\lambda_S = 26T - 160$$

where:

λ_S	=	surface loading rate (kg/ha d);
T	=	temperature (°C).

Mara (1987) reviewed the situation and produced an equation for calculation of appropriate design loading rates at different temperatures. Using Gloyna's concept of the relationship of pond performance to temperature, he noted that the temperature coefficient is only constant over a short range and therefore used the term 'b-cT' to replace it. From experience with pond systems in Brazil and France, he knew that the respective loading rates of 350kg/ha.day at 25°C and 100kg/ha.day at 10°C both produced good performance. He then selected an arbitrary upper limit of 500kg/ha.day for 35°C.

These relationships were incorporated into the general equation:

$$\lambda_S = a(b - cT)^{T - T_{Ref}}$$

where:

a,b,c = constants (a in kg/ha day);
T_{ref} = reference temperature (°C).

Mara (1987) then solved this equation simultaneously to establish values for the unknown constants and so derived the final equation:

$$\lambda_S = 350(1.107 - 0.002T)^{T-25}$$

This equation sits safely inside the McGarry and Pescod equation and has now become widely used for design. Further details of facultative pond design based on this equation are given in Chapter 9.

8.2 EMPIRICAL DESIGN EQUATIONS

These equations attempt to account for numerous variables that may have an influence on pond performance, but essentially they still treat the pond as a 'black box'. They are derived from regressions of pond performance data, rather than from studies of actual pond treatment mechanisms.

Larsen (1974) published a typical example of this form of design equation. Using a pond in New Mexico, he analysed the data and developed an equation that incorporates variables such as BOD, solar radiation, wind speed, temperature and relative humidity. The design variables were incorporated in the following parameters given below:

$$\mathrm{MOT} = \frac{\text{Surface Area (Solar Radiation)}^{1/3}}{\text{Influent Flow Rate (Influent BOD)}^{1/3}}$$

$$\mathrm{RED} = \frac{\text{Influent BOD - Effluent BOD}}{\text{Influent BOD}}$$

$$\mathrm{TTC} = \frac{\text{Wind Speed (Influent BOD)}^{1/3}}{\text{(Solar Radiation)}^{1/3}}$$

$$\mathrm{TEMPR} = \frac{\text{Lagoon Liquid Temperature}}{\text{Air Temperature}}$$

$$\mathrm{DRY} = \text{Relative Humidity}$$

After analysis of the experimental data, a design equation that incorporates these parameters was produced:

$$\text{MOT} = (2.47^{\text{RED}} + 2.47^{\text{TTC}} + 24.9/\text{TEMPR} + 150.0/\text{DRY}) \times 10^6$$

Larsen claimed that this equation could then be used to back-calculate the pond surface area required to "accomplish any desired biochemical oxygen demand reduction from easily obtained climatic data for the geographical area in which the lagoon is to be located".

While regression will give an equation of best fit for the data from which it was derived, it is questionable how applicable this is to different ponds. With regard to hydraulics, for example, this equation is again incapable of differentiating between different pond shapes, inlet designs, etc. Prats and Llavador (1994) stated that the validity of this approach when applied to different locations is debatable. Finney and Middlebrooks (1980, pg 142) concluded that Larsen's equation was "totally useless".

8.3 POND DESIGN USING REACTOR THEORY

This approach attempts to apply standard reactor theory derived from the process-engineering field. The mechanisms that act to provide stabilisation of the pollutants in a pond system are complex and numerous. Instead of attempting to model these mechanisms individually, this approach attempts to quantify their overall combined effect. The reaction rates for organic and pathogen removal are normally assumed to follow first-order kinetics. If the first-order-rate law is incorporated into an appropriate mass balance and integrated using boundary conditions that reflect the hydraulic regime of the reactor, an equation suitable for design is derived (Tchobanoglous and Schroeder, 1985). The simplest examples of these are the ideal flow equations.

8.3.1 Ideal flow

There are two extremes of ideal flow: plug flow and completely mixed flow. The concept of plug flow assumes there is no mixing or diffusion of the wastewater in the reactor:

$$\frac{C_e}{C_i} = e^{-kt}$$

Alternatively, if the wastewater is assumed to be instantaneously completely mixed upon entering the reactor the 'completely stirred tank reactor' (CSTR) equation can be derived:

$$\frac{C_e}{C_i} = \frac{1}{1 + kt}$$

where in both cases:

C_e = effluent concentration (mg/l);
C_i = influent concentration (mg/l);
k = first order reaction rate constant (d^{-1});
t = time (d).

Marais and Shaw (1961) proposed the use of the completely mixed model for the predication of both BOD and faecal bacterial reduction in waste stabilisation ponds. Marais (1974) suggested that, in order to achieve maximum bacterial die-off, each pond in a series of ponds should be equally sized. When a series of equally sized ponds are used, the overall reduction can be described by the following equation:

$$\frac{C_e}{C_i} = \frac{1}{(1 + kt)^n}$$

where:

n = the number of ponds in series.

Alternatively Thirumurthi (1974) stated that a completely mixed flow model "should never be recommended for the rational design of stabilization ponds" and argued for the use of the *plug flow equation* instead. Preul and Wagner (1987) warned against the use of the ideal flow models in general, stating that their accuracy "may vary substantially with actual pond conditions and therefore their application is limited." However, despite these differing opinions the completely mixed equation has now become widely used to design a pond series for the required degree of faecal bacterial removal and a practical guide to doing this is outlined in Chapter 9.

8.3.2 Non-ideal flow

Ideal flow is, of course, only a theoretical concept. In practice the flow through ponds will always exist somewhere between the two extremes of plug and completely mixed flow and is referred to as non-ideal (or dispersed) flow. As an alternative to the ideal flow equations, Thirumurthi (1969) proposed the use of the Wehner-Wilhelm equation.

Wehner and Wilhelm (1956) started with the dimensionless steady-state differential equation for a plug flow reactor incorporating substrate consumption according to first order kinetics and its axial transport by diffusion (molecular and/or turbulent). They undertook an analysis of the boundary conditions and solved the equation analytically. This equation is valid for reactors with any kind of entry or exit configurations and has most commonly been denoted in the form below, as given in Levenspiel (1972):

$$\frac{C_e}{C_i} = 4a \frac{e^{\frac{1}{2d}}}{(1+a)^2 e^{\frac{a}{2d}} - (1-a)^2 e^{\frac{-a}{2d}}}$$

in which:

$$a = \sqrt{1 + 4ktd}$$

where:

C_e, C_i	=	effluent and influent concentration (g/m^3);
d	=	dispersion number;
k	=	first order reaction rate constant (d^{-1});
t	=	retention time (d).

As an approximation, Thirumurthi noted that the second term in the denominator, which is small, could be neglected, thereby simplifying the equation to:

$$\frac{C_e}{C_i} = 4a \frac{e^{\frac{1-a}{2d}}}{(1+a)^2}$$

This simplified equation is considered suitable for design until the value of d is >2, after which the error may be significant. Thirumurthi (1969), however, noted that d is seldom likely to exceed a value of 1 in waste stabilisation ponds.

Polprasert and Bhattarai (1985) evaluated the use of the Wehner-Wilhelm equation against the completely mixed flow equation for predicting the total and faecal coliform die-off in a number of ponds in hot climates. They found that the results obtained by use of the Wehner-Wilhelm equation "had significantly higher correlation coefficient values than those of the completely-mixed equations".

Conversely, Ferrara and Harleman (1981) noted that the dispersion number needed in this equation can be difficult to predict, and even if obtained by tracer study, is only representative of the environmental conditions that prevailed for the period over which the tracer study was conducted. Given this, they suggested that the simpler plug flow and completely mixed models might provide results that are just as relevant.

8.3.3 Combined pond models

In these models the pond is represented as a number of separate but interconnected regions with flow exchange between them. These different regions are modelled as discrete reactors behaving as plug flow, completely mixed flow, dispersed flow or as simple dead flow, retention zones (Watters *et al.*, 1973). The combined model used by Watters *et al.* (1973) is known as the 'finite stage model'. As seen in Figure 8.1, the model consists of a series of modules each containing a completely mixed flow unit (F_a), a dead flow unit (F_b) and a plug flow unit (F_c), each of which represent the behaviour of a defined fraction of the total pond volume. The dead flow unit interchanges flow with the completely mixed unit at a fraction of the main flow rate (Q) defined as K_H. Any number (n) of these modules are combined to characterise the pond and, therefore, this model requires five parameters – Fa, F_b, F_c, K_H and n. Watters *et al.* (1973) noted that, although it would be desirable to be able to predict these parameters, this is not possible and instead they must be determined by undertaking a tracer study.

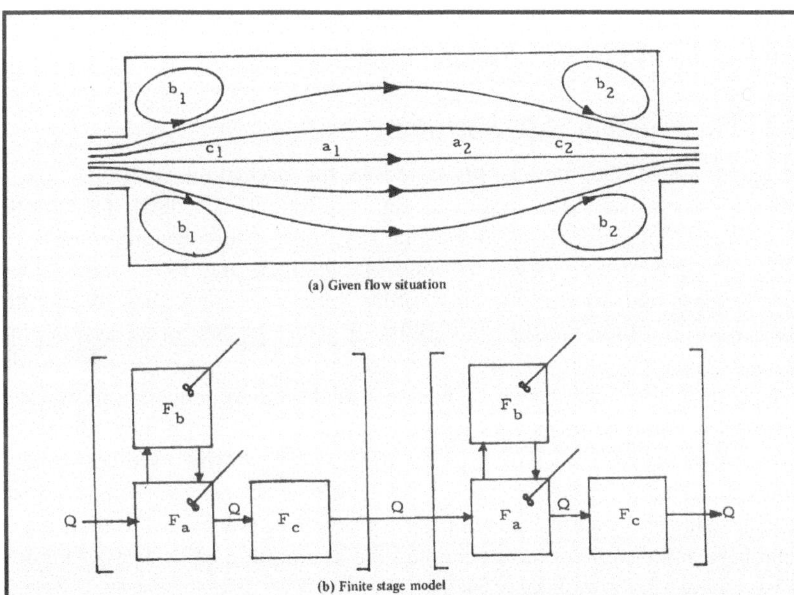

Figure 8.1: The finite stage model (Watters *et al.*, 1973, pg. 16)

Ferrara and Harleman (1981) attempted to improve on this approach using a pond model consisting of a central 'active' zone, where the flow moves from the inlet to outlet, and back via 'return' zones down the sides. They claimed that their model "reliably represents the transport process" and has the benefits of reducing the required input parameters to three: the dispersion coefficient of the active zone, the size of the active zone and the dilution ratio. Unfortunately, the latter two parameters still have to be determined by calibrating the model against experimental data.

Preul and Wagner (1987) sought to extend the work of Ferrara and Harleman. Instead of representing the active zone as a single reactor they divided it into separate plug flow and completely mixed zones. Further, they introduced top flow and bottom flow options for the model, which they claimed could be used to account for stratification effects during different seasons.

Overall, the use of combined pond models has produced some very good correlations with experimental data. However, it is essential to remember that this method is not predictive. Experimental data must first be collected to allow the model parameters to be calculated. Unless extensive work is done to develop predictive equations for the unknown parameters, it is unlikely that this approach will ever be useful for design purposes.

8.3.4 The reaction rate constant

One thing all of the models presented in the preceding sections have in common is their use of the first order rate coefficient, k. Indeed, Thirumurthi (1974) stated that evaluation of k was the key to the whole design process.

As mentioned previously, the rates of pathogen and BOD removal are typically assumed to follow first order kinetics. There has been little discussion in the literature of the validity of using the first order assumption; however, given its significance, it would seem to warrant greater interest. Thirumurthi (1991) discussed a laboratory-scale experiment that showed the reaction rate could be proportional to the substrate concentration to the power of 1.1. Wood (1987) also questioned the validity of this assumption as it implies the rates of processes such as oxygen mass transfer and algal growth are such that they are not rate limiting. In practice, however, the majority of researchers and designers have accepted the assumption of first order kinetics and have gone on to implement its use.

There are a large number of predictive equations for estimating the first order rate constant, k, for the removal of BOD and faecal coliforms.

One of the better documented studies, is that of Thirumurthi (1974) who published a relatively involved method of determining k:

$$k = k_s C_{Te} C_O C_{Tox}$$

where:

k_s	=	a 'standard' value of k (d^{-1});
C_{Te}	=	correction factor for temperature;
C_o	=	correction factor for organic load;
C_{Tox}	=	correction factor for industrial toxic chemicals.

Using data from the literature combined with results from a pond in Canada, Thirumurthi used the plug flow equation to back-calculate 'field k' values using the average influent and effluent BOD and the pond's theoretical retention time. These 'field k' values were then adjusted for temperature and organic load to produce k_s values.

Thirumurthi reported that the average k_s value for all the ponds studied was 0.056 day^{-1} and that the range was from 0.042 to 0.07 day^{-1}. However, these numbers were themselves based on averages of k_s values calculated for ponds at different times and averages from multiple ponds at different sites. An example of this is a pond that had k_s values ranging from 0.0026 to 0.0968 day^{-1} over the 19 dates that data were collected. Overall the raw field values actually had a range of 0.0017 to 0.128 day^{-1}!

This method of using field data to back-calculate k via one of the ideal flow equations is the usual method for determining k values. But as illustrated in the example given above, this method has its shortcomings. For example, Thirumurthi noted that fluctuations in the k_s values were partly due to the variation of the BOD over time. Additionally, the actual hydraulic characteristics of the different ponds were ignored by use of the theoretical retention time.

There are many alternative publications to Thirumurthi's work that also involve back-calculation from field data. For example, for BOD removal, Marais (1966) found that the best fit for experimental data was given by:

$$k_T = 1.2(1.085)^{T-35}$$

and Mara (1975) proposed:

$$k_T = 0.3(1.05)^{T-20}$$

where:

k_T	=	first order reaction rate constant (day^{-1});
T	=	temperature (°C).

In addition to the removal of BOD, equations are also available for pathogen decay. For example Marais (1974) analysed faecal coliform data from a pond series in South Carolina and found that the first-order rate constant for faecal coliform removal in facultative and maturation ponds was given by:

$$k_T = 2.6(1.19)^{T-20}$$

Von Sperling (1999, 2002, 2003) analysed the data from 33 facultative and maturation systems in tropical and subtropical Brazil (latitude 7–24° South). From this he derived the following equation for the first-order faecal coliform removal rate constant using the Wehner-Wilhelm equation for non-ideal flow in which he used the reciprocal of the pond's length-to-breadth ratio as a simple estimate of the pond dispersion number (see section 8.3.5):

$$k_T = 0.92D^{-0.88}\theta^{-0.33}(1.07)^{T-20}$$

where:

D = pond depth (m);
θ = pond retention time (d).

As an alternative to using field data, a number of researchers have considered the use of laboratory-based studies for the determination and study of the first order reaction rate constant.

Thirumurthi and Nashashibi (1967) undertook a laboratory study to determine the reaction rate constant for a synthetic wastewater using small bench-scale reactors under artificial lighting. This work was limited to three experimental runs undertaken at a combination of different loading and lighting regimes. In a more comprehensive study, Uhlmann (1979) examined the treatment performance of small model ponds as a function of organic loading, retention time and temperature. These were also fed on a synthetic wastewater and held under controlled artificial lighting. The reaction rate constants were then back calculated via the plug flow equation. In a subsequent paper, Uhlmann *et al.* (1983) undertook a regression analysis of the data to produce an equation for prediction of the reaction rate constant based on organic loading, mean retention time and mean temperature.

Wood (1987) reviewed the research of Thirumurthi and Nashashibi (1967) and Uhlmann (1979, 1983). He was particularly critical of the way these studies used the ideal flow models to back-calculate the reaction rate constants while in practice the model ponds were fed in discrete daily additions. Using a semi-continuous flow model he recalculated the reaction rate constants and showed this yielded significantly different results. Wood also went on to conclude that there was a need to determine the rate limiting steps and their kinetic parameters.

Brissaud *et al.* (2000) noted that rate constants given in the literature vary widely as a function of the water depth, temperature, solar radiation, organic load and the hydraulic model used. Because of this variation they used pilot-scale experimental ponds to determine the reaction rate constant for faecal coliform removal in a maturation pond. Two pilot-scale ponds were used, each of 1-m depth. They were filled with lagoon water and left exposed to the climatic conditions. The derived reaction rate constant of 0.6 day^{-1} was then used with tracer data from a full-scale pond to predict the theoretical treatment efficiency of the full-scale ponds under study. These results compared very favourably with the actual treatment efficiencies measured for these ponds. Although further studies are required to confirm the reliability of this technique, the work of Brissaud *et al.* (2000) does appear to offer an appropriate method for determining values of the reaction rate constant. It represents a compromise between the problem of back-calculating field data through an equation for ideal flow, thereby disregarding the influence of the actual hydraulic regime of the pond, and the problem of the artificial conditions present in laboratory-scale experiments.

It is worth noting that any set of pond performance data can be analysed to yield a value of the first-order reaction rate constant (k) for BOD or faecal coliform (or other parameter) removal on the assumption of plug flow, complete mixing or dispersed flow. Provided subsequent designers always use this value of k (adjusted for temperature as necessary) in the equation from which it was derived, then it obviously will recreate the same treatment performance data from which it was back calculated. However the choice of equation is not irrelevant. If the hydraulic behaviour of the pond being designed is different to that of the ponds from which the first-order reaction rate constant was derived then likewise so will the performance be different. It is possible to more accurately predict a pond's performance using integrated hydraulic and reaction modelling (see Section 8.4.3) which takes the actual hydraulic behaviour of a particular pond into account, but this is simply not always going to be practical for many pond designers. The general practice for this type of pond design is therefore to assume 'completely mixed' hydraulics which is more conservative than assuming 'plug flow' hydraulics and pragmatically allows a simple design method that has been found to work reasonably effectively in a wide range of practical designs.

8.3.5 The Dispersion Number

Fick's Law describes molecular diffusion. If general dispersion in, say, the x-direction is considered to have equivalent behaviour, then the dispersion of a tracer, C, can be described by:

$$\frac{\partial C}{\partial t} = D\frac{\partial^2 C}{\partial x^2}$$

Where D is the coefficient of axial dispersion that defines the degree of back-mixing.

If u and L are the velocity component and the length in the x-direction, then a dimensionless form of the equation can be derived as:

$$\frac{\partial C}{\partial \theta} = d\frac{\partial^2 C}{\partial z^2} - \frac{\partial C}{\partial z}$$

where:

θ = $t/t_{mean} = tu/L$;

z = $(ut + x)/L$;

d = (D/uL).

The dimensionless constant d is known as the dispersion number and can be experimentally derived from the results of a tracer study. In reality the dispersion number is effectively a single overall factor that attempts to account for the wide range of physical influences that can affect fluid movement in a pond. These influences include:

- the flowrate and its variation over time;
- the inlet size, position and orientation;
- the outlet position and design;
- wind shear and its variation over time;
- pond geometry (including influences of baffles); and
- temperature/density effects.

For the design of new ponds an accurate method of predicting the dispersion number has been sought in a number of research studies. Arceivala (1981), using data from the literature, proposed four simple empirical equations for the prediction of the coefficient of axial dispersion, D, from which the dispersion number can be determined:

- wider than 30 m with baffles, D = 33W;
- wider than 30 m without baffles, D = 16.7W;
- narrower than 10 m with baffles, D = $11W^2$;
- narrower than 10 m without baffles, D = $2W^2$;
- where W is the pond width (m).

Polprasert and Bhattarai (1985) built on work by Fischer (1967) into the prediction of dispersion in streams and rivers to propose the following predictive equation for the dispersion number in ponds:

$$d = \frac{0.184[\theta v(W+2Z)]^{0.489} W^{1.511}}{(LZ)^{1.489}}$$

where:

θ	=	hydraulic retention time (s);
v	=	kinematic viscosity (m^2/s);
W	=	pond width (m);
Z	=	pond depth (m);
L	=	length of fluid travel from inlet to outlet (m).

Marecos do Monte (1985) undertook tracer studies on two facultative waste stabilisation ponds in Portugal. She compared the dispersion numbers obtained with those predicted by the Polprasert and Bhattarai equation. There was little resemblance between the predicted and the measured results, leading her to state that the predictive equation cannot be considered to be valid for all ponds. She concluded that, for design, the completely mixed reactor equation should be applied as it yields the more conservative pond sizing.

Agunwamba (1991) published a review of dispersion number prediction equations. He wrote that the existing equations had yet to prove useful due to the disparity between experimental and predicted results. To explain this problem he suggested that omission of factors such as "wind speed, dead zones, secondary currents and seasonal effects; sampling time after tracer release; pond breadth-to-depth ratio and Reynolds number" could be to blame. Agunwamba *et al.* (1992) presented an alternative predictive equation for the dispersion number (d):

$$d = 0.10201\left(\frac{u^*}{u}\right)^{-0.81963}\left(\frac{H}{L}\right)\left(\frac{H}{W}\right)^{-\left(0.98074+1.38485\frac{H}{W}\right)}$$

where:

u	=	flow velocity (m/d);
u*	=	shear velocity (m/d);

H = pond depth (m);
L = pond length (m);
W = pond width (m).

In the same year Agunwamba (1992) also published a new method of dispersion number determination requiring only data on the bacterial variation along the pond as input. This method was claimed to be "simple, accurate and economical" in comparison to the use of tracer studies. However, to date there have been no other publications that have documented use of this technique.

Nameche and Vasel (1998) reviewed the work of a number of researchers on the prediction of the dispersion number. These authors compared these against tracer study data from more than thirty existing pond and aerated lagoon systems and used linear and multi-linear regressions to develop their own predictive equations. For waste stabilisation ponds they proposed:

$$P = 0.1\left(\frac{L}{W}\right) + 0.01\left(\frac{L}{Z}\right)$$

where P is the Peclet number (the inverse of the dispersion number).

Von Sperling (2003) used the technique of Monte Carlo simulation to investigate the sensitivity of predicting the dispersion number in comparison to the uncertainly of estimating other design variables such as flowrate, reaction rate constant and so on. In conclusion he found that when accounting for the uncertainly in other such variables, very simple models for predicting the dispersion number such as $d = (\text{Length/Breadth})^{-1}$, which he proposed in von Sperling (1999), could be used without significantly affecting the overall accuracy of the design.

8.4 MATHEMATICAL MODELLING

8.4.1 Reaction modelling

The first order reaction rate constant (k), discussed in Section 8.3.4, represents the overall effect of the many physical, biological and chemical processes that contribute to waste stabilisation in a pond. Marais and Shaw (1961) recognised this and stated that "the theory does not concern itself with the biological agencies responsible for the degradation action but only in the results they produce which give rise to the value k."

In (1979) Fritz *et al.* modelled the process dynamics occurring in ponds.

Their work had the objective of linking mass balance equations for twelve of the key biomass/biochemical variables to develop a non-steady-state mechanistic model for a typical facultative waste stabilisation pond. A conceptual summary of their pond model is shown in Figure 8.2. The model accounted for the inflow and outflow concentrations of soluble organics (represented by COD), dissolved oxygen, bacterial cell mass, algal cell mass, inorganic carbon, organic nitrogen, ammonia, nitrate, organic and inorganic phosphorus, and alkalinity.

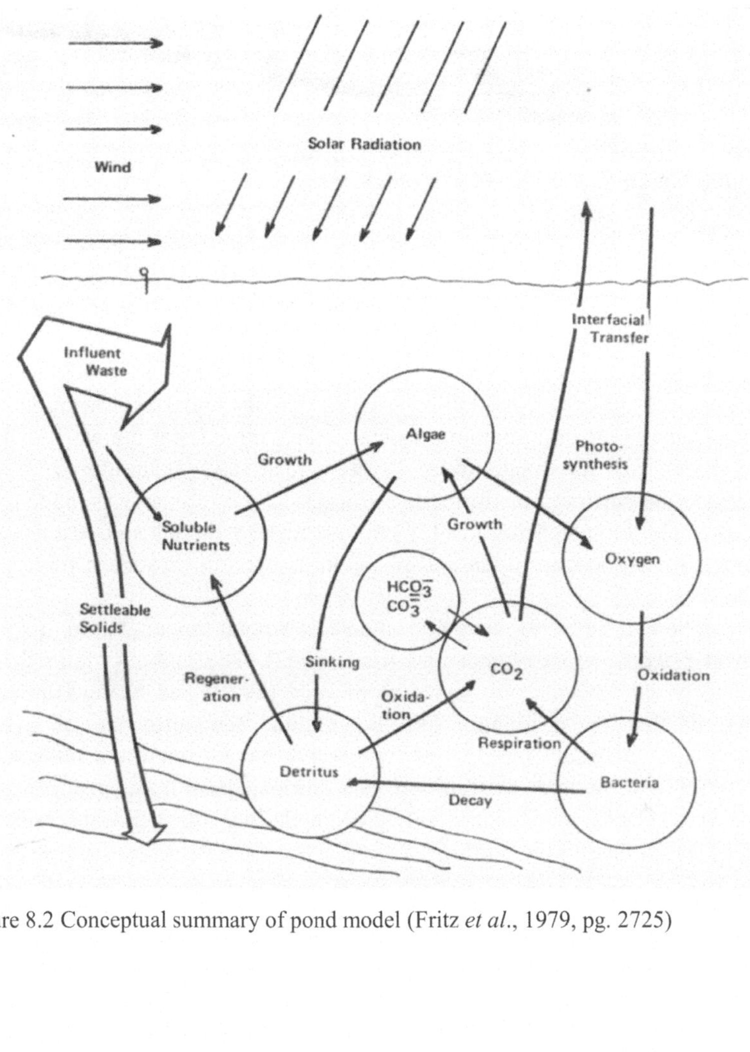

Figure 8.2 Conceptual summary of pond model (Fritz *et al.*, 1979, pg. 2725)

In addition to these external inputs and outputs, the model incorporated internal mass transfer of bacterial and algal cell mass into, and carbon dioxide, methane, ammonia and inorganic phosphorus out of, a 'black-box' sludge layer undergoing anaerobic digestion. A set of differential equations was proposed to represent the rate of change of these variables within the pond. The influence of dynamic external factors such as solar radiation, temperature and wastewater characteristics, which affect the reaction rates, were also included in the model. A fourth order Runge-Kutta technique was then used to solve the set of differential equations, giving the resultant concentrations for an annual cycle. Data published by Larsen (1974), from an existing pond in New Mexico, was used to evaluate the model. As the 'first attempt' at developing a mechanistic pond model it gave reasonable results and provided insight into the process dynamics within pond systems. However, the authors made a range of conclusions and recommendations suggesting the model required some reasonable degree of further development. One specific example of this was with regard to the lack of understanding of pond hydraulics and their effect on the biological processes.

Colomer and Rico (1992) sought to improve on the Fritz model. Their revised model was evaluated in comparison to field data for a facultative pond receiving primary effluent. Error analysis for each parameter indicated it was an improvement on the Fritz model, with better predictions for all parameters except for nitrate-nitrogen.

Xiang-Hua *et al.* (1994) studied nutrient transformation in a pond system. This work was based on modelling work originally undertaken a number of years earlier as part of a doctorate thesis. They claimed that the work of Fritz and others had not undertaken "systematical and quantitative studies concerning the nutrient transformation process". No reference was made to the work of Colomer and Rico (1992). A 12-equation model and results were presented for carbon, nitrogen and phosphorus cycling through a three-pond system.

Kayombo *et al.*, (1999), have presented a new pond model. Again this work made no reference to Colomer and Rico (1992). The authors claimed their model to be an advance on the works of Fritz (1979) and Xiang-Hua *et al.* (1994), in that these authors did not include the influence of pH on heterotrophic bacterial and algal activities and that nitrogen and phosphorus were assumed to be non-rate-limiting. Another paper from this group was by Senzia *et al.* (1999) and focused solely on modelling nitrogen transformation and removal in facultative ponds.

8.4.2 Hydraulic modelling

Mathematical models can also be used to study a wide range of hydraulic behaviours. However, their application to wastewater stabilisation ponds has only started to be more widely researched in recent times as computing speed and software applications have dramatically improved.

Fares (1993) developed a unique numerical model, based on the shallow water equations, for simulating circulation patterns and mass transport in large basins driven by wind and thermocline effects. Fares and Lloyd (1995) then adapted this model, with the addition of point sources and sinks, representing the inlet and outlet, to undertake a study of the flow behaviour of a waste stabilisation pond system on Grand Cayman in the British West Indies. Their analysis confirmed the presence of short-circuiting, which they attributed to wind affects. Fares *et al*., (1996) continued the work with an objective of using the model to investigate the effects of alternative inlet/outlet configurations under the influence of differing wind speeds and directions.

Wood *et al*., (1995) published the first journal paper describing the application of a commercial computational fluid dynamics (CFD) package to the design of waste stabilisation ponds. This consisted of a two-dimensional, laminar model. Wood *et al*., (1998) then incorporated the k-ε turbulence model. This paper also presented the technique of introducing a virtual tracer to simulate hydraulic retention time distribution curves. These studies were, however, limited to two-dimensional flow and the authors reported that this leads to difficulties in representing the pond inlet. In conclusion, they stated that two-dimensional models could not be used to adequately describe flow in ponds. The work of Wood then continued with three-dimensional modelling.

Following on from the work of Wood *et al*., two new papers on the application of CFD to pond design using three-dimensional CFD models incorporating turbulence were presented by Salter *et al*., (2000) and Shilton (2000). Since this time the potential of CFD has become increasingly recognised and is attracting a growing amount of research.

In addition to solving the equations of fluid flow, CFD modelling also allows incorporation of other equations within its solution domain. The next logical development was, therefore, the integration of the hydraulic model with a reaction model. This development is still in its infancy but has the potential to become the future of pond design as discussed in the following section.

8.4.3 Integrated models

In Section 8.3.4 above it was highlighted that while something like the ideal flow equation is a simple and generally effective method of sizing ponds it has the shortcoming of assuming ideal hydraulic behaviour in all ponds. This shortcoming was highlighted by Wood *et al.*, (1995, pg.112) who stated "it is currently impossible to reliably predict how various modifications of pond design, such as placement and number of inlets, use of baffles, etc., might affect pond performance".

Shilton and Harrison (2002) embedded the reaction rate model for first order kinetics within a CFD model of a pond to simulate the decay of coliforms as a direct function of the hydraulic flow pattern. To demonstrate this they presented a design example of a facultative pond sized for a flow of 10,000 m^3/d. As seen below, the standard pond was modelled (Figure 8.3) along with a design incorporating two baffles (Figure 8.4). In all cases the inlet is located in the bottom left corner, while the outlet is located in the top right corner.

As is typically found in pond systems, the standard design suffered from severe short-circuiting with the model predicting a value of 6.15×10^6 cfu/100ml at the outlet while the baffled design improved treatment efficiency to reduce this to 5.60×10^3 cfu/100ml.

Figure 8.3 CFD model of coliform decay – standard pond design (Shilton and Harrison, 2002)

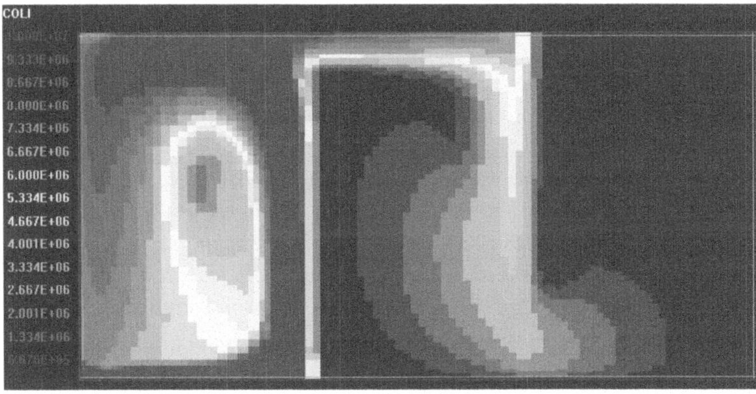

Figure 8.4 CFD model of coliform decay – 2 baffle system (Shilton and Harrison, 2002)

This development is clearly a significant step forward in pond design as it allows a designer to test a range of sizes and physical designs to optimise a final design. Although not attempted at this time, future researchers will undoubtedly add a further level of sophistication to the modelling by moving to integrate more mechanistic description of reaction kinetics, as described in Section 8.4.1 into CFD models. At the present time, however, even the more simple approach taken by Shilton and Harrison (2002) is too resource intensive for all but large engineering consultants. Therefore a more pragmatic approach is to follow simplified guidelines for improved hydraulic design of ponds such as produced by Shilton and Harrison (2003), and summarised in chapter 10.

8.5 SUMMARY

In a general review of pond design, Metcalf and Eddy (1991, pg. 438) stated that "The amount of effort that has been devoted to the characterization of facultative ponds is staggering, and an equal amount has probably been spent trying to develop appropriate design equations."

In particular there have been a large number of publications that have attempted to apply reactor theory to pond design. Debate over the use of the ideal flow assumption has led to work on the non-ideal dispersed flow model and various combined flow models. Although apparently more sophisticated, there has been difficulty in obtaining reliable prediction of the input parameters required for these models. This has limited their application and led several researchers to recommend the return to use of the simpler ideal flow models.

The design manuals by Mara and Pearson (1998) and Mara *et al.* (1992) use a loading rate adjusted for temperature to size ponds for organic loading and the completely mixed ponds-in-series reactor model for pathogen removal. These methods provide today's engineers with a safe and consistent design methodology.

However, they cannot directly account for the effect of design variables such as inlet, outlet, shape, baffles and so on; however, guidelines such those by Shilton and Harrison (2003) can assist in this particular regard.

This chapter has reviewed a large amount of research that has been conducted to improve pond design. However, for an engineer wanting to design a new pond system this can be confusing and somewhat overwhelming. Therefore in the following chapter (Chapter 9) a practical guide to pond sizing is outlined along with several design examples. Then in Chapter 10 hydraulic design guidelines are presented that allow further refinement of the pond design for optimisation of hydraulic behaviour.

REFERENCES

Agunwamba, J. (1992b). A new method for dispersion number determination in waste stabilization pond. *Water, Air and Soil Pollution* **63**, 361-369.

Agunwamba, J., Egbuniwe, N. & Ademiluyi, J. (1992). Prediction of the dispersion number in waste stabilization ponds. *Water Research* **26**(1), 85-89.

Agunwamba, J. (1991). Dispersion number determination in waste stabilization ponds. *Water Air and Soil Pollution* **59**, 241- 247.

Arcelivala, S. (1981). Hydraulic modeling for waste stabilization ponds. *Journal of the Environmental Engineering Division, ASCE.*

Brissaud, F., Lazarova, V., Ducoup, C., Joseph, C., Levine, B. & Tournoud, M. (2000). Hydrodynamic behaviour and faecal coliform removal in a maturation pond. *Water Science and Technology* **42**(10-11), 119-126.

Colomer, F. & Rico, D. (1992). Mechanistic model for facultative stabilization ponds. *Water Environment Research* **65**(5), 679-685.

Fares, Y. (1993). Circulation pattern in long narrow lakes based on shallow water equations. *Advances in Hydro-Science and Engineering.* Wang, S. (Ed). Volume 1: 1142-1147.

Fares, Y., Frederick, G., Vorkas, C. & Lloyd, B. (1996). Hydrodynamic effects on performance of waste stabilisation lagoons. Unpublished copy obtained from author.

Fares, Y. and Lloyd, B. (1995). Wind effects on residence time in waste stabilisation lagoons. *HYDRA 2000*; Thomas Telford; London.

Ferrara, R. & Harleman, D. (1981). Hydraulic modelling for waste stabilization ponds. *Journal of the Environmental Engineering Division, ASCE*, **107**(EE4), 817-830.

Fischer, H. (1967). The mechanics of dispersion in natural streams. *Journal of the Hydraulics Division, ASCE*, **93**(HY6), 187-216.

Finney, B. & Middlebrooks, E. (1980). Facultative waste stabilization pond design. *Journal of the Water Pollution Control Federation*, 52(1): 134-147.

Fritz, J., Middleton, A. & Meredith, D. (1979). Dynamic process modelling of wastewater stabilization ponds. *Journal of the Water Pollution Control Federation* **51**(11), 2724-2743.

Gloyna, E. (1976). Facultative waste stabilization pond design. *Ponds as a Wastewater Treatment Alternative.* Gloyna. E., Malina, J. and Davis, E. (Ed.). University of Texas, Austin; 143-157.

Kayombo, S., Mbwette, T., Mayo, A., Katima, J. & Jorgensen, S. (1999). Development of a holistic ecological model for design of facultative waste stabilization ponds in tropical climates. *Proceedings of the 4th IAWQ Specialist Group Conference on Waste Stabilisation Ponds.* Pearson, H. (Ed). Marrakech, Morocco.

Larsen, T. (1974). *A Dimensionless Design Equation for Sewage Lagoons.* Doctorate Thesis; University of New Mexico; Albuquerque, USA.

Levenspiel, O. (1972). *Chemical Reaction Engineering.* John Wiley & Sons; New York, USA.

McGarry, M. & Pescod, M. (1970). Stabilization pond design criteria for tropical Asia. *Proceedings of the 2nd International Symposium for Waste Treatment Lagoons.* McKinney, R. (Ed). University of Kansas; Kansas City, Kansas, USA; 114-132.

Mara, D. & Pearson, H. (1998). *Design Manual for Waste Stabilization Ponds in Mediterranean Countries.* Lagoon Technology International; Leeds, UK.

Mara, D., Alabaster, G., Pearson, H. & Mills, S. (1992). *Waste Stabilisation Ponds - A Design Manual for Eastern Africa.* Lagoon Technology International; Leeds, UK.

Mara, D. (1975). Proposed design for oxidation ponds in hot climates. *Journal of the Environmental Engineering Division, ASCE,* **101**(EE2), 296-300.

Mara, D (1987). Waste stabilization ponds: problems and controversies. *Water Quality International* **1**, 20–22.

Marais, G. (1974). Faecal bacterial kinetics in waste stabilization ponds. *Journal of the Environmental Engineering Division, ASCE,* **100**(EE1), 120-139.

Marais, G. (1970). Dynamic behaviour of oxidation ponds. *Proceedings of the 2nd International Symposium for Waste Treatment Lagoons.* McKinney, R. (Ed). University of Kansas; Kansas City, Kansas, USA; 15-46.

Marais, G. (1966). New factors in the design, operation and performance of waste-stabilization ponds. *Bulletin of the World Health Organisation* **34**, 737-763.

Marais, G. & Shaw, V. (1961). A rational theory for the design of sewage stabilization ponds in Central and South Africa. *Transactions of the South African Institution of Civil Engineers* **3**, 205-227.

Marecos do Monte, M. (1985). *Hydraulic Dispersion in Waste Stabilization Ponds in Portugal.* MSc Thesis; Department of Civil Engineering, University of Leeds; Leeds, UK.

Metcalf and Eddy, Inc. (1991). *Wastewater Engineering: Treatment, Disposal and Reuse.* McGraw-Hill; New York, USA.

MWD (Ministry of Works and Development) (1974). *Guideline for the Design, Construction and Operation of Oxidation Ponds.* Public Health Engineering Section, Ministry of Works and Development; Wellington, New Zealand.

Nameche, T. & Vasel, J. (1998). Hydrodynamic studies and modelization for aerated lagoons and waste stabilization ponds. *Water Research* **32**(10), 3039-3045.

Polprasert, C. & Bhattarai, K. (1985). Dispersion model for waste stabilization ponds. *Journal of the Environmental Engineering Division, ASCE* **111**(EE1), 45-59.

Prats, D. & Llavador, F. (1994). Stability of kinetic models from waste stabilization ponds. *Water Research* **28**(10), 2125-2132.

Preul, H. & Wagner, R. (1987). Waste stabilization pond prediction model. *Water Science Technology* **19**(12), 205-211.

Salter, H., Ta, C., Ouki, S. & Williams, S. (2000). Three-dimensional computational fluid dynamic modelling of a facultative lagoon. *Water Science and Technology* **42**(10-11), 335-342.

Shilton, A. N. & Harrison, J. (2003). *Guidelines for the Hydraulic Design of Waste Stabilisation Ponds*. Palmerston North, New Zealand: Institute of Technology and Engineering, Massey University.

Shilton, A. & Harrison, J. (2002). Integration of coliform decay within a CFD model of a waste stabilisation pond. *Water Science and Technology* **48**(2), 205-210.

Shilton, A. (2000). Potential application of computational fluid dynamics to pond design. *Water Science and Technology* 42(10-11), 327-334.

Senzia, M., Mayo, A., Mbwette, T., Katima, J. & Jorgensen, S. (1999). Modelling nitrogen transformation and removal in facultative ponds. *Proceedings of the 4th IAWQ Specialist Group Conference on Waste Stabilisation Ponds*. Pearson, H. (Ed). Marrakech, Morocco.

Tchobanoglous, G. & Schroeder, E. (1985). *Water Quality Characteristics, Modeling, Modification*. Addison-Wesley; Reading, Massachusetts, USA.

Thirumurthi, D. (1991). Biodegradation in waste stabilization ponds (facultative lagoons). *Biological Degradation of Wastes*. Elsevier; London, England; 231-246.

Thirumurthi, D. (1974). Design criteria for waste stabilization ponds. *Journal of the Water Pollution Control Federation* **46**(9), 2094- 2106.

Thirumurthi, D. (1969). Design principles of waste stabilization ponds. *Journal of the Sanitary Engineering Division, ASCE*, **95**(SA2), 311-330.

Thirumurthi, D. & Nashashibi, O. (1967). A new approach for designing waste stabilization ponds. *Water and Sewage Works*, **114**(R), 208-218.

Uhlmann, D., Recknagel, F., Sandring, G., Schwarz, S. & Eckelmann, G. (1983). A new design procedure for waste stabilization ponds. *Journal of the Water Pollution Control Federation* **55**(10), 1252-1255.

Uhlmann, D. (1979). BOD removal rates of waste stabilization ponds as a function of loading, retention time, temperature and hydraulic flow pattern. *Water Research* **13**, 193-200.

von Sperling, M. (1999). Performance evaluation and mathematical modeling of coliform die-off in tropical and subtropical waste stabilization ponds. *Water Research* **33**(6), 1435-1448.

von Sperling, M. (2002). Relationship between first-order decay coefficients in ponds, for plug flow, CSTR and dispersed flow regimes. *Water Science and Technology* **45**(1), 17–24.

von Sperling, M. (2003). Influence of the dispersion number on the estimation of coliform removal in ponds. *Water Science and Technology* **48**(2), 181–188.

Watters, G., Mangelson, K., & George, R. (1973). *The Hydraulics of Waste Stabilization Ponds*. Research Report; Utah Water Research Laboratory, College of Engineering, Utah State University; Utah, USA.

Wehner, J. & Wilhelm, R. (1956). Boundary conditions of flow reactor. *Chemical Engineering Science* **6,** 89-93.

Wood, M., Howes, T., Keller, J. & Johns, M. (1998). Two-dimensional computational fluid dynamic models for waste stabilisation ponds. *Water Research* **32**(3), 958-963.

Wood, M., Greenfield, P., Howes, T., Johns, M. & Keller, J. (1995). Computational fluid dynamic modelling of wastewater ponds to improve design. *Water Science and Technology* **12**, 111- 118.

Wood, T. (1987). Interpretation of laboratory-scale waste stabilization pond studies. *Water Science and Technology* **19**(12), 195- 203.

Xiang-Hua, W., Yi, Q. & Xia-Sheng, G. (1994). Graphical presentation of the transformation of some nutrients in a wastewater stabilization pond system. *Water Research* **28**(7), 1659-1669.

9

Pond process design – a practical guide

Duncan Mara

9.1 INTRODUCTION

In the previous chapter, it has been seen that over the years a wide range of design criteria/equations have been proposed. For the design engineer needing to size a pond system this can be confusing and so in this chapter a practical and well-recognised approach to the process design of waste stabilisation ponds (i.e. the calculation of pond volumes, areas and retention times) is described. The pond types considered are anaerobic ponds, facultative ponds and maturation ponds. They are arranged in series, such that each series comprises a single anaerobic pond, a single facultative pond and one or more maturation ponds. Whether or not maturation ponds are included depends on the characteristics of the raw wastewater and on the required effluent quality. At any one pond site there will generally (except at very small installations serving fewer than around 2000 people) be more than one series. The use of several series of ponds in

ISBN: 1843390205. Published by IWA Publishing, London, UK.

parallel allows greater flexibility in operation and maintenance and, in some cases, for dealing with high seasonal variations in loading.

Prior to entering the ponds, the wastewater should receive preliminary treatment by screening and grit removal. Screening is particularly important to reduce the risk of discharging visually offensive material, such as condoms or sanitary tampons, to receiving waters. Furthermore, effective screening at the front end of a pond system prevents the objectionable appearance of such material building up as scum mats within the ponds and avoids its accumulation in the pond sludge. Land application/disposal of contaminated pond sludge obviously becomes a problem if farmers refuse to accept condom/tampon-laden sludge's on their land.

Grit accumulation can create small islands near the inlet of a pond and more generally can reduce the effective treatment volume of a pond. While grit removal has often been considered a lower priority for ponds because of their large storage volume, in a number of cases accumulation of grit has caused significant problems. Examples of this are often seen in areas such as the Middle East where high amounts of sand enter the wastewater. In New Zealand silt from a food processing plant led to the loss of a significant amount of effective pond volume and a subsequent loss of performance efficiency.

For a fully worked design example and a case study, see sections 9.8 and 9.9 of this chapter.

9.2 EFFLUENT QUALITY

The effluent quality required by the local regulatory agency would normally cover some of the following parameters: BOD (or COD), suspended solids, ammonia-nitrogen, faecal coliform bacteria and human intestinal nematode eggs. Other parameters may also be specified, such as total nitrogen and human trematode eggs. The precise selection of parameters and parameter values set by the regulator should reflect the intended destination of the effluent – surface or coastal water discharge, or reuse in agriculture and/or aquaculture. Typically BOD and suspended solids are always regulated. Increasingly nutrients, particularly nitrogen, are also a focus. If waterways have recreational use for bathing and/or other water-based sports, or if there are local shellfisheries, faecal coliform bacteria (or faecal enterococci) are commonly specified. For agricultural reuse faecal coliforms and human intestinal nematode eggs, and for aquacultural reuse faecal coliforms and human trematode eggs, are relevant.

Pond designers should examine very carefully the effluent qualities specified by regulators. This is extremely important because not all regulators set sensible effluent quality requirements. Negotiation with the regulator can result in a more appropriate set of effluent quality requirements. A good example of this is whether pond effluent samples should be filtered prior to BOD analysis. As ponds essentially convert much of the raw wastewater BOD into algae, the question to be answered is whether the algal BOD be included or excluded. The EU Urban Waste Water Treatment Directive (CEC, 1991) requires sample filtration for pond effluents in order to exclude the algal BOD. The arguments in favour of this are that the pond algae in the receiving watercourse are quickly consumed by the aquatic biota, and that they produce oxygen in the receiving water, during daylight hours at least.

9.3 ANAEROBIC PONDS

Anaerobic ponds receive such a high organic loading (usually >100g BOD/m^3d, equivalent to >3000 kg/ha d for a depth of 3 m) that they contain no dissolved oxygen and no algae (although occasionally a thin film of *Chlamydomonas* may be seen at the surface). They work extremely well: a properly designed anaerobic pond will achieve at least 40 percent BOD removal at 10°C, 60 percent at 20°C and 70 percent at 25°C. Retention times are short: for wastewater with a BOD of 300 mg/l, 3 days are required at 10°C and below, and 1 day is sufficient at 20°C and above.

Designers have in the past been too afraid to incorporate anaerobic ponds in case they cause odour. Hydrogen sulphide, formed mainly by the anaerobic reduction of sulphate by sulphate-reducing bacteria such as *Desulfovibrio,* is the principal potential source of odour. However, at the pH values normally found in anaerobic ponds (around 7.5), most of the sulphide is present as the odourless bisulphide ion.

Many practical applications have shown that odour is *not* a problem if the recommended design loadings are not exceeded and if the sulphate concentration in the raw wastewater is <500 mg SO_4^{2-}/1 (Gloyna and Espino, 1969). A small amount of sulphide is beneficial as it is lethal to *Vibrio cholerae* (Oragui *et al.* 1993), and it reacts with heavy metals to form insoluble metal sulphides, which precipitate out (Moshe *et al.* 1972). However, concentrations of 50–150 mg/l can inhibit methanogenesis (Pfeffer, 1970).

9.3.1 Design

Anaerobic ponds are typically designed on the basis of volumetric BOD loading (λ_v, g/m^3 d), given by:

$$\lambda_v = L_i Q/V_a \tag{9.1}$$

where:

Q = influent wastewater flow, m^3d;
L_i = influent wastewater BOD, mg/l;
V_a = anaerobic pond volume, m^3 (= pond mid-depth area, A_a, m^2 × pond depth, D_a, m).

The design value of λ_v varies with temperature, which is taken as the mean air temperature of the coldest month. BOD removal also varies with temperature (Table 9.1). D_a is normally in the range 2–5 m. Ideally the deeper the better, but depth may be limited by the height of the water table and more practically the economics of excavation, the cost of which increases with depth. Once the depth is chosen, the pond mid-depth area can be calculated. The mean hydraulic retention time (θ_a, d) is then calculated from:

$$\theta_a = V_a/Q = L_i/\lambda_v \qquad (9.2)$$

subject to a practical minimum value for θ_a of 1 day.

Table 9.1 Variation of design volumetric BOD loading on, and BOD removal in, anaerobic ponds

Temperature *(°C)*	Design loading *(g/m³ d)*	BOD removal *(%)*
<10	100	40
10 – 20	20*T* – 100*	2*T* + 20*
20 – 25	10*T* + 100*	2*T* + 20*
>25	350	70

**T* = temperature, °C

9.4 FACULTATIVE PONDS

Facultative ponds receive either raw wastewater (after screening and grit removal), when they are termed 'primary' facultative ponds, or the effluent from anaerobic ponds, when they are termed 'secondary' facultative ponds. Except at small installations serving less than around 2000 people, the latter option is preferred; due to the high efficiency of anaerobic ponds, the combined areas for an anaerobic pond and a secondary facultative pond are much lower (by 30–50 percent) than the area for a primary facultative pond. Facultative ponds are also typically designed for BOD removal on the basis of a relatively low surface loading (100–400 kg BOD/ha d). This permits the development of a healthy algal population.

While the microbiological, physical and chemical environment in the pond has been extensively reviewed in Chapters 2 and 3, it is useful to reiterate some key aspects again before focusing on the design.

In the case of a facultative pond, the algae generate the bulk of the oxygen required for BOD removal by the pond bacteria. Due to the algae, facultative ponds are coloured dark green, although they may occasionally appear red or pink (especially when slightly overloaded) due to the presence of anaerobic purple sulphide-oxidising photosynthetic bacteria. The algae that predominate in the turbid waters of facultative ponds are the motile genera, such as *Chlamydomonas, Pyrobotrys* and *Euglena*, as these can optimise their vertical position in the pond in relation to incident light intensity and temperature more easily than non-motile forms (such as *Chlorella,* although these are fairly common in facultative ponds). The concentration of algae in a healthy facultative pond depends on loading and temperature, but is usually in the range 500–2000 μg chlorophyll *a* per litre.

As a result of the photosynthetic activities of the pond algae, there is a diurnal variation in the concentration of dissolved oxygen. After sunrise, the dissolved oxygen level gradually rises to a maximum in the mid-afternoon, after which it falls to a minimum during the night. The position of the oxypause (the depth at which the dissolved oxygen concentration reaches zero) similarly changes, as does the pH since at peak algal activity carbonate and bicarbonate ions react to provide more carbon dioxide for the algae, so leaving an excess of hydroxyl ions with the result that the pH can rise to above 9.4, which, as discussed in Chapter 6, significantly assists disinfection processes.

9.4.1 Design

Facultative ponds are designed on the basis of surface BOD loading (λ_s, kg/ha d), given by Mara's (1987) equation:

$$\lambda_s = 350\,(1.107 - 0.002T)^{T-25} \tag{9.3}$$

where T = mean air temperature of the coldest month, °C.

For $T \leq 8$°C a minimum design value of 80kg/ha d is used (CEMAGREF *et al.* 1997). The pond area (A_f, m^2) is given by:

$$A_f = 10\,L_i\,Q/\lambda_s \tag{9.4}$$

where L_i = influent BOD (i.e., the BOD of effluent from anaerobic pond in the case of designing a secondary facultative pond), mg/l.

Pond depth varies from 1 to 2m, with 1.5m being typical. The pond retention time (θ_f, d) is given by:

$$\theta_f = A_f D_f / [\tfrac{1}{2}\,(Q_i + Q_e)] \tag{9.5}$$

where D_f = facultative pond depth, m;
Q_i = influent wastewater flow, m^3/d;
Q_e = effluent wastewater flow, m^3/d.

If seepage is negligible, then:

$$Q_e = Q_i - 0.001eA_f \tag{9.6}$$

where e = net evaporation (= evaporation – rainfall), mm/day.

Thus:

$$\theta_f = 2A_fD_f/(2Q_i - 0.001eA_f) \tag{9.7}$$

subject to a practical minimum value for θ_f of 4 days.

9.4.2 Treatment efficiency

At this point in the design process, it is appropriate to check the treatment efficiency obtained thus far so as to determine whether the design is now adequate in its own right or if further treatment is needed. In the sections below this is reviewed for both stream discharge and crop irrigation, using criteria from the literature and several key agencies.

BOD

In assessing BOD it is important to determine whether the BOD sample should be filtered or not. For example, the EU Urban Waste Water Treatment Directive (CEC, 1991) specifies that the effluent BOD from a pond system should not be greater than 25 mg per litre of *filtered* effluent.

Assuming that BOD removal is reasonably well described by first order kinetics, the *un*filtered BOD of the effluent of the facultative pond (L_e, mg/l) is given by:

$$L_e = L_i/(1 + k_1\theta_f) \tag{9.8}$$

where k_1 = first-order rate removal constant for BOD removal, d^{-1}.

The value of k_1 varies with temperature, as follows (Mara, 1976):

$$k_{1(T)} = k_{1(20)}(1.05)^{T-20} \tag{9.9}$$

Typical values of $k_{1(20)}$ are 0.3 d^{-1} for primary facultative ponds and 0.1 d^{-1} for secondary facultative ponds. Since approximately 70–90 percent of the effluent BOD from a facultative pond is due to the algae it contains, the filtered effluent BOD can be conservatively estimated as:

$$L_e\,(\text{filtered}) = 0.3L_e\,(\text{unfiltered}) \tag{9.10}$$

Pathogens

WHO (1989) recommends that effluents used for restricted irrigation should not contain more than one human intestinal nematode egg per litre (these nematodes are *Ascaris lumbricoides*, the human roundworm; *Trichuris trichiura*, the human whipworm; and *Ancylostoma duodenale* and *Necator americanus*, the human hookworms). Mara and Pearson (1987) and Blumenthal *et al.* (2000) also recommend that the effluent should contain no more than 10^5 faecal coliform bacteria per 100 ml.

The design equation for egg removal is (Ayres *et al.* 1992):

$$R = 100\,[1 - 0.41\exp(-0.49\theta + 0.0085\theta^2)] \qquad (9.11)$$

where R = percentage egg removal.

Equation 9.11 is applied to the anaerobic and facultative ponds in turn.

The design equation for faecal coliform removal proposed by Marais (1974) is used here:

$$N_e = N_i / (1 + k_B\theta_a)\,(1 + k_B\theta_f) \qquad (9.12)$$

where N_e = number of faecal coliforms per 100 ml of facultative pond effluent;
N_i = number of faecal coliforms per 100 ml of raw wastewater; and
k_B = the first-order rate constant for faecal coliform removal, d^{-1}.

The value of k_B varies with temperature, as follows (Marais, 1974):

$$k_{B(T)} = 2.6\,(1.19)^{T-20} \qquad (9.13)$$

The value of N_i in raw domestic wastewater is commonly 10^7–10^8 per 100 ml, with 5×10^7 per 100 ml often being a suitable design value. The temperature used in equation 9.13 is the mean air temperature in the coolest month of the irrigation season.

9.5 MATURATION PONDS

If the assessment of the treatment efficiency, as discussed in 9.4.2, indicates further treatment is necessary, then one or more maturation ponds are required after the facultative pond. The primary function of maturation ponds is the removal of excreted pathogens, and this can be efficiently achieved by a properly designed series of ponds. Maturation ponds also provide some additional polishing of BOD, although this is usually limited to only around 10–25 percent in each pond.

The size and number of maturation ponds is typically determined by the required bacteriological quality of the final effluent. For example, the World Health Organisation suggests that for unrestricted irrigation this should be ≤1000 faecal coliforms per 100 ml, and ≤200 per 100 ml for watering public parks, sports fields and hotel lawns and gardens (WHO, 1989).

Maturation ponds usually show less vertical biological and physicochemical stratification than facultative ponds and are well oxygenated throughout the day. Their algal population is thus more diverse than that of facultative ponds, with non-motile genera tending to be more common; algal diversity increases from pond to pond along the series.

9.5.1 Design

The retention time in the first maturation pond (θ_{ml}, d) is subject to the following three constraints:

(1) $\theta_m \not> \theta_f$

(2) $\theta_m \not< \theta_m^{min}$,

where θ_m^{min} is the minimum retention time to prevent algal washout and minimise hydraulic short-circuiting. Marais (1974) suggests a value of 3 days for θ_m^{min}, but 5 days are more appropriate when temperatures are <10° C; and

(3) $\lambda_{s(ml)} \not> 0.75\ \lambda_{s(f)}$,

where $\lambda_{s(ml)}$ and $\lambda_{s(f)}$ are the surface BOD loadings on the first maturation pond and the facultative pond, respectively.

The last constraint sets the minimum retention time in the first maturation pond, as follows. Since $\theta = AD/Q$, equation 9.4 can be rewritten for the first maturation pond as:

$$\lambda_{s(ml)} = 10 L_i D_m / \theta_{ml} \qquad (9.14)$$

where

L_i = the BOD of the influent to the first maturation pond (i.e. the effluent from the facultative pond calculated from equations 9.8 and 9.9);

D_m = the depth of the maturation ponds, m (usually 1–1.5 m).

Application of constraint no. 3 above permits equation 9.14 to be rewritten as:

$$\theta_{ml} = 10 L_i D_m / 0.75 \lambda_{s(f)} \qquad (9.15)$$

Equation 9.15 therefore determines the minimum retention time required in the first maturation pond of the series so as to be in compliance with constraint no. 3. Now that these constraints have been quantified, the calculations to determine the actual retention times required to achieve the required final pathogen removal are undertaken.

In an extension to equation 9.12 terms are added to those of the anaerobic pond and the facultative pond to account for the first maturation pond (as determined by 9.15) and a series of n subsequent equally sized maturation ponds:

$$N_e = N_i / (1 + k_B\,\theta_a)\,(1 + k_B\,\theta_f)\,(1 + k_B\,\theta_{ml})\,(1 + k_B\,\theta_m)^n \tag{9.16}$$

where

N_e = faecal coliform count per 100 ml of the final effluent;
θ_m = retention time in the second to nth maturation ponds, d;
n = number of maturation ponds additional to the first maturation pond.

Because the design is now focused on determining the retention time (and therefore the size) of the second to nth maturation ponds (θ_m), equation 9.16 is rearranged to give:

$$\theta_m = \{[N_i / N_e(1 + k_B\theta_a)(1 + k_B\theta_f)(1 + k_B\theta_{m1})]^{1/n} - 1\}/k_B \tag{9.17}$$

subject to $\theta_m \not< \theta_m^{\,min}$

Note: θ_m refers to the retention time in *each* of the second to nth maturation ponds. By calculating the value of θ_m for different values of n (1, 2, 3, ...), several different combinations of pond size/number are determined that will achieve the required faecal coliform reduction. By considering constraints 1 and 2 and then by multiplying n by θ_m to minimise the total retention time (and therefore size) of the series, the final design is determined. See the worked example in Section 9.8 for further guidance.

Ammonia removal

In anaerobic ponds there is no removal of ammonia; indeed the effluent ammonia concentration is generally greater than the influent concentration as some of the organic nitrogen is converted to ammonia. For ammonia removal in facultative and maturation ponds, the following equations can be used (Pano and Middlebrooks, 1982):

- for temperatures <20 °C:

$$C_e = C_i / \{1 - [(A/Q)(0.0038 + 0.000134T)\exp((1.041 + 0.044T)(\mathrm{pH} - 6.6))]\} \quad (9.18)$$

- for temperatures >20 °C:

$$C_e = C_i / \{1 + [5.035 \times 10^{-3}(A/Q)][\exp(1.540)(\mathrm{pH} - 6.6)]\} \quad (9.19)$$

where:

C_e and C_i = effluent and influent concentrations of ammonia-N, respectively, mg/l.

The pH may be estimated from:

$$\mathrm{pH} = 7.3\exp(0.0005\mathbf{A}) \quad (9.20)$$

where:

$\mathbf{A}$ = alkalinity, mg $CaCO_3$/l.

Equation 9.18 or 9.19 is applied sequentially to the facultative pond and each maturation pond, so that the final effluent concentration of ammonia-N can be determined.

9.6 PHYSICAL SIZING

The process design equations given in sections 9.3–9.5 give the mid-depth area of each pond type. At the physical design stage the results of the process design have to be 'translated' into the actual design that will be implemented. First, the number of series of anaerobic, facultative and maturation ponds must be determined. Only at small WSP installations will there only be one series. At installations serving more than a few thousand persons two or more series are needed to increase operational flexibility. Additionally, since the site is never quite flat, several separate ponds help minimise earthworks and thus construction costs. The number of series to be used in any one situation is left to the designer's judgement, but must take into account the site conditions.

The designer must place the divided areas in *series* and not in *parallel*: this means that if the designer decides to have N series, then each series comprises (1) an anaerobic pond with an area (A_a/N), where A_a is the required total anaerobic pond area calculated in the process design for the total flow; (2) a facultative pond with an area (A_f/N); and (3) n maturation ponds, the first with an area (A_{m1}/N) and the others each with an area of (A_m/N). Arrangements must be made to split the flow into N equal parts, so that each series is equally loaded. All this should be obvious, but mistakes do occur, most commonly when it is (for example, to minimise earthworks), an anaerobic pond followed by, say, two facultative ponds – the mistake occurs if these two facultative ponds are constructed in series, rather than in parallel. This error means that the load received by the first facultative pond is twice what it should be, and thus it will unquestionably fail and may cause odour nuisance.

Once the number of series has been decided and the correct layout of each series selected, the actual dimensions of each pond can be determined. The areas calculated in the process design are 'mid-depth' areas and the resulting mid-depth dimensions must be converted to dimensions at the pond base and at the embankment top. This conversion requires selection of three parameters: the pond length-to-breadth ratio, the embankment slope, and the required freeboard. The pond depth assumed in the process design must also be confirmed (or, if site constraints so require, altered and the process design redone).

The mid-depth dimensions are determined from the calculated pond mid-depth area and the chosen length-to-breadth ratio. For anaerobic ponds the length-to-breadth ratio is commonly 1–2 to 1 for anaerobic ponds, and 2–3 to 1 for primary facultative ponds. For secondary facultative and maturation ponds much higher values can be used (up to 10 to 1, depending on the site topography). Embankment slopes are most commonly 1 in 3 internally and 1 in 2 externally; steeper slopes can be used if the soil is suitable providing slope stability has been assessed by standard geotechnical methods.

Freeboard, the vertical height of the top of the embankment above top water level in the pond, is provided to prevent wind-induced waves overtopping the embankments and to allow for build-up during high flow periods. For small ponds (up to 0.5 ha) a minimum freeboard of 0.5 m is required, and for larger ponds at least 1 m should be provided. The area where the pond water surface meets the soil embankment requires some protection against wave action that would otherwise erode it away. Installing a concrete 'wave band' or rock 'rip-rap' protection around the whole perimeter of the pond typically provides this.

Once the mid-depth values of length, width and depth have been determined, the actual pond dimensions are calculated as follows:

$$\text{Base dimension} = (\text{mid-depth dimension}) - nD$$
$$\text{Embankment-top dimension} = (\text{mid-depth dimension}) + n(D + 2F)$$

where n = the embankment slope factor (ie, a slope of 1 in n);
D = pond liquid depth, m; and
F = freeboard, m.

A more complete discussion of WSP physical design (rather than just sizing) is given in Mara (1997).

9.7 POND EFFLUENT REUSE

9.7.1 Agricultural reuse

It is relatively straightforward to design WSP systems that will comply with the recommendations of WHO (1989) and Blumenthal *et al.* (2000) for the microbiological quality of treated wastewaters used for either restricted or unrestricted irrigation. Effluents from WSP systems treating mainly domestic wastewater also comply with the United Nations Food and Agriculture Organisation's recommendations for the physicochemical quality of irrigation waters (Ayers and Westcot, 1985), although effluents (including domestic wastewater effluents) should be routinely monitored for electrical conductivity, sodium absorption ratio, total nitrogen, boron and pH.

9.7.2 Aquacultural reuse

Mara *et al.* (1993) and Mara (1997) present a rational design for wastewater-fed fishponds based on the minimal wastewater treatment in anaerobic and facultative ponds for the maximal production of microbiologically safe fish. The procedure is: (a) design an anaerobic pond and a facultative pond as described in sections 9.3.1 and 9.4.1; (b) design the fishpond, which receives the facultative pond effluent, on the basis of a total nitrogen loading of 4 kg total N/ha day; and (c) check that the fishpond contains < 1000 faecal coliforms per 100 ml and <0.5 mg free ammonia-N/l. Steps (b) and (c) are described in the subsections below.

Total nitrogen

Reed's (1985) equation is used to estimate total nitrogen removal in the facultative pond (assuming no net removal occurs in the anaerobic pond):

$$C_e = C_i \exp\{-[0.0064(1.039)^{T-20}][\theta + 60.6(\mathrm{pH} - 6.6)]\} \qquad (9.21)$$

where:

C_e and C_i = effluent and influent concentrations of total nitrogen, respectively, mg/l. The pH is estimated from equation 9.20.

The fishpond area is determined from the following version of equation 9.4:

$$A_{fp} = 10C_eQ / \lambda_{s(TN)} \qquad (9.22)$$

where:

A_{fp} = area of the fishpond, m^2; C_e = total nitrogen concentration in the effluent from the facultative pond given by equation 9.21;

Q = effluent flow from the facultative pond, m^3/day; and $\lambda_{s(TN)}$ = total nitrogen loading rate, kg/ha day (= 4).

Faecal coliforms

The retention time in the fishpond (θ_{fp}, d) is calculated from equation 9.7, and the number of faecal coliforms in the fishpond (N_{fp}, per 100 ml) is calculated from the following version of equation 9.12:

$$N_{fp} = N_i / (1 + k_B\theta_a)(1 + k_B\theta_f)(1 + k_B\theta_{fp}) \qquad (9.23)$$

WHO (1989) recommends that N_{fp} should be less than 1000 per 100 ml. If it is >1000 per 100 ml, then either increase θ_{fp} or consider having a maturation pond between the facultative pond and the fishpond.

WHO (1989) also recommends that the influent to the fishpond should contain no viable human trematode eggs per litre (the human trematodes are *Schistosoma* spp., the human blood flukes; *Clonorchis sinensis*, the Oriental liver fluke; and *Fasciolopsis buski*, the giant intestinal fluke). These eggs are removed in ponds more quickly than the human intestinal nematode eggs (equation 9.11), and they rapidly become unviable within the few days retention in the anaerobic and facultative ponds.

Free ammonia

Equation 9.18 or 9.19 is used to determine the concentration of ammonia-N. This is firstly done for the facultative pond effluent (assuming that the conversion of total nitrogen produces an effluent ammonia concentration in the effluent of the anaerobic pond, the influent to the facultative pond, equal to 75% of the total nitrogen concentration in the raw wastewater). This is then repeated for the fishpond.

The ammonia concentration is the total concentration of NH_3 and NH_4^+, sometimes termed 'free and saline ammonia'. In order to protect the fish from free ammonia (NH_3) toxicity, the concentration of NH_3 should be <0.5 mg N/l. The percentage (p) of free ammonia in aqueous ammonia solutions depends on the absolute temperature ($\boldsymbol{T}$, K; K = °C + 273) and pH, as follows (Emerson *et al.* 1975):

$$p = 1/[10^{(\mathrm{pKa} - \mathrm{pH})} + 1] \qquad (9.24)$$

where $\mathrm{pK_a}$ is given by:

$$\mathrm{pK_a} = 0.09018 + (2729.92/\boldsymbol{T}) \qquad (9.25)$$

Equations 9.24 and 9.25 should be used to determine the free ammonia concentration in the fishpond, assuming a pH of 7.5 (the pH range in wastewater-fed fishponds is usually 6.5 – 7.5). If the NH_3 concentration in the fishpond is >0.5 mg N/l, consideration should be given to installing a maturation pond between the facultative pond and the fishpond.

9.8 DESIGN EXAMPLE

A WSP system is to be designed for a tourist resort in North Africa. Known design parameter values are given in Table 9.2. The treated wastewater is used for hotel garden and lawn watering and unrestricted crop irrigation during April – October.

9.8.1 Solution

Assume the raw wastewater contains 5×10^7 faecal coliform bacteria per 100 ml and 300 human intestinal nematode eggs per litre. The effluent must contain ≤200 faecal coliforms per 100 ml (the WHO, 1987 recommendation for hotel lawn and garden watering) and ≤1 egg per litre (the WHO recommendation for unrestricted irrigation). Thus a pond series comprising an anaerobic pond, a secondary facultative pond and, at this stage, an unknown number of maturation ponds, is required.

Table 9.2 Design parameter values for a WSP system for a tourist resort in North Africa

Month	Mean temperature *(°C)*	Net evaporation *(mm)*	Wastewater flow *(m^3/d)*	BOD concentration *(mg/l)*
January	11.2	– 60	10 000	300
February	12.5	– 84	10 000	300
March	14.8	50	10 000	300
April	18.1	78	14 000	320
May	21.4	109	16 000	330
June	23.6	134	21 000	345
July	27.1	155	26 000	380
August	25.5	150	26 000	380
September	22.1	126	21 000	345
October	18.0	83	14 000	320
November	14.2	44	10 000	300
December	11.4	12	10 000	300

Using Table 9.1 the permissible design volumetric BOD loading for the anaerobic pond and equation 9.3 for the permissible design surface BOD loadings for the secondary facultative pond are determined. For each month (and each temperature), the BOD removals from the anaerobic pond are estimated using the appropriate equation in Table 9.1 and the parameter values in Table 9.2. The anaerobic pond volume and the secondary facultative pond area are then calculated from equations 9.1 and 9.4, respectively. The calculated values are given in Table 9.3, which shows that in this case peak summer conditions control the anaerobic pond design, and winter conditions control the secondary facultative pond design.

Table 9.3 Calculated required anaerobic pond volumes and secondary facultative pond areas for each month of the year for the North Africa design example

Month	Anaerobic pond volume *(thousand m^3)*	Secondary facultative pond area *(ha)*
January	25.0	15.5
February	20.0	14.0
March	15.0	9.0
April	17.2	9.1
May	17.0	7.4
June	20.3	7.0
July	28.2	7.6
August	28.2	8.2
September	22.6	8.9
October	17.2	9.1
November	16.7	10.2
December	25.0	15.5

Specimen calculations for these controlling conditions are as follows:

(a) Anaerobic pond for summer:
For July, $T = 27$ °C and therefore from Table 9.1 $\lambda_v = 350$ g/m^3 d. Thus the anaerobic pond volume is given by:

$$
\begin{aligned}
V_a &= L_i Q/\lambda_v \\
&= 380 \times 26\,000/350 \\
&= 28\,200 \text{ m}^3
\end{aligned}
$$

(b) Secondary facultative pond for winter:
For January, T = 11°C and therefore from Table 9.1 the BOD removal in the anaerobic pond is 42 percent. The resultant anaerobic pond effluent (i.e., secondary facultative pond influent) BOD is 0.58 × 300, = 174 mg/l.

From equation 9.3, $\lambda_v = 112$ kg/ha day for 11°C and so the secondary facultative pond area is calculated by:

$$
\begin{aligned}
A_f &= 10 L_i Q/\lambda_s \\
&= 10 \times 174 \times 10\,000/112 \\
&= 155\,000 \text{ m}^2 \text{ (15.5 ha)}
\end{aligned}
$$

The next step is to calculate the retention times in these two ponds in April or October, the critical (coolest) months for effluent reuse:

$$\begin{aligned}\theta_a &= V_a/Q \\ &= 28\,200/14\,000 \\ &= 2 \text{ days}\end{aligned}$$

Taking the depth of the secondary facultative pond as 1.5 m:

$$\begin{aligned}\theta_f &= 2A_fD_f/(2Q_i - 0.001eA_f) \\ &= [2 \times 155\,000 \times 1.5/[(2 \times 14\,000) - (0.001 \times 78/30 \times 155\,000)] \\ &= 16.8 \text{ days}\end{aligned}$$

The retention time in the first maturation pond is determined from equation 9.15:

$$\theta_{ml} = 10L_iD_m/0.75\lambda_{s\ (fac)}$$

Taking $D_m = 1$ m and $\lambda_{s\ (fac)} = 217$ kg/ha day for 18 °C, and assuming that a 70% BOD removal is achieved in the first two ponds:

$$\begin{aligned}\theta_{m1} &= 10 \times (0.3 \times 320) \times 1/(0.75 \times 217) \\ &= 6 \text{ days}\end{aligned}$$

From equation 9.13, $k_B = 1.83\ d^{-1}$ for 18 °C. Using this value in equation 9.17 (with $N_e = 200$ per 100 ml):

$$\begin{aligned}\theta_m &= \{[N_i/N_e(1+k_B\theta_a)(1+k_B\theta_f)(1+k_B\theta_{m1})]^{-1/n} - 1\}/k_B \\ &= \{[5\times10^7/200[1+(1.83\times2)][1+(1.83\times16.8)][1+(1.83\times6)]]^{1/n} - 1\}/1.83\end{aligned}$$

Thus for $n = 2$, $\theta_m = 6$ days; and, for $n = 3$, $\theta_m = 2.3$ days, which is below the minimum value of 3 days. However, the combination $n = 3$ and $\theta_m = \theta_m^{min} = 3$ days is valid, and in this case better than the combination of $n = 2$ and $\theta_m = 6$ days, as the total retention time is 9 days, rather than 12 days. Therefore the design adopted is three maturation ponds (additional to the first maturation pond), each with a retention time of 3 days.

The area of the first maturation pond is given by:

$$A_{m1} = Q\theta_{m1}/D_m$$

Taking D_{m1} as 1 m:

$$
\begin{aligned}
A_m &= 14\,000 \times 6/1 \\
&= 84\,000 \text{ m}^2 \text{ (8.4 ha)}
\end{aligned}
$$

The area of each of the second–fourth maturation ponds is:

$$
\begin{aligned}
A_m &= 14\,000 \times 3/1 \\
&= 42\,000 \text{ m}^2 \text{ (4.2 ha)}
\end{aligned}
$$

Equation 9.11 could be used to calculate the number of eggs in the final effluent, but the count will be <1 egg per litre.

Thus the whole WSP system comprises:

- Anaerobic pond: 2.8 ha, 3 m deep;
- Facultative pond: 15.5 ha, 1.5 m deep;
- First maturation pond: 8.4 ha, 1 m deep;
- Second–fourth maturation ponds, each 4.3 ha, 1m deep.
- A total mid-depth area of around 40 ha.

9.9 CASE STUDY

This is a case study of a situation not uncommon in, for example, North Africa and the Middle East, where the winter design temperature is 10 °C and the irrigation season design temperature is 23 °C. As shown below, the system can be operated in different modes during the different seasons.

A WSP system is required to treat 10 000 m^3/day of a wastewater which has a BOD of 300 mg/l and 5×10^7 faecal coliforms per 100 ml. The effluent is to be used for unrestricted irrigation.

9.9.1 Solution

The required anaerobic pond volume and retention time are determined for winter conditions:

$$
\begin{aligned}
V_a &= L_i Q/\lambda_v \\
&= 300 \times 10\,000/100 \\
&= 30\,000 \text{ m}^3
\end{aligned}
$$

$$
\begin{aligned}
\theta_a &= V/Q \\
&= 30\,000/10\,000 \\
&= 3 \text{ days}
\end{aligned}
$$

The required facultative pond area and retention time (for a depth of 1.5 m) are also determined for winter conditions, taking the BOD removal in the anaerobic pond as 40 percent:

$$\begin{aligned} A_f &= 10\, L_i Q/\lambda_s \\ &= 10\,(0.6 \times 300)\, 10\,000/100 \\ &= 180\,000\ \text{m}^2 \end{aligned}$$

$$\begin{aligned} \theta_f &= A_f D_f/Q \\ &= 180\,000 \times 1.5/10\,000 \\ &= 27\ \text{days} \end{aligned}$$

There are two series of ponds operating most of the year, but during the irrigation season when higher temperatures improve the pond system's capacity to handle higher organic loadings, the effluents from both anaerobic ponds are switched to discharge into one of the facultative ponds, which in turn then discharges into the other facultative pond which now acts as a maturation pond. Calculate the faecal coliform numbers in the final effluent:

$$N_e = N_i/(1 + k_B\theta_a)(1 + k_B\theta_f/2)^2$$

with $k_B = 4.38\ \text{d}^{-1}$ for $T = 23$ °C:

$$\begin{aligned} N_e &= 5 \times 10^7/[1 + (4.38 \times 3)][1 + (4.38 \times 27/2)]^2 \\ &= 980 \text{ per } 100 \text{ ml.} \end{aligned}$$

The egg count will be <<1 per litre.

9.10 FUTURE DESIGN DIRECTIONS

Waste stabilisation ponds can be satisfactorily designed using the process design equations given in sections 9.3–9.5, but of course the resulting designs are only as good as the values of the design parameters used. A more sophisticated approach, recommended by von Sperling (1996) for facultative ponds, would be to adopt ranges of values for the design parameters, such as flow, BOD and faecal coliform numbers, and determine the pond volumes, areas and retention times required to achieve the required effluent quality (which might be specified on a 95-percentile basis) using a multi-trial Monte Carlo simulation program. Such an approach, combined with a physical design, which optimizes pond hydraulics (Chapter 10), is likely to produce a design of greater reliability in terms of effluent quality, and possibly one with greater cost-effectiveness. This approach certainly deserves more detailed investigation.

REFERENCES

Ayers, R.S. and. Westcot D.W (1985). Water Quality for Agriculture. Irrigation and Drainage Paper No. 29, Rev. 1. Rome, Italy: Food and Agriculture Organization of the United Nations.

Ayres, R.M., Alabaster G.P., Mara D.D. and Lee D.L. (1992). A design equation for human intestinal nematode egg removal in waste stabilization ponds. *Water Research* **26**(6), 863–865.

Blumenthal, U.J., Mara D.D., Peasey A., Ruiz-Palacios G. and Stott R. (2000). Guidelines for the microbiological quality of treated wastewater used in agriculture: recommendations for revising WHO guidelines. *Bulletin of the World Health Organization*, **78**(9), 1104–1116.

CEMAGREF, SATESE, Ecole National de la Santé Publique and Agences de l'Eau (1997). Le Lagunage Naturel: Les Leçons Tirées de 15 Ans de Pratique en France. Lyon, France: Centre National du Machinisme Agricole, du Génie Rural, des Eaux et des Forêts.

Council of the European Communities (1991). Council Directive of 21 May 1991 concerning urban waste water treatment (91/271/EEC). *Official Journal of the European Communities*, L135/40 (30 May).

Emerson, K., R.C. Russo, R.E. Lund and R.T. Thurston (1975). Aqueous ammonia equilibrium calculations: effect of pH and temperature. *Journal of the Fisheries Research Board of Canada* **32**(12), 2379–2383.

Gloyna, E.F. and Espino E. (1969). Sulfide production in waste stabilization ponds. *Journal of the Sanitary Engineering Division*, American Society of Civil Engineers, **95**(SA3), 607–628.

Mara, D.D. (1976). Sewage Treatment in Hot Climates. Chichester, England, UK: John Wiley and Sons.

Mara, D.D. (1997). Design Manual for Waste Stabilization Ponds in India. Leeds, UK: Lagoon Technology International (available at: http://www.leeds.ac.uk/civil/ceri/water/tphe/publicn/list.html).

Mara, D.D. and Pearson H.W. (1987). Waste Stabilization Ponds: Design Manual for Mediterranean Europe. Copenhagen, Denmark: World Health Organization Regional Office for Europe.

Mara, D.D., Edwards P., Clark D. and Mills S.W. (1993). A rational approach to the design of wastewater-fed fishponds. Water Research, 27 (12), 1797–1799.

Marais, G.v.R. (1974). Faecal bacterial kinetics in waste stabilization ponds. *Journal of the Environmental Engineering Division*, American Society of Civil Engineers, **100**(EE1), 119–139.

Moshe, M., Betzer N. and Kott Y. (1972). Effect of industrial wastes on oxidation pond performance. *Water Research* **6**, 1165–1171.

Oragui, J.I., Arridge H., Mara D.D., Pearson H.W. and Silva S.A. (1993). Vibrio cholerae O1 removal in waste stabilization ponds in northeast Brazil. *Water Research* **27**(4), 727–728.

Pano, A. and Middlebrooks E.J. (1982). Ammonia nitrogen removal in facultative wastewater stabilization ponds. *Journal of the Water Pollution Control Federation* **54**(4), 344–351.

Pfeffer, J.T. (1970). Anaerobic lagoons: theoretical considerations. In: *Proceedings of the Second International Symposium on Waste Treatment Lagoons* (ed. R.E. McKinney), pp. 310-320. Laurence, KS: University of Kansas.

von Sperling, M. (1996). Design of facultative ponds based on uncertainty analysis. *Water Science and Technology* **33**(7), 41–47.

WHO (1989). Health Guidelines for the Use of Wastewater in Agriculture and Aquaculture. Technical Report Series No. 778. Geneva, Switzerland: World Health Organization [see Note below].

Note:

WHO (1989) is currently being revised, and the new guidelines are expected to be published in 2005. It is likely that the following changes will be made (details will be given at www.who.int/water_sanitation_health):

(a) restricted irrigation: an additional guideline of ≤105 E. coli per 100 ml; and a reduction to ≤0.1 egg per litre when children under 15 are exposed;

(b) unrestricted irrigation: ≤0.1 egg per litre when children under 15 are exposed; and

(c) aquaculture: a relaxation of the number of E. coli in the fishpond to ≤104 per 100 ml.

10

Hydraulic design

Andy Shilton and David Sweeney

10.1 INTRODUCTION TO POND HYDRAULICS

Despite the large amount of research that has been undertaken into various aspects of waste stabilisation pond technology, the treatment efficiency of these systems is still often compromised by hydraulic problems. Before getting into more detail, it is important to firstly introduce and review some of the common jargon, and performance parameters used in this area.

10.1.1 The Theoretical Hydraulic Retention Time (HRT)

The theoretical HRT is simply the ratio of the pond volume and rate of inflow:

$$\theta_{Theo} = \frac{V}{Q}$$

ISBN: 1843390205. Published by IWA Publishing, London, UK.

where:

θ_{Theo} = theoretical hydraulic retention time, (d);
V = pond volume, (m^3);
Q = average (or design) flowrate, (m^3/d).

However, in reality ponds rarely operate at their theoretical HRT because:

- The flowrate is not steady and continuously varies from the value used to calculate the theoretical HRT;
- Sludge accumulation decreases the pond volume.

To determine the actual HRT of a pond a tracer study can be conducted, although this will only be representative of the flow conditions encountered during its duration. Tracer studies are discussed further in Section 10.1.6.

10.1.2 Ideal flow - plug-flow and completely mixed flow

The concept of 'plug flow' assumes that there is no mixing or diffusion as the wastewater moves through the pond. Alternatively, 'completely mixed flow' assumes the wastewater is instantaneously and completely mixed upon entering the pond. By assuming that the pond hydraulics conform to either of these types of flow behaviour, and assuming that the decay of pollutants follows first order kinetics (see Section 10.3.1), equations can be derived which enable calculation of the treatment efficiency achieved as a function of time. These are known as the ideal flow equations.

The plug flow equation:

$$\frac{C_e}{C_i} = e^{-\mathrm{kt}}$$

The completely mixed equation (sometimes called the continuous stirred tank reactor, or CSTR equation):

$$\frac{C_e}{C_i} = \frac{1}{1+\mathrm{kt}}$$

In both cases:

C_e = effluent pollutant concentration, (mg/l or counts/mL);
C_i = influent pollutant concentration, (mg/l or counts/mL);
k = first order reaction rate constant, (1/d);
t = time, (d).

In 1961, Marais and Shaw proposed the use of the completely mixed model for predicting the extent of faecal bacteria deactivation in waste stabilisation ponds. Marais later expanded this model in 1966 and 1970 to incorporate the effect of anaerobic conditions on the bacterial death rate, and again in 1974 to account for the influence of temperature on the first order rate constant (Marais, 1974). Marais suggested that in order to achieve maximum bacterial die-off, a series of ponds of equal size should be used. This being the case, the overall reduction can be described by the following equation:

$$\frac{C_e}{C_i} = \frac{1}{(1+\mathrm{kt})^n}$$

where:

n = the number of equal sized ponds in series.

This equation is often used when designing for pathogen removal in a pond series (see Chapters 8 and 9 for its application).

While the hydraulic behaviour of a pond may approach plug flow or completely mixed conditions under special circumstances, in practice its behaviour will always be somewhere between these extremes and this is termed as 'non-ideal' flow. Non-ideal flow is discussed further in Section 10.1.5, but firstly we need to review the main reasons that ponds are hydraulically inefficient – short-circuiting (Section 10.1.3) and dead space (Section 10.1.4).

10.1.3 Hydraulic short-circuiting

When wastewater enters a pond it does not simply move uniformly from the inlet to the outlet. It mixes with the rest of the pond water. Most will mix into the main body of the pond and then steadily discharge over a reasonable period. However, some water will enter and leave the pond in a very short period of time, often just a matter of a few hours in ponds that have theoretical hydraulic retention times measured in weeks. This is called hydraulic 'short-circuiting', because it has 'short-circuited' the treatment process.

Numerous examples of short-circuiting in ponds have been identified from tracer studies. This effect has been blamed on a number of possible causes including thermal stratification (MacDonald and Ernst, 1986; Uluatam and Kurum, 1992; Pedahzur *et al.*, 1993; Salter, 1999), channelling from inlet to outlet (Marecos do Monte and Mara, 1987) and wind effects (Fares and Lloyd, 1995; Fares *et al.*, 1996; Frederick and Lloyd, 1996; Wood, 1997; Salter, 1999; Vorkas and Lloyd, 2000). However, it has also become clear that the momentum from the inflow, especially if the inlet is horizontally aligned, will cause the influent to swirl around the pond. Should this influent circulate around past the outlet then short-circuiting will occur, resulting in the discharge of only partially treated wastewater (Shilton, 2001).

10.1.4 Dead space and flow velocities

The volume occupied by settled sludge is 'physical' dead space. It reduces the physical volume of the pond, thereby reducing the hydraulic retention time and treatment efficiency.

However, zones of low flow within ponds can also have a negative impact on the treatment efficiency. For example, back-eddies in corners are slow to mix and interchange fluid with the main flow of the pond, and are often referred to as 'dead zones'. The small fraction of the wastewater that lingers in these dead zones has a long retention time and consequently receives a high degree of treatment. Meanwhile, most of the wastewater passes through the remaining 'active volume' of the pond, which provides the majority of the treatment to the bulk of the flow.

To explain this concept, let's consider a theoretical example of an irregularly shaped pond with part of the volume (say 10%) connected to the main pond by a narrow 'neck'. Let's assume that the water that enters this isolated volume remains there for a year before finally mixing back into the main body of the pond. This isolated volume is obviously a hydraulic 'dead zone'. There is so little flow in and out of this zone that it provides negligible treatment to the main wastewater flow. In essence the effective pond volume has been reduced by 10% and it is the other 90% of 'active volume' that is providing the treatment to practically all the wastewater.

If we did a tracer study on the above pond example for an infinite period of time, and if there was no sludge accumulation, no decay of tracer and a constant flow, we could show that the entire tracer quantity that went in does eventually come out. We would then determine from the tracer data that the mean HRT equalled the theoretical HRT. However, in practice tracer studies have a limited duration and a common 'rule of thumb' has been that tracer studies should be monitored for three theoretical HRTs. For example, if the theoretical pond retention time is 20 days, then this means that tracer data should be collected for 60 days. While the concentration of tracer still emerging over the final period seems small, the fact that this 'tail' does extend over so such a long time means that this quantity of tracer is not insignificant. Analysis of the tracer data to determine the mean HRT assumes that all of the tracer has emerged. If the tracer study is completed too soon then this leads to a somewhat reduced value of the mean HRT derived from the data. Recent modelling by one of the authors (unpublished) has actually indicated that a minimum of five HRTs of tracer monitoring is needed to accurately determine the mean HRT from tracer data.

Because this period of monitoring is not generally undertaken then it seems that at least some of the 'dead space' determined in tracer studies is more a result of experimental technique rather than sludge accumulation.

10.1.5 Non-ideal flow and the dispersion number

An alternative to the ideal flow models discussed in Section 10.1.2 is the Wehner-Wilhelm model for dispersed (non-ideal) flow. The hydraulic nature of the flow is characterised using the dispersion number, *d*, which defines the extent of longitudinal mixing of the flow as it passes through the pond (Thirumurthi, 1969). The resulting treatment in the pond is given by:

$$\frac{C_e}{C_i} = 4a \frac{e^{\frac{1}{2d}}}{(1+a)^2 e^{\frac{a}{2d}} - (1-a)^2 e^{\frac{-a}{2d}}}$$

in which:

$$a = \sqrt{1 + 4\mathrm{kt}d}$$

where:

C_e, C_i = effluent and influent pollutant concentration, (mg/l or counts/mL);
d = dimensionless dispersion number;
k = first order reaction rate constant, (1/d);
t = retention time, (d).

For non-ideal flow regimes, the extent of dispersion as denoted by the dispersion number is greater than zero (ideal plug flow), but less than infinity (ideal completely mixed flow). The dispersion number is calculated from the results of a tracer study and is therefore a function of all the physical influences that affected flow within the pond during the study period, for example, wind, variable flowrates and so on. However, in order to be a useful tool for pond design, the dispersion number must be predicted in advance. A number of researchers have presented design equations to predict the dispersion number as a function of pond geometry and flow, including Ferrara and Harleman (1981), Arceivala (1983), Polprasert and Bhattarai (1985), Agunwumba *et al.* (1992) and von Sperling (2003).

The accuracy of these predictions has been assessed by a number of authors including Marecos do Monte (1985) and Dorego and Leduc (1996). Ferrara and Harleman (1981) noted that the dispersion number is difficult to predict and, even when obtained by tracer study, is only a representative of the environmental conditions that prevailed for the period over which the tracer study was conducted. In a comparison of tracer study results with the Polprasert and Bhattarai equation, Marecos do Monte found there was little resemblance between the measured and predicted value of *d*, leading her to conclude that this predictive equation cannot be considered to be valid for all ponds. Considerable uncertainty surrounds any prediction of dispersion numbers and, as a consequence, the use of the dispersed flow model for design is not widespread.

In addition to the dispersion number, the reaction rate parameter, k, in the Wehner-Wilhelm model is also subject to a degree of uncertainty. Using a technique called sensitivity analysis, Von Sperling (2003) demonstrated that, despite the limitations of the dispersion number, its impact on the accuracy of the final model prediction was not significant compared to the uncertainty caused by variation in the factors that influenced the reaction rate constant, k.

A further limitation of the Wehner-Wilhelm model is that the dispersion number is unable to provide specific information about where zones of short-circuiting and mixing are occurring in the pond. Rather than attempting to describe the extent of flow dispersion using a single parameter, researchers now focus less on this approach, and more on the application of computer modelling which can predict the actual flow patterns within a pond and quantify its resultant hydraulic and treatment efficiencies – see Section 10.1.7 below.

10.1.6 Tracer studies

A tracer is a substance that is added to the pond inflow to track the movement of flow/pollutants through the pond. While chemical tracers such as rhodamine WT are most commonly used, microbial and, occasionally, radioactive tracers have also been used (e.g. Frederick and Lloyd, 1996; Olsen and Tjomsland, 1998). Tracer studies are applied in many different applications throughout the fields of water and chemical engineering, and Levenspiel (1972) presents a comprehensive review of this technique.

The large majority of tracer studies undertaken on waste stabilisation ponds have used what is known as the stimulus-response tracer technique. This involves adding a 'pulse' or 'slug' of tracer to the pond inlet and then monitoring the concentration of the tracer at the outlet over time. Previous studies include Mangelson and Watters (1972); Racault *et al.* (1984); Chapple (1985); MacDonald and Ernst (1986); Marecos do Monte and Mara (1987); Moreno (1990); Uluatam and Kurum (1992); Pedahzur *et al.* (1993); Frederick and Lloyd (1996); Wood (1997); Salter (1999); Brissaud *et al.* (2000); Shilton *et al.* (2000); and Vorkas and Lloyd (2000) amongst others. The response of tracer concentration at the outlet, when plotted relative to time, is known as the hydraulic retention time distribution curve. The tracer concentration must be monitored at the outlet until negligible amounts of tracer remain in the system, as discussed earlier this can take up to five theoretical HRTs. It is common practice to adjust the tracer response data so that the total area under the curve is equal to one (unity). In this way, response curves can be directly compared regardless of the quantity of tracer used, and/or the units in which they were measured.

Using the data obtained from a tracer study, various hydraulic parameters can be determined, including the:

- Mean retention time;
- Dispersion number;
- Time to start of short-circuiting;
- Time for 10% and 90% (the t10 and t90 fractions) of tracer discharge, etc.

It is also possible to integrate a first order model of removal kinetics with the tracer data to quantify the ponds' performance in terms of treatment efficiency. Levenspiel (1972) generally explains this approach while Shilton *et al.* (2000) gives an example of its application to waste stabilisation ponds.

10.1.7 Computational fluid dynamics

Computational fluid dynamics (CFD) is a computer-based method of solving the equations that govern the conservation of mass, momentum, and energy throughout a body of fluid. This technique can be used to model the flow of fluid, gas and solid phases with a high degree of accuracy in one, two, or three dimensions. Because the equations of state covering fluid motion are applied on a local basis, it is necessary to divide the region being modelled into a grid containing a large number of cells. By assigning boundary conditions to the grid, and using an iterative solution technique, mass flow can be approximated at each point throughout the computational grid.

This technique is increasingly being used to model a range of water and wastewater treatment processes including single phase applications (e.g. membrane filters and UV reactors) and multiple phase flows, such as ozone contactors, and clarification (Levecq *et al.*, 2001) and has in recent years become increasingly applied to waste stabilisation ponds. Apart from quantifying the flow velocities and pattern throughout the pond, CFD can also simulate tracer studies. Furthermore, it is also possible to integrate and solve reaction kinetics within the CFD model allowing changes in hydraulic design to be evaluated in terms of impact on treatment efficiency (Shilton and Harrison, 2002).

Clearly, CFD provides a very valuable tool for rapidly advancing this research area. Indeed, a large amount of the recommendations in the rest of this chapter are drawn from the "Guidelines for the Hydraulic Design of Waste Stabilisation Ponds" (Shilton and Harrison, 2003), which made wide use of this technique.

10.2 INPUTS AND INFLUENCES ON HYDRAULICS

In order to fully understand pond hydraulics, inputs and influences on the flow momentum within a pond must be considered. These include:

- Flowrate—higher flowrates increase inlet momentum;
- Inlet size—smaller inlets increase the inlet velocity and hence also the inlet momentum;
- Inlet position and orientation—these determine the way in which the inlet momentum is introduced into the main body of the pond, and the resulting influence on the main flow pattern;
- Outlet position—sets the distance from the inlet and, therefore, the time for short-circuiting flow to reach the outlet;
- Pond geometry and baffles—these have a strong influence on flow patterns and define the degree of 'channelling';
- Temperature/density effects—these may influence the channelling and circulation of the main flow;
- Wind shear—higher wind velocities and greater pond surface areas increase the momentum imputed, and as a result influence the main flow pattern;
- Mechanical aerators—if present, these constitute a significant momentum input, and as a result can have a strong influence on the main flow pattern.

This chapter provides guidance on improving pond design through more complete consideration of these effects, however, it must be recognised that the current understanding of some of these inputs and influences is still very limited.

10.3 RELATING HYDRAULICS TO TREATMENT

While this chapter focuses on improving pond hydraulics, the ultimate objective is improved treatment efficiency. It is, therefore, important that the relationship between pond hydraulics and the kinetics of treatment is understood. It is also essential that we don't ignore the realities of solids and organic loading limitations.

10.3.1 The treatment relationship

It is typical to assume that the decay of water quality indicators such as BOD and coliforms in a waste stabilisation pond can be predicted using first order reaction kinetics.

This relationship is:

$$\begin{bmatrix} \text{Rate} \\ \text{of} \\ \text{treatment} \end{bmatrix} = \begin{bmatrix} \text{Reaction} \\ \text{rate} \\ \text{constant} \end{bmatrix} \times \begin{bmatrix} \text{Concentration} \\ \text{of pollutant} \\ \text{remaining} \end{bmatrix}$$

Because the concentration of pollutant in the pond decreases with time, there is a non-linear relationship between the rate of treatment and time. Simply put, this means that when the wastewater is initially highly polluted, the rate of treatment is high, but once the wastewater is stabilised to low pollutant concentrations then the rate of treatment is much lower, and little further treatment is achieved. This relationship explains why it is important to ensure that short-circuiting is prevented in a pond. If only a small fraction of the total flow is short-circuiting without adequate treatment, then this still contributes a disproportionately large amount of the pollutant remaining in the effluent. This is particularly the case when we are considering a water quality parameter such as coliforms in a strict regulatory environment, where a reduction of several orders of magnitude is normally required (i.e. more than 99.9%).

It is important to clarify that the first order reaction model does not 'explain' pathogen disinfection behaviour in ponds. It is merely an empirical model that can be used to represent the net resultant effect of a very complex set of mechanisms dependent on factors such as dissolved oxygen, pH and light conditions (Curtis *et al.*, 1992; Davies-Colley *et al.*, 1999). However, in the absence of a more widely validated reaction model, this assumption of first order kinetics is still widely used in pond design equations.

10.3.2 Integrating hydraulic and treatment efficiency

Research into pond hydraulics has tended to use hydraulic parameters such as 'mean retention time', 'dispersion number', 'dead space' and so on. However, as discussed previously, the rate of treatment in a pond is non-linear. So how do such terms relate to actual pond treatment efficiency? The simple answer is that by using hydraulic parameters alone we cannot be sure! For example, it is possible to have two ponds with the same 'mean hydraulic retention time', but with different treatment efficiencies if one has a greater degree of short-circuiting.

As discussed in Chapter 8, there are a number of ways of sizing ponds. Regardless of which approach is used, they all have a common weakness: they take no account of the physical configuration. For example, is the inlet pipe directing the flow straight towards the outlet? Does adding a couple of baffles into a pond improve its treatment efficiency? (If so, by how much?)

As mentioned in Section 10.1.6, it is possible to integrate reaction kinetics with tracer data to yield the result of a tracer study in terms of the efficiency of a particular treatment. However, tracer studies are very time consuming and subject to changing conditions in the field such as wind and variable flowrates. As discussed in Section 10.1.7 there now exists the capability to undertake computer simulations of ponds hydraulics with integrated reaction kinetics.

10.3.3 Why not just design for plug flow?

Assuming first order kinetics, from a treatment perspective the most effective hydraulic design will always be plug flow. While perfect plug flow conditions cannot be obtained in actual pond systems, there are certainly ways of making the flow through a pond system more plug in nature. These include:

- Designing a number of smaller ponds in series rather than just one large pond;
- Constructing long, narrow ponds or ponds fitted with many baffles (thus creating a high length-to-width ratio);
- Using inlets that dissipate inflow momentum to reduce jetting and short-circuiting.

Practical considerations may, however, not always make 'plug flow' designs the best option. For example, long narrow ponds or multiple smaller ponds may be more expensive to construct. However, the most important consideration is with regard to loading. The first pond in the series will be subjected to a much higher organic loading rate than the subsequent ponds. This same concern also applies at the front end of a long, narrow or baffled pond. Elevated loading rates change the whole nature of the chemical and biological environment in the pond and, in extreme cases, may lead to organic overloading.

This restricts application of the 'plug flow design approach' to maturation ponds where the organic loading has already been substantially reduced, and indeed the use of a number of ponds in series is a common practice for maturation pond design (see Chapter 9).

Similar caution is needed when using inlets that act to dissipate the inlet momentum (e.g. vertically orientated inlets). For example, in a pond receiving wastewater containing a significant organic and/or solids loading, the use of a vertical inlet will slow the velocity of the fluid in the inlet region. This could create problems of sludge build-up around the inlet and again create the potential for localised organic overloading. Additionally, if the horizontal momentum is minimised, wind effects alone may dominate the flow pattern and this may also lead to poor hydraulic efficiency.

Ideally, the best general behaviour for a pond, especially if receiving raw wastewater, is the aim for a design in which the influent is rapidly mixed into the main body of the pond. This distributes the solids and organics load more evenly. But at the same time the design must also prevent jetting of the influent rapidly around past the outlet creating short-circuiting problems. Techniques by which these objectives can be achieved, by appropriate inlet/outlet design and the use of baffles, are discussed in the following sections.

10.4 INLET DESIGN

Recent research suggests that the inlet position and its relation to the outlet are more important than previously thought. Pearson *et al.* (1995) concluded "...the positioning and depths of the inlet and outlets may have a greater beneficial impact on treatment efficiency than pond shape" (page 137). Wood (1997), Persson (2000), and Shilton (2001) all noted that the position and design of the inlet does indeed have a significant impact on the hydraulic efficiency of a pond. However, previous design manuals have provided little practical guidance regarding the design and positioning of inlets.

There has been uncertainty in the literature regarding what flow patterns prevail within waste stabilisation ponds. A number of researchers have assumed that fluid moves reasonably directly from the inlet towards the outlet. However, it has been found that horizontally oriented inlets can drive the pond contents to circulate in large cells, with much higher velocities than if the flow was simply moving from the inlet to the outlet in a plug flow manner.

It is useful to think of the inlet as a small drive on a large flywheel where, in the case of the pond, the flywheel is the bulk volume. Although the jetting effect from an inlet pipe is quite localised, it provides a consistent source of momentum inputted in a fixed direction at a fixed point. This momentum is transferred into the bulk volume and thereby drives the main circulation (Shilton 2001). The flow conditions that result from this driving force depend to a large extent on the magnitude, location and orientation of this momentum source.

The results of laboratory experiments, computer modelling and fieldwork have all repeatedly highlighted this jetting effect that a horizontal inlet creates in a pond. The picture of a tracer study in Figure 10.1 shows the jetting effect for wastewater flowing from a primary pond into a secondary pond via a pipe in the embankment.

Figure 10.1 'Jetting' effect of the inlet as seen in a tracer study on an operational pond

10.4.1 Use of large horizontal inlets

The use of large pipe diameters and a large inlet channel to reduce the jetting effect associated with horizontal inlet pipes have been tested (Shilton, 2001; Shilton and Harrison, 2003). While larger inlets did decrease the velocity of the main flow circulation, the bulk pattern of flow in the pond was just the same. Short-circuiting was delayed but the net effect, in terms of improving treatment efficiency, was not particularly significant.

10.4.2 The jet attachment technique

Rather than attempting to reduce jetting created by a horizontal inlet pipe, in some cases this effect could be utilised to improve treatment in the pond. Instead of directing a horizontal pipe straight out into the main body of the pond, the inlet flow can be oriented in along a sidewall, which causes the jet from the inlet to 'cling' to the side. This is known as 'jet attachment'.

Previously it was noted that a common problem with pond hydraulics is that the influent can circulate too rapidly from the inlet to an outlet around the edge of a pond. So why would we want to use an inlet that encourages this effect? In some circumstances, jet attachment can be used to control the flow pattern in the pond. For example, if an outlet is located in the centre of a pond the influent could be encouraged to swirl around the edge before slowly spiralling into the centre.

10.4.3 Vertical inlet

If a horizontal inlet causes short-circuiting problems, then a relatively inexpensive method of avoiding this would appear to be to modify the inlet to discharge vertically. Hydraulic testing on ponds with vertical inlets has, however, produced variable results, with the method working well in some cases but not in others (Shilton, 2001; Shilton and Harrison, 2003). It seems fair to assume that, in the case of a vertical inlet, the tracer will be discharged and then slowly spread out evenly across the pond. However, in some cases the tracer appeared to move out in two plumes along either adjacent wall. This led to the testing of a vertical inlet with short (stub) baffles placed on either adjacent wall to block the circulation along these walls. This approach was tested in both the laboratory and in computer modelling of the full-scale pond. Both cases gave excellent results. The tracer was rapidly mixed within the baffled inlet area and then moved uniformly out into the main body of the pond through the wide gap between the two baffles.

10.4.4 Diffuse (manifold) inlet

Early pond design guidelines recommend the use of multiple inlets, or an inlet manifold to distribute the inlet flow evenly across the pond (Hermann and Gloyna, 1958; Canter and England, 1970). These recommendations were later supported by results from studies of scale model systems (Mangleson *et al.*, 1973) and computer models of full-scale ponds (Fares *et al.*, 1996; Perrson, 2000).

Shilton and Harrison (2003) conducted laboratory testing using an inlet pipe running the width of a pond, which contained eight equally spaced small diffuser holes facing downward towards the base of the pond. This configuration created an even distribution of the tracer that then spread down the length of the pond. The movement of the tracer down the pond was relatively uniform, and yielded an improvement in the theoretical treatment capacity of the pond, compared to the single inlet case. However, this improvement was not as significant as might have been expected.

While a manifold inlet can be used to provide improved treatment, installing and maintaining this sort of inlet on a large full-scale pond may not always be practical or cost effective compared to other options.

10.4.5 Inflow dropping from a horizontal pipe

Many ponds are currently fitted with a horizontal inlet pipe that discharges a short distance above the pond surface. So is this working like a horizontal inlet or a vertical inlet?

As the water plunges down into the pond it will certainly pick up a significant vertical velocity. However, the horizontal component of momentum remains after discharge. Tracer testing (Shilton and Harrison, 2003) confirmed that for this type of inlet, the influent swirls around to the outlet at a rate very similar to that for an equivalent submerged horizontal inlet.

10.4.6 Inlet type – practical considerations and recommendations

While a range of alternative inlet designs exists, most of these are simply methods for avoiding or minimising the jetting effect that results from the use of a horizontal inlet pipe.

In Section 10.3 we highlighted the need to consider organic and solids loading. While from a purely hydraulic viewpoint it may be useful to dissipate the inlet momentum, in so doing we lose the useful effect that this momentum has in rapidly distributing the organic and solids loading out into the main body of the pond. This is not, however, an issue for maturation ponds where organic and solids loads have already been significantly reduced by prior treatment.

A second practical consideration is wind. If the inlet design acts to dissipate the driving force of the inlet then it is more likely that, on a windy day, the flow pattern in a pond will be driven by the wind. In certain cases this may drive the influent rapidly towards the outlet, leading to short-circuiting problems and poor hydraulic efficiency. Each scenario will have its own considerations for the design engineer to take into account, but as a general guide the following recommendations are given:

- For ponds receiving wastewater, which has significant organic and/or solids loadings, overloading in the vicinity of the inlet can be avoided by using a horizontal inlet pipe, ensuring good mixing and distribution of the influent out into the pond. However, attention must be given to avoid the inflow swirling quickly around past the outlet creating short-circuiting. Careful consideration of the location of the outlet and the placement of baffles could be used to achieve this (see Sections 10.5 and 10.7).
- For ponds with baffles receiving pre-treated wastewater with low organic and/or solids loadings, a horizontal or a vertical inlet may be used. If a vertical inlet is used, a minimum of two adjacent stub baffles is strongly recommended.

10.4.7 Inlet position

Since the inlet is often an important driving force on the main pond circulation, engineers need to assess the broad flow pattern that will result from inlet positioning as part of the design process. The ideal approach is to model this on a computer but, at the time of writing, this is still a relatively specialist application that many practitioners do not have access to, and is expensive to commission.

An alternative approach is to develop an intuitive estimation of the flow pattern that will be set up, using a plan diagram of the pond design. Consider a horizontal jet to be a source of momentum that will then drive the larger bulk circulation around the pond just like a small mixer would.

It has been observed that ponds which have a length to width ratio of roughly 1:1 to 2:1 tend to circulate in a single large cell, typically with small counter-current circulations tucked in the corners (back-eddies). Note that the jet attachment technique can also be used to improve the predictability of the flow path.

10.4.8 Effect of varying flowrate

The flowrate entering a pond varies both through a diurnal cycle, and with more extremity during periods of wet weather. Laboratory testing was undertaken on a range of different pond geometries at different flowrates, and it was found the resulting flow patterns were largely independent of the inlet flowrate in each case. This is good news for the designer, as it would be difficult to optimise pond hydraulics if the prevailing flow pattern changed at different flowrates. The only time that this finding may not hold is when wind effects are able to dominate, which is most likely when inlet flow (and so the momentum) is reduced. This scenario is discussed in Section 10.6.

10.5 OUTLET DESIGN

10.5.1 Outlet depth

The Design Manual of Mara & Pearson (1998) recommends the following depths for outlets:

- Anaerobic Ponds 300mm;
- Facultative Ponds 600mm.

In anaerobic ponds, the outlet should be deep enough to be clear of any surface crust. In facultative ponds the depth is selected so as to discharge from below the maximum depth of the surface algal band. If an outlet weir is to be used, as opposed to a simple outlet pipe, then this should incorporate a scum guard that extends to the indicated depth (Mara & Pearson 1998), however, in maturation ponds where "algal bands are irrelevant" (Mara & Pearson, 1998, p. 62), the outlet should be located close to the surface to provide the best microbial quality.

10.5.2 Outlet position – influence on efficiency

There is no doubt that the positioning of the outlet is critical in terms of hydraulic efficiency. If the outlet is located in such a position that flow from the inlet circulates directly around past the outlet, then short-circuiting will occur and treatment efficiency will be compromised. It is, therefore, important to ensure that the outlet is kept out of the main flowpath of the incoming wastewater.

The engineer needs to assess what the likely flow pattern will be, and then select an outlet position in a 'sheltered' spot. Some practical suggestions for achieving this are given at the end of this section, but first we need to consider what effect moving the outlet position might have on the overall flow pattern.

10.5.3 Outlet position – influence on flow pattern

Does the positioning of the outlet affect the main circulation pattern? Observations made during laboratory and modelling work have indicated that, for ponds with inlet driven recirculation patterns, the outlet has only a localised influence, and its relocation does not alter the bulk flow pattern that exists in a pond. This is a useful observation as it means that after the flow pattern has been optimised by design of the inlet in conjunction with the shape/baffles, then the outlet can be relocated for maximum efficiency, without the likelihood that it will significantly alter the flow pattern.

10.5.4 Outlet manifolds

Some designers have advocated the use of outlet manifolds. These can consist of a weir running down the width of the pond at the opposite end from the inlet. However, as our understanding of flow behaviour has improved, we now realise that wastewater in a pond doesn't simply move slowly from one end to the other. Hence, although such a structure will incur additional expense to construct, it is unlikely to have a significant positive impact on the hydraulic efficiency.

Because an outlet manifold collects flow from a number of locations, using a manifold may actually compromise pond performance. This is because, compared to a single outlet, it is more difficult to protect a long outlet manifold from an inflow jet that short-circuits around the pond.

10.5.5 Outlet position - design suggestions

It has generally been considered that the best position for an outlet is at the opposite end of the pond to the inlet. However, we now realise that wastewater can swirl rapidly around from the inlet past the outlet. Picking the best spot for the outlet in any pond is still going to require some reasonable degree of judgment from the designer, however the following rules should help:

Hydraulic dead spots

The outlet should be located close to a corner or, if the pond is irregularly shaped, into a zone that is obviously out of the main flow path.

Use of baffles to control flow

Baffles can be used to improve hydraulic efficiency throughout a pond, including near the outlet. Outlet baffles can be used to reduce short-circuiting by effectively 'shielding' the outlet. The use of baffles is discussed further in Section 10.7.

Central outlets

By positioning the outlet in the middle of the pond and using the inlet to promote a swirling action around the outer edge, a flow pattern is established in which the flow slowly spirals into the centre. This idea has been modelled on computer without wind effects, with excellent results. However, the possibility that the wind could drive the flow over into the central zone needs to be carefully considered. Although this sort of a design appears to offer potential, it has not been tested in the field, and a full-scale research study of its performance is needed before it can be generally recommended.

Use of flow deflectors

Further protection of a central outlet can be achieved by placing small walls or sheets of baffle material around it to deflect flow away from this area. In this way a separate 'sub-pond' is established inside the pond.

Distance between inlet and outlet

In Section 10.6.3 we discuss the relative significance of the inlet versus wind. However, for now it should be realised that on a windy day the wind will certainly have some effect and therefore a sensible separation distance is required between the inlet and outlet. This same point applies when using a central outlet. For example, on a rectangular pond, if the outlet is to be placed in the centre zone it might be better kept towards the opposite end of this zone, away from the inlet.

10.6 WIND

10.6.1 Just how important are wind effects?

It has generally been believed that ponds operate more effectively at higher wind speeds due to a perceived improvement in aeration and mixing. There are two main mechanisms of oxygenation in pond systems: mass diffusion from the atmosphere and oxygen production by algae within the pond. However, of these, the oxygen produced by algal photosynthesis is the most significant, with oxygen transfer due to surface aeration generally limited to windy periods during the hours of darkness (Davies and Cornell, 1991).

With regard to mixing, a number of researchers have recently found evidence that wind may create flow patterns in ponds that encourage short-circuiting problems (Mara and Pearson, 1998; Lloyd *et al.*, 2003). Because of this there is a growing belief that wind may have more of a negative influence on pond performance than a positive one. However, due to the difficulties associated with quantifying this effect on flow conditions in ponds, very little experimental work has been undertaken to validate these views.

10.6.2 Wind induced circulation

Using computer modelling it has been shown that, in ponds with a large surface area at high wind speeds, wind fetch will create a three-dimensional circulation pattern, with short-circuiting across flow across the surface of the pond, and a return flow across the pond floor (Fares, 1998; Sweeney *et al.*, 2003). To date, attempts to validate these models using experimental drogue tracking have proven inconclusive.

By contrast, experimental drogue tracking work presented by Shilton (2001) determined that, in a smaller pond with a relatively higher inlet power input, the flow pattern was predominantly two-dimensional, circulating in the horizontal rather than the vertical plane. Shilton (2001) also conducted mathematical modelling of the wind effect on this pond, which provided evidence that a wind induced reverse bottom current was only present near the very bottom of the pond. At the depths of 0.5 metres and 1.0 metre, which corresponded to the depths of the experimental drogues, the model confirmed the flow to be circulating two-dimensionally in the horizontal plane.

Predicting the circulation patterns in a pond subjected to wind shear is a complex problem, but nevertheless it is important that the engineer has at least a basic understanding of these behaviours. Intuitively, as the wind dominance increases, the circulation in the vertical plane also increases, but when a horizontal inlet dominates then the flow predominantly swirls around in the horizontal plane.

10.6.3 Wind versus inlet mixing

If engineers are to improve pond hydraulic design, then it is important that the power inputs that drive the flow circulation are understood.

By using a power analysis approach (detailed in the following section), Shilton (2001) evaluated two ponds, sized using a modern design manual, and found the power input via the inlet to be more dominant than the power input due to the wind, except at high wind speed or if a large inlet was used. In another evaluation Shilton (2001) undertook a tracer study on a waste stabilisation pond and then used a computer model to simulate the pond hydraulics both with and without wind effects. When the wind shear was incorporated into the computer simulation this improved the agreement of the simulated tracer results and the experimental tracer response, but not by a substantial amount. In these cases it would appear that the power input from the inlet jet was much more significant than the power input from wind and that the pond hydraulics were essentially 'inlet driven'. This is, however, not always going to be the case.

There are several reasons why the inlet power may not always be so dominant in all pond systems and in such cases the wind power may be significantly larger than the inlet power and the pond will be 'wind driven':

1. Overly large inlets are often used, which means that the inlet velocity (and its power input) is significantly reduced. This is often the case for small ponds where the inlet pipe is kept large to avoid blockage.
2. The use of an inlet manifold will distribute the inflow energy more evenly across the pond, lessening the concentrated jetting effect of a single inlet.
3. A significant number of ponds in current use are oversized with larger surface areas than modern designs. This increases the relative influence of the wind.

While wind speed and direction can be highly variable at a site, the energy input at the inlet is relatively consistent over time, and always as a concentrated point source in a fixed position and direction. While wind effects must be considered in some circumstances, it seems that engineers have tended to underestimate the influence that the inlet has (or could have) on the pond hydraulics. In the following section a method is outlined for estimating the relative significance of wind versus inlet power.

10.6.4 Approximate analysis of mixing power input

In order to provide a broad comparison of the power inputs into a pond from an inflow pipe and from wind shear across the pond surface, theoretical approximations can be used for both the inlet and wind energy inputs (Larsen, 1999; Shilton and Harrison, 2003). It should be noted, however, that these estimates are approximate, as they do not account for mechanisms such as the internal transfer and dissipation of energy within the pond.

The power input from an inlet (P_{inlet}) can be estimated from:

$$P_{inlet} = 0.5\,\rho_w\,v^3\,A$$

where:

- ρ_w = density of water, (kg/m^3);
- v = velocity of water, (m/s);
- A = cross-sectional area of inlet, (m^2).

If this inflow enters via a circular pipe with a given flowrate Q (m^3/s), and assuming a value of 1000kg/m^3 for water density, the relationship between the power input and the pipe diameter ϕ (m) is given by:

$$P_{inlet} = \frac{811\,Q^3}{\phi^4}$$

The input of wind power (P_{wind}) can be estimated using:

$$P_{wind} = u_s\,\tau_{wind}\,A_{pond}$$

where:

- u_s = surface water velocity, (m/s);
- τ_{wind} = wind shear stress on the water surface, (kg/m.s^2);
- A_{pond} = surface area over which wind shear is exerted, (m^2).

The wind shear stress can be estimated from:

$$\tau_{wind} = k\,\rho_a\,v_{wind}^2$$

where:

- k = empirical constant;
- ρ_a = density of air, (kg/m^3);
- v_{wind} = velocity of wind, (m/s).

Larsen (1999) noted that the entrained surface velocity (u_s) of a water body is approximately equal to 3% of the wind velocity (v_w), a finding that is supported by a range of other studies Wood (1997). By substituting in this relationship and the general empirical equation for wind induced shear stress, τ_w, the equation for wind power becomes:

$$P_{wind} = 0.03\, k\, \rho_a\, v_{wind}^{\ 3}\, A_{pond}$$

For a pond of given area this equation can be used to provide an estimation of the power input for a range of wind velocities.

Selection of the empirical constant, k, is important and depends on the height at which the wind velocity is measured. From studies on a model yacht pond 60m wide, 240m long and 2m deep, Van Dorn (1953) cites values for this empirical constant that are dependant on the height at which the wind speed is measured. These range from 0.0037 for a measurement height of 0.25 metres to 0.0011 for a measurement height of 10 metres.

10.6.5 Controlling the effect of wind on pond hydraulics

Whilst the occurrence of wind at a pond cannot be practically controlled, the inlet pipe is a physical structure that can be easily manipulated. By selecting an inlet pipe diameter that ensures that inlet effects dominate the power input into a pond, it may be possible to force the pond flow into a predetermined pattern, rather than allowing it to fluctuate with changes in wind speed and direction.

The approximate power input analysis technique outlined in 10.6.4 offers engineers a practical method of controlling the flow pattern so as to optimise the hydraulic efficiency of a pond. These correlations are approximate and as a design tool this approach is novel, but in the absence of any other approach it provides the design engineer with a tool for controlling pond hydraulic behaviour.

10.7 BAFFLES AND SHAPE

Baffles are commonly used as a means of improving the hydraulic and treatment efficiency of ponds. In cases where inlet-energy driven flow prevails, baffles can be used to optimise the hydraulic flow patterns in the pond. For ponds with wind-energy driven hydraulics, baffles can be used to reduce any short-circuiting that occurs under certain wind conditions.

There are numerous materials from which baffles can be built. It is generally easier to build baffles as part of the original design, in which case a concrete block or earthen wall is often used. However, retro-fitting baffles into existing ponds is also commonly undertaken typically using plastic fabric/sheeting with a seam at the top containing floats (such as sealed plastic pipes) and a seam at the bottom containing weights (such as heavy chain). Regardless of the material used, it is important to ensure that the baffle is well sealed, especially at the base, and fully impermeable to avoid any 'leakage' or short-circuiting through the baffle.

10.7.1 Horizontal baffling across the pond (transverse)

Watters *et al.* (1973) tested three different widths of baffles: 50%, 70% and 90% of the width of the pond. Each width was tested using 2, 4, 6 and 8 baffles spaced evenly down the length of the pond. For the 50% width baffles, short-circuiting problems actually increased in the pond as the flow tracked directly down the middle of the pond and, in effect, the baffle cells created hydraulic 'dead space'. 90% width baffles were found to give a lower hydraulic efficiency than was seen for the 70% width baffles. This was probably due to a channelling effect. It was concluded that the 70% width baffles were the most hydraulically efficient option out of the three widths tested.

10.7.2 Vertical baffling through the pond depth

Watters *et al.* (1973) undertook four further experiments using vertical baffling: two with four baffles and two with six baffles. An example of the type of baffling used is shown in Figure 10.2. Surprisingly, the four-baffle cases proved to be more efficient than the six-baffle cases. Again this was attributed to channelling effects.

When the results were compared against the horizontal baffle experiments, it was found that, for a comparable amount of baffling, the horizontal configuration was more efficient.

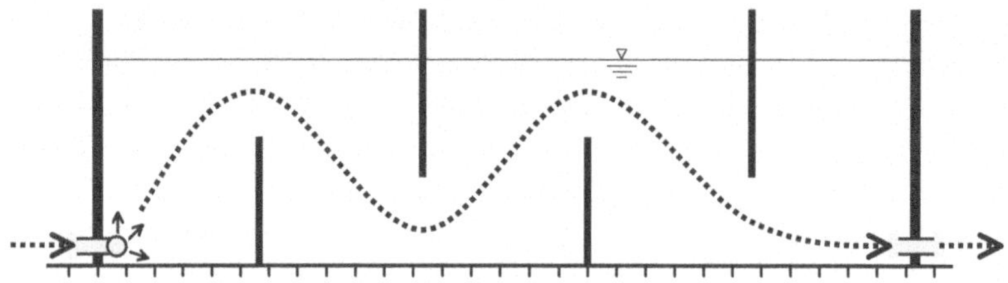

Figure 10.2 Experimental set-up for vertical baffle runs (Watters et al., 1973)

10.7.3 Longitudinal versus transverse baffling

Horizontal baffles can also be oriented along the length of the pond, instead of the width, to provide longitudinal baffling. Watters *et al.* (1973) found that for a comparable length of baffling, a similar efficiency was achieved as for transverse baffling across the pond width. It is noted, however, that this experiment used a manifold inlet. If a simple horizontal inlet was used (which has a 'jetting' effect) then it might be expected that transverse baffling would be more effective.

10.7.4 Interactions of baffles and inlets

Using computer flow modelling, Persson (2000) investigated the use of sub-surface berms or islands (in effect, submerged baffles) located close to the inlet. The use of berms resulted in a reduction in short-circuiting. This improvement results from the improved dissipation and redirection of the inlet jet(s). Work has been done using full-depth baffles located close to the inlet with similar results (Shilton and Harrison, 2003).

Shilton (2001) tested the effect of installing a single baffle (67% of the pond width) located halfway down the length of a laboratory model. The baffle was tested with three different inlets. In the first two tests, horizontal inlets were used. It was found that, after baffle addition, the time taken for the tracer front to reach the outlet was lengthened considerably. However, when using a vertical inlet, the baffle addition gave no further improvement over the unbaffled case.

10.7.5 Number of baffles

Typically transverse baffles that extend across a large part of the pond width are used. A number of studies have concluded that using more baffles gives better hydraulic efficiency, but with cost in mind it is important to consider the relative impact the number of baffles has on treatment efficiency.

Shilton and Harrison (2003) studied a series of evenly spaced, 70% width baffles using CFD modelling. It was found that while a single baffle does give an improvement over the unbaffled case, stepping up to a two-baffle system was far superior. The study showed a further but smaller improvement was achieved by shifting to four baffles, and where high treatment efficiency is required this may be warranted. However, the use of six baffles did not offer any substantial further improvement, and the extra gain provided by using eight baffles does not seem to warrant the extra cost in most applications.

In facultative ponds it may be advisable to limit the number of baffles to just two. This is because for a pond with a geometric length to width ratio of, say, 2:1, two baffles will increase the effective length to width ratio of the flow path to approximately 5:1 and any further increase would increase the possibility of organics or solids overloading at the front end of the pond.

10.7.6 Length to width ratio

From experiences with unbaffled laboratory and field ponds, Shilton and Harrison (2003) observed that at length to width ratios of 1:1 to 2:1, the flow tends to circulate right around the circumference of the pond if a horizontal inlet is positioned in a corner so as to discharge down the pond length. However, in laboratory testing on a pond with a length to width ratio of 3:1, it was found that, when the inlet was aligned to discharge along the shorter width, the behaviour changed dramatically. The flow moved across the width and around the corner, but then instead of travelling to the far end, it travelled about one third of the pond length and then turned back quite sharply into the middle of the pond (it appeared as if an 'invisible' baffle was in place!) The reason for this phenomenon is quite obvious in retrospect. As the pond becomes narrower then a series of counter-current circulation cells that are roughly the diameter of the pond width can be created. A series of these circulation cells could actually work very effectively, but further research is needed to better define this effect before it can be recommended as a design technique.

Clearly, the traditional assumption that in a narrow pond, the influent simply flows slowly, in a plug flow manner, from one end to the other is not necessarily correct. In experiments with baffles, Shilton and Harrison (2003) observed that the flow behaviour changed from a series of circulating cells at lower length to width ratios, for example at a 4:1 ratio, to become more channelled (or plug flow) at higher length to width ratios such as at over 10:1.

10.7.7 Alternative baffle positioning

Usually, transverse or longitudinal baffles are evenly spaced along the length or across the width of the pond. To determine whether other, less conventional, baffle positioning could offer any advantages, Shilton and Harrison (2003) examined a range of different single and twin baffle arrangements, including moving the baffles closer towards the inlet and/or the outlet.

If only a single baffle is to be used, it was found that shifting the baffle from the mid-length of the pond closer to the outlet provided a shielding effect that improved flow efficiency. However, no single baffle configuration can come close to providing the efficiency improvements achieved by using two baffles.

After testing a wide variety of alternative baffle placements for a twin baffled pond, the conventional approach of using equal spacing along the pond length was confirmed as the preferred arrangement.

10.7.8 The stub baffle

While long baffles (70% of the pond width) have traditionally been used in ponds, Shilton and Harrison (2003) have also investigated whether strategically positioned stub baffles (15% of the pond width or length) may provide the benefits associated with the long baffles, but at lower cost.

Upon testing these designs by computer modelling and in the laboratory, it was found that the stub baffling performed extremely well. The stub baffle set up a tight circulation localised in the inlet corner and then dispersed the flow evenly out into the larger main pond area. The longer baffles were more effective in channelling the flow, but overall both configurations provided similar treatment efficiency (as determined by computer modelling) and similar degrees of short-circuiting (as confirmed in the laboratory). However, testing undertaken on the application of the stub baffles in other configurations did not produce the same level of treatment efficiency as their long baffle counterparts. It seems that while the stub baffle can work extremely well in some cases, hydraulic performance is quite sensitive to changes in pond configuration and potentially other external factors such as wind.

While stub baffles have a lot of potential, they cannot yet be considered a substitute to long baffles in general application. However, their application for improving the operation of vertical inlets, or for shielding outlets is effective and has been described in Sections 10.4.3 and 10.5.5.

10.8 AERATORS, MIXERS AND TEMPERATURE

10.8.1 Aerators and mixers

Comparative power input analysis, as discussed in Section 10.6.4, can be used to demonstrate that if a pond contains aerators and other types of mixers these are likely to dictate the flow pattern in a pond (Shilton and Harrison, 2003). This also highlights that haphazard placement of aerators could have quite negative effects if they act to promote short-circuiting of the inflow towards the outlet. For design purposes, the previous comments made in regard to placement of horizontal inlets and shielding of the outlet also apply.

10.8.2 High rate algal ponds

Paddle wheel mixers are an integral part of the design of high rate algal ponds (see Chapter 13). The gentle circulation (typically 0.15 m/s) of wastewater around the baffled pond maintains algae in a mixed suspension (Craggs, 2002, pers. comm.).

The influent is normally added downstream of the paddlewheel near the base of the pond, while the outlet is located at the surface on the other side, thereby ensuring that the influent wastewater must make at least one circulation around the pond before any discharge (Craggs, 2002, pers. comm.). However the disadvantage of this circulating flow is that it does quickly move influent around the pond and back past the vicinity of the outlet, which facilitates short-circuiting.

Very little research has been published on the hydraulics of these systems, but it would seem likely that these systems could benefit from the use of flow shields or deflectors, as discussed previously, to shelter the outlet and prevent wastewater short-circuiting the treatment process after only the first few circulations.

10.8.3 Temperature effects

Thermal stratification may be detrimental to the hydraulic behaviour of a pond system, by facilitating short-circuiting of the inflow across the surface or floor of a pond. This effect can be a particular problem if the influent flow has a significantly different temperature to that of the main body of the pond, and is not well mixed upon entry.

Wastewater that is confined to one layer will have a significantly reduced retention time and hence treatment efficiency. It has been proposed from the results of a number of tracer studies that thermal stratification has been the cause of short-circuiting in ponds (MacDonald and Ernst, 1986; and Pedahzur *et al.*, 1993). However, it is important to note that, while these studies identified short-circuiting at the outlet and linked it with stratified condition in the pond, no direct measurements were taken within the pond to confirm this behaviour.

Potential solutions to this problem may involve:

1. The use of vertical baffling to ensure the vertical mixing of the flow.
2. Ensuring adequate mixing of the influent into the main body of the pond.
3. Provision of mixing in the vertical profile.

The build-up and breakdown of stratification are frequently assumed to entail some degree of convective mixing. However, it is important to note that the two are not necessarily linked. Convective mixing will only occur in a pond if it becomes thermally unstable. This results from a rapid cooling, such that the lower layers do not thermally equalise quickly enough by conduction with the cooler upper layers. In this case the warmer lower layer convects upwards in exchange with the cooler (and so denser) upper layer. Because convection currents act immediately to equalise any thermal imbalance, this effect is very difficult to study experimentally. Extremely accurate temperature measurements taken simultaneously throughout the pond's depth are required, and to date only preliminary work in this area has been undertaken (Sweeney *et al.*, 2005). What is occasionally observed, however, is the incidence of pond turnover.

Overturn has a serious impact on pond operation. Traditionally, this has been blamed on convective mixing of the stratified pond liquid layers. It is possible, however, that the mechanism is somewhat different and may be more directly linked to the sludge layer. Two separate studies (unpublished) in New Zealand have found that the sludge layer frequently has higher temperatures than the water column above it. Furthermore the sludge layer generates and contains the gaseous products of anaerobic or anoxic stabilisation. Both these factors decrease the density of the sludge layer and, rather than pond overturn being due to convection of the lower liquid layer, it may actually be due to the rising of the sludge layer.

10.9 SUMMARY AND RESEARCH RECOMMENDATIONS

While poor hydraulic performance is a major cause of poor pond performance, designing for optimal hydraulic conditions in a pond is not an area that is well understood.

When considering hydraulic design it is first necessary to abandon the assumption that flow is either entirely plug or completely mixed in the pond. All ponds have some degree of short-circuiting and improving treatment performance requires minimisation of this problem. Developing, at least, a basic understanding of the hydraulic conditions, which are likely to prevail in a pond must be seen as the key to optimising the design. The physical design, positioning and orientation of pond inlets and outlets; the pond loading; the basic pond geometry; the location, length and orientation of baffles; and external driving forces such as wind are all key factors that must be assessed in combination.

The impact that computational fluid dynamics has made in advancing this research area is significant but to date these models are still relatively basic in many areas, integrated pathogen decay, for example, and so this area offers much potential for ongoing research work. Throughout this chapter a range of practical design modifications have been suggested for optimising pond hydraulics. However, in many areas it was noted that ideas, such as the use of central outlets, have not been rigorously tested in full-scale applications and further research is needed before such techniques can be practically applied with confidence. The same issue applies to the 'power analysis' concept, which potentially offers engineers a very valuable assessment tool.

Optimising pond hydraulic design can seem complex, especially if computer modelling is contemplated, but in reality many ponds are designed with very little regard to the hydraulics and even simple application of some of the ideas contained in this chapter have the potential to yield significant improvements in hydraulic and treatment efficiency.

REFERENCES

Agunwamba J. C., Egbuniwe N. and Ademiluyi J. O. (1992) Prediction of the dispersion number in waste stabilization ponds. *Water Research* **26**(1), 85–89.

Arceivala S. J. (1983) Hydraulic modeling for waste stabilization ponds. *Journal of the Environmental Engineering Division, ASCE* **109**(EE1), 265–268.

Brissaud F., Lazarova V., Ducoup C., Joseph C., Levine B. and Tournoud M. (2000) Hydrodynamic behaviour and faecal coliform removal in a maturation pond. *Water Science and Technology* **42**(10–11), 119–126.

Canter L. W. and Englande A. J. (1970) States' design criteria for waste stabilization ponds. *Journal of the Water Pollution Control Federation* **42**(10), 1840–1847.

Chapple L. G. (1985) *A Study of Bacterial Kinetics and Hydraulic Short Circuiting in Sewage Lagoons*. Masters Thesis, Department of Civil Engineering, University of Queensland.

Craggs R. J. (2002) National Institute of Water and Atmospheric Research, Hamilton, New Zealand. Personal Communication.

Curtis T. P., Mara D. D. and Silva S. A. (1992) Influence of pH, oxygen and humic substances on ability of sunlight to damage fecal coliforms in water stabilization pond water. *Applied and Environmental Microbiology* **58**(4), 1335–1343.

Davies M. L. and Corwell D. A. (1991) *Introduction to environmental engineering*, 2nd Edn. McGraw-Hill, New York.

Davies-Colley R. J., Donnison A. M. and Speed D. J. (1999) Towards a mechanistic understanding of pond disinfection. *Water Science and Technology* **48**(2), 149–158.

Dorego N. C. and Leduc R. (1996) Characterisation of hydraulic flow patterns in facultative aerated lagoons. *Water Science and Technology* **34**(11), 99–106.

Fares Y. R., Frederick G. L., Vorkas C. A. and Lloyd B. J. (1996) Hydrodynamic effects on performance of waste stabilisation lagoons. In: *Hydrodynamics—Theory and Applications ICHD-96*, A. T. Chwang, J. H. W. Lee and D. Y. C. Leung (eds), Balkema, Rotterdam, 341–348.

Fares Y. R. and Lloyd B. J. (1995) Wind effects on residence time in waste stabilisation lagoons. In: *Hydra 2000: Proceedings of the 26th Congress of the International Association for Hydraulic Research*, 11th–15th September, London, UK, 205–211.

Ferrara R. A. and Harleman D. R. F. (1981) Hydraulic modeling for waste stabilization ponds. *Journal of the Environmental Engineering Division, ASCE* **107**(EE4), 817–830.

Frederick G. L. and Lloyd B. J. (1996) An evaluation of retention time and short-circuiting in waste stabilisation ponds using *Serratia marcescens* bacteriophage as a tracer. *Water Science and Technology* **33**(7), 49–56.

Hermann E. R. and Gloyna E. F. (1958) Waste stabilization ponds III. Formulation of design equations. *Sewage and Industrial Wastes* **30**(8), 963–975.

Larsen T. (1999) Department of Civil Engineering, University of Aalborg, Aalborg, Denmark. Personal Communication.

Levecq C., de Traversay C. and Essemiani K. (2001) Computational fluid dynamics applications in water treatment. In: *Water Software Systems: Theory and Applications,* B. Ulanicki, B. Coulbeck and J. P. Rance (eds), Research Studies Press, Baldock, 257–267.

Levenspiel O. (1972) *Chemical Reaction Engineering*, 2nd Edn. John Wiley and Sons, New York, USA.

MacDonald R. and Ernst A. (1986) Disinfection efficiency and problems associated with maturation ponds. *Water Science and Technology* **18**(10), 19–29.

Mangleson K. A. and Watters G. Z. (1972) Treatment efficiency of waste stabilization ponds. *Journal of the Sanitary Engineering Division, ASCE* **98**(SA2), 407–425.

Mara D. D. and Pearson H. W. (1998) *Design Manual for Waste Stabilization Ponds in Mediterranean Countries*. Lagoon Technology International; Leeds, UK.

Marais G. v. R. (1974) Faecal bacterial kinetics in stabilization ponds. *Journal of the Environmental Engineering Division, ASCE* **100**(EE1), 119–139.

Marecos do Monte M. H. F. and Mara D. D. (1987) The hydraulic performance of waste stabilization ponds in Portugal. *Water Science and Technology* **19**(12), 219–227.

Moreno M. D. (1990) A tracer study of the hydraulics of facultative stabilization ponds. *Water Research* **24**(8), 1025–1030.

Olsen N. R. B. and Tjomsland T. (1998) 3D CFD modelling of wind induced currents and dispersion of radioactive tracer in a lake. Presented at: *3rd International Conference on Hydroscience and Engineering*, 31st August – 3rd September, Cottbus, Germany.

Pearson H. W., Mara D. D. and Arridge, H. A. (1995) The influence of pond geometry and configuration on facultative and maturation waste stabilisation pond performance and efficiency. *Water Science and Technology* **31**(12), 129–139.

Pedahzur R., Nasser A. M., Dor I., Fattal B. and Shuval H. I. (1993) The effect of baffle installation on the performance of a single-cell stabilization pond. *Water Science and Technology* **27**(7–8), 45–52.

Persson J. (2000) The hydraulic performance of ponds of various layouts. *Urban Water* **2** pp. 243–250.

Polprasert C. and Bhattarai K. K. (1985) Dispersion model for waste stabilization ponds. *Journal of Environmental Engineering* **111**(1), 45–58.

Racault Y., Boutin P. and Douat J. (1984) Study by tracer experimentation of the behaviour of a waste stabilization pond: influence of the basin geometry. *Revue Francaise des Sciences de L'eau* **3,** 197–218.

Salter H. E. (1999) *Enhancing the Pathogen Removal Performance of Tertiary Lagoons.* Doctorate Thesis, Centre for Environmental Health Engineering, University of Surrey.

Shilton, A. N., Wilks, T., Smyth, J. and Bickers, P. (2000) Tracer studies of a New Zealand waste stabilisation pond, analysis of treatment efficiency. *Water Science and Technology* **42**(10–11), 343–348.

*Shilton A. N. (2001) *Studies into the Hydraulics of Waste Stabilisation Ponds.* Doctorate Thesis. Institute of Technology and Engineering, Massey University; Palmerston North, New Zealand.

Shilton A. N. and Harrison J. (2002) Integration of coliform decay within a CFD model of a waste stabilisation pond. *Water Science and Technology* **48**(2), 205–210.

*Shilton A. N. and Harrison J. (2003) *Guidelines for the Hydraulic Design of Waste Stabilization Ponds.* Palmerston North, New Zealand: Institute of Technology and Engineering, Massey University.

Sweeney D. G., Cromar N. J., Nixon J. B., Ta C. T. and Fallowfield H. J. (2003) The spatial significance of water quality indicators in waste stabilization ponds—Limitations of residence time distribution analysis in predicting treatment efficiency. *Water Science and Technology* **48**(2), 211–218.

Sweeney D. G., Nixon J. B., Cromar N. J. and Fallowfield H. J. (2005) Profiling and modelling of thermal changes in a large waste stabilisation pond. *Water Science and Technology*, in press.

Thirumurthi D. (1969) Design principles of waste stabilization ponds. Journal of the Sanitary Engineering Division, ASCE SA2 (Apr) pp. 311–330. Thirumurthi D. (1969) Design principles of waste stabilization ponds. *Journal of the Sanitary Engineering Division, ASCE* **SA2** (Apr) pp. 311–330.

Uluatam S. and Kurum Z. (1992) Evaluation of the wastewater stabilisation pond at the METU treatment plant. *International Journal of Environmental Studies* **41**(1–2), 71–80.

Van Dorn W. (1953) Wind stress on an artificial pond. *Journal of Marine Research*, **12** (3), 249–276.

von Sperling M. (2003) Influence of the dispersion number on the estimation of coliform removal in ponds. *Water Science and Technology* **48**(2), 181–188.

Vorkas C. and Lloyd B. (2000) The application of a diagnostic methodology for the identification of hydraulic design deficiencies affecting pathogen removal. *Water Science and Technology* **42**(10–11), 99–110.

Watters G., Mangelson K, and George R. (1973) *The Hydraulics of Waste Stabilization Ponds.* Research Report; Utah Water Research Laboratory, College of Engineering, Utah State University; Utah, USA.

Wood M. (1997) *Development of Computational Fluid Dynamic Models for the Design of Waste Stabilisation Ponds.* Doctorate Thesis. Department of Chemical Engineering, University of Queensland; Brisbane, Australia.

* = These two references can be downloaded from either of the following sites:

- http://ite.massey.ac.nz/staff/shiltona.htm
- http://www.leeds.ac.uk/civil/ceri/water/tphe/publicat/publicat.html

11

Solids removal and other upgrading techniques

E. Joe Middlebrooks, V. Dean Adams,
Stuart Bilby and Andy Shilton

11.1 INTRODUCTION

Sometimes pond effluent quality is inadequate to meet environmental objectives in the receiving waters. For example, parameters that may need improving are pathogens (see Chapter 6), nutrients and ammonia (see Chapter 5), suspended solids and BOD (see Chapter 4) (Hickey *et al.,* 1989).

Removing algae is one of the most challenging aspects of upgrading pond systems because of its tendency to clog conventional filter systems. In New Zealand ponds, blue-green algae blooms are common in late summer and regularly result in effluent suspended solids levels of over 150 mg/L. (Hickey *et al*, 1989; Hutchinson & Oliver, 1978).

ISBN: 1843390205. Published by IWA Publishing, London, UK.

The appropriate level of upgrading and the appropriate method is strongly dependent on the needs of the receiving environment.

The solids are primarily composed of algae and other pond detritus rather than wastewater solids. These high concentrations are usually limited to 2 to 4 months during the year.

11.2 INTERMITTENT SLOW SAND FILTRATION

Intermittent slow sand filtration is capable of polishing pond effluents at relatively low cost and is similar to the practice of slow sand filtration in potable water treatment. Intermittent sand filtration of pond effluents is the application of pond effluent on a periodic or intermittent basis to a sand filter bed. As the wastewater passes through the bed, suspended solids and other organic matter are removed through a combination of physical straining and biological degradation processes. The particulate matter collects in the top 5 to 8 cm of the filter bed, and while regular raking helps maintain the filter, eventually this accumulation clogs the surface and prevents effective infiltration of additional effluent. When this happens, the bed is taken out of service, the top layer of clogged sand removed, and the unit is put back into service. The removed sand can be washed and reused or discarded.

11.2.1 Summary of performance

Summaries of the performance of intermittent sand filters treating pond effluents presented in Table 11.1 show that it is possible to produce an effluent with TSS and BODs less than 15 mg/L from anaerobic, facultative ponds and aerated lagoons followed by intermittent sand filters with effective sizes less than or equal to 0.3 mm. Effective nitrification can also be achieved with intermittent sand filters.

Rich and Wahlberg (1990) evaluated the performance of five facultative pond-intermittent sand filter systems located in South Carolina and Georgia, and found that the systems provided superior performance when compared with ten aerated lagoon systems.

Truax and Shindala (1994) reported the results of an extensive evaluation of facultative pond-intermittent sand filter systems using four grades of sand with effective sizes of 0.18 to 0.70 mm and uniformity coefficients ranging from 1.4 to 7.0. Performance was directly related to the effective size of the sand and hydraulic loading rate. With effective size sands of 0.37 mm or less and hydraulic loading rates of 0.2 $m^3/m^2.d$, effluents with BODs and TSS of less than 15 mg/L were obtained. TKN concentrations were reduced from 11.6 mg/L to 4.3 mg/L at the 0.2 $m^3/m^2.d$ loading rate. The experiments were conducted in a mild climate, and it is unlikely that similar nitrogen removal would occur during cold months or more severe climates.

Melcer *et al.* (1995) reported the performance of a full-scale aerated lagoon-intermittent sand filter system located in New Hamburg, Ontario, Canada, that had been in operation since 1980. Results for 1990 and from January to August of 1991 are presented in Table 11.1. Surface loading rates for both periods were 3.24 m^3/m^2.d with influent BODs, TSS and TKN concentrations of 12, 16 and 19 mg/L respectively. Filter effluent quality was exceptional with BODs, TSS and TKN concentrations less than 2 mg/L.

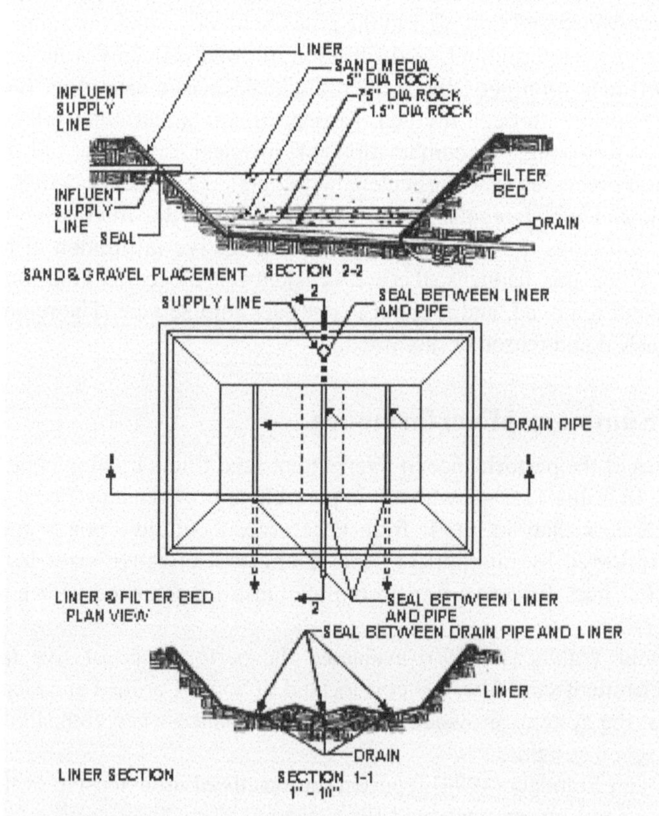

Figure 11.1 Cross-sectional and plan views of typical intermittent sand filter (Middlebrooks *et al.*, 1983)

Table 11.1 Performance of an aerated lagoon and intermittent sand filter system, Hamburg Plant (Melcer *et al.*, 1995)

Location	Parameter	1990	1991 (Jan-Aug)
Raw sewage	Average flow rate, m^3/d	1676	1673
	Max. flow rate, m^3/d	4530	3990
	BOD, mg/L	186	120
	TSS, mg/L	314	171
	TKN, mg/L	45	44
	TP, mg/L	9.3	9.5
Aerated cell	HRT, d	7	7
	BOD loading, $kg/m^3.d$	0.03	0.02
Aerated cell effluent	BOD, mg/L	34	36
	TSS, mg/L	44	44
	TP, mg/L	6	5
Facultative pond	HRT, d	165	165
	Avg BOD loading, kg/100 m^2.d	0.51	0.55
Cell 2 effluent	BOD, mg/L	12	11
	TSS, mg/L	16	18
	TKN, mg/L	19	18
	NH_3-N, mg/L	15	14
	NO(T)-N, mg/L	1.1	0.8
	TP, mg/L	1.2	0.7
Filter	Annual surface loading, m^3/m^2	195	153
	Surface loading, $L/m^2.d$	3240	3240
		Mar-Dec	Mar-Aug
Filter effluent	BOD, mg/L	2	2
	TSS, mg/L	1.7	1.1
	TKN, mg/L	2	1.1
	NH_3-N, mg/L	1.2	0.6
	NO(T)-N, mg/L	7	9
	TP, mg/L	0.5	0.4

11.2.2 Operating periods

The length of filter run is a function of the effective size of the sand and the quantity of solids deposited on the surface of the filter. The EPA Design Manual - Municipal Wastewater Stabilisation Ponds (1983) and several publications (Marshall and Middlebrooks, 1974; Messinger, 1976; 1979; Harris *et al.*, 1978; Hill *et al.*, 1977; Bishop *et al.*, 1977; Tupyi *et al.*, 1979; Russell *et al.*, 1983) contain extensive information on the relationship between solids deposited on the surface of a filter and the length of run time. Truax and Shindala (1994) also reported run times very similar to those in the above studies, and their results are presented in Table 11.2.

Table 11.2 Pond & intermittent sand filtration - run lengths (Truax and Shindalla, 1994)

Sand Characteristics	Hydraulic loading rate $m^3/m^2.d$	Days of filter operation before initial clogging
Sand #1	0.2	469
• e.s. = 0.70 mm	0.9	335
• U.C. = 2.1	1.1	106
Sand # 2	0.2	468
• e.s. = 0.35 mm	0.7	259
• U.C. = 1.4	0.9	16
Sand #3	0.2	130
• e.s. = 0.37 mm	0.4	305
• U.C. = 7.0	0.6	159
	0.7	27
	0.9	9
Sand #4	0.2	131
• e.s. = 0.18 mm	0.4	130
• U.C. = 2.7	0.7	35
	0.9	5

11.2.3 Maintenance requirements

Maintenance is directly related to the quantity of solids applied to the surface of the filter, and this is related to the concentration of solids in the influent to the filter and the hydraulic loading rate. Filters with low hydraulic loading rates tend to operate for extended periods. With such extended operating periods, maintenance consists of routine inspection of the filter, removing weeds, raking to break up the accumulated surface layer, and an occasional cleaning by removing the top five to eight centimetres of sand after allowing the filter to dry out.

11.2.4 Hydraulic loading rates

Typical hydraulic loading rates on a single stage filter range from 0.37 to 0.56 m^3/m^2.d. If the suspended solids in the influent to the filter will routinely exceed 50 mg/L, the hydraulic loading rate should be reduced to 0.19 to 0.37 m^3/m^2.d to increase the filter run. Experience at Pauanui, a small New Zealand community, over 20 years has been to use 0.4 m^3/m^2.d in summer and 0.2 m^3/m^2.d in winter.

In cold weather locations, the lower end of the range is recommended during winter to avoid the possible need for bed cleaning during the winter months.

Pond algal blooms and suspended solids levels above 150 mg/L can lead to rapid clogging of intermittent sand filters and generous storage or a rock filter pre-treatment may be desirable if such conditions are anticipated.

It is prudent to carefully assess peak wet-weather flows during design and to use a conservatively low loading rate because intermittent sand filters are unable to easily take increased flows. Under-sizing can rapidly result in an operational spiral of overloaded filters, increased clogging, decreased capacity, inadequate storage and more overloaded filters.

11.2.5 Design of intermittent sand filters

Algae removal from pond effluent is almost totally a function of the sand gradation used. If an effluent BOD and SS of below 30 mg/L is sufficient, a single-stage filter with medium sand (effective size = 0.3 mm) will produce a reasonable filter run. If better effluent quality is required, finer sand (effective size = 0.15 to 0.2 mm) is necessary or a two-stage filtration system with the finer sand in the second stage can be used.

The total filter area required for a single-stage operation is obtained by dividing the anticipated influent flow rate by the hydraulic loading rate selected for the system. One spare filter unit should be included to permit continuous operation since the cleaning operation may require several days. An alternate approach is to provide temporary storage in the ponds. Three filter beds are the preferred arrangement to permit maximum flexibility. In small systems that depend on manual cleaning, the individual bed should not be bigger than about 90 m^2. Larger systems with mechanical cleaning equipment might have individual filter beds up to 5000 m^2 in area.

The sand used as the filter media is generally described by effective size (es) and uniformity coefficient (U.C.). The es is the 10 percentile size, i.e. only 10 percent of the filter sand, by weight, is smaller than that size. The uniformity coefficient is the ratio of the 60th percentile size to the 10th percentile size. The sand for single-stage filters should have an es ranging from 0.20 to 0.30 mm and a U.C. of less than 7.0, with less than 1 percent of the sand smaller than 0.1 mm. The U.C. value seems to have little effect on performance, and values ranging from 1.5 to 7.0 are acceptable. In the general case, clean, pit-run concrete sand is suitable for use in intermittent sand filters providing that the es, U.C., and minimum size are suitable.

The design depth of sand in the bed should be a minimum of 45 cm plus additional depth to allow for at least 1 year of cleaning cycles. A single cleaning operation may remove 2.5 to 5 cm of sand. A 30-day filter run would then require an additional 30 cm of sand. In the typical case an initial bed depth of about 90 cm of sand is usually provided. A graded gravel layer 30 to 45 cm separates the sand layer from the under drains. The bottom layer is graded so that its es is four times as great as the openings in the under drain piping. The successive layers of gravel are progressively finer to prevent intrusion of sand. A common and cheaper alternative is to use gravel around the under-drain piping and then a permeable geotextile membrane to separate the sand from the gravel. Further details on design and performance are presented in the EPA Design Manual - Municipal Wastewater Stabilisation Ponds (1983). A design example for an intermittent sand filter treating a pond effluent is presented below.

11.2.6 Design example – intermittent sand filters

Determine dimensions of filters

Design Flow (Q) = 379 m^3/day
Hydraulic Loading Rate (HLR) = 0.29 m^3/m^2.day
Filter Area Required = Q/HLR = 1307 m^2
Split above into 3 separate cells of 436m^2, L:W = 2:1
Individual filter cell size: width = 14.8 m, length = 29.5 m
Could also add another spare cell of equal size to allow for filter cleaning.

Influent distribution system

Dosing siphon used to gravity feed filters. Electrically activated valves may also be used. Loading sequence will be designed to deliver one-half the daily flow rate to filter unit in two equal doses (or more frequent smaller doses).
Daily flow to each filter unit = 379/3 = 126 m^3/day
Size of dose = 126/2 = 63 m^3

Minimum freeboard required for filters

Should be adequate to hold three doses.
Water depth (assuming no passage through filter) =
No. doses x dose / (W x L) = 3 x (63) / (14.8 x 29.5) = 0.43m

11.3 ROCK FILTERS

Rock filters have been installed throughout the United States and the world, with variable performance (EPA, 1983; Middlebrooks *et al.*, 1982; Saidam *et al.*, 1995). A rock filter operates by allowing pond effluent to travel through a submerged porous rock bed, causing algae to settle out on the rock surface as the liquid flows through the void spaces. The accumulated algae are then biologically degraded. Algae removal with rock filters has been studied extensively at Eudora, Kansas; California, Missouri; and Veneta, Oregon (EPA, 1983). A diagram of the Veneta, Oregon rock filter is shown in Figure 11.2.

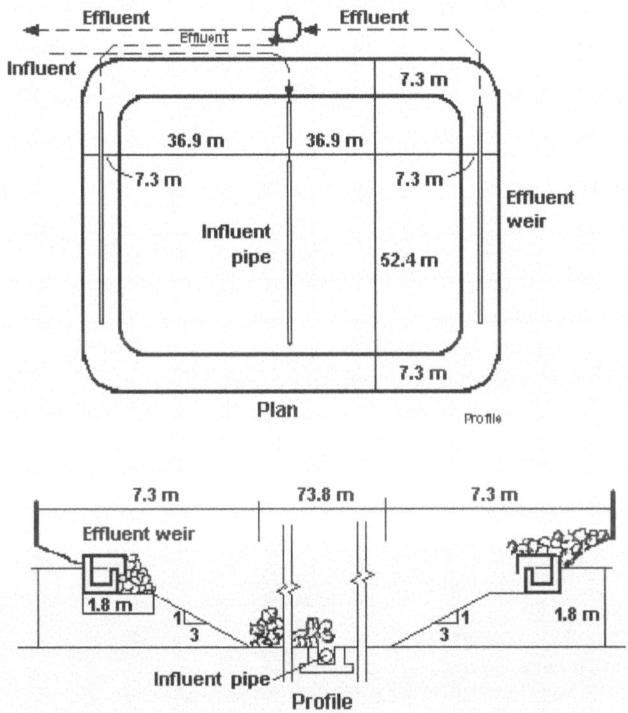

Figure 11.2 Rock filter at Veneta, Oregon (Williamson and Swanson, 1978)

The principal advantages of the rock filter are its relatively low construction cost and simple operation. Odour problems can occur, and the design life for the filters and the cleaning procedures has not yet been firmly established; however, several units have operated successfully for over 20 years.

11.3.1 Performance of rock filters

Mixed results have been obtained with rock filters. Performance data from one of the most effective filters, the Veneta system, which treats approximately 1,000m^3/day (Williamson and Swanson, 1978), are shown in Table 11.3. After approximately 20 years of operation, the system was still producing an effluent meeting secondary standards with regard to BOD, TSS and FC. However, during the winter months, high ammonia nitrogen concentrations were observed in the effluent.

Table 11.3 Mean and range of performance data for Veneta, OR Wastewater Treatment Plant – 1994

Constituent	Influent	Effluent
BOD_5, mg/L	138 (50-238)	17 (5-30)
TSS, mg/L	124 (50-202)	9 (2-27)
FC, #/100 mL	-	<10 (<10-20)

Stamberg *et al* (1984) presented performance results for two rock filters operating in West Monroe, LA. The systems were loaded at higher hydraulic loading rates than that used at the Veneta facility and produced effluent BOD_5 and TSS concentrations less than 30 mg/L. However, there were occasional concentrations of up to 40 mg/L in both BOD_5 and TSS.

Saidam *et al.* (1995) performed a series of studies of rock filters treating pond effluent in Jordan, but even though the rock sizes and loading rates of several of the filters were similar to those at Veneta and West Monroe, the quality of the effluents was much less

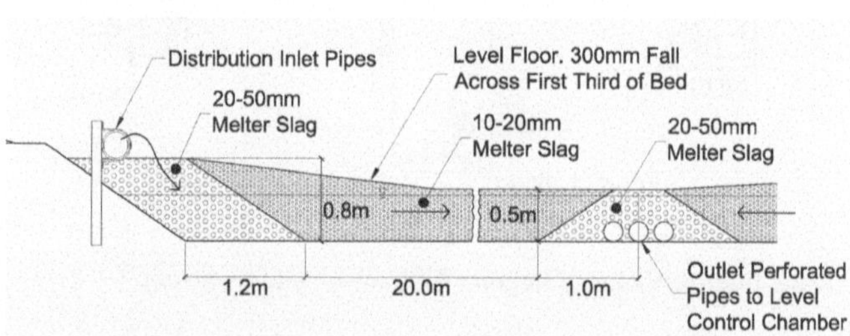

Figure 11.3 Cross-section through rock filter at Paeroa, New Zealand

In New Zealand rock filters have been successfully used for removing high levels of algae from pond effluents. The systems were developed from sub-surface flow wetlands designs but in this case, without any plants. The rock used was 10 mm to 20 mm diameter with coarser stone at the inlet and outlet to provide even flow distribution

In three of the systems the rock used was steel melter slag. The steel making slag has the advantage of a high porosity, low cost, less hydrogen sulphide odour and also provides some phosphorous removal, at least for the first years of operation. These rock filters follow aerated facultative ponds. The rock filters have provided a consistent removal of suspended solids to below 25 mg/L and averages of below 12 mg/L even when pond suspended solids levels were well above 100 mg/L.

Because the rock filters are generally anoxic, little nitrification occurs. However, the gravel filters are well suited to denitrification. The effluent from the filters can be anaerobic and issues with hydrogen sulphide odours can occur. When the filter is removing high concentrations of algae, the organic nitrogen in the algae will be released as ammonia in the effluent.

Table 11.4 Design parameters and performance of some New Zealand rock filters

		Waiuku	**Paeroa**	**Ngatea**	**Clarks Beach**
Design flow (average)	m^3/day	3,000	2,067	460	375
Current flow (average)	m^3/day	1,800	2,100	250	290
Width	m	29.6	22	26.3	32
Length	m	97.4	131	136.0	62
No. of beds		10	8	2	2
Total rock filter area	m^2	28,868	23,056	7,154	3,875
Rock size	mm	20/10	20/10	20/10	20/10
Rock type		slag	slag	slag	greywacke
Rock depth	m	0.5	0.5-0.8	0.5-0.8	0.5-0.65
Rock filter loading rate (average)	mm/day	62	91	35	75
Rock filter loading rate (average)	m^3/m^3.day	0.14	0.20	0.08	0.17
Average water depth	m	0.45	0.45	0.45	0.45
Hydraulic retention time (average)	days	3.3	2.2	5.8	1.5
Year constructed		1993	2000	2002	1998
$CBOD_5$	average	6	4	3	
$CBOD_5$	95 percentile	11	19	6	
Suspended solids	average	12	9	6	
Suspended solids	95 percentile	24	17	9	
Ammonia	average	5	7	15	
Ammonia	95 percentile	24	12	27	
Total nitrogen	average	8	10	19	
Total nitrogen	95 percentile	20	17	36	

11.3.2 Inter-pond rock filters

Inter-pond rock filters have been used in New Zealand recently to improve suspended solids and nitrogen removal. (Archer and O'Brien, 2005). These consist of permeable rock embankments across the ponds that provide filtering, reduce short-circuiting and provide a surface to grow nitrifying bacteria. Where additional nitrification is desirable, air can be bubbled through perforated pipes beneath the submerged rock embankments.

Alternatively nitrification filter beds (NFB) (Reed *et al.*, 1995) where pond effluent is sprayed onto a vertical rock filter placed at the side of a pond or wetland have been used successfully. Based on work with attached growth systems the suggested specific surface area of rock material required is:

$$A_v = \frac{2713 - 1115C_e + 204(C_e)^2 - 12(C_e)^3}{K_T}$$

Where:

A_v = specific surface area, $m^2/kg\ NH_4.d$
C_e = desired NFB effluent ammonia concentration, mg/L
K_T = temperature dependent coefficient:

- $0.626(1.15)^{(T-10)}$ at 1-10°C
- $1.048^{(T-20)}$ at >10°C

This formula is applicable for effluent ammonia concentrations (C_e) in the range 0-6 mg/L and requires that BOD be low (BOD/TKN <1), that the surface be continually wet, that there is an adequate oxygen supply and that there is sufficient alkalinity (≈ 10 g alkalinity/L g ammonia).

11.3.3 Design of rock filters

Rock filters have been designed using a number of varying parameters. The critical factor in the design of rock filters appears to be the hydraulic loading rate. Rates less than 0.3 $m^3/m^3.d$ appear to give the best results with rocks in the size range of 8 to 20 cm and with a depth of 2 m with the water applied in an up-flow pattern. Rates of 0.15 to 0.30 $m^3/m^3.d$ can work well with 1 to 2 cm rock.

An adjustable outlet level is highly desirable to allow for different head losses caused by seasonal flows, flow increases or clogging of the beds. Correct hydraulic design can be critical. It is important to keep the flow below the surface of the rock otherwise growth of algae and insect nuisance may occur.

11.3.4 Maintenance requirements

Pond systems are often selected where land is relatively cheap and where there is a need to have robust, simple to operate treatment processes. Lightly loaded rock filters require almost no maintenance and are a useful complement to ponds. Most of the clogging occurs, when it does occur, in the first third of the rock filter. The system at Waiuku, New Zealand has been operating for twelve years without clogging problems. If solids ever did accumulate to the point of causing difficulty, the rock at the inlet of the filter would need to be replaced or lifted and washed.

Occasional weed removal by flooding, hand weeding or spraying is necessary.

11.4 RAPID SAND FILTRATION

When rapid sand filtration has been used for the removal of algae from wastewater stabilisation pond effluents, very poor results have been obtained. Improved efficiency can be obtained when chemicals are added prior to filtration or when the wastewater is treated by coagulation and flocculation prior to filtration. Because of the variable nature of algae species and pond effluent parameters even with coagulant and flocculent dosing, operation of rapid sand filtration to achieve consistent effluent quality is difficult.

Diatomaceous earth filtration is capable of producing a high-quality effluent when treating wastewater stabilisation pond water, but the filter cycles are generally less than 3 hours. This results in excessive usage of backwash water and diatomaceous earth, which leads to very high cost and eliminates this method of filtration as an option for upgrading wastewater stabilisation pond effluents.

Use of microscreens and textile screens for algae removal have generally been unsuccessful. Fabric openings of 1 micrometre are required to effectively remove the algae and back washing is, again, excessive.

11.5 COAGULATION-FLOCCULATION

Coagulation followed by sedimentation has been applied extensively for the removal of suspended and colloidal materials from water. Lime, alum, and ferric salts are the most commonly used coagulating agents. Floc formation is sensitive to parameters such as pH, alkalinity, turbidity, and temperature. Most of these variables have been studied, and their effects on the removal of turbidity of water supplies have been evaluated. In the case of the chemical treatment of wastewater stabilisation pond effluents, however, the data are not comprehensive.

Shindala and Stewart (1971) investigated chemical treatment of stabilisation pond effluents as a post-treatment process to remove the algae and to improve the quality of the effluent. They found that the optimum dosage for best removal of the parameters studied was 75-100 mg/L of alum. When this dosage was used, the removal of phosphate was 90% and the COD was 70%.

Tenny (1968) has shown that at a pH range of 2 to 4, algal flocculation was effective when a constant concentration of a cationic polyelectrolyte (10 mg/L of C.31) was used. Golueke and Oswald (1965) conducted a series of experiments to investigate the relation of hydrogen ion concentrations to algal flocculation. In this study, only H_2SO_4 was used, and only to lower the pH. Golueke and Oswald (1965) found that flocculation was most extensive at a pH value of 3, with which Tenney's results agree. They obtained algal removals of about 80-90%. Algal removal efficiencies were not affected in the pH range of 6-10 by cationic polyelectrolytes.

McGarry (1970) studied the coagulation of algae in waste stabilisation pond effluents, and reported the results of a complete factorial designed experiment using the common jar test. Tests were performed to determine the economic feasibility of using polyelectrolytes as primary coagulants alone or in combination with alum. He also investigated some of the independent variables that affected the flocculation process, such as concentration of alum, flocculation turbulence, concentration of polyelectrolyte, pH after the addition of coagulants, chemical dispersal conditions, and high rate algal pond suspension characteristics. Alum was found to be effective for coagulation of algae from high rate algal pond effluents, and the polyelectrolytes used did not reduce the overall costs of algal removal. The minimum cost per unit algal removal was obtained with alum alone (75-100 mg/L).

Al-Layla and Middlebrooks (1975) evaluated the effects of temperature on algae removal using coagulation-flocculation-sedimentation. Algae removal at a given alum dosage decreased as the temperature increased. Maximum algae removal generally occurred at an alum dosage of approximately 300 mg/L at 10°C. At higher temperatures alum dosages as high as 600 mg/L could not produce equivalent removals. Settling time required was found to vary adversely as the temperature of the wastewater increased.

Coagulation-flocculation is not easily controlled and requires skilled operating personnel at all times. A large volume of sludge is produced, and this introduces additional operational problems. For a small community accustomed to minimal operation and maintenance of its wastewater, coagulation-flocculation may not often be appropriate.

11.6 DISSOLVED AIR FLOTATION

Several studies have shown the dissolved air flotation process to be an efficient and a relatively cost effective means of algae removal from wastewater stabilisation pond effluents. The performance obtained in several of these studies is summarised in Table 11.5.

Table 11.5 Summary of typical dissolved air flotation performance (Parker, 1976)

Investigator and Location	Coagulant and Dose (mg/L)	Overflow Rate (m/hour)	Detention Time (minutes)	BOD_5			Suspended Solids		
				Influent (mg/L)	Effluent (mg/L)	% Removal	Influent (mg/L)	Effluent (mg/L)	% Removal
Parker (1976) Stockton, CA	Alum: 225 mg/L. Acid added to pH 6.4	6.6[a]	17[a]	46	5	89	104	20	81
Ort (1972) Lubbock, TX	Lime[c]: 150 mg/L	NA	12[b]	280-450	0-3	>99	240-360	0-50	>79
Komline-Sanderson (1972) El Dorado, Arkansas	Alum: 200 mg/L	9.8[c]	8[c]	93	<3	>97	450	36	92
Bare (1971) Logan, UT	Alum: 300 mg/L	3.2-5.9[d]	NA	NA	NA	NA	100	4	96
Stone *et al.* (1975) Sunnyvale, CA	Alum: 175 mg/L. Acid added - pH 6.0 - 6.3	4.9[e]	11[e]	NA	NA	NA	150	30	80

[a] Including 33% pressurised (240-415 kPa) recycle. [b] Including 30% pressurised (345 kPa) recycle. [c] Including 100% pressurised recycle. [d] Including 25% pressurised (310 kPa) recycle. [e] Including 27% pressurised (380-480 kPa) influent

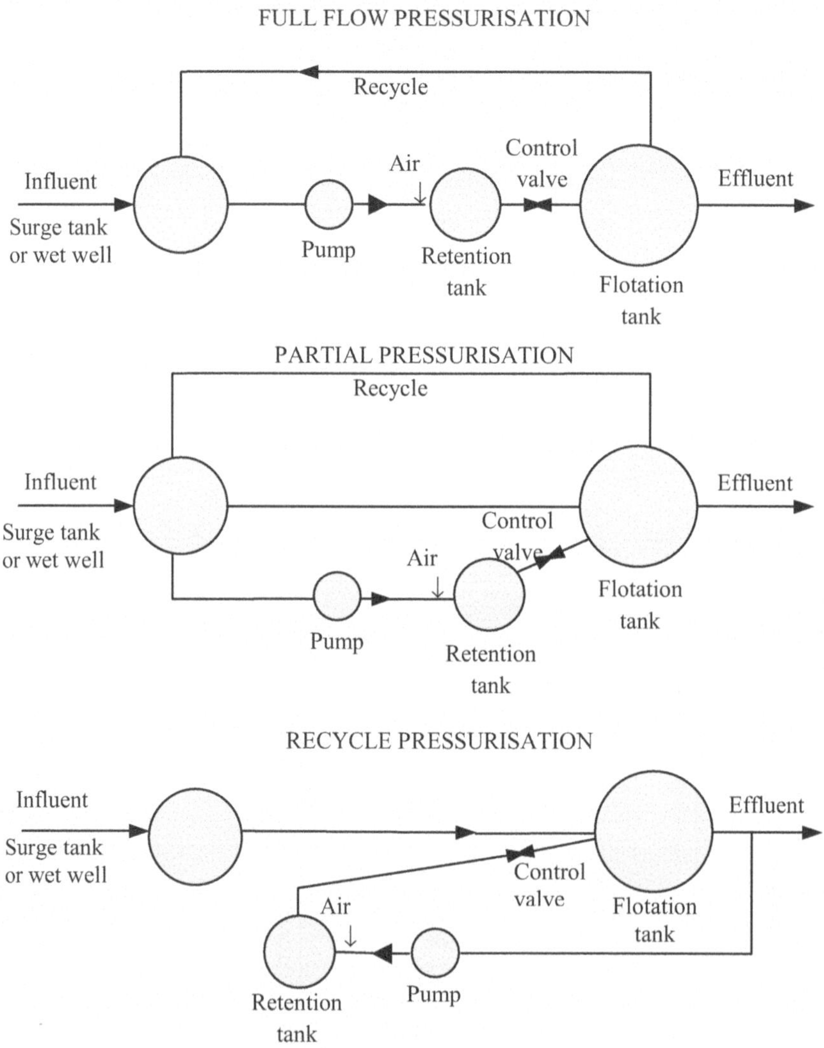

Figure 11.4 Types of dissolved air flotation systems (Snider, 1976)

Three basic types of dissolved air filtration are employed to treat wastewaters: full flow pressurisation, partial pressurisation, and recycle pressurisation. These three types of dissolved air flotation are illustrated by flow diagrams in Figure 11.4. In the full flow pressurisation system the entire wastewater stream is injected with air and pressurised and held in a retention tank before entering the flotation cell. The flow is direct, and all recycled effluent is repressurised. In partial pressurisation only part of the wastewater stream is pressurised, and the remainder of the flow bypasses the air dissolution system and enters the separator directly. Recycling serves to protect the pump during periods of low flow, but it does hydraulically load the separator. Partial pressurisation requires a smaller pump and a smaller pressurisation system. In recycle pressurisation, clarified effluent is recycled for the purpose of adding air and then is injected into the raw wastewater. Approximately 20-50% of the effluent is pressurised in this system. The recycle flow is blended with the raw water flow in the flotation cell or in an inlet manifold.

Important parameters in the design of a flotation system are hydraulic loading rate including recycle, concentration of suspended solids contained within the flow, coagulant dosage, and the air-to-solids ratio required to effect efficient removal. Pilot-plant studies by Stone *et al.* (1975) and Snider (1976) have shown the maximum hydraulic loading rate (also known as the 'overflow rate') to range between 81.5 and 101.8 L/min.m^2 (4.9 and 6.1 m/hour). An efficient air-to-solids ratio was found to be 0.019:1 by Bare (1971). Solids concentrations during Bare's studies were 125 mg/L. However, experimental results with the removal of algae indicate that lower hydraulic rates and air-to-solids ratios than those recommended by the manufacturers of industrial equipment should be used.

Dissolved air flotation with the application of coagulants performs essentially the same function as coagulation-flocculation-sedimentation, except that a much smaller system is required with flotation. Flotation will occur in shallow tanks with hydraulic residence times of 7-20 minutes compared with the several hours retention time in deep sedimentation tanks. Overflow rates of 81.5-101.8 L/min.m^2 (4.9 and 6.1 m/hour) can be employed with flotation; whereas, a value of less than 40 L/min.m^2 (2.4 m/hour) is recommended with sedimentation. However, it must be pointed out that the sedimentation process is much simpler than the flotation process, and when applied to small systems, consideration must be given to whether this increase in system complexity is an appropriate solution.

The flotation process does not require a separate flocculation unit, and it has been found that it is best to add alum at the point of pressure release where mixing occurs and a good dispersion of the chemicals occurs. Brown and Caldwell (1976) designed two tertiary treatment plants that employ algae flotation and have developed several design features not included in standard units, which should be incorporated for good algae removal (Parker, 1976). The Brown and Caldwell study recommended that the tank surface be protected from wind to prevent float movement to one side of the tank. It was also recommended that the flotation tank be covered in rainy climates to prevent the breakdown of the floc by rain. Another alternative proposed has been to store the wastewater in stabilisation ponds during the rainy season and then operate the flotation process at a higher rate during dry weather.

Recently induced air flotation (IAF) systems using a falling flow of water to create fine air bubbles have shown some promise for algae removal from pond effluents and are being used in full-scale systems. A mixture of polymer, coagulant and surfactant to enhance algae and phosphorus removal is mixed with the flow before it enters the flotation tank. With a tank retention time of only five minutes, BOD_5 and suspended solids levels of less than 10 mg/L were consistently achieved in full-scale trials (Conway *et al.*, 2002; Duncan, 2002).

Alum-algae sludge has been returned to the wastewater stabilisation ponds for over 3 years at Sunnyvale, California with no apparent detrimental effect (Farnham, 1981). Neither sludge banks, floating mats of material, nor increased suspended solids concentrations in the pond effluent have been observed. Return of the float to the pond system is also a possibility. Estimates of a period of time that sludge can be returned could range from 10 to 20 years.

Sludge disposal from a dissolved air flotation system can impose considerable difficulties. Alum-algae sludge is very difficult to dewater and discard. Centrifugation and vacuum filtration of unconditioned-algae-alum sludge have produced marginal results. Indications are that lime coagulation may prove to be as effective as alum and produce more easily dewatered sludge.

Brown and Caldwell (1976) evaluated heat treatment of alum-algae sludge using the 'Porteous, Zimpro low-oxidation' and 'Zimpro high-oxidation processes' and found relatively inefficient results. The 'Purifa process', using chlorine stabilisation, produced a sludge dewaterable on sand beds or in a pond, however, the high cost of chlorine eliminates this alternative. If algae are killed before entering an anaerobic digester, volatile matter destruction and dewatering results are reasonable. But as with the other sludge treatment and disposal processes, additional operations and costs are incurred, and the option of dissolved air flotation loses its competitive position.

11.7 MODIFICATIONS AND ADDITIONS TO TYPICAL DESIGNS

11.7.1 Controlled discharge

Controlled discharge is defined as limiting the discharge from a pond system to those periods when the effluent quality will satisfy existing discharge requirements. This practice prevents discharge from the pond during the winter period and during the spring and autumn overturn periods and algal bloom periods. Indeed, in a number of areas, pond discharges during winter months have not been permitted.

Pierce (1974) reported on the quality of pond effluent obtained from 49 pond installations in Michigan, USA that practices controlled discharge. Of these 49 pond systems, 27 have two cells, 19 have three cells, 2 have four cells, and one has five cells. Discharge from these systems is generally limited to late spring and early autumn. However, several of the systems discharged at various times throughout the year. The period of discharge varied from fewer than 5 days to more than 31 days. The ponds were emptied to a minimum depth of approximately 0.46 m during each controlled discharge to provide storage capacity for the non-discharge periods.

During the discharge period, the pond effluent was monitored for BODs, SS, and FC. The results of the study indicated that the most probable (median) effluent BOD_5 concentration for controlled discharge systems was 17 mg/L for the two-cell pond systems and 14 mg/L for three-or-more cell pond systems. There was a 90% probability that the effluent BOD_5 concentration from both two-cell and three- or more cell pond systems would not exceed 27 mg/L. The most probable effluent suspended solids concentration was found to be 30 mg/L for two-cell pond systems and 27 mg/L for three or more cell pond systems, while the 90% probability for effluent SS was 46mg/L for two cells and 47mg/L for three or more.

A similar study of controlled discharge pond systems was conducted in Minnesota (Pierce, 1974). The discharge practices of the 39 installations studied were similar to those employed in Michigan. The results of that study from the autumn discharge period indicated that the effluent BODs concentrations for 36 of 39 installations sampled were less than 25 mg/L, and the effluent SS concentrations were less than 30 mg/L. In addition, effluent FC concentrations were measured at 18 of the pond installations studied. All the installations reported effluent FC concentrations of less than 200/100 mL. During the spring discharge period 49 municipal pond installations were monitored. Effluent BODs concentrations exceeded 30 mg/L at only three installations, while the maximum effluent BODs concentration reported was only 39 mg/L. Effluent suspended solids concentrations ranged from 7 mg/L to 128 mg/L, with 16 of the 49 installations reporting effluent suspended solids concentrations greater than 30 mg/L. Only 3 of the 45 installations monitored for effluent FC concentrations exceeded 200 per 100 mL.

The controlled discharge of pond effluent is a simple, economical, and practical method of achieving a high degree of treatment. Experience indicates that routine monitoring of the pond effluent is necessary to determine the proper discharge period. However, these discharge periods can extend throughout the major portion of the year. While it would be necessary to increase the storage capacity of some pond systems to employ controlled discharge, many pond systems already have additional freeboard and storage capacity that could be utilised without significant additional modification.

11.7.2 Hydrograph-controlled release

The hydrograph-controlled release (HCR) pond is a variation of the Controlled Discharge pond. In this technique the discharge periods are controlled by a gauging station in the receiving stream and are allowed to occur during high-flow periods. During low-flow periods the effluent is stored in the HCR pond. The process design uses conventional facultative or aerated ponds for the basic treatment, followed by the HCR pond for storage/discharge. No treatment allowances are assumed during design for the residence time in the HCR pond, its sole function is storage.

Depending on stream flow conditions, storage needs may range from 30 to 120 days. The design maximum water level in the HCR pond is typically about 2.4 m, with the minimum water level at 0.6 m. Other physical elements are similar to conventional pond systems. The major advantage of HCR systems is the possibility of utilising lower discharge standards during high-flow conditions as compared to a system designed for very stringent low-flow requirements, which is then operated in that mode on a continuous basis. A summary of the design approach used in the USA is shown in Table 11.6.

Table 11.6 Overview of hydrograph-controlled release pond design in the USA

a) Basic principle: At critical low river flow, BOD and SS loadings are reduced by restricting effluent discharge rates rather than decreasing concentration of pollutants.
b) Must be sized to retain wastewater during low flow (Q10/7- once in 10-year low flow rate for 7-day period). Use existing ponds or build storage ponds.
c) Assimilative capacity of receiving stream must be estimated or established by studying historical data

Zirschsky and Thomas (1987) performed a nationwide (USA) assessment of HCR systems, and showed that the HCR system is an effective, economical and easily operated system. It was also found to be an effective means of upgrading a pond system. Several simple effluent release structures are illustrated in the article.

11.7.3 Complete-retention ponds

In areas of the world where the moisture deficit, evaporation minus rainfall, exceeds 75 cm annually, a complete-retention wastewater pond may prove to be the most economical method of disposal if low-cost land is available. The pond must be sized to provide the necessary surface area to evaporate the total annual wastewater volume plus the precipitation that would fall on the pond. The system should be designed for the maximum wet year and minimum evaporation year of record if overflow is not permissible under any circumstance. Less stringent design standards may also be appropriate in situations where occasional overflow is acceptable or alternative disposal is available under emergency conditions. Monthly evaporation and precipitation rates must be known in order to size the system properly. Complete-retention ponds usually require large land areas.

11.8 AUTOFLOCCULATION AND PHASE ISOLATION

Autoflocculation of algae has been observed during some studies (Golueke and Oswald, 1965; McGriff and McKinney, 1971; McKinney *et al.*, 1971; Hill *et al.*, 1977). Chlorella was the predominant algae occurring in most of the cultures. Laboratory-scale continuous experiments with mixtures of activated sludge and algae have produced large bacteria-algae flocs with good settling characteristics (Hill *et al.*, 1977; Hill and Shindala, 1977).

Floating algae blankets have been reported in some cases in the presence of chemical coagulants (Shindala and Stewart, 1971; van Vuuren and van Durren, 1965). The phenomenon may be caused by the entrapment of gas bubbles produced during metabolism or by the fact that in a particular physiological state the algae have neutral buoyancy. In an 11,355 L/hr pilot plant (combined flocculation and sedimentation), a floating algal blanket occurred with alum doses of 125-170 mg/L. About 50% of the algae removed were skimmed from the surface (van Vuuren and van Durren, 1965).

Because of the infrequent occurrence of conditions necessary for autoflocculation, it is not a usable alternative to remove algae from wastewater stabilisation ponds.

Phase isolation experiments to remove algae from pond effluents are based on this concept and some success has been reported, however, full-scale operation of a phase isolation system did not produce consistent results (McGriff, 1981).

11.9 ATTACHED GROWTH

Although baffles are considered useful primarily to ensure good hydraulic efficiency, they also provide a structure for bacteria, algae, and other microorganisms to grow on and this has been attributed to improved performance (Reynolds *et al.*, 1975; Polprasert and Agarwalla, 1995). In ponds with baffling or some other media supporting attached growth, the microbiological community consists of a gradient of algae to photosynthetic, chromogenic bacteria and, finally to nonphotosynthetic, nonchromogenic bacteria (Reynolds *et al.*, 1975).

Polprasert and Agarwalla (1995) presented a model for substrate utilisation in facultative ponds using first-order reactions for both suspended and biofilm biomass, which demonstrated the significance of biofilm biomass growing on the sidewalls and bottoms of ponds.

Details of attached growth pond systems can also be found in Chapter 2.

11.10 LAND APPLICATION/TREATMENT

Land application techniques offer another alternative for upgrading pond systems, again predominantly using filtration type mechanisms. The design and operation of land treatment systems is described in detail in Reed *et al.*, (1995) and Crites *et al.*, (2000).

Land application systems receiving wastewater effluents can be classified in three major categories: slow rate, rapid infiltration and overland flow. The rate of application and the pathway for treated water determine the type of system.

11.10.1 Slow rate irrigation (SR)

Slow rate irrigation is the most commonly used form of land treatment. Wastewater may be applied with sprinklers or surface flooding. Application rates vary from 0.61 to 6.1 m per year. Vegetation is a critical component in the process, and all types of crops have been used: forests, greenbelts, golf courses, and general agriculture. While a large amount of the water is recycled back to the atmosphere via evapotranspiration, the remainder filters down through the root zone and is collected in under-drains or more typically, enters the groundwater. Soil types are ideally medium to fine textured with moderate permeability. Winter storage is required in cold/wet climates. The key design objectives are to avoid runoff, organic/solids overloading at the surface, and leaching of pathogens and/or nitrogen to the groundwater.

11.10.2 Rapid infiltration (RI)

Rapid infiltration usually consists of intermittent flooding of shallow basins in relatively coarse soils with rapid permeability. Application rates vary from 6.1 to 122 m per year, and climate is not critical. Vegetation is not a critical factor in the process and the water percolates to under drains, wells, or groundwater. The potential for contamination of the groundwater needs careful consideration.

11.10.3 Overland flow (OF)

Overland flow is similar to that used for SR, but the wastewater is applied to gently sloping fields and the soils are generally impermeable. Sheet flow down the slope is assumed and the effluent is collected in ditches at the toe of the slope. Vegetation is an essential component and generally systems use water tolerant grasses. Application rates vary from 3 to 21 m of water per year. Winter wastewater storage is needed in cold/wet climates. This process has been successfully demonstrated with municipal and industrial wastewaters, although more recently appears to have become less favoured due to problems with "channelling".

Table 11.7 Site characteristics, design features and expected quality for land treatment (Crites, Reed and Bastian, 2000)

Parameter	Slow Rate	Rapid Infiltration	Overland Flow
Grade	20% cultivated site 40% uncultivated site	Not critical	2 to 8% for final slopes
Soil permeability	Moderate	Rapid	Slow to none
Groundwater depth	0.6 – 3 m	0.9 m during application	Not critical
Climate	Winter storage in cold climates	Not critical	Winter storage in cold climates
Application method	Sprinkler or surface	Usually surface	Sprinkler or surface
Annual loading, m	0.6 – 6	6 – 122	3 – 21
Minimum preliminary treatment	Primary	Primary	Grit removal and comminution
Need for vegetation	Required	Grass sometimes used	Water-tolerant grasses
BOD_5, mg/L	<2	5	10
TSS, mg/L	<1	2	10
NH_3/NH_4 (as N), mg/L	<0.5	0.5	<4
Total N, mg/L	3	10	5
Total P, mg/L	<0.1	1	4
Faecal coli (number/100 ml)	0	10	200+

11.10.4 Design concepts

A summary of site characteristics, design features, and expected effluent from these techniques is presented in Table 11.7. System selection is dependent on the availability of land with the characteristics summarised in Table 11.7.

In humid climates slow rate systems are generally designed for the maximum possible hydraulic loading to minimise land requirements. Systems of this type have been successfully designed for forests, pastures, forage grasses, corn and other crop production. In arid climates in which water conservation is more critical, wastewater is often applied at rates that just equal the irrigation needs of the crop.

The design of pond systems to meet treatment standards required for 'restricted' and 'unrestricted' irrigation is outlined in Chapter 9.

11.11 PARTIAL-MIX AERATED PONDS

In the partial-mix aerated pond system the aeration serves only to provide an adequate oxygen supply, and there is no attempt to keep all of the solids in suspension in the pond as is done with complete-mix and activated sludge systems. Some mixing obviously occurs and keeps portions of the solids suspended, however, an anaerobic degradation of the organic matter that settles does occur. The system is sometimes referred to as a facultative aerated pond system. In the USA suspended solids in partial mix aerated ponds followed by a settling pond are generally less than 30 mg/L assuming that a settling pond following the aerated lagoons is designed for a HRT of 2 days to limit growth of algae.

Even though the pond is only partially mixed, it is conventional to estimate the BOD removal using a simple complete mix model based on first order reaction kinetics. Studies have shown that a plug flow model and first order kinetics more closely predict the performance of these ponds when either surface or diffused aeration is used (Middlebrooks, *et al.*, 1982). However, most of the ponds evaluated in this study were lightly loaded and the reaction rates calculated are very conservative. As illogical as it sounds, because of the lack of better design reaction rates, it is still necessary to design partial-mix ponds using complete-mix kinetics.

11.11.1 Partial-mix design model

The design model using first order kinetics and operating "n" number of equal sized ponds in series is given by Equation 11.3.

$$\frac{C_n}{C_o} = \left[\frac{1}{1+(kt/n)^n}\right] \tag{11.3}$$

Where:

C_n	=	effluent BOD concentration in pond n, mg/L
C_o	=	influent BOD concentration, mg/L
k	=	first order reaction rate constant, $days^{-1}$
	=	0.276 day^{-1} at 20°C (see section 11.11.2)
t	=	total hydraulic residence time in pond system, days
n	=	number of ponds in the series

If other than a series of equal volume ponds are to be employed, it is necessary to use the following general equation:

$$\frac{C_n}{C_o} = \left(\frac{1}{1+k_1t_1}\right)\left(\frac{1}{1+k_2t_2}\right)....\left(\frac{1}{1+k_nt_n}\right) \tag{11.4}$$

Where k_1, k_2 ... k_n are the reaction rates in ponds 1 through n (all usually assumed equal for lack of better information) and t_1, t_2 ... t_n are the hydraulic residence times in the respective ponds.

Mara (1976) has shown that a number of equal volume reactors in series is more efficient than unequal volumes, however, due to site topography or other factors there may be cases where it is necessary to construct ponds of unequal volume.

11.11.2 Selection of reaction rate constants

The selection of the k value is the critical decision in the design of any pond system. A design value of 0.276 day^{-1} is recommended by the Ten States Standards (1978, 1997) at 20°C and 0.138 d^{-1} at 1°C. Boulier and Atchinson (1975) recommended values of k of 0.2 to 0.3 at 20°C and 0.1 to 0.15 at 0.5°C. Reid (1970) suggested a k value of 0.28 at 20°C and 0.14 at 0.5°C based on research with partial mix ponds aerated with perforated tubing in central Alaska. These values are essentially identical to the Ten States Standards (1978, 1997) recommendations. Similar values of k are reported in textbooks.

11.11.3 Influence of number of ponds

When using the partial mix design model, the number of ponds in series has a pronounced effect on the retention time, and therefore, the size of the pond system required to achieve a specified degree of treatment. The effect can be demonstrated by rearranging Equation 11.3 and solving for t:

$$t = \frac{n}{k}\left[\left(\frac{C_o}{C_n}\right)^{1/n} - 1\right] \tag{11.5}$$

All terms in this equation have been defined previously.

11.11.4 Design example

Compare detention times for the same BOD removal levels in partial-mix aerated ponds having one to five ponds. Assume $C_o = 200$ mg/L, $C_e = 30$ mg/L, $k = 0.28\ d^{-1}$, $T_w = 20°C$.

Solution:

1. Solve Equation 11.3 for a single pond system

$$t = \frac{n}{k}\left[\left(\frac{C_o}{C_n}\right)^{1/n} - 1\right]$$

For n = 1

$$t = \frac{1}{0.28}\left[\left(\frac{200}{30}\right)^{1/1} - 1\right]$$

$$t = 20.2\ days$$

2. Repeat for n = 1, 2, 3, …etc

n	t (each cell)	t (total series)
2	5.65 days	11.3 days
3	3.15 days	9.5 days
4	2.17 days	8.7 days
5	1.65 days	8.3 days

3. Continuing to increase n will result in the detention time trending towards the detention time in a plug flow reactor. It can be seen from the tabulation above that the advantages diminish after the third or fourth pond.

11.11.5 Temperature effect

The influence of temperature on the reaction rate is defined by Equation 11.6.

$$k_T = k_{20}\,\theta^{T_w-20} \tag{11.6}$$

Where:

k_T = reaction rate at temperature T, day^{-1}
k_{20} = reaction rate at 20°C, day^{-1}
θ = temperature coefficient = 1.036
T_w = temperature of pond water, °C

The pond water temperature (T_w) can be estimated using the following equation developed by Mancini and Barnhart (1976).

$$T_w = \frac{AfT_a + QT_i}{Af + Q} \tag{11.7}$$

Where:

T_w = pond water temperature, °C
T_a = ambient air temperature, °C
T_i = influent water temperature, °C
A = surface area of pond, m^2
F = proportionality factor = 0.5
Q = wastewater flow rate, m^3/day

Because the pond's surface area is not yet known, an estimate is made, the temperature is calculated using Equation 11.7 and then the pond is sized. After several iterations, the determination of the detention time, and thus the size of the system is completed.

11.11.6 Pond configuration

Even though partial mix ponds are designed using the complete mix model, it is recommended that the ponds be configured with a length to width ratio of 3:1, this is done to minimise hydraulic inefficiency.

11.11.7 Mixing and aeration

The oxygen requirements control the power input required for partial mix pond systems. A complete-mix system would require approximately ten times the power as a system designed to satisfy the oxygen requirements only. In most cases partial mix system design is based on the BOD entering the system to estimate the oxygen requirements.

11.12 MACROPHYTE SYSTEMS

Macrophyte systems are discussed in Chapter 2 (Microbiology) and in particular, Chapter 15 (Wetlands).

11.13 AQUACULTURE

Aquaculture ponds are discussed in Chapter 16.

11.14 UASB

UASB (Upflow Anaerobic Sludge Blanket) systems offer a relatively compact technique for significantly reducing the organic loading before a waste stabilisation pond system therefore reducing the total land area requirement. UASB's are essentially an alternative to anaerobic ponds and have been used effectively in warm climates, particularly Brazil.

UASB's are approximately 20 to 30 times smaller than anaerobic ponds. As part of their design UASB's typically incorporate biogas capture providing for energy recovery and odour control. While the construction cost of a UASB per unit volume is greater than that for an anaerobic pond, it has been argued that because of its reduced size it offers a cost effective alternative to an anaerobic pond (Frassinetti *et al.*, 1996).

11.15 ULTRAVIOLET DISINFECTION

Ultraviolet disinfection can be used to reduce microbiological pathogen levels in waste stabilisation pond effluents. Nelson (2000) found that effluent from pond systems have very low numbers of coliforms associated with its effluent suspended solids compared to effluents from other forms of wastewater treatment. This means that despite the fact that ponds might have relatively high suspended solids concentrations in their effluents they are still easily disinfected by ultraviolet irradiation. For example, a facultative pond effluent containing 170 mg/l of suspended solids was disinfected to produce a residual coliform concentration of 5.6 MPN/100ml. This compared against an activated sludge system that had a suspended solids concentration of 6.8 mg/l and achieved a residual coliform concentration of 42.3 MPN/100ml. Because algal cells have a surface charge and are normally resistant to flocculation it appears that they do not shield pathogens cells from UV radiation to the extent that other wastewater particles tend to do.

11.16 PERFORMANCE COMPARISONS WITH OTHER REMOVAL METHODS

Designers and owners of small systems are strongly encouraged to use as simple technology as feasible. Experience has shown that small communities, or large ones without properly trained operating personnel and access to spare parts, using sophisticated technology inevitably encounter serious maintenance problems and can fail to meet effluent standards. Discussed in this section, are methods that require good maintenance and operator skills, such as dissolved air flotation, coagulation-flocculation and granular media filtration (rapid sand or mixed media filters with chemical addition). At locations where operation and maintenance are available, these processes can be made to work well.

Small communities with limited resources and untrained operating personnel should select as simple a system as applicable to their site situation. In rural areas with adequate land, systems such as controlled discharge ponds or hydrograph controlled release ponds are an excellent choice. Performance by these types of treatment is controlled by selecting the time of discharge and can be controlled to produce an excellent effluent BOD_5 and suspended solids. In arid areas, total containment ponds should be considered.

Where land is limited but resources and personnel are unavailable, it is again best to utilise relatively simple methods to control algae in effluents. Intermittent sand filters, application of effluent to farmlands, overland flow, rapid infiltration, constructed wetlands and rock filters may serve well. Intermittent sand filters with low application rates and a warm climate will provide nitrification. Land application to farmland offers the potential for nutrient recycling.

Energy savings with these type processes are substantial, while still maintaining excellent effluent quality (Middlebrooks *et al.*, 1982). Table 11.8 shows a comparison of expected effluent quality and energy consumption in relatively simple processes up to the most sophisticated used in wastewater treatment. Obviously there are many variables to be considered in design, but energy consumption and the appropriateness of the technology to its location are essential considerations for achieving a well-engineered pond upgrade solution.

Table 11.8 Total annual energy for typical 3785 m^3/day (1 mgd) system including electrical plus fuel (Middlebrooks, *et al.*, 1982)

	EFFLUENT QUALITY				
TREATMENT SYSTEM	**BOD mg/L**	**SS mg/L**	**P mg/L**	**N mg/L**	**ENERGY (1000 kWh/yr)**
Rapid infiltration (facultative pond)	5	1	2	10	150
Slow rate, ridge + furrow (facultative pond)	1	1	0.1	3	181
Overland flow (facultative pond)	5	5	5	3	226
Facultative pond + intermittent sand filter	15	15	-	10	241
Facultative pond + microscreens	30	30		15	281
Aerated lagoon + intermittent sand filter	15	15		20	506
Extended aeration + sludge drying	20	20	-	-	683
Extended aeration + intermittent sand filter	15	15	-	-	708
Trickling filter + anaerobic digestion	30	30	-	-	783
RBC + anaerobic digestion	30	30			794
Trickling filter + gravity filtration	20	10	-	-	805
Trickling filter + N removal + filter	20	10	-	-	805
Activated sludge + anaerobic digestion	20	20	-	-	889
Activated sludge + anaerobic digestion + filter	15	10	-	-	911
Activated sludge + nitrification + filter	15	10	-	-	1,051
Activated sludge + sludge incineration	20	20	-	-	1,440
Physical chemical advanced secondary	30	10	1	-	4,464

REFERENCES

Al-Layla, M. A. and Middlebrooks E. J. (1975) Effect of Temperature on Algal Removal from Wastewater Stabilisation Ponds by Alum Coagulation. *Water Research* **9**, 873-879.

Archer, H. E. and O'Brien, B. M. (2005) Improving Nitrogen Reduction in Waste Stabilisation Ponds. *Water Science and Technology* **51**(12), 133-138.

Bare, W. F. R. (1971) *Algae Removal from Waste Stabilization Lagoon Effluents using Dissolved Air Floatation*. M.S. thesis, Utah State University, Logan, USA.

Bishop, R. P., Reynolds J. H., Filip D. S., and Middlebrooks E. J. (1977) *Upgrading Aerated Lagoon Effluent with Intermittent Sand Filtration*. PRWR&T 167-1, Utah Water Research Laboratory, Utah State University, Logan, UT, 1977.

Boulier, G.A. and Atchinson T.J. (1975) *Practical Design and Application of the Aerated-Facultative Lagoon Process*. Hinde Engineering Company, Highland Park, IL.

Brown and Caldwell. (1976) *Draft Project Report, City of Davis-Algae Removal Facilities*. Walnut Creek, California.

Conway, C., Atkinson, B. and Earnshaw, C. (2002). Returning pond effluent to sensitive waters, a cost effective process for effluent polishing. *5th International IWA Specialist Group Conference on Waste Stabilisation Ponds, Conference Papers Vol. 2*, April 2002, Auckland, New Zealand.

Crites, R. W., Reed S. C., and Bastian R. K. (2000) *Land Treatment Systems for Municipal and Industrial Wastes*. ISBN 0-07-061040-1. McGraw-Hill, New York, USA.

Duncan, C B M. (2002) *Upgrading Processes for Municipal Oxidation Pond Effluents*. Report for MengSt. Department of Civil and Environmental Engineering, The University of Auckland, New Zealand.

EPA. (1983) *Design Manual: Municipal Wastewater Stabilization Ponds*. EPA 625/1-83-015, Environmental Protection Agency, Center for Environmental Research Information, Cincinnati, OH, USA.

Farnham, Helen. (1981) Personal Communication, Sunnyvale, California Wastewater Treatment Plant, Sunnyvale, CA, USA.

Frassinetti, P., Catunda, C. and van Haandel, A. (1996) Improved performance and increased applicability of waste stabilisation ponds by pretreatment in a UASB reactor. *Water Science and Technology*. **33**(7), 147–156.

Golueke, C. and Oswald W. J. (1965) Harvesting and Processing Sewage-Grown Planktonic Algae. *Journal Water Pollution Control Federation* **37**(4), 471-498.

Harris, S. E., Filip D.S., Reynolds J.H., and Middlebrooks E.J. (1978) Separation of Algal Cells from Wastewater Lagoon Effluents, Volume I: Intermittent Sand Filtration to Upgrade Waste Stabilization Lagoon Effluent. EPA-600/2-78-033, NTIS No. PB 284925, U.S. Environmental Protection Agency, Municipal Environmental Research Laboratory, Cincinnati, OH.

Hickey C.W., Quinn J.M., and Davies-Colley R. J. Effluent Characteristics of domestic sewage oxidation ponds and their potential impacts on rivers. *New Zealand Journal of Marine and Freshwater Research* (1989) Volume 23, 585-600.

Hill, D.O. and Shindala A. (1977) Performance Evaluation of Kilmichael Lagoon. US Environmental Protection Agency, EPA-600/2-77-109, Municipal Environmental Research Laboratory, Cincinnati, OH, USA.

Hill, F.E., Reynolds J.H., Filip D.S., and Middlebrooks E.J. (1977) Series Intermittent Sand Filtration to Upgrade Wastewater Lagoon Effluents. PRWR 153-1, Utah Water Research Laboratory, Utah State University, Logan, UT, USA.

Hutchinson E.G. and Oliver D.A. (1978) A Survey of Oxidation Ponds in the Auckland Region. Auckland Regional Authority. Works Division.

Mancini, J.L., and Barnhart E.L. (1976) Industrial Waste Treatment in Aerated Lagoons, Ponds as a Wastewater Treatment Alternative, *Water Resources Symp. No. 9*, University of Texas, Austin, USA.

Mara, D.D. (1976) Sewage Treatment in Hot Climates, John Wiley, New York, USA.

Marshall, G.R., and Middlebrooks E.J. (1974) Intermittent Sand Filtration to Upgrade Existing Wastewater Treatment Facilities. *PRJEW 115-2*, Utah Water Research Laboratory, Utah State University, Logan, UT, USA.

McGarry, M. G. (1970) Algal Flocculation with Aluminum Sulfate and Polyelectrolytes. Journal Water Pollution Control Federation, **42**(5), R191.

McGriff, E.C. and McKinney R.E. (1971) Activated Algae? A Nutrient Removal Process. *Water and Sewage Works* **118**, 337.

McGriff, E.C. (1981) Facultative Lagoon Effluent Polishing Using Phase Isolation Ponds. EPA-600/2-81-084, NTIS No. PB 81-205965, U. S. Environmental Protection Agency, Municipal Environmental Research Laboratory, Cincinnati, OH, USA.

McKinney, R.E., et al. (1971) Ahead: Activated Algae? *Water Wastes Engineering*, **8**(51).

McNabb, C.D. (1976) The Potential of Submerged Vascular Plants for Reclamation of Wastewater in Temperate Zone Ponds, in Biological Control of Water Pollution, University of Pennsylvania Press, Philadelphia, pp123-132.

Messinger, S.S. (1976) Anaerobic Lagoon-Intermittent Sand Filter System for Treatment of Dairy Parlor Wastes. M.S. Thesis, Utah State University, Logan, UT, USA.

Melcer, H., Evans B., Nutt S.G. and Ho A. (1995) Upgrading Effluent Quality for Lagoon-Based Systems. *Water Science and Technology* **31**(12), 379-387.

Middlebrooks, E.J., Middlebrooks C.H., Reynolds J.H., Watters G.Z., Reed S.C. and George D.B. (1982) Wastewater Stabilization Lagoon Design, Performance, and Upgrading, Macmillan, New York, USA.

Nelson, K. (2000). Ultraviolet light disinfection of wastewater stabilization pond effluents. *Water Science and Technology* **42**(10-11), 165-170.

Parker, D. S. (1976) Performance of Alternative Algae Removal Systems. In Ponds as a Wastewater Treatment Alternative, edited by E. F. Gloyna, J. F. Malina, Jr., and E. M. Davis, Center for Research in Water Resources, College of Engineering, The University of Texas at Austin, USA.

Pierce, D. M. (1974) Performance of Raw Waste Stabilization Lagoons in Michigan with Long Period Storage Before Discharge. In Upgrading Wastewater Stabilization Ponds to Meet New Discharge Standards. PRWG151. Utah Water Research Laboratory, Utah State University, Logan, Utah, USA.

Polprasert, C. and Agarwalla B.K. (1995) Significance of Biofilm Activity in Facultative Pond Design and Performance. *Water Science and Technology* **31**(12), 119-128.

Rich, L. G. and Wahlberg, E. J. (1990). Performance of Lagoon-Intermittent Sand Filter Systems. *J. Water Pollut. Control Fed.* **62**, 697-699.

Reed, S.C., Crites R.W., and Middlebrooks E.J. (1995) Natural Systems for Waste Management and Treatment, 2nd Edition, ISBN 0-07-060982-9, McGraw-Hill, New York.

Reid, L.D., Jr. (1970) Design and Operation for Aerated Lagoons in the Arctic and Subarctic, Report 120, U.S. Public Health Service, Arctic Health Research Center, College, AK.

Reynolds, J.H., Nielson S.B., and Middlebrooks E.J. (1975) Biomass Distribution and Kinetics of Baffled Lagoons. *Journal of the Environmental Engineering Division*, ASCE, 101, EE6, 1005-1024.

Russell, J.S., Middlebrooks E.J., Lewis R.F., and Barth E.F. (1983) Lagoon Effluent Polishing with Intermittent Sand Filters. *Environmental Engineering Journal*, ASCE, **109**(6), 1333-1353.

Saidam, M.Y., Ramadan S.A. and Butler D. (1995) Upgrading Waste Stabilization Pond Effluent by Rock Filters. *Water Science and Technology* **31**(12), 369-378.

Shindala, A., and Stewart J.W. (1971) Chemical Coagulation of Effluents from Municipal Waste Stabilization Ponds. *Water and Sewage Works*, **118**(4) 100-103.

Snider, E.F., Jr. (1976) Algae Removal by Air Flotation. In Ponds as a Wastewater Treatment Alternative, edited by E. F. Gloyna, J. F. Malina, Jr., and E. M. Davis, Center for Research in Water Resources, College of Engineering, The University of Texas at Austin.

Stamberg, J. B., et al. (1984). Simple Rock Filter Upgrades Lagoon Effluent to AWT Quality in West Monroe, Louisiana, paper presented at 57th Conf. Water Pollution Control Federation, New Orleans, LA, USA.

Tenney, M.W. (1968) Algal Flocculation with Aluminum Sulfate and Polyelectrolytes. *Applied Microbiology* **18**(6), 965.

Ten States Recommended Standards for Sewage Works. (1978, 1997) A Report of the Committee of Great Lakes-Upper Mississippi River Board of State Sanitary Engineers, Health Education Services Inc., Albany, NY, USA.

Truax, D.D. and Shindala A.. (1994) A Filtration Technique for Algal Removal from Lagoon Effluents. *Water Environment Research* **66**(7), 894-898.

Tupyi, B., Reynolds J.H., Filip D.S. and Middlebrooks E.J. (1979) Separation of Algal Cells from Wastewater Lagoon Effluents, Volume II: Effect of Sand Size on the Performance of Intermittent Sand Filters. EPA-600/2-79-152, NTIS No. PB 80-120132, U.S. Environmental Protection Agency, Municipal Environmental Research Laboratory, Cincinnati, OH.

van Vuuren, L.R.J. and van Duuren F.A. (1965) Removal of Algae from Wastewater Maturation Pond Effluent. *Journal Water Pollution Control Federation*, **37**, 1256.

Williamson, K.J. and Swanson, G.R. (1978) Field Evaluation of Rock Filters for Removal of Algae from Lagoon Effluents. In Proceedings of Conference on Performance and Upgrading of Wastewater Stabilization Ponds. August 23-25 1978, Utah State University, Logan, Utah, USA.

Zirschsky, J. and Thomas R.E. (1987) State of the Art Hydrograph Controlled Release (HCR) Lagoons. *Journal Water Pollution Control Federation* **59**(7), 695-698.

12

Operation, maintenance and monitoring

Barry Lloyd

12.1 INTRODUCTION

The operation, maintenance and monitoring of Waste Stabilisation Pond (WSP) systems ranges from excellent practice to total neglect. Smaller systems tend to be the most neglected for infra-structural and financial reasons. Although WSPs are promoted for their simplicity in construction and operation and maintenance (O and M), it does not necessarily follow that they work effectively if neglected. The long-term success or failure of a pond system depends upon adequate and sustained financial support for O and M activities.

ISBN: 1843390205. Published by IWA Publishing, London, UK.

12.2 OPERATION

12.2.1 The operating manual

Although WSPs are considered to be simple to operate it is still essential that the designer of the pond system should provide an operating manual.

Design concepts and operational procedures, and associated estimates of operating costs and staffing requirements, should have been discussed at an early stage in the project so that the operating authority clearly understands and budgets for system operation. This may seem obvious for large schemes but also applies to WSP systems serving small communities and towns, particularly in developing countries, where systems are failing due to lack of attention to basic operating guidelines.

The completed draft of the manual should be placed in the hands of the operating authority at least three months before a system is commissioned.

The operating manual must cover:

1) design criteria and their implications, with technical detail including ratings and significant dimensional parameters;
2) operating procedures, supported by manufacturers' instructions for all machinery and equipment;
3) maintenance procedures, supported by manufacturers' instructions and parts lists, and lists of essential tools.

After the first year of operating experience the manual should be reviewed and amended by the operating authority in consultation with the operators. Subsequent periodic amendments should be made to take account of any changes (extension, up-grading or rehabilitation) of the system.

12.2.2 Records

Records refer to:

1) full engineering drawings of the system as constructed;
2) summarised design criteria and calculations of performance characteristics;
3) records required for operation and maintenance.

The engineering drawings are required to locate all the components of the system on the ground and under the ground (e.g. de-sludging drainage pipes). The drawings aid in the understanding of the design and hence how the system is intended to operate.

The operation of a system is dependent upon an understanding of the layout of the component parts. A flow chart may be used to show sequentially the treatment stages in a system. A series of scale drawings showing increasing detail, starting with the layout of the whole WSP system and ending with detail of individual components is required. Multiple sets of the drawings should be held in the office of the managing authority and at least one set should be permanently stored at the treatment plant if it is large enough to warrant an on-site office/maintenance building. The layout of the whole treatment plant should be provided at an appropriate scale. Include all control valves and pipe dimensions. Provide details on service buildings, and any additional land available for extension of the pond system. The site layout should extend at least to the security perimeter fencing. It is also advisable to include features external to the treatment plant such as access roads, sewer pipes, pumping stations and power supply feeding the plant and final discharge point to receiving water bodies.

Records required for operation and maintenance

A record is required for every item of mechanical and electrical plant in the system for identification and for recording performance and maintenance carried out. In one developing country study, 100% of pumped WSP systems were not operational from 5 months up to 12 months of the study year due to operation and maintenance problems in pumping (Lloyd *et al.* 2002).

12.2.3 Operations staff

Recommended staffing levels for WSP systems were defined by Arthur (1983) according to the population served and hence the size and complexity of the plant. In the 1960s and 1970s most medium sized sewage works in developed countries had their own on-site laboratory and the level of supervisory and technical staffing indicated in Table 12.1 was a fair representation of the situation then. Privatisation within the water sector in the 1990s had a marked impact on staffing levels and in a number of countries, particularly in Western Europe, staffing levels have been cut by around 25-30%. The tendency to 'rationalise' operations and 'down size' their staffing has extended to quality monitoring so that few medium sized works now operate an on-site laboratory, and quality monitoring increasingly tends to be conducted in centralised laboratories. In some countries, such Australia and New Zealand, the staffing is less than a third of the totals in Table 12.1 .

Mara and Pearson (1998) noted the importance of adequate staffing levels for routine O and M, and indicated that the level of staffing also depends on:

- the type of inlet works (manual screen raking; maintenance of grit removal or mechanical screens);

- on-site laboratory facilities (requiring a technician);
- site maintenance, particularly grass cutting (manually requiring more labourers than mechanical mowers).

Table 12.1 Staffing levels for WSP systems proposed by Arthur (1983)

Population served	**10,000**	**25,000**	**50,000**	**100,000**	**250,000**
Foreman/Supervisor	-	-	1	1	1
Mechanical engineer	-	-	-	1	1
Laboratory technician	-	1	1	1	2
Assistant foreman	-	1	2	2	2
Labourers	1	2	4	6	10
Driver	-	1	1	1	2
Watchman	1	1	2	3	5
Total	2	6	11	15	23

Table 12.2 Daily hours of operator (hours/day) devoted to O and M, and the proportion of pond systems with either fair or poor O and M in Catalonia (Garcia *et al.* 1999).

Parameter	Aerated pond systems [n = 13]	Stabilisation pond systems [n = 7]	WWTP with maturation ponds [n = 4]
Daily hours			
• average	3.2	0.6	5.9
• range	(0.6 – 8)	(0.2 – 1)	(1.0 – 12)
Proportion with fair to poor O & M	8 %	85 %	0 %

Garcia *et al.* (1999) reported the hours/day worked by operators at 24 treatment plants in Catalonia where all plants had at least one operator. Table 12.2 shows that the lowest average hours worked per day (0.6 h d^{-1}) were at small stabilisation pond systems where O and M inputs were done only weekly or less frequently. The highest requirements for O and M (5.9 h d^{-1}) were for conventional Wastewater Treatment Plants (WWTPs) incorporating subsequent maturation ponds, due to their electro-mechanical equipment and large size. All of the stabilisation pond systems except one suffered from problems of preliminary treatment failures, plant growth on embankments and failure to meet European effluent standards as a result of the minimal and poor level of O and M. By contrast the aerated pond systems and the conventional WWTPs met the physicochemical standards thus benefiting from the increased time invested. This study emphasises the importance of attention to O and M even in WSP systems, which are promoted for their low maintenance requirements.

Table 12.3 Factors influencing job satisfaction (adapted from Hertzberg, 1968)

	Motivators	Dissatisfaction
1	Achievement	Company policy/administration
2	Recognition	Lack of supervision
3	Good facilities	Working conditions, including security
4	Responsibility	Poor interpersonal relations
5	Training	Low status
6	Advancement	Low wage, no prospect of promotion

The effective operation of even the smallest system depends upon trained, interested and motivated operators. However, WSP system operators in many countries are unskilled or semi-skilled. Their remuneration is low, their motivation is low and therefore their performance and work attendance may also be low.

Table 12.3 allows the manager to consider what factors might improve operators' motivation, and hence job satisfaction and performance. Although it is well known that managerial interest is a powerful motivator, personnel management and supervision are still key areas for improvement worldwide. This applies equally to the construction and rehabilitation of works as to the operation, maintenance and monitoring of systems. Gloyna (1971) pointed out that 'people accept responsibility and work most effectively when there are well-defined lines of communication'. However, in many developing countries the responsibility for the day-to-day operation of systems has been assigned to small municipalities that lack the management infrastructure to operate and maintain them in a professional manner. Furthermore, the lines of communication are often poorly defined. Only irregular contacts with the professional regional management authority occur and the WSP system supervisor in small municipalities is also often multi-tasked. These are fundamental problems that should have been addressed in the operating manual and at the commissioning stage of the system. It must be added that some of the most critical operational problems could be avoided if supervisors regularly checked the work done by their operators and ensured that they were provided with the facilities and tools for the job.

Training

The importance of training cannot be over-emphasised in contributing to job satisfaction and enrichment. It is axiomatic that operators should attend an initial training workshop, which should include:

- their specific duties;

- the basic principles of the WSP system for which they are responsible;
- the significance of WSPs for human health and environmental quality.

Periodic refresher courses should be organised, e.g. when the system is upgraded or rehabilitated. This should be organised so as to bring together operators from other systems for information exchange.

Operator advancement is problematic in that the opportunities are limited. However a progressive authority will set in place on-the-job training and grading systems so that unskilled staff can progress to semi-skilled and on to supervisory level. One possible grading system could be: 1) Basic training. 2) Operators certificate. 3) Higher operators certificate. 4) Supervisors certificate.

12.2.4 Facilities, safety and security

Even in small pond systems it is desirable to provide basic facilities for operators and visitors including:

1) Operators' office with desk, separate toilet, wash basin, shower and first aid kit;
2) Protective clothing, including life jackets for sampling from a boat;
3) Lifebuoys (lifebelt) placed at strategic locations adjacent to each pond;
4) Secure storage for tools and equipment.

The operating manual should include an inventory of tools required for maintenance. Overall plant security requires that a high fence must surround the entire plant with lockage access gates. Warning notices must be attached to the fence clearly stating that it is a "wastewater treatment plant, hazardous to health; no person may enter without the authorisation of the (named) authority".

Hygiene standards must be defined to protect workers' with health and safety notices displayed at key points on the works such as the main gate and office.

In addition to the facilities mentioned above, protective clothing for staff should include the items listed in Table 12.4.

Table 12.4 Protective clothing for O and M staff

Work clothes	Overalls	Raincoat
Work boots (steel capped)	Rubber boots	Harness
Protective gloves	Rubber gloves	Nose and mouth mask
Hard hat	Plastic protective glasses	

12.3 MAINTENANCE

12.3.1 Maintenance duties and procedures

Table 12.5 attempts to summarise the range of duties necessary for the adequate maintenance of a WSP system and the frequency with which these activities should be carried out. However, it is difficult to be prescriptive regarding the frequency of every activity, since every works is different.

Table 12.5 Summary of principal activities for maintenance of WSP systems

Location	Activity	Frequency
Inlet works	1. Intake screens: removal of solids	Daily
	2. Grit channels: removal of grit	Daily
	3. Transfer of solids and grit to a designated storage or disposal area	Daily
	4. Pumps: greasing/oiling and inspection	As required
	5. Maintenance of pumping records	Daily
Ponds	1.Maintaining flow through plant by cleaning and removal of solids accumulated at inlet and outlet structures and connecting canals and overflow weirs	Daily
	2.Removal and dispersion of floating scum	As required
	3.Maintaining pond embankments by cutting grass and repairing damage caused by burrowing animals and wave action	As required + weekly inspection
	4. Control insect breeding by oscillating pond levels and removing emergent macrophytes from the ponds	As required
	5. Operation and greasing of valves for pond drainage	Monthly
General site	1. Cleanliness and safety of the plant	Weekly
	2. Inspection and repair of perimeter fence and hedges and access roads	Quarterly
	3. Painting of corrodible structures	Yearly
	4. Record keeping	Daily and weekly

The less skilled the operators are, the greater the need for them to have a programme of duties to follow during their shift. Gloyna's (1971) advice is as valid today as it was then: 'Management must establish an effective operation and maintenance programme by formulating and adopting an organisational chart that clearly defines a chain of authority and responsibility....operators must be given regular instruction in maintenance; each operator must be thoroughly acquainted with the planned or "preventive" maintenance that must be routinely performed, and with any "curative" maintenance that may be required in an emergency'.

The operating manual should define standards for the duties to be undertaken including their timing and safety protocols.

12.3.2 Inlet works and intake structures

Some pond installations do not have inlet screening, but this results in items such as condoms and sanitary rags accumulating in the pond and in its sludge. While removal of floating items can be made from the pond once this material is in the sludge, it makes it less desirable for land application when it comes time to desludge the pond. Therefore screening at the inlet is strongly recommended. A common form of intake structure for small WSP systems is illustrated in Figure 12.1. This shows the coarse bar screens, sometimes called bar racks, at the entrance of a small WSP system. The primary bar screens illustrated must be 'hand-cleaned' by rake at least daily. Typically bar screens are constructed from parallel steel bars with clear openings between each bar of 15mm which allows sand and grit, but not coarse gravel or pebbles to pass. In larger more sophisticated plants the screens are mechanically cleaned, or a perforated rotating drum may undertake the screening, and the screenings are automatically removed to a skip for disposal to landfill. An account of a comprehensive range of mechanical screening devices may be found in Metcalf and Eddy (2003).

Figure 12.1 shows two parallel grit channels behind the bar screens. These should be designed to allow grit to settle out by reducing the velocity of flow to approximately 0.3m per second. The constant velocity can be achieved by parabola-shaped channels or placing a Parshall flume immediately after the grit channel. The two channels can be operated independently and shut off by gates (as shown) to allow drainage before cleaning. The grit can then be manually removed to be barrowed for storage and disposal. The adze is an appropriate tool for manual grit removal since its blade is at a right angle to the handle, facilitating the lifting of grit from the channel. In large mechanical systems grit removal may be achieved by power-operated suction hoses, but these are expensive to install and maintain and therefore not generally appropriate for small installations in developing countries.

Figure 12.1 Manual maintenance of inlet screen (left) ahead of parallel grit channels showing independent operation for grit removal (right). Photos: A. Leitner.

In addition to horizontal channels, other methods of grit removal exist and may be considered for larger plants. Again, general wastewater treatment textbooks such as Metcalf and Eddy (2003) provide further information on these techniques. There is some debate as to whether grit removal is always needed on pond systems at all because of their large storage volume. But, as discussed in Chapter 9, lack of grit removal can create problems, such as build-up near inlets.

12.3.3 Pond maintenance

In order to maintain a uniform flow through the ponds and to help avoid short-circuiting the operator must routinely remove any solids accumulating in inlet and outlet channels and pipes, from outlet baffles and from overflow weirs.

The removal of floating scum from ponds is important, as this is sometimes the source of odours (as described in section 12.6). Scum removal can often be achieved using long handled skimmers and pressure hoses may disperse larger accumulations of scum and solids. In advanced facultative ponds, Green *et al.* (1996) have described the incorporation of down-wind concrete beaches or scum ramps where floatable trash can be cast up by the wind to dry. This material can then be more easily collected for burial or disposal as a solid waste.

The routine maintenance of pond embankments is essential to their long-term use. Grass embankments have to be regularly cut, and damage caused by burrowing animals and wave action repaired. Weed growth is a particular problem on unprotected soil embankments as this creates quiescent conditions, which favour insect breeding, such as mosquitoes, so emergent plants should be removed at regular intervals. For concrete and rip-rap embankments this is a lesser problem, but weed growth and seedling trees should still be regularly removed from joints between concrete sections.

In some countries it has been found that rock filled pond embankments provide the ideal breeding site for indigenous nuisance insects just below the water line. For example, at the large WSP system of Mangere in Auckland, New Zealand, (now decommissioned) chironomid midge infestations occurred in the main water body of the ponds when organic loadings were below 40kg BOD_5 ha^{-1} d^{-1}, but with higher loadings, midge breeding was limited to the ponds edges. Lawty *et al.* (1996) reported that the midge larvae could be controlled by injecting malathion just below the water surface at the pond edges. Elsewhere the development of biological control using the insect pathogen *Bacillus thuringiensis* has progressed to the commercial branding of strains of this organism.

12.3.4 Maintenance records

Because routine maintenance procedures are straightforward in WSP systems this is not a justification for a lack of record keeping. Good maintenance records complement reliable monitoring records and taken together are indicative of professional standards and a responsible work ethic. Conversely, the lack of records is a recipe for failure. It is recommended that a weekly worksheet is completed that records the shifts worked and by whom and provides checkboxes for a list of regular maintenance tasks and space for general observations to be recorded. Overall responsibility for ensuring that daily and less frequent activities are done is the responsibility of the plant supervisor who should therefore sign off the report sheet as a true record of maintenance carried out each week.

12.4 MONITORING

12.4.1 Definitions and objectives

In the general context of the water sector, monitoring has been defined by Meybeck and Helmer (1996) as "the long term, standardised measurement and observation of the aquatic environment, in order to define status and trends".

In WSP monitoring, the overall objective is to ensure that at the point of final disposal, the effluent conforms to established quality standards. These standards are set with regard to the characteristics of the receiving water body, taking into consideration the need for:

- protection of human health;
- conservation of aquatic life; and

- potentially, multiple forms of reuse.

The simplest form of routine WSP monitoring is when only the effluent is monitored to check compliance. Unfortunately in many countries monitoring data are produced and filed, and nothing further is done with the information. The World Health Organisation has therefore attempted to promote additional investigative and evaluative techniques to encourage and promote remedial action, under the umbrella term of surveillance. Surveillance was defined by WHO (1976) as: "The continuous and vigilant public health assessment and overview of the safety and acceptability of drinking-water supplies".

Although originally applied primarily to drinking water and water resources, Lloyd (2002) has extended the application of this terminology to the monitoring and evaluation of the water cycle, including sewage treatment plants. In the specific case of WSPs, surveillance may be taken to mean "the monitoring, investigation and evaluation of WSPs for the purpose of effluent quality (and sludge) management and operational control".

When WSP systems do not meet mandatory standards more complicated monitoring and evaluations are called for to determine whether this is due to under-performance of the treatment plant and/or to factors associated with qualitative and quantitative changes in the raw sewage. Thus additional objectives of surveillance and monitoring are to:

- identify the cause(s) of under-performance of the treatment plant;
- check if that hydraulic and physicochemical loadings are within the design specification;
- predict future upgrading requirements;
- check the physical, chemical and biological characteristics of raw sewage sources.

The last of the bullet points (above) allows for the consideration of a rapidly increasing array of industrial chemicals. Some authors claim that WSP systems are relatively more resilient to many types of trade effluent when compared with e.g. conventional biological treatment processes. However, a discussion of the numerous contaminants found in industrial effluents is outside the scope of this section. Industrial chemicals should be dealt with by trade effluent regulation, and thus be prevented from entering the foul sewer.

When many developing countries are struggling to develop the most basic of monitoring programmes, it seems more important to identify a short list of critical parameters. Consequently the focus of this section of the chapter will be placed on those fundamental parameters identified as being of key significance to the receiving environment, and presenting the greatest risks to human health when reuse is practised.

12.4.2 Regulatory monitoring programmes

The qualitative and quantitative design of basic monitoring programmes are dictated by the regulatory authority, which prescribes the parameters and standards which effluent discharge must meet. As an example, for purposes of statutory control the Urban Wastewater Treatment Directive for the European Economic Community (EEC, 1991) listed only five effluent parameters. The Directive states 'it is necessary to monitor treatment plants, receiving waters and the disposal of sludge to ensure that the environment is protected from the adverse effects of the discharge of wastewaters' and that this information 'is made available to the public in the form of periodic reports'. Three of the parameters (BOD_5, COD and Total Suspended Solids) are required to meet the standards, presented in Table 12.6, by ensuring that 'secondary (conventional) treatment, or an equivalent (WSP) alternative' is in place.

Table 12.6 EEC quality requirements for all discharges from urban wastewater treatment plants

Parameters	Effluent concentration	Minimum reduction (%)	Reference method of measurement
BOD_5 at 20^0C, without nitrification	25 mg/l	70 - 90	Homogenised, unfiltered, undecanted sample. Determination of dissolved oxygen before and after 5 days incubation at 20^0C in darkness. Addition of a nitrification inhibitor.
COD	125 mg/l	75	Homogenised, unfiltered, undecanted sample. Potassium dichromate method.
Total suspended solids	35 mg/l	90	Filtering of a representative sample through a 0.45 μm membrane. Drying at 105°C.

The EEC Directive also defined the minimum percentage reduction in each parameter that the treatment plants must achieve, and the reference method prescribed for the analysis, as indicated in Table 12.6. Whereas the first three parameters listed apply to all discharges to water bodies, two additional parameters defined in the Directive are intended to protect water bodies, which are sensitive to eutrophication. These parameters are total phosphorus and total nitrogen. Table 12.7 specifies different standards for different population equivalents (p.e.) reflecting the greater difficulty for smaller, less well managed systems to achieve the more stringent standards.

Table 12.7 EEC requirements for discharges from urban wastewater treatment plants to sensitive water bodies

Parameters	Effluent concentration	Minimum reduction (%)	Reference method of measurement
Total phosphorus	2 mg/l P (10,000–100,000 p.e.) 1 mg/l P (>100,000 p.e.)	80	Molecular absorption spectrophotometry
Total nitrogen	15 mg/l N (10,000 –100,000 p.e.) 10 mg/l N (>100,000 p.e.)	70 - 80	Molecular absorption spectrophotometry

Standards have also been traditionally influenced by the quantitative flow to the plant, the degree to which the effluent is diluted by receiving waters, and also the uses for which waters are required.

Some governments and in some cases, supra-governmental bodies, prescribe the minimum frequency with which samples must be taken and analysed. Thus the Council of the European Communities (EEC, 1991) requires samples to be taken according to the population equivalents (p.e.) served.

It is emphasised that the frequency of sampling indicated in Table 12.8 should be regarded as the minimum required for statutory purposes. There is no way that one or two samples per month can be considered as effluent quality control. Responsible authorities are encouraged to increase sampling frequency but rarely do so for reasons of cost. The fact that increasing the number of samples increases the statistical probability of identifying a failure can also be a disincentive to increasing sampling frequency.

Table 12.8 Prescribed effluent monitoring frequency (EEC, 1991)

Population equivalents (p.e.)	Samples per year
2,000 - 49,999	12
≥ 50,000	24

Other factors that should be considered in deciding on the frequency of sampling include the following:

- *Seasonal*: Pond performance is strongly influenced by climate and therefore seasonal changes in temperature, rainfall and wind patterns should be considered.
- *Diurnal*: Raw sewage generally shows greater daily variation in its strength and flow than subsequent stages of treatment and, at least,

at the beginning of a monitoring programme this should be accounted for by increased sampling.

- *Under-performance*: If the plant fails to meet discharge standards a review of operation, maintenance and monitoring must be undertaken by a knowledgeable investigator. If necessary monitoring is intensified in order to identify the causes of failure and rectify them.

12.4.3 WSP evaluations

Pearson *et al.* (1987) presented guidelines for the minimum *evaluation* of full-scale ponds, to obtain a reasonable estimate of *performance*. For this purpose they listed 7 important parameters: BOD_5, COD, SS, NH_3-N, NO_3-N, total P and faecal coliforms.

Pearson and his colleagues drew a clear distinction between *effluent quality monitoring* and *evaluation of pond performance*. They recognised that 'the technical man-power, equipment and materials required even for (effluent monitoring) may be partly or totally lacking' and therefore recommended the 'absolute minimum that must be done in order to gain ... a limited estimate of pond performance'.

Pearson *et al.* (1987) and Mara and Pearson (1998) recommended two levels of *effluent monitoring programme*, and a third level for more advanced 'minimum' evaluations which are often beyond the capabilities of local organisations:

- *Level 1*: In accordance with the EEC Directive (EEC, 1991) shown in Tables 12.6 -12.8.
- *Level 2*: To be applied when level 1 showed that a pond effluent is failing to meet its discharge or reuse quality.
- *Level 3*: To include additional sample locations, as indicated in Table 12.9.

For Level 2 effluent monitoring, Mara and Pearson (1998) recommended all 15 parameters listed in Table 12.9, although parameters 8-14 are only required for assessing the suitability of effluents for reuse in crop irrigation. They specified that parameter 6) pH, 7) Temperature, and 15) Faecal coliforms, should be taken as grab samples and that two samples for pH and temperature should be taken, one in the morning between 8.00 am and 10.00 am, the other between 2.00 pm and 6.00 pm. The faecal coliform sample should be taken between 8.00 am and 10.00 am. They also recommended that the other 12 quality parameters should be taken as 24-hour flow-weighted composite samples. If resources permit, these can be collected using automatic samplers, which take samples every one or two hours. Otherwise grab samples have to be taken every one to three hours, and subsequently combined.

The Level 3 monitoring proposed by Mara and Pearson (1998) extends sampling from final effluent, to all inlet-outlet points in the WSP series, for eight of the parameters (Table 12.9). This means that the performance of each stage of treatment can be calculated for relevant parameters and compared with the design specification. Additional parameters (chlorophyll, algal identification, column sampling, sulphide, dissolved oxygen) may also be monitored. If this strategy fails to identify the cause(s) of under-performance, then more advanced investigations including hydraulic evaluation may have to be undertaken (Lloyd *et al.* 2002).

Table 12.9 WSP level 2 effluent monitoring parameters and level 3 evaluation parameters (Adapted from Pearson *et al.* 1987)

Level 2 effluent parameters	Sample locations for Level 3 evaluation	Sample container	Preservation	Maximum storage time
1) Flow	RS and final effluent	-	-	-
2) BOD_5	RS, all pond effluents	PE	Cool 4°C	4 h
3) COD	RS, all pond effluents	PE	Cool 4°C	24 h
4) Suspended Solids	RS, all pond effluents	PE	Cool 4°C	7 days
5) Ammonia-N	RS, all pond effluents	PE	Cool 4°C acidified	24 h
6) pH	RS, all pond effluents	PE	None	6 h
7) Temp. ^{0}C	RS, all pond effluents	-	-	On site
8) Total Nitrogen-N	RS and final effluent *	PE	Cool 4°C	24 h
9) Total Phosphorus	RS and final effluent *	PE	Cool 4°C	1 month
10) Chloride	RS and final effluent *	PE	Cool 4°C	7 days
11) Electrical Conductivity	RS and final effluent *	PE	Cool 4°C	24 h
12) Ca, Mg, Na	Final effluent *	PE	Cool 4°C	7 days
13) Boron	Final effluent *	PE	Cool 4°C	6 months
Microbiology				
14) Helminth Eggs	RS, all pond effluents *	PE	None	
15) Faecal Coliforms	RS, all pond effluents	Glass	Cool 4°C	< 6 h

RS = Raw sewage. * = Samples required only when the effluent is for crop irrigation. PE = Polyethylene.

Lloyd and Vorkas (1999) used the review of locally available routine monitoring data as the starting point (Step 1) for the development of a more advanced and systematic evaluation method, which is summarised in Figure 12.2.

The desk study of monitoring data from a collection of systems may be used to determine that monitoring is adequate to assess whether WSPs are meeting national standards, and whether the standards are being met. If these criteria are met then all is well; if they aren't then Step 2 is initiated for selected systems.

Step 2 (preliminary diagnostic), involves a rapid on-site assessment to identify obvious design and operational problems, and requires access to the full engineering drawings of the plant and to the operation and maintenance records. The findings of the preliminary diagnostic work lead to Step 3 (intensive evaluation). This involves setting up a properly designed, advanced monitoring programme and a series of additional hydraulic investigations to determine the cause of under-performance.

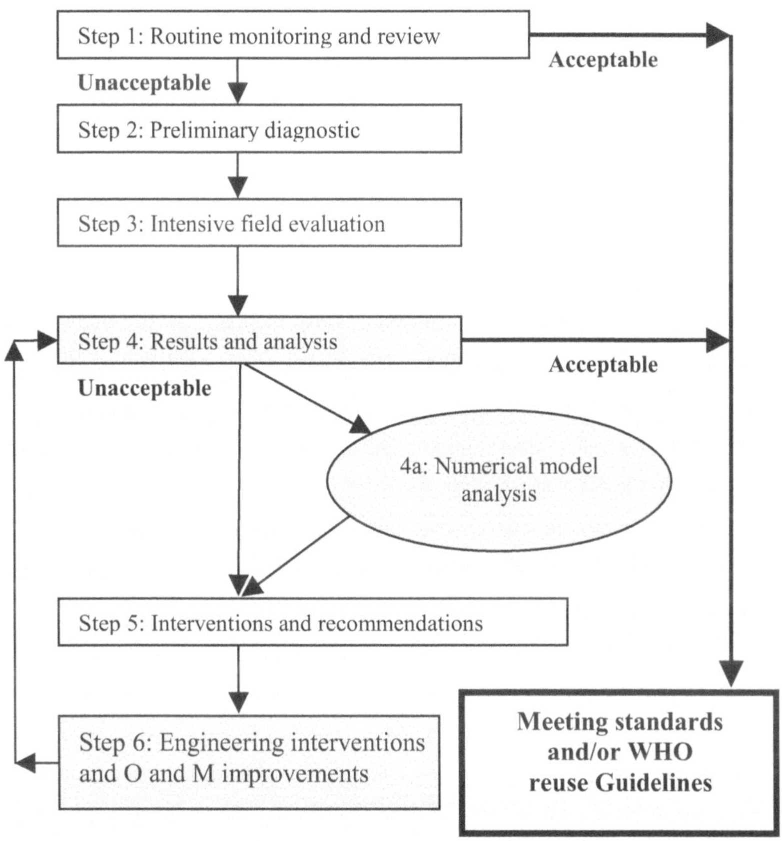

Figure 12.2 Evaluation methodology leading from routine WSP monitoring, via evaluation to performance improvement (Adapted from Lloyd and Vorkas, 1999)

The methodology summarised above has been applied by Lloyd *et al.* (2002) to identify the causes of under-performance in a collection of eight WSP systems as a pre-requisite for rehabilitation and upgrading. In this study Step 1 revealed that none of the WSP systems met any of the national standards for protection of the aquatic environment or for reuse. Step 2 identified fundamental operational and performance problems in all stages of treatment. Step 3 revealed low hydraulic efficiency and short-circuiting in all ponds in the WSP series using a variety of tracer and flow tracking techniques. For information on hydraulic assessments the reader is referred to Chapter 10.

12.4.4 Flow

Sewage normally enters the works in an open channel and this limits the range of devices available for metering the flow of sewage. Most commonly this is done at the works inlet by the installation of a Parshall flume, whilst a V-notch or flat weir is useful for monitoring the flow leaving the system. While weirs are cheaper, they cannot be used at the inlet due to fouling with solids.

Both flumes and weirs operate by creating hydraulic conditions where flow is proportional to height (or water depth), which is easily measured. To be accurate, weirs and flumes need careful design but there are numerous textbooks on hydraulics/fluid mechanics that fully detail their design.

Once a flume or a weir is installed, flow depth can be manually or automatically recorded and flowrate deduced. In large treatment works it is usual to have a continuous record of flow. In small works routine flow measurements may have to be taken manually to check hydraulic loading, and in pumped systems such measurements should also be used to check pump performance against ratings.

12.4.5 Analytical methods

Full descriptions of the standard methods for analysis of water and wastewater are detailed in specialist monographs such as “Standard Methods for the Examination of Water and Wastewater” (APHA, 1992).

12.5 SLUDGE

12.5.1 Predicting sludge accumulation

Rules of thumb are sometimes provided as to the frequency with which sludge removal should take place. Mara and Pearson (1998) suggest that ‘anaerobic ponds require desludging when they are around one third full of sludge (by volume)'. They also provide a formula to calculate when this is likely to occur:

$$n = V_a / 3\,Ps$$

where:

n	=	the number of years to reach 1/3 of the volume available;
V_a	=	volume of the anaerobic pond in m^3;
P	=	population served;
s	=	sludge accumulation rate, m^3/ capita year.

They suggest 'a slightly conservative value for s at 0.1 m^3/ capita year. This value is suitable for winter temperatures below ~10°C; at higher temperatures lower values may be used, e.g. 0.04 m^3 / capita year at 20°C.'

Saqquar and Pescod (1995) developed a model to describe sludge accumulation in primary anaerobic ponds. They argue that sludge accumulation per capita can vary considerably from place to place, and cite Gloyna (1971) who estimated accumulation as 0.03–0.05 m^3/capita as contrasted with Arceivala (1986) reporting 0.08 m^3 /capita. Saqquar and Pescod (1995) apply the principle that the volume of sludge accumulating in primary ponds is controlled by the non-biodegradable portion of the settled solids that either enter the system or are produced as a result of bacterial activity. Using this principle they derived an equation to estimate the volume when only limited information is available:

$$V_{AS} = 2.1\,[(F_{XSS,0}) / (1000)]$$

where:

V_{AS}	=	the predicted volume of accumulated sludge in m^3;
$F_{XSS,0}$	=	flow rate of the suspended solids at the pond inlet (kg/day).

12.5.2 Measuring sludge accumulation

Table 12.10 below gives a range of sludge accumulation rates from the literature. Saqquar and Pescod (1995) noted that sludge accumulation is far from uniform. In their study of the Alsamra WSPs in Jordan the sludge was scoured away from the inlet and outlet areas of the anaerobic pond due to the higher velocities in these regions. Middlebrooks *et al.* (1982) concluded that the prevailing wind did not significantly affect the pattern of sludge accumulation. However, Frederick and Lloyd (1996), Vorkas and Lloyd (2000) and Lloyd *et al.* (2002) carried out detailed bathymetric surveys of the distribution of sludge and found that inlet, outlet locations and prevailing wind direction, measured on-site, strongly influenced the distribution of sludge in ponds. They also found that sludge bathymetry is related to short-circuiting flow paths.

The considerable variation in average accumulation rates is not particularly informative in the absence of loading rates, retention times and details of the ponds design. However, in Table 12.10, the highest rate of accumulation shown for facultative ponds (4.86 cm/yr) in the Cayman Islands example is due to the fact that this is a primary facultative pond receiving raw sewage with a high saline content and reduced anaerobic digestion rates. Consequently, maturation ponds used as a second stage at the Cayman ponds also experienced significantly higher accumulation rates (1.7cm /year; 12 cm in 7 years) (Frederick, 1995).

Table 12.10 WSP Sludge accumulation rates reported from various regions

Anaerobic		Facultative	
Rate (cm/yr)	Total accumulation reported [**Region** – Author]	Rate (cm/yr)	Total accumulation reported [**Region** – Author]
1.5	15 cm in 10 years [**France**-Schetrite Racault]	0.6 - 1.1 56 d.ret	5.5 cm in 5 years [**France** Schetrite Racault]
6.1 5.7 d. ret	61 cm in 10 years [High altitude – **Mexico**]	1.5 8 d. ret	15 cm in 10 years [High altitude – **Mexico**]
20	40 cm in 2 years [**Australia** – Parker *at al.*]		
46.4	170 cm in 3.66 years [**Jordan** – Saqquar Pescod]	4.86	34 cm in 7 years [**Cayman Islands** – Frederick]

d. ret = Nominal pond retention time in days

Pond sludge survey methods

Paing *et al.* (1999) consider that the most sensitive method for the measurement of sludge depth is by the use of a pH electrode fixed on a graduated pole. The pH at the liquid sludge interface drops suddenly as the electrode enters the sludge. The total pond depth is measured by the graduated pole, without the pH probe, and the sludge depth obtained from the difference.

The more commonly used method described below has been adapted from the white towel technique originally described by Malan (1964). This requires a pole with a diameter of 5–7cm. A white cloth/towel is wrapped around the bottom 2m of the pole attached together with a tape measure, at 20 cm intervals.

The following procedures and safety precautions are included here from Lloyd and Vorkas (1999). The depth of the sludge is measured from a boat by lowering the pole vertically into the pond until it reaches the bottom; it is then slowly withdrawn. Sludge particles are trapped on the towelling material so that the sludge depth is clearly visible. The sludge depth and full depth of the overlying liquid column are reported. The disadvantage of the white towel method is that the towel surface becomes fouled quite quickly and the adhering particles must be removed by rinsing, rubbing or brushing. The withdrawal of the pole is illustrated in Figure 12.3, which also shows several important safety features including the use of gloves, mask and safety jacket. A safety line is in use, which also serves as a guide for the sampling grid as shown in Figure 12.4.

Safety notes. The survey procedure is extremely hazardous, particularly for the sampler, and therefore health and safety procedures should be strictly adhered to! The task should be undertaken by a team of 4 competent technical staff. The sludge sampler is required to use protective clothing, including elbow length gloves and face mask (see Figure 12.3). The sampler should remain seated during the survey and be assigned the exclusive task of handling the measuring pole and calling the readings. The readings should be recorded by another team member, on a pre-prepared grid representing the pond.

Figure 12.3 Anaerobic pond sludge measurement and safety procedures

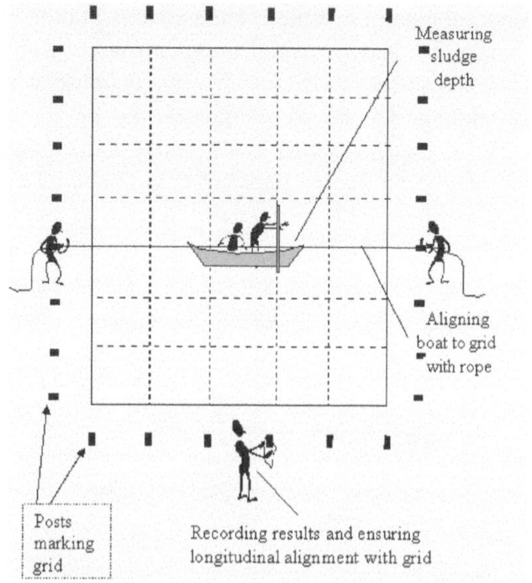

Figure 12.4 Procedures for collecting sludge survey data from a WSP

Sludge survey data and analysis

The survey data are entered into a Microsoft EXCEL spreadsheet for analysis of pond liquid volume, sludge volume, identification of dead zones, and topographic display using software such as SURFER™. Sludge volumes can be calculated fairly precisely from bathymetric surveys, providing sufficient sample points are used.

Results from sludge surveys at a 3-stage, parallel series WSP system, which had been operated for 10 years without desludging are summarised in Table 12.11. The plant was located at a high altitude (2,500 m amsl) in the State of Mexico. The results demonstrated the critical situation in the primary, so-called 'bio-digester', with 41% by volume occupied by sludge, reflecting the lack of routine maintenance of the first stage of treatment. The effect of this was to reduce the nominal retention time of the north bio-digester from 0.55 days to (177 m^3/605 m^3d^{-1}) = 0.3 days. This reduced its performance from 75% to <20% solids removal.

In the anaerobic pond 24% by volume was occupied by sludge. With a total volume of 3,437 m^3 and an average flow of 605 m^3d^{-1} this gave 5.7 days nominal retention time without the reduced volume. However with the volume reduced to 2,612 m^3 the retention time was reduced to 4.3 days. The facultative pond's nominal retention time of 8.15 days was reduced by 6.5% to 7.6 days.

The overall impact of the accumulated sludge on the performance of the pond system was to reduce FC removal efficiency by about 1 log and the plant failed all the regulatory effluent discharge standards.

Table 12.11 Summary of results from the sludge surveys of the northern series of WSPs in the plant of Mexicaltzingo, Mexico

Treatment stage	Total pond vol. (m^3)	Ave pond depth from top water level (m)	Average sludge depth (m)	Sludge vol. (m^3)	Sludge vol. (%)
1. Bio-digester	300	3.2	1.7	123	41
2. Anaerobic	3,437	2.71	0.61	825	24
3. Facultative	4,932	1.65	0.15	345	7

Plots of sludge bathymetry

Contour maps and 3-dimensional plots can be generated from sludge bathymetric data using SURFER™ and ARCVIEW ™ software. Data mapping is useful for locating submerged mounds of sludge, since some ponds are desludged blind, without draining the pond. Bathymetric data can also be used for the preparation of accurate physical grids in the hydraulic modelling of ponds to study short-circuiting in calibrated models such as HYDRO-3D (Guganesharajah *et al.* 2002), although it can be argued that it is the flow pattern that defines where the sludge is deposited not the other way around (Shilton and Harrison, 2003).

12.5.3 Sludge removal

According to Mara and Pearson (1998) sludge removal can be readily achieved by using a raft-mounted sludge pump, however this procedure means that desludging is conducted blind and the result should be checked by a further sludge survey. In the case of small WSP treatment plants with only a single line of ponds and without by-pass channels, it has the advantage that the pond system can continue working whilst desludging proceeds. This operation tends to remove a lot of liquid which means a lot of 'low solids sludge' has to be handled and disposed of, making this operation expensive.

At sites designed with parallel series of ponds, a pond in one series may be completely desludged by first draining down the pond to the sludge layer. The drained supernatant should be returned to the works inlet, then the sludge layer dewatered to the point where the sludge slurry can be moved to a sludge storage area or sludge drying beds.

In some locations the sludge can be allowed to dry out in situ as shown in Figure 12.5 and the dry sludge removed manually. The production of a dry sludge cake may take several months depending on how well the sludge has already digested and climatic conditions.

Figure 12.5 Dry sludge removal 2/3 complete from a facultative pond [Photo: A. Leitner, 2002]

It is clearly important for designers of new pond systems to develop a strategy design for pond desludging in advance, and design the system accordingly to allow for this.

12.5.4 Sludge disposal

One of the advantages of WSPs is their ability to digest the organic solids accumulating in anaerobic and facultative ponds. However, the fact that anaerobic digestion is proceeding efficiently also means that gases, principally CH_4 and CO_2, are released from ponds and, if not collected, are wasted or contribute to the greenhouse effect. Few investigators (Green *et al.* 1995; DeGarie *et al.* 2000) have addressed this issue.

Ultimately the solids from ponds must also be disposed of, and the disposal of sewage sludge has become a major problem in recent years in densely populated countries of Western Europe.

There are two major environmental concerns regarding the application of sludge to agricultural land:

- health of the soil and hence of plant crops associated with the uptake of heavy metals and other chemical constituents;

- human and animal health hazards associated with pathogens.

Chemicals

While regulations regarding heavy metals vary around the world, as a guide, Table 12.12 highlights key changes to an EEC directive on heavy metals in sewage sludge.

Table 12.12 Key changes to the 1986 EEC directive on principal heavy metals in sewage sludge

Metal	Heavy metal limits in sludge (mg kg^{-1}, dry matter)			
	1986 Directive	Proposed (initial)	Proposed (medium term)	Proposed (long term)
Cd	20-40	10	5	2
Cu	1,000-1,750	1,000	800	600
Hg	16-25	10	5	2
Ni	300-400	300	200	100
Pb	750-1200	750	500	200
Zn	2,500-4,000	2,500	2,000	1,500

(Adapted from ENDS, 2000)

In addition to the careful long-term surveillance of metals content, the pH of the soils and sludges must also be monitored. Most metals are more readily available for uptake by crops under acidic conditions.

The main value of sewage sludge lies in its nutrients rather than its organic content. The principal nutrients in sludge are nitrogen and phosphorus, both of which are necessary for the health and growth of crops. However, an excess of nitrogen over the requirements of crops may reduce yields and make them more susceptible to disease and pest attack. In some soils excess application may increase the risk of nitrates reaching rivers or groundwater.

Pathogens

Whereas the amounts of metals entering sewage, principally by way of industrial effluents can be, to a large extent, regulated by trade waste control, there is no practical means of preventing the presence of pathogens in sewage. Whilst it is recognised that pathogens, like metals, are concentrated in sludge, it should be expected that, given the long retention times of sludge in ponds, all but the most recently deposited sludge will have minimal pathogens and parasites surviving, and hence the risk of transmission of disease through the application to land is relatively lower than that from sludges from other treatment plants.

12.6 EMISSIONS

12.6.1 Sources of odour

Well-designed, well-maintained WSPs that are correctly loaded should not give rise to any objectionable odour nuisance. The principal sources of odour, if they do arise, are typically from the inlet works, anaerobic ponds and/or overloaded facultative ponds.

A common source of unpleasant odours at sewage works is hydrogen sulphide, which is detectable at very low thresholds (0.5 ppb). There are however a variety of other compounds which develop in domestic sewage that can be detected at similarly low levels as indicated in Table 12.13.

Hydrogen sulphide and mercaptans are produced from the heterotrophic bacterial breakdown of sulphur containing amino acids, but hydrogen sulphide is also produced from inorganic sulphate. The bacterial putrefaction of organic nitrogen can also lead to the production of unpleasant odours when methylamine, scatole and indole are released. Many bacteria ferment sugars via pyruvic acid to a broad range of volatile organic compounds, such as aldehydes and acids, including acetic, butyric and valeric acids; these also have the potential to produce nuisance odours at low concentrations.

Many types of bacteria can fulfil their sulphur requirements from sulphate by reducing it to sulphide prior to its incorporation into an amino acid, and hence into microbial protein. This assimilatory reduction is however distinct from the dissimilatory reduction of sulphate when it is used as a terminal electron acceptor in anaerobic respiration (Postgate, 1984). The latter process is confined to two groups of strict anaerobes: Desulfotomaculum and Desulfovibrio. Their typical habitats are anaerobic sediments containing organic matter and adequate sulphate. The combination of sewage organics and sulphate may permit high levels of hydrogen sulphide to form from the metabolism of sulphate reducing heterotrophs.

The metabolism of the sulphate-reducers may be described by two half-reactions of an oxidation-reduction pair. Sulphide is generated from the reduction of sulphate, which serves as the final electron acceptor for the oxidation of organic substrate (Poduska and Anderson, 1981). During this process 8 electrons are transferred to the sulphur atom in the reduction half reaction as shown:

$$SO_4^{2-} + 8H^+ + 8e^- \rightarrow S^{2-} + 4H_2O$$

Table 12.13 Olfactory thresholds for some nuisance odour compounds associated with domestic wastewater treatment (Vincent and Hobson, 1998)

GROUP	Compound	Odour description	Odour threshold (parts per billion)
Sulphurous	Hydrogen sulphide	Rotten eggs	0.5
	Methyl mercaptan	Decayed cabbage	0.0014 – 18
	Ethyl mercaptan	Decayed cabbage	0.02
	Dimethyl sulphide	Decayed vegetables	0.12 - 0.4
	Dimethyl disulphide	Putrefaction	0.3 – 11
Nitrogenous	Methylamine	Fishy, rotten	0.9 – 53
	Scatole	Faecal, repulsive	0.002- 0.06
	Indole	Faecal, repulsive	1.4
Acids	Acetic	Vinegar	16
	Butyric	Rancid	0.09 – 20
	Valeric	Sweat	1.8 – 2630
Aldehydes and Ketones	Formaldehyde	Acrid, suffocating	370
	Acetaldehyde	Fruit, apple	0.005 – 2
	Butyraldehyde	Rancid, sweaty	4.6

Using methyl alcohol as the organic substrate for oxidation, the equation representing the overall reaction is:

$$4CH_3OH + 3SO_4^{2-} + 6H^+ \rightarrow 3H_2S + 4CO_2 + 8H_2O$$

The activity of Desulfotomaculum and Desulfovibrio can sometimes result in a massive generation of H_2S. In some areas, particularly low lying coastal areas, where saline intrusion into sewers has raised sulphate levels, microbial conversion of sulphate to sulphide is the major cause of odour problems.

12.6.2 Odour control

Odour control techniques should consider both prevention of generation and minimisation of release. The principal design methods for prevention of odour generation have been summarised by Vincent and Hobson (1998), so it is more appropriate here to consider minimisation of release by good operational practices.

Inlet works

Many odour producing substances, including the important ones listed in Table 12.13, are formed in sewers. This problem is increased if the sewage is not moved quickly to the treatment plant as it will become septic and start decaying while still in the pipeline. Consequently odorous sewer gases may first be vented at the inlet (head works) as sewage enters the treatment works. Where head works or pumping stations are identified as a source of complaint, it may be necessary to cover the inlet area and treat the odour by chemical or biological scrubbing.

Safety note: H_2S *is a highly toxic gas. The occupational exposure limit (OEL) is 10 ppm at which concentration it is extremely odorous. Concentrations of* H_2S *can rise above the OEL in confined spaces. Operators and works visitors should never enter a confined space e.g. a covered inlet works, without appropriate training and equipment.*

Covering ponds

The area of WSPs is generally considered too large to make totally covering them a practical proposition and it is not in keeping with their low cost ethos. However, DeGarie *et al.* (2000) have described the design, installation and commissioning of two, 3.9 hectare floating, self draining, geo-membranes covering the anaerobic section of two of their pond systems in Melbourne, Australia. The project was designed for odour control and also included the design and installation of a biogas collection and control system.

For other types of treatment works the subject of covers for odour containment is reviewed by Koe (2001). One special case that may be of relevance to odour control in the future is the Advanced Integrated Pond System (AIPS) developed by Oswald and his colleagues at the University of Berkeley. The AIPS primary pond comprises a deep pit digester within a surrounding facultative pond. The deep pit serves to confine anaerobic processes, including H_2S release, to a much smaller volume and area than in normal anaerobic ponds. Experimental work conducted by Green *at al.*, (1995) demonstrated that gaseous emissions (methane, carbon dioxide and traces of hydrogen, nitrogen and hydrogen sulphide) could be collected in a submerged gas canopy and collector. Although the principal objective of this work was to increase the efficiency of carbon removal and potential recovery of biogas, a spin-off from this would also be odour control.

Anaerobic and facultative ponds

Most odour complaints are associated with anaerobic ponds, which are overloaded, but since overloaded facultative ponds effectively become anaerobic ponds they are dealt with here together. Yanez (1980) reported that in Peru, facultative pond loadings between 200-400 kg BOD_5/Ha d^{-1} result in occasional slight odour problems; at 400-700 kg BOD_5/Ha d^{-1} there were frequent light and occasional strong odour problems; at higher loadings light or strong odour problems were permanently present. Veenstra *et al.* (1995) reported on poorly functioning ponds in the Yemen where low water use results in high BOD_5 concentrations (800 mg/l) and has led to serious odour problems in the early morning. This coincided with the variation of sulphide over the day, and the fact that the photosynthetic purple sulphur bacteria were actively converting the sulphide to elemental sulphur during daylight hours. This phenomenon was earlier reported in India where the worst H_2S emissions occurred during the night.

Gloyna (1971) noted that 'during periods of high water temperatures in shallow ponds, sludge mats may rise from the bottom. Usually the bacterial activity is intense and the odours are overpowering'. He attributed this problem to the growth of blue-green algae (Cyanobacteria) and recommended the immediate dispersal of these floating mats by creating turbulence by applying a jet of water from a hose to break up the mats and resettle them. An alternative way of agitating the surface to break up the mats is by the use of aerators. Aerators should also be effective in dealing with the BOD_5 overload/sulphide generation problem referred to in the previous paragraph. Since the maintenance of aerobic conditions will inhibit the action of sulphate reducing bacteria, it has been argued that the operation of aerators will be an effective measure to control H_2S release.

Aerators, however, are not always effective in dealing with odour problems as demonstrated by Frederick (1995) in WSPs treating saline sewage in the Cayman Islands. In her study, four aerators were installed in each of two facultative ponds to deal specifically with odour complaints. Prior to the installation of the aerators the concentration of H_2S in the raw sewage had reached an average of 9.6 mg/l as a result of saline intrusion into the sewers. The facultative pond effluents averaged 1.5 mg/l H_2S and the final maturation pond effluent contained 0.2 mg/l of H_2S. For the first 2 months after installation the aerators were operated 24 hours/day and major reductions in H_2S concentrations were achieved.

Due to energy costs the aeration was then reduced to 16 hours/day and H_2S in the facultative pond effluents rose to almost 6 mg/l and to almost 1.5 mg/l in the final effluent. It is noteworthy that during the initial study period SO_4 concentrations rose steadily by more than an order of magnitude, from 47 to 511 mg/l. In the following 2 years the problem of saline intrusion continued to worsen resulting in a doubling of SO_4 concentrations to greater than 1,000 mg/l in the raw sewage.

Ultimately the saline groundwater leakage into the sewers contributed more than 40% of the total sewage flow and conductivity attained almost 25,0000 μS/cm. This resulted in further rises in H_2S levels to 14–16 mg/l in the facultative ponds and 2.5 mg/l in the final maturation pond effluent and numerous odour complaints.

Because of high electricity costs the aerators could only be used for 2 short periods in response to periods of maximum complaint. The odour problem was finally resolved by repairing 3.7 km of sewer pipe; this also reduced the flow by 40% and electricity pumping costs by 22%. Only then were the aerators removed from the ponds.

Maturation ponds

These do not usually give rise to complaints if they are truly aerobic. However some WSP systems treating highly saline wastewaters may result in such high concentrations of H_2S developing in the anaerobic or facultative ponds that this is carried right through the pond system, destroying the aerobic pond community, and resulting in odour emissions from all stages of treatment.

12.6.3 Aerosols

Aerosols have been defined as particles in the size range 0.01µm to 50 µm suspended in air.

Compared with activated sludge plants and even percolating filters, basic WSP systems are not generally considered to produce significant health hazards associated with aerosols. The exceptions worth consideration are aerated ponds, and spray irrigation since WSP effluent reuse is on the increase. Additionally there are several operational activities where staff are exposed to air-borne droplets and sprays including:

- Inlet screen cleaning and grit removal;
- Sampling and desludging anaerobic and facultative ponds.

It is recommended that operators should wear facemasks, in addition to normal protective clothing for these tasks.

12.7 FUTURE DEVELOPMENTS

Monitoring reveals that the effluents from WSP systems in many countries fail to meet regulatory standards. Although some of the important reasons why failures are widespread include inadequate design and inadequate numbers of stages of treatment, lack of attention to operation and maintenance is also a common cause of failure. In some instances technological add-ons are required. In those developing regions where under-performance is widespread, this problem should be addressed before WSP technology acquires a reputation as being unreliable.

REFERENCES

APHA American Public Health Association Federation. (1992). Standard methods for the examination of water and wastewater. 18th Edition New York, *American Public Health Association.*

Arceivala, S.J. (1986). Wastewater treatment for pollution control. Tata McGraw-Hill Publishing Co Ltd, Delhi, India.

Arthur, J.P. (1983). Notes on the Design and Operation of Waste Stabilization Ponds in Warm Climates of Developing Countries. Technical Paper No. 7. Washington, DC:The World Bank.

DeGarie, C.J., Crapper, T., Howe, B.M., Burke, B.F. and McCarthy, P.J. (2000). Floating geo-membranes for odour control and biogas collection/utilization in Municipal lagoons. Paper presented at 4th International Specialist WSP conference in Marrakech, 1999.

EEC. (1991). Council of the European Communities directive of 21 May 1991 concerning urban wastewater treatment (91/271/EEC). *Official Journal of the European Communities* **L135**/40 (30 May).

Frederick, G.L. (1995). The performance of full scale WSPs treating saline wastewater. Ph D Thesis, University of Surrey, England.

Frederick, G.L. and Lloyd, B.J. (1996). An evaluation of retention time and short-circuiting in waste stabilisation ponds using Serratia marcescens bacteriophage as a tracer. *Water Science and Technology* **33**(7), 49-56.

Garcia, J., Mujeriego, R., Bourrouet, A., Peñuelas, G. and Freixes, A. (1999). Wastewater treatment by pond systems: Catalonia experience. Presented at the *4th IAWQ conference on Waste Stabilisation Ponds: Technology and Applications.*

Gloyna, E.F. (1971). Waste stabilisation ponds. WHO Monograph series No. 60 World Health Organisation, Geneva.

Green, F.B., Lundquist, T.J., Muir, J., Tresan, R.B. and Oswald, W.J. (1995). Methane fermentation, submerged gas collection, and the fate of carbon in advanced integrated wastewater pond systems. *Wat. Sci. Tech.* **31**(12), 55-65.

Green, F.B., Bernstone, L.S., Lundquist, T.J. and Oswald, W.J. (1996). Advanced integrated wastewater pond systems for nitrogen removal. *Water Science and Technology* **33**(7) 207-217.

Guganesharajah R.K., Lloyd B.J, and Vorkas C.A (2002). The development of HYDRO-3D; a computational hydraulic model for assessing and designing waste stabilisation ponds. Paper presented at the *IWA 5th Int. Specialist Conf. on Waste Stabilisation Ponds: Pond Technology for the New Millenium* 2-5th April. Auckland, New Zealand.

Hertzberg, F. (1968). Work and the nature of man. Staple Press, London

Koe, L. (2001). Process covers for odour containment, in *Odours in Wastewater Treatment*. Eds., Stuetz and Frechen. IWA Publishing, London, UK.

Lawty, R., Ashworth, J. and Mara, D.D. (1996). Waste stabilisation pond decommissioning: a painful but necessary decision. *Water Science and Technology* **33**(7), 107-115.

Lloyd, B.J. and Vorkas, C.A. (1999). A diagnostic methodology for the evaluation and hydraulic optimisation of waste stabilisation pond design for pathogen removal using field assessments and modelling. Paper presented at the *IAWQ 4th Specialist Int.Conf. on Waste Stabilisation Ponds: Technology and the Environment*. 20th-23rd April 1999, Marrakech, Morocco.

Lloyd, B.J., Leitner, A.R., Vorkas, C.A. and Guganesharajah, R.K. (2002). Under-performance and rehabilitation strategy for waste stabilization ponds in Mexico. Paper presented at the *IWA 5th Int. Specialist Conf. on Waste Stabilisation Ponds: Pond Technology for the New Millenium 2-5th April*. Auckland, New Zealand

Lloyd, B.J. (2002). Water and sanitation for environmental protection and human development. Paper presented at the *PAHO IAHR Centennial meeting PAHO/WHO*, Washington D.C. June 2002.

Malan, W.M. (1964). A guide to the use of septic tank systems in South Africa. CSIR report N0. 219. Pretoria, S Africa: National Institute for Water Research.

Mara, D.D and Pearson, H.W. (1998). Design manual for waste stabilisation ponds in Mediterranean Countries. European Investment Bank.

Metcalf and Eddy. (2003). Wastewater engineering: treatment, disposal and reuse. 4th Edition, McGraw Hill.

Meybeck, M. and Helmer, R. (1996). An introduction to water quality, in Water Quality Assessments 2nd Edition edited by D Chapman, Publ. E and FN Spon, ISBN 0 419 21590 (HB).

Middlebrooks, E.J., Middlebrooks, C.H., Reynolds, J.H., Watter, G.Z., Reed, S.C. and George, D.B. (1982). Waste stabilisation lagoon design, performance and upgrading. Macmillan Publishing Co Ltd. New York, USA.

Paing, J., Picot, B., Sambuco, J.P. and Rambaud, A. (1999). Sludge accumulation and methanogenic activity in anaerobic lagoon. Paper presented at *4th International IAWQ Specialist Conference on Waste Stabilisation Ponds; Technology and Environment*, Marrakech, 20-23 April 1999.

Poduska, R.A. and Anderson, B.D. (1981). Successful storage lagoon odour control *J.Water Poll. Control Fed.* **53**(3), 299-310.

Pearson, H.W., Mara, D.D. and Bartone, C.R. (1987). Guidelines for the minimum evaluation of the performance of full-scale waste stabilization pond systems. *Water Research* **21**(9), 1067-1075.

Postgate, J.R. (1984). The sulphate reducing bacteria. 2nd Edition. Cambridge University Press.

Saqquar, M.M. and Pescod, M.B. (1995). Modelling sludge accumulation in anaerobic wastewater stabilisation ponds. *Wat. Sci. Tech.* **31**(12), 185-190.

Shilton, A. and Harrison, J. (2003). Guidelines for the hydraulic design of waste stabilisation ponds. Palmerston North, New Zealand: Institute of Technology and Engineering, Massey University.

Veenstra, F.A., Al-Nozaily and Alaerts, G.J. (1995). Purple non-sulphur bacteria and their influence on waste stabilisation ponds in the Republic of Yemen. *Water Science and Technology* **31**(12), 141-149.

Vincent, A. and Hobson, J. (1998). Odour control. CIWEM Monographs on Best Practice No. 2, Terence Dalton Publishers, London.

Vorkas, C.A. and Lloyd, B.J. (2000). The application of a diagnostic methodology for the evaluation of hydraulic design deficiencies affecting pathogen removal. *Water Science and Technology* **44**(10/11), 99-109.

WHO. (1976). Surveillance of drinking-water quality. Monograph series No. 63, World Health Organisation, Geneva.

Yanez, F. (1980). Evaluation of the San Juan stabilisation ponds. Final research report. CEPIS/PAHO/IDRC, Lima.

13

Advanced integrated wastewater ponds

Rupert Craggs

13.1 INTRODUCTION

Advanced Integrated Wastewater Pond Systems (AIWPS) are designed to achieve both efficient wastewater treatment and recovery of resources from the wastewater through capture of biogas as an energy source, harvest of algal biomass as a fertiliser or feed and reuse of treated effluent (Oswald 1991; El Hamouri *et al.* 1995; Green *et al.* 1995b; Oswald 1996).

AIWPS retain many of the benefits of conventional ponds including cost-effective construction, easy operation and tolerance of shock loads (Belsare and Belsare 1987; Agunwamba 1991; Oswald 1991). Moreover, AIWPS have been developed to incorporate many design improvements over conventional stabilisation pond systems to overcome problems such as: sludge accumulation, odour release and poor effluent quality in terms of high concentrations of algal

ISBN: 1843390205. Published by IWA Publishing, London, UK.

suspended solids, nutrients and faecal indicators (WHO 1987; Oswald 1988a; Oswald 1990; Oswald 1991; Oswald 1996).

AIWPS are specially designed to promote particular natural wastewater treatment processes in a series of four types of ponds that are arranged in the most favourable sequence for wastewater treatment (Figure 13.1): Advanced Facultative Ponds (AFPs), High Rate Ponds (HRPs), Algae Settling Ponds (ASPs) and Maturation Ponds (MPs) (Oswald 1990; Oswald 1991;). AIWPS usually consist of two AFPs in parallel, a HRP, two ASPs in parallel, and one or more MPs in series. However, the four ponds of AIWPS have an overall land requirement similar to that of conventional stabilisation pond systems (Oswald 1996).

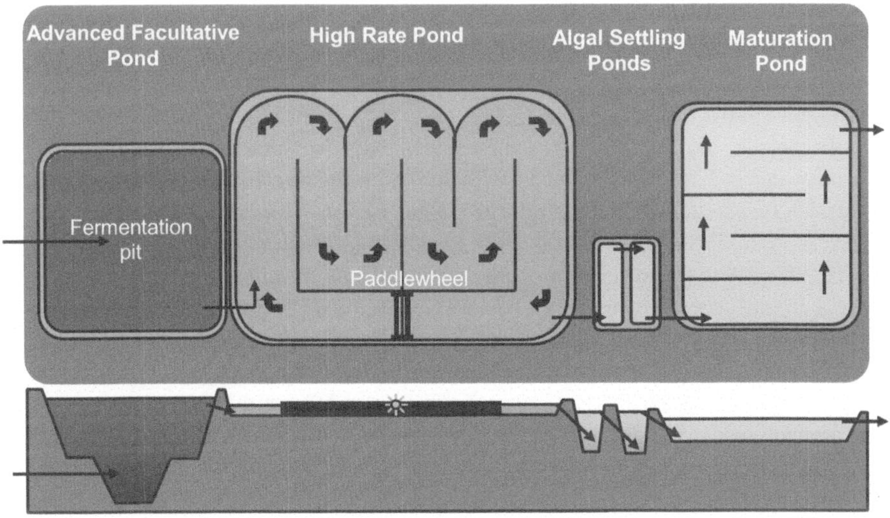

Figure 13.1 Schematic diagram of an Advanced Integrated Wastewater Pond System (AIWPS)

13.1.1 Historical development

AIWPS have been developed by Oswald and co-workers over the last five decades at the University of California at Berkeley, USA. Oswald began to study the role of microalgae in sewage ponds in 1949. Early research focused on the symbiosis between algae and bacteria, which Oswald called 'photosynthetic oxygenation' (Ludwig *et al.* 1950; Ludwig and Oswald 1952; Oswald *et al.* 1953; Oswald and Gotaas 1955).

In 1957 Oswald developed the High Rate Pond for combined wastewater treatment and recovery of nutrients as algal biomass (Oswald and Gotaas 1957; Oswald *et al.* 1957). Several small-scale HRPs were tested at research sites in Richmond and Concord, California (Oswald 1963). At the time, the Richmond HRP was the largest outdoor algae cultivation pond in the world.

In the early 1960's research focused on methane fermentation in pond systems. Ponds that had deep anaerobic pits were found to provide better treatment than shallower ponds (Oswald *et al.* 1963). The Advanced Facultative Pond, a deep facultative pond incorporating a deep fermentation pit was first used in the full-scale AIWPS built in 1967 at St Helena, California, which still operates today (Oswald 1996).

Research on algal growth and productivity in High Rate Ponds continued in the 1970's (Oswald 1970; Oswald 1978). In 1978 the propeller pumps used to mix the Richmond HRP were replaced with more efficient paddlewheels. The second full-scale municipal AIWPS was built in 1980 at Hollister, California (Oswald 1996).

In the late 1970's and 1980's the use of HRPs for the treatment of domestic wastewater in Israel was refined by Shelef and co-workers in collaboration with German researchers (Shelef *et al.* 1977; Shelef *et al.* 1978a; b; Oron *et al.* 1979; Berend *et al.* 1980; Shelef *et al.* 1980; Shelef 1982; Shelef and Azov 1987).

In the 1980's and 1990's work continued on nutrient removal, resource recovery and energy efficiency of AIWPS systems including: methane recovery and utilization from Advanced Facultative Ponds (Von Hippel and Oswald 1985; 1989; Oswald *et al.* 1994; Green *et al.* 1995a,b; Green *et al.* 1996b) and algae harvest from HRPs (Hall 1989; Nurdogan and Oswald 1995).

Recent research has focused on the use of AIWPS technology to remove selenium from agricultural drainage waters (Oswald *et al.* 1989; Gerhardt *et al.* 1991; Green *et al.* 2003) and enhancing AIWPS effluents using dissolved air flotation, sand filtration and reverse osmosis (Downing *et al.* 2002).

13.1.2 AIWPS implementation

Several wastewater stabilisation pond systems based on the AIWPS research at Berkeley are presently operating in California and as can be seen in Table 13.1, AIWPS concepts have been implemented or researched in many countries around the world.

Table 13.1 Application of AIWPS concepts around the world

Country	Ponds	WW	Research	Reference
Australia	HRP	Abattoir	O_2 dynamics	Evans *et al.* 2003
Belgium	HRP	Piggery	Marine algae	De Pauw and Vaerenbergh 1983
Brazil	HRP	Sewage	Treatment/Algae	Kawai *et al.* 1984
	AP/HRP /MP	Piggery	Treatment	Costa *et al.* 2000
Canada	HRP	Piggery	Algal flocculation	Buelna *et al.* 1990
China	HRP	1^{ry} Sew	Macrophytes	Chen *et al.* 2003
Egypt	HRP	Sewage	Treatment	El-Gohary *et al.* 1991
Ethiopia	AFP	Tannery	Treatment	Tadesse *et al.* 2003
France	HRP	Sewage	Disinfection	Bahlaoui *et al.* 1997
	HRP	Aquaculture - Macroalgae		Pagand *et al.* 2000
Germany	HRP	Piggery	Rotifer /Algae	Schluter *et al.* 1987
	HRP	Sewage	O_2/Model	Grobbelaar *et al.* 1988
India	HRP	Poultry	Algae biomass	Mahadevaswamy and Venkatamaran 1986
	HRP	2^{ry} sewage	Disinfection	Sebastian and Nair 1984
Ireland	HRP	Piggery	Treatment/Algae	Fallowfield and Garrett 1985
Israel	HRP	Sewage	Review	Shelef and Azov 1987
Italy	HRP	Sewage	Energy	Balloni *et al.* 1983
Kuwait	FP/HRP	Sewage	Algal settling	Al-Shayji *et al.* 1994
Malaysia	HRP	Goat, duck, palm, rubber		Phang 1990
Mexico	HRP	Sewage	Aquaculture	Paniagua-Michel *et al.* 1987
	Dig/ HRP	Mixed	Resources	Rovirosa *et al.* 1995
Morroco	AP/HRP	Sewage	COD Model	El Hamouri *et al.* 2003
New Zealand	AIWPS	Sewage	Treatment	Craggs *et al.* 2003
Netherlands	HRP	Sewage	Model	Kroon *et al.* 1989
Philippines	HRP	Sewage	Algae biomass	Oswald 1980
Portugal	HRP	Sewage	Overloading	Pinheiro *et al.* 1987
	HRP	Tomato	Treatment/Algae	Rodrigues and Oliveira 1987a
	HRP	Piggery	Treatment/Algae	Rodrigues and Oliveira 1987b
Scotland	HRP	Piggery	Batch feed	Fallowfield *et al.* 1999
Singapore	HRP	Piggery	Treatment/Algae	Taiganides 1997
South Africa	HRP	Cattle	Fish harvest	Cloete *et al.* 1984
	Wet/HRP	Sewage	Filamentous algae	Wood *et al.* 1989
	AFP/ FP /TF	Sewage	Petro	Meiring and Oellermann 1995
	HRP	Tannery	Spirulina/Use	Rose, *et al.* 1995
	HRP	Acid mine	Integrated ponds	Rose, *et al.* 1998
Spain	HRP	Sewage	Cryptosporidium	Araki *et al.* 2000
Switzerland	HRP	Nutrients	Algae production	Schanz and Dubinsky 1988
Thailand	HRP	Sewage	Fish harvest	Edwards and Sinchumpasak 1981
USA	AP/HRP	Poultry	Treatment/Algae	Duggan *et al.* 1972
	HRP	Piggery	Algae/flotation	Koopman and Lincoln 1983
	HRP	Cattle	Bioflocculation	Lincoln and Koopman 1986
	AIWPS	Winery, Cannery		Oswald 1995
	AIWPS	Sewage	DAF, RO	Downing *et al.* 2002
	AIWPS	Drainage	Selenium, NO_3	Green *et al.* 2003
Vietnam	HRP	Rubber	Hyacinth	Bich *et al.* 1999

Dig – Digestor; AP – Anaerobic Pond; FP – Facultative Pond; DAF Dissolved Air Flotation; RO Reverse Osmosis

13.2 ADVANCED FACULTATIVE PONDS

Advanced Facultative Ponds (AFPs) promote sedimentation and anaerobic treatment processes (Oswald 1996). AFPs can be likened to a *deep* (3-4 m) conventional facultative pond overlying a conventional anaerobic pond (the fermentation pit) which can be a further 3-4 m deep (Oswald *et al.* 1994) (Figure 13.2.). A major problem with conventional (1.0-1.5 m deep) facultative ponds is that efficient methane fermentation is limited by intermittent mixing and oxygenation of the anaerobic bottom zone, resulting in acidic conditions and odour release (Oswald 1991; Green *et al.* 1995b).

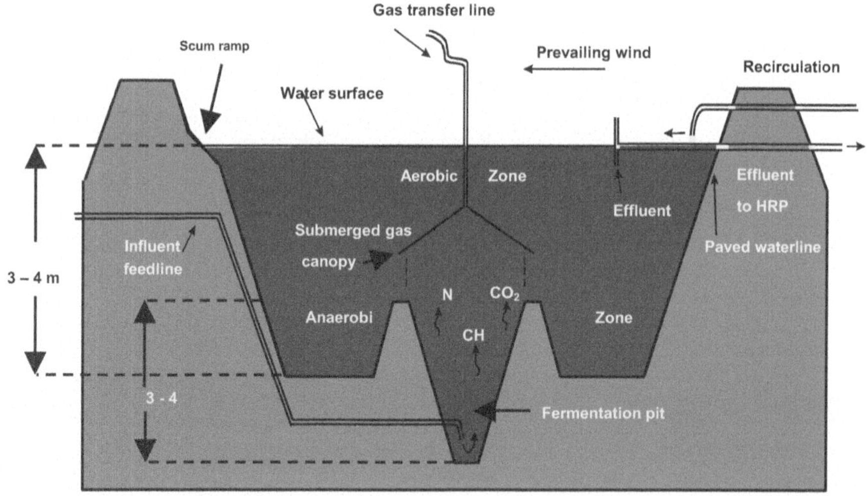

Figure 13.2 Side elevation of an Advanced Facultative Pond (not to scale)

The deep anaerobic fermentation pit of the AFP is surrounded by a wall or berm at the level of the AFP floor to prevent wind mixing and intrusion of oxygenated water (Oswald *et al.* 1994). Short-circuiting of inflow to outflow is reduced by piping the influent wastewater to the centre of the fermentation pit, which minimises the formation of circulation currents (Green *et al.* 1996b). The "upflow velocity" in the fermentation pit (typically below 2.0 m d^{-1}) is designed to be less than the settling rate of most sewage solids including helminth (parasitic worm) ova and cysts (Oswald *et al.* 1994; Green *et al.* 1996b). AIWPS systems in the United States have been operating for as long as 30 years without sludge removal (Green *et al.* 1996b).

Fermentation pits are designed by matching the rate of sedimentation of wastewater solids to the rate of anaerobic digestion so that only the non-organic component of the incoming wastewater solids accumulates at the bottom of the pit (Oswald 1996). The permissible volumetric loading rate (kg BOD_U m^{-3} d^{-1}) of fermentation pits is calculated using temperature dependent reaction rates which vary with the solid content (24 hr settled solids) of the wastewater. The typical residence time is 1 - 3 days (Oswald 1996).

Settled solids with adhering bacteria are temporarily resuspended in the fermentation pit by bubbles of biogas that are formed during the fermentation process (Green *et al.* 1996b). These solids rise and fall within the fermentation pit as the gas bubbles expand and then break away from the bottom sludge layer, which creates an anaerobic sludge blanket through which all the wastewater must flow (Green *et al.* 1996b). When initially formed in the fermentation pit, the biogas consists mainly of methane, nitrogen gas and carbon dioxide (50%, 40% and 7% respectively) (Green *et al.* 1996b). Nitrogen gas is released by heterotrophic nitrification of organic nitrogen compounds (Verstraete and Alexander 1973; Tate 1980; Castignetti and Hollocher 1984; van de Graaf *et al.* 1995; Diab *et al.* 1993) followed by denitrification, which may account for over 30% of the total nitrogen removal from the wastewater (Green *et al.* 1996b). As the biogas rises through the overlying water column of the AFP much of the nitrogen gas and carbon dioxide dissolve, increasing the biogas methane concentration to greater than 80% (Green *et al.* 1995b).

The AFP is designed to have adequate residence time for treatment of the remaining wastewater BOD_U to a level that is acceptable for treatment in the HRP (Oswald 1996). For domestic sewage, the fermentation pit will typically remove 60% of the BOD_U and the optimal BOD_U concentration of HRP influent is 100 g m^{-3} (Oswald 1996). The surrounding pond residence time is calculated using a temperature dependent first order decay rate for the remaining BOD_U and is typically between 15 and 30 days (Oswald 1996).

The oxygen-rich, aerobic surface layer of the AFP oxidises any malodorous gases (e.g. H_2S) that are generated in the fermentation pit preventing the release of offensive pond odours (Green *et al.* 1996b). Supplementary surface aeration may be necessary for odour control during prolonged overcast periods and should be operated whenever the surface dissolved oxygen concentration of the AFP falls below 2 g m^{-3} (Oswald 1996). A low level of permanent aeration may be required in locations where water supplies contain more than 100 g m^{-3} of sulphate to prevent H_2S release (Oswald 1991). Only aerators that aerate the pond surface and do not introduce oxygen into the fermentation pits should be used. A paved, gently sloped *scum ramp* along the water line of the downwind corner can be used to collect floatable solids for simple removal

(Green *et al.* 1996b) if the influent wastewater is unscreened. The AFP outflow pipe should draw horizontally from 0.5 m below the pond surface to avoid carry-over of grease and floatables to the HRP and ensure that any influent that is less dense than the pond contents will remain in the pond (Oswald 1996).

AFPs provide sludge and grease removal and can remove 60 to 80% of the influent BOD_5 and virtually all of the original wastewater solids (Oswald 1991; Green *et al.* 1996b). Additionally, toxic organic compounds such as pesticides can be degraded and heavy metals are precipitated as sulphides in the fermentation pit (Oswald 1996).

13.3 HIGH RATE PONDS

The second pond in the system is the High Rate Pond (HRP) or High Rate Algal Pond (HRAP). Algae in High Rate Ponds provide 'photosynthetic oxygenation' which promotes aerobic breakdown of the remaining dissolved organic matter in the wastewater by heterotrophic bacteria (Oswald 1996). Nutrients are removed by assimilation into algae biomass, or by volatilisation (ammonia) and precipitation (phosphate) at the high pH that occurs as a result of intense algal photosynthesis (Oswald 1996). Elevated rates of disinfection occur as a result of high exposure to UV in solar radiation in the presence of high dissolved oxygen (DO) and pH levels in the pond (Oswald 1988a). Refer to Chapters 5 and 6 for further information on nutrient and pathogen removal. The HRP is a *shallow* (0.2 to 1.0 m deep) continuously mixed raceway or meandering channel (Figure 13.3.) with a residence time between 2 and 8 days (Oron and Shelef 1982; Oswald 1996).

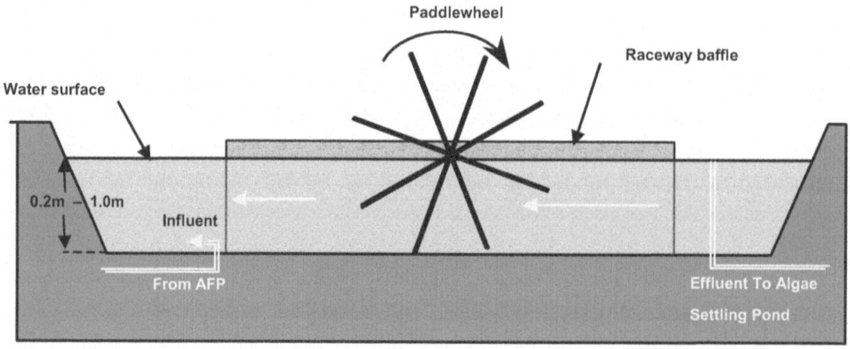

Figure 13.3 Side elevation of a High Rate Pond (not to scale)

The low, mean water velocity (15 cm s^{-1}) in the HRP prevents thermal stratification and keeps algae uniformly suspended, while larger aerobic bacterial flocs tend to "tumble" along the pond bottom (Abeliovich and Azov 1976; Shelef *et al.* 1978a; Grobbelaar *et al.* 1988; Oswald 1988a;b; Oswald 1991). This gentle mixing is most economically provided by a slowly rotating paddle-wheel which is used to separate the influent and effluent pipes of the HRP to minimise any short-circuiting (Oswald 1996).

HRPs enhance the symbiosis between microalgae and heterotrophic bacteria (Oswald 1996). Microalgal photosynthetic oxygenation facilitates degradation of dissolved organic compounds by aerobic bacteria which release nutrients and carbon dioxide, which, in turn, are assimilated into more algae biomass (Oswald *et al.* 1953; Richmond and Grobbelaar 1986; Oswald 1988a;b). Thus, oxidation of wastewater is achieved inexpensively using solar energy, and nutrients are recycled into algal biomass in the same pond (de la Noüe and De Pauw 1988).

Algal photosynthetic efficiency is improved by the gentle mixing and by maintaining an optimum depth for light penetration to ensure that the algal cells receive maximum exposure to solar radiation (Fallowfield and Garrett 1985; Terry and Raymond 1985; Dodd 1986; Richmond 1986; Oswald 1988a;b). Since the ratios of oxygen production to algal production and oxygen production to BODu decomposition are both approximately 1.6, HRPs are designed by matching the algal cell concentration to the wastewater influent BOD_U concentration (Oswald *et al.* 1953; Shelef *et al.* 1980; Grobbelaar *et al.* 1988; Belsare and Belsare 1987; Oswald 1988a; Oswald 1996). The pond depth is calculated using a modified Beer-Lambert Law to predict the depth of light penetration based on the required algal concentration and the minimum total daily insolation (Oswald 1996). Oswald (1978) found that the algal concentration (C_c, in g m^{-3}) and depth (d, in m) are reciprocally related to a constant that depends on the insolation and light attenuation over the pond depth. This constant varies from 60 g m^{-2} (in areas with cloud cover and for ponds with large amounts of non-algal suspended matter) to 90 g m^{-2} (in sunny areas and for ponds mainly containing algae), thus:

$$d = \frac{60}{C_C} \text{ to } \frac{90}{C_C}$$

HRP depths between 0.3 - 0.5 m appear to be most suitable for most wastewaters (Oron *et al.* 1981a; De Pauw and Van Vaerenbergh 1983; Richmond 1983; Dodd 1986; Richmond 1986), although shallower (0.2 m) depths are required in highly pigmented (light absorbing) wastewaters.

The HRP residence time is determined by the time required to grow the algal concentration needed to release the oxygen to degrade the remaining BOD (Oswald 1996). This is based on the available solar insolation and an estimated photosynthetic efficiency (Oswald and Gotaas 1957):

$$\theta = \frac{h\,C_c\,d}{SF}$$

In which h is the energy content of algae (normally about 0.106 Mol g^{-1} dry wt, where 1 Mol is equal to 217 kJ), C_c is the desired algal concentration (g m^{-3} dry wt), S is the total solar insolation at the pond surface (Mol m^{-2} d^{-1}), d is the depth of the pond (m), F is the photosynthetic efficiency (typically 2 - 3 % of total insolation) and θ is the residence time (d) (Oswald 1996).

A design model for HRPs based on climatic and seasonal conditions has been developed by Fallowfield and colleagues using experimental data from HRPs operated in Scotland and France (Martin and Fallowfield 1989; Fallowfield *et al.* 1992). This was based on a model of algal productivity in HRPs (Grobbelaar *et al.* 1988; Grobbelaar *et al.* 1990). Recently Jupsin *et al.* (2003) have proposed a dynamic model for HRPs, based on the River Water Quality Model 1.

HRPs actually produce a surplus of dissolved oxygen (~ 1.9 times the applied BOD_5; Grobbelaar *et al.* 1988) and some of the HRP effluent (typically 25%) can be recirculated during the day to augment the aerobic surface layer of the AFP with warm, slightly alkaline, oxygen rich-water (Oswald 1988a; Oswald 1996). The oxygen balance in HRPs has recently been modelled by El Ouarghi *et al.* (2000).

The gentle mixing in HRPs prevents thermal stratification, which limits the depth of the aerobic zone in conventional facultative ponds, decreasing their treatment efficiency (Oswald 1978; Ayala and Bravo 1984; Fallowfield and Garrett 1985; Terry and Raymond 1985; de la Noüe and De Pauw 1988; Grobbelaar *et al.* 1988). Mixing also promotes the growth of non-motile, colonial microalgal species such as *Scenedesmus* and *Micractinium* (Benemann *et al.* 1977; Oron *et al.* 1981a;b; Esen *et al.* 1987; Oswald 1996), which unlike the small or free-swimming algal species (such as *Chlorella* and *Euglena*) that dominate in conventional stabilisation ponds, produce extracellular polymers and flocculate and so are more easily settled in the subsequent Algae Settling Pond (Oswald 1988a; Hall 1989; Nurdogan and Oswald 1996).

Nutrient removal results not only from assimilation into algal biomass but from the tendency of intense algal photosynthesis to raise the daytime pH of the HRP water, typically, to 9.5 or 10 (Azov *et al.* 1982; Oswald 1996). At this high pH, ammonium is converted to ammonia gas, which volatilises to the air (Azov and Shelef 1987; Gomez *et al.* 1995; Garcia *et al.* 2000a), and phosphorus may be precipitated with polyvalent cations such as iron, calcium and magnesium (Picot *et al.* 1991; Moutin *et al.* 1992; Nurdogan and Oswald 1995; Mesple *et al.* 1995; Garcia *et al.* 2002). Nitrification does not generally occur in domestic wastewater HRPs, but has been observed in HRPs treating high strength (BOD and organic-N) wastewaters (Cromar *et al.* 1996; Fallowfield *et al.* 1999; Chen *et al.* 2003; Costa *et al.* 2000).

Disinfection in HRPs is improved by the shallow depth and mixing (increasing exposure of the wastewater to the UV in solar radiation) which is augmented by the high pH and DO levels in the pond water (Parhad and Rao 1974; Sebastian and Nair 1984; Curtis *et al.* 1992; El Hamouri *et al.* 1994; 1995; Davies-Colley *et al.* 2000; Shelef and Azov 2000; Davies-Colley *et al.* 2003). Faecal coliform disinfection rate constants in HRPs have been found to vary with both season and location. For example, they range from 6.1 – 42.6 d^{-1} in Israel (Shelef *et al.* 1980) 1.0 – 24.7 d^{-1} in Morocco (El Hamouri *et al.* 1995) to 0.4 – 2.3 d^{-1} in Scotland (Fallowfield *et al.* 1996). However, these rates are much higher than typical rates for conventional maturation ponds under the same conditions (Bahlaoui *et al.* 1997). HRPs have also been shown to provide efficient removal of viruses, nematode eggs and cryptosporidium (Sobsey and Cooper 1973; El Hamouri *et al.* 1994; 1995; Araki *et al.* 2000; Araki *et al.* 2001).

Mixing reduces problems with invertebrates grazing the pond algae by promoting resuspension and regrowth of large, grazer resistant species (Schluter *et al.* 1987). Thus, although conventional ponds are often troubled by algal grazers, mature HRPs rarely have severe problems (Oswald 1989; Oswald 1996).

13.3.1 HRP operation

Depth

Depth is a very important variable in ponds, with deep ponds suffering from self-shading light-limitation of algae, whereas very shallow ponds have thermal instability due to temperature fluctuation (Abeliovich 1980). Studies to optimise the performance of HRPs have shown that the two main factors affecting pond performance are solar radiation and water temperature (Oswald 1996). In tropical regions, where insolation and water temperature are fairly constant year-round, HRPs may be operated with constant depth and residence time (Oswald 1996). HRPs in temperate regions are best operated with seasonally variable retention times (Oswald 1996). This is most economically achieved by varying the pond depth, with greater depth in the winter when the algal concentration is lower, and therefore light penetration deeper, and shallower depth in the summer when both algal concentration and light attenuation are higher (Goldman 1979; Buhr and Miller 1983; Shelef and Azov 1987; Picot *et al.* 1991; Oswald 1996). Recent research in New Zealand has shown little difference in the annual treatment performance of HRPs operated with retention times of 7.5 days and constant depths of 0.3 m and 0.45 m (Craggs *et al.* 2003).

Paddlewheel mixing

Paddle wheels provide very efficient mixing of ponds (Groeneweg *et al.* 1980; Lincoln and Hill 1980; Oswald 1980; Fallowfield and Garrett 1985; Oswald 1988a;b). HRP water velocities as low as 5 cm s^{-1} will prevent thermal stratification and maintain algae in suspension (Oswald 1988a), however in wide shallow channels, average velocities of ~15 cm s^{-1} are generally required to ensure at least 5 cm s^{-1} in all locations (Shelef *et al.* 1978a; Moraine *et al.* 1979; Fallowfield and Garrett 1985; Richmond and Grobbelaar 1986; Dodd 1986; Oswald 1988a;b). The energy required for HRP mixing is dependent upon water velocity, channel geometry and surface roughness, and can be calculated using classical hydraulic equations for kinetic head loss, Mannings equation for open channel flow and a hydraulic power equation (Oswald 1988a; Green *et al.* 1995a; Oswald 1996).

Discharge

High Rate Ponds normally discharge continuously; however, better overall treatment may be achieved by only discharging during the daytime (Green *et al.* 1996b) or by batch operation (Sebastian and Nair 1984; Bontoux and Picot 1994; Fallowfield *et al.* 1999).

13.4 ALGAE SETTLING POND

The third pond is the Algae Settling Pond (ASP), which promotes natural sedimentation and removal of the algae-bacterial biomass from the HRP effluent and enables recovery of the algal solids (Oswald 1996) (Figure 13.4.). The ASP may be up to 3 m deep at the inflow end and the bottom of this pond slopes up along its length to a depth of 0.5 – 1.0 m at the outflow end. ASPs typically have a length to width ratio of 5:1 (Oswald 1996). The inflow pipe discharges 0.5 – 1.0 m from the pond bottom and the outflow is a weir at the pond surface (Oswald 1996). A HRP may produce from 75 to 250 kg ha^{-1} d^{-1} of algae-bacterial biomass that will readily settle under the quiescent conditions in an ASP (Eisenberg *et al.* 1981; Oswald 1990; Nurdogan and Oswald 1996). ASPs are designed to have a residence time to enable maximum settling of algae but limit further algal growth and they incorporate sufficient volume to store the settled algae solids that accumulate in the time between algae removal (Oswald 1996). A one to two day residence time is usually sufficient to remove from 50 to 80% of the algae by sedimentation (Nurdogan and Oswald 1996; Garcia *et al.* 2000b).

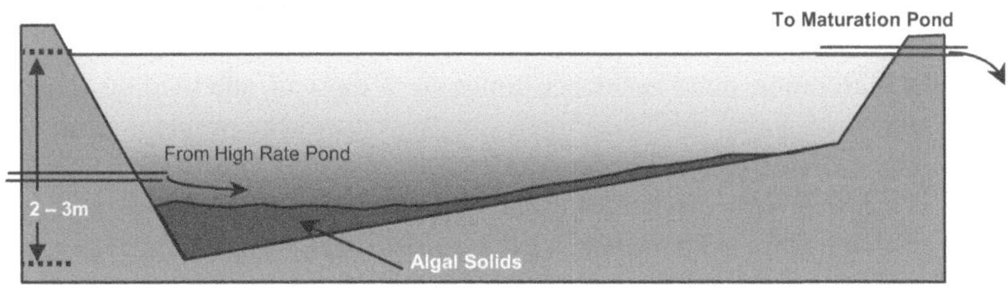

Figure 13.4 Side elevation of an Algae Settling Pond (not to scale)

13.4.1 ASP operation

Compared to the bacterial sludge of primary and secondary in-tank mechanical treatment systems, the settled algae biomass is relatively innocuous and thus can remain at the bottom of the ASP for many months without releasing significant amounts of nutrients and BOD back into the pond effluent (Oswald 1996). Algal solids should however be removed from the ASP at intervals of 4 to 6 months and with two settling ponds in parallel, one can be taken off line and simply decanted down to the layer of settled algae by pumping the supernatant back to the High Rate Pond (Green *et al.* 1996b). The settled algae, which are typically 3-4 % solids, (Eisenberg *et al.* 1981; Green *et al.* 1995a) are either left to dry out and removed as a solid cake or are pumped from the pond to be spread onto land directly or dewatered on a sand or geotextile drying bed.

13.5 MATURATION POND

The fourth pond is the Maturation Pond (MP), which is designed for effluent "polishing" by promoting zooplankton grazing of the remaining unsettled algae and natural disinfection by a combination of solar-UV radiation, sedimentation and protozoan grazing (Oswald 1996) (Figure 13.5.). MPs can also provide some storage to allow for controlled discharge or irrigation of the effluent (Oswald 1991).

MPs can have a depth of 1 to 3 m and are designed with sufficient residence time to provide a high level of disinfection (Oswald 1996). MPs are traditionally designed using temperature dependent decay rates for faecal indicator bacteria (Marais 1974) to determine required residence times. These tend to give conservative predictions of faecal indicator disinfection in APS systems (Oswald 1996). In general, MPs with a total residence time of 10 to 20 days can easily reduce faecal coliform levels to below 1000 MPN $(100\ \text{ml})^{-1}$ (Oswald

1988a, Oswald 1989) and levels below 100 MPN can be consistently achieved (Craggs et al. 2003).

Limiting maturation pond residence times to 3 days or subdividing larger MPs into cells with 3 day residence times greatly reduces the problem of algal regrowth (algae divide every 2-3 days) that can cause TSS and BOD levels to increase. The inflow pipe discharges at the bottom of the pond and the outflow pipe takes effluent from the pond surface (Oswald 1996).

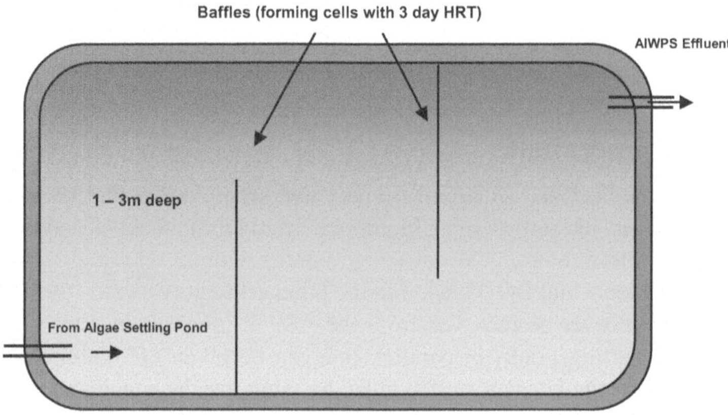

Figure 13.5 Plan view of a Maturation Pond (not to scale)

13.6 TREATMENT PERFORMANCE

Over 90% of the wastewater BOD_5, 80% of nitrogen and 70 % of phosphorus may be consistently removed by AIWPS systems (Oswald 1970; Oswald 1991; Green *et al.* 1995a; Oswald 1996). Typical MP effluent quality for AIWPS systems treating domestic wastewater is shown in Table 13.2.

Table 13.2 MP effluent quality for AIWPS systems treating domestic wastewater (compiled from Oswald 1991; Green *et al.* 1995a; Green *et al.* 1996a; Craggs et al. 2003)

Parameter	AIWPS Effluent
BOD_5 (g m^{-3})	30
$fBOD_5$ (g m^{-3})	7
TSS (g m^{-3})	40
Total N (g m^{-3})	20
Nitrate-N (g m^{-3})	2
Ammoniacal-N (g m^{-3})	4
Total P (g m^{-3})	5
Dissolved Reactive P (g m^{-3})	2
*Faecal coliforms (cfu / 100 ml)	100

*or *E. coli* (MPN / 100ml)

13.6.1 Lime Addition

If the wastewater is soft (i.e. with low concentrations of cations such as calcium, magnesium and iron), treatment can be improved by addition of lime (e.g. 60 g m^{-3} of calcium oxide) to the HRP to enhance algal flocculation and phosphate precipitation at the high pH naturally attained in the pond (Nurdogan and Oswald 1995). Much less lime is required than for conventional lime coagulation of phosphorus, which requires concentrations of 250-300 g m^{-3} to elevate the wastewater pH as well as precipitate phosphates (Oswald 1996). Algae removal may be increased to >90% and ammoniacal-N and dissolved reactive phosphorus removal to 85% and 95% respectively by lime addition to HRPs (Nurdogan and Oswald 1995).

13.6.2 Effluent Disposal

The high quality effluent of AIWPS systems should have minimal impact on most large receiving waters. Gloyna and Tischler (1981) have suggested that pond discharges may actually enhance the receiving water ecosystem, provided appropriate dilution is achieved. Remaining algae will provide oxygen during daylight hours and are a food source for aquatic invertebrates (Oswald 1988b; Quinn and Hickey 1993).

AIWPS effluent may also be discharged to natural or constructed wetlands or rock filters if additional effluent "polishing" is desired. If effluent reuse or land disposal is preferred, AIWPS effluent is sufficiently low in BOD_5 and suspended solids to percolate readily into the ground (Oswald 1989; Puskas and Esen 1989).

13.7 AIWPS COSTS

13.7.1 Capital Costs

AIWPS provide primary, secondary and partial tertiary (in terms of nutrient and faecal indicator removal) wastewater treatment, at capital costs that are considerably lower than those of conventional in-tank mechanical treatment plants (Oswald 1991). The capital cost of an AIWPS is similar to that of a conventional pond system using supplementary aeration and is typically a third to a half of the cost of an in-tank mechanical treatment plant designed for the same flow and organic load (Soeder 1980, Middlebrooks *et al.* 1981; Oswald 1991; Green *et al.* 1996a).

Land costs

AIWPS require the same or less land area as conventional stabilisation ponds but can provide a more consistent and higher quality effluent (Green *et al.* 1995a). Picot *et al.* (1992) showed that conventional facultative ponds require approximately five times more land than HRPs to produce an effluent of similar quality. However, like conventional pond systems, the main limitation for the use of AIWPS to treat the wastewater from large cities is the availability of a large area of affordable land (Oswald 1996).

Construction costs

The simplicity of AIWPS means that they can be designed, constructed and commissioned more economically and in less time than in-tank mechanical treatment plants. Construction costs are primarily for civil works and mechanical equipment (Green *et al.* 1996a). AIWPS ponds are constructed from low-cost formed, compacted earth. Higher costs are incurred if the ponds require a clay or plastic liner, however, the construction cost of lined ponds is still considerably less than that of steel-reinforced concrete reactors (such as anaerobic digesters, activated sludge tanks and clarifiers) (Green *et al.* 1996a). AIWPS have only four major mechanical components: a paddle-wheel to mix the HRP, a pump to recirculate HRP effluent to the AFP surface, a supplementary aerator for the AFP (if required), and a pump to remove settled algal solids from the bottom of the ASP (if used) (Green *et al.* 1996a). In contrast, in-tank mechanical treatment plants require a wide range of specialised equipment.

13.7.2 Operation and Maintenance Costs

The operation and maintenance costs of AIWPS are typically a third of those of in-tank mechanical treatment plants (Soeder 1980; Oswald 1991; Green *et al.* 1996a). Operation and maintenance costs are associated with sludge removal and disposal, power requirement, equipment maintenance, personnel and disinfection costs (Green *et al.* 1996a).

Sludge removal and disposal

The cost of daily removal and disposal of sewage sludge from in-tank mechanical treatment plants is considerable. It can amount to over 50% of the operation costs of combined primary and secondary in-tank mechanical treatment and includes between 15% and 25% of the total electrical costs of the plant (Green *et al.* 1996a). Sludge removal and disposal costs and land requirements are often not accounted for in economic comparisons between in-tank mechanical plants and pond systems (Green *et al.* 1996a). Sludge removal has *never been required* at AIWPS in Hollister and St. Helena, California, which have operated for over 20 and 30 years respectively (Green *et al.* 1996a). The rate of sludge accumulation in the St Helena AFPs was found to be only 0.017 metric tons (of 35% solids) / 1,000 m^3 of wastewater treated (Green *et al.* 1996a).

Power requirement

Unlike conventional WSPs, AIWPS do require some power, primarily for operating the HRP paddlewheel and recirculation pump (Green *et al.* 1996a). AIWPS have a similar energy requirement to conventional pond systems that have supplementary mechanical aeration, but this is still a fraction of that required for in-tank mechanical treatment systems such as activated sludge (Green *et al.* 1996a). Essentially, pond systems replace electrical energy used for mechanical aeration with solar-powered photosynthetic oxidation by algae. A fundamental difference between photosynthetic oxidation of wastewater in HRPs and mechanical aeration of wastewater in activated sludge systems is that HRPs operate at dissolved oxygen levels at or often above saturation (for example up to 20 to 30 g m^{-3} during the day), while for optimal oxygen transfer from air to water, mechanical aeration must operate at levels considerably *below* saturation (Oswald 1991; Green *et al.* 1996a). HRPs can typically produce between 6 and 13 kg of dissolved oxygen per kWh used to operate the paddle wheel compared to mechanical aerators, which can provide between 0.9 and 2.5 kg of dissolved oxygen per kWh (Metcalf and Eddy 1991; Green *et al.* 1995a). Recirculation (up to 25% of the total flow) of highly aerobic effluent from the HRP to the surface of the AFP during daylight hours reduces the need for supplementary aeration of the AFP surface to prevent odour release (Oswald 1991).

Personnel costs

AIWPS, like all pond systems, are less complex, generally more robust and less sensitive to toxic substances and shock loads than in-tank mechanical treatment systems, and can operate for extended periods without wastewater inflow while maintaining wastewater treatment capability (Belsare and Belsare 1987; WHO 1987; Oswald 1988a; Agunwamba 1991; Green *et al.* 1994). AIWPS can therefore be operated and maintained by fewer and less-skilled personnel than in-tank mechanical treatment plants (Oswald 1989; Green *et al.* 1995a; Green *et al.* 1996a).

13.8 ADDITIONAL TREATMENT

The economic advantage of AIWPS over in-tank mechanical treatment further increases when the costs of conventional disinfection and tertiary treatment are considered (Green *et al.* 1996a). Complete tertiary treatment to remove ammoniacal-N, nitrate and phosphate by chemical, mechanical and biological processes is extremely expensive, typically costing twice as much as that of primary and secondary in-tank mechanical treatment combined (Oswald 1988a; Green *et al.* 1996a).

Since AIWPS already provide significant tertiary treatment and disinfection, the overall cost of achieving a higher level of wastewater treatment (e.g. complete algal suspended solids removal, further nutrient removal, disinfection or even salt removal) would still be considerably less than the cost of an in-tank mechanical tertiary treatment plant with disinfection (Green *et al.* 1996a ; Green *et al.* 1996b).

Additional treatment methods including enhanced settling, filtration, dissolved air flotation and electrolytic flocculation for the removal of algae, reverse osmosis for salt removal and UV disinfection have all been evaluated (Sandbank and Shelef 1987; Esen *et al.* 1991; Cromar *et al.* 1992; Rose *et al.* 1992; Al-Shayji *et al.* 1994; Meiring and Oellermann 1995; Middlebrooks 1995; Nurdogan and Oswald 1996; Poelman *et al.* 1997; Downing *et al.* 2002).

13.9 RESOURCE RECOVERY

AIWPS are specifically designed to enable potential recovery of resources (energy, nutrients and water) from wastewater, which could partially offset the cost of wastewater treatment (Green *et al.* 1996; Oswald 1996).

13.9.1 Methane

Efficient methane fermentation is usually inhibited in conventional facultative ponds due to lack of having consistently anaerobic conditions on the pond bottom (Green *et al.* 1995b). Furthermore, because of their large surface area, capture of biogas for energy production is impractical. AFPs can be designed to capture the methane-rich biogas by incorporating a submerged gas collector, which only covers the fermentation pit and can therefore be installed at modest additional cost (Green *et al.* 1995b). Submerged gas collectors are also far less susceptible to damage by wind and solar-UV than surface collectors and enable the biogas to be "scrubbed," and the carbon dioxide and nutrients recycled in microalgal biomass (Green *et al.* 1994). Estimated average annual energy production from the proposed gas collection and electrical generation at the St Helena AIWPS is 53 kWh / 1000 m^3 of wastewater treated (Green *et al.* 1995b; Green et al. 1996).

13.9.2 Algae Biomass

Algae biomass is rich in nutrients such as nitrogen, phosphorus and potassium and as such is an excellent fertiliser (Metting *et al.* 1988). The comparative resistance of the algal cell wall to bacterial decay means that the fertiliser is released slowly, minimising nutrient loss from the soil. Algal biomass also has potential as high-protein feed for aquaculture (Edwards and Sinchumpasak 1981; Groeneweg and Schluter 1981; Gaigher 1982; Cloete *et al.* 1984; Rose *et al.* 1995; Costa *et al.* 2000) and livestock such as poultry, pigs and ruminants (Soeder 1980; Azov and Shelef 1982; McGarry 1982; Shelef 1982; Oswald 1988a;b).

There are possible uses of algal biomass for extraction of industrially valuable oils, pigments, pharmaceutical agents and vitamins or as a selective adsorption material for heavy metals (Oswald 1996). Many of these uses are reviewed by Aaronson and Dubinsky (1982), Metting and Pyne (1988), de la Noüe and De Pauw (1988) and Edwards (1992).

Algal biomass can also be fermented to produce methane or alcohol for energy generation. One kilogram of algae will produce sufficient methane to generate about one kilowatt-hour of electricity (Oswald 1996). If the settled algae biomass from the ASP were to be recycled to the Advanced Facultative Pond fermentation pits, energy generation from methane could potentially exceed the total energy requirement for the whole AIWPS (Green *et al.* 1995b).

13.10 UPGRADING CONVENTIONAL WSP'S

Since AIWPS are modular systems they can be implemented in a number of ways to upgrade existing pond systems. The extent of the pond upgrade will depend upon the effluent quality required to meet effluent or receiving water guidelines, the size and performance of the existing pond system and the site characteristics. A conventional primary and secondary facultative pond system can be subdivided to form Advanced Facultative Ponds (with deeper fermentation pits to facilitate methane fermentation), High Rate Ponds, Algae Settling Ponds and Maturation Ponds, with little additional land requirement. Simpler upgrades excluding the AFP may be possible for existing pond systems that are already removing wastewater BOD and solids adequately, but now require improved disinfection and nutrient removal.

13.11 TREATMENT OF OTHER WASTES

As shown in Table 13.1, AIWPS in part or full have been successfully used to treat various agricultural wastes including: animal manure (cattle, poultry, piggeries, dairy cows), agro-industrial waste (pineapple processing, fruit and vegetable canning, palm oil, rubber), slaughterhouse and meat-packing waste, agricultural drainage water for selenium removal and acid mine drainage water (see Shelef (1989) for a review). Typically agricultural and industrial wastewaters have much higher BOD than domestic sewage and may require more supplementary surface aeration of the AFP or substitution with an anaerobic pond (Oswald 1996). Furthermore, the HRP may need to operate at shallower depths with high light-attenuating wastewaters to ensure adequate sunlight exposure for algal growth and photosynthesis (Oswald 1996). However, the value of the algal biomass reclaimed either as fertiliser, which could be spray irrigated directly from the HRP, or used as a livestock feed supplement, could offset the costs of the system (Pieterse *et al.* 1982; Oswald 1991). HRPs have also been used to treat sewage seawater mixtures (Ryther 1983), saline wastewaters (tannery; Rose *et al.* 1992; Rose *et al.* 1995; Tadesse *et al.* 2003) and polluted seawater from aquaculture facilities (Pagand *et al.* 1998; 2000).

13.12 SUMMARY

AIWPS are a simple upgrade to conventional WSPs that provide reliable, nuisance free and efficient wastewater treatment at less than half the cost of activated sludge in-tank mechanical wastewater treatment. The high treatment performance of AIWPS is due to integration of AFPs for sedimentation and methane fermentation of wastewater solids; paddlewheel-mixed HRPs for low-cost wastewater oxidation, nutrient assimilation and disinfection; ASPs for simple algae removal by sedimentation; and MPs for natural wastewater disinfection and polishing. AIWPS are designed by matching the wastewater-loading rate to the rates of natural wastewater treatment processes (including: sedimentation, methane fermentation, algal photosynthetic oxygenation and disinfection), which are calculated for local climatic conditions.

Since AIWPS already provide significant tertiary treatment and disinfection, the overall cost of achieving higher levels of wastewater treatment with supplementary treatment would still be considerably less than the cost of an in-tank mechanical tertiary treatment plant with disinfection.

AIWPS retain the simple construction, operation and aesthetics of conventional ponds and enable opportunities for resource recovery through nutrient reclamation as algae and use of biogas as an energy source.

13.13 FUTURE RESEARCH NEEDS

Although much research has been conducted throughout the world on components of AIWPS systems more is required on the complete system, and particularly the AFP. AIWPS systems also need to be evaluated for treatment and recovery of resources from other organic wastewaters, including saline/seawater systems.

AFPs

Research is required to further elucidate the processes of sludge digestion and nitrogen removal in fermentation pits, and verify the removal of heavy metals, hydrocarbons and other persistent organic compounds. Methods of incorporating fermentation pits in shallow ponds (where depth is limited by a high water table) are needed. Gas production, capture and use from AFPs needs to be studied further and demonstrated at full-scale.

HRPs

Research is needed on HRP hydraulics, to minimise short-circuiting of inflow to outflow, and determine how pond dimensions influence eddy formation. More data on the diurnal performance of HRPs needs to be reported to determine whether treatment performance could be improved by only discharging during the day. The use of well-mixed HRPs as a tool to evaluate particular aerobic treatment and disinfection processes that occur in all WSPs should be explored.

ASPs

Techniques are needed to improve settling of algae such as providing shade through the addition of surface covers, biofilm screens or floating aquatic plants. Research is required on the practical removal, handling and use of the algae/bacterial biomass at large scale.

MPs

Design equations based on solar insolation or solar-UV dose (incorporating the influence of fluctuating DO and pH) rather than temperature are required. Research is needed on a "whole ecosystem approach" to MP design, incorporating other components of natural freshwater communities to further improve water quality. For example, integrating aspects of wetlands for filtration, buffering, effluent polishing and increasing habitat complexity.

REFERENCES

Aaronson, S. and Dubinsky, Z. (1982) Mass production of microalgae. *Experientia* **38**(1), 36-40.

Abeliovich, A. (1980). Factors limiting algal growth in high rate oxidation ponds. In *Algae Biomass*, (eds G. Shelef and C.J. Soeder), pp. 205-215, Elsevier, Nth Holland, Amsterdam.

Abeliovich, A. and Azov, Y. (1976) Toxicity of ammonia to algae in sewage oxidation ponds. *Applied Environmental Microbiology* **31**, 801-806.

Agunwamba, J.C. (1991) Simplified optimal design of the waste stabilization pond. *Water Air and Soil Pollution* **59**, 299-309.

Al-Shayji, Y.A., Puskas, K., Al-Daher, R. and Esen, II. (1994) Production and separation of algae in a high-rate ponds system. *Environment International* **20**(4), 541-550.

Araki, S., Gonzalez, J.M., de Luis, E. and Becares, E. (2000) Viability of nematode eggs in high rate algal ponds: the effect of physico-chemical conditions. *Water Science & Technology* **42**(10-11), 371-374.

Araki, S., Martin-Gomez, S., Becares, E., De Luis-Calabuig, E. and Rojo-Vazquez, F. (2001) Effect of high-rate algal ponds on viability of *Cryptosporidium parvum* oocysts. *Applied and Environmental Microbiology* **67**(7), 3322-3324.

Ayala, F.J. and Bravo, R.B. (1984) Animal wastes for *Spirulina* production. *Arch. Hydrobiol. Suppl.* **67**(3), 349-355.

Azov, Y. and Shelef, G. (1982) Operation of high-rate oxidation ponds: theory and experiments. *Water Research* **16**(7), 1153-1160.

Azov, Y. and Shelef, G. (1987) Effect of pH on the performance of high-rate oxidation ponds. *Water Science and Technology* **19**(12), 381-383.

Azov, Y., Shelef, G. and Moraine, R. (1982) Carbon limitation of biomass production in high-rate oxidation ponds. *Biotechnology and Bioengineering* **24**(3), 579-594.

Bahlaoui, M.A., Baleux, B. and Troussellier, M. (1997) Dynamics of pollution-indicator and pathogenic bacteria in high-rate oxidation wastewater treatment ponds. *Water Research* **31**(3), 630-638.

Balloni, W., Florenzano, G., Materassi, R., Tredeci, M., Soeder, C.J. and Wagener, K. (1983). Mass Cultures of algae for energy farming in coastal deserts. In *Energy From Biomass, Second European Community Conference.*, (eds A. Straub, P. Chartier and G. Schleser), pp. 291-295, Applied Sci. Publ., N.Y.

Belsare, D.K. and Belsare, S.D. (1987) High rate oxidation pond for water and nutrient recovery from domestic sewage in urban areas in the tropics. *Arch. Hydrobiol. Beih. Ergebn. LImnol.* **28**, 123-128.

Benemann, J.R., Weissman, J.C., Koopman, B.L. and Oswald, W.J. (1977) Energy production by microbial photosynthesis. *Nature* **268**, 19-23.

Berend, I., Simovitch, E. and Ollian, A. (1980). Economic aspects of algal animal food production. In *Algae Biomass: Production and use*, (eds G. Shelef and C.J. Soeder), vol , pp. 799-818, Elservier/Nth Holland Public Corp, Amsterdam.

Bich, N.N., Yaziz, M.I. and Bakti, N.A.K. (1999) Combination of *Chlorella vulgaris* and *Eichhornia crassipes* for wastewater nitrogen removal. *Water Research* **33**(10), 2357-2362.

Bontoux, J. and Picot, B. (1994) Possibilities and limits of stabilization ponds in wastewater treatment. *Water Pollution Research Journal Canada* **29**(4), 545-556.

Buelna, G., Bhattarai, K.K., de la Noue, J. and Taiganides, E.P. (1990) Evaluation of various flocculants for the recovery of algal biomass grown on pig-waste. *Biological Wastes* **31**(3), 211-222.

Buhr, H.O. and Miller, S.B. (1983) A dynamic model of the high-rate algal-bacterial wastewater treatment pond. *Water Research* **17**(1), 29-38.

Castignetti, D. and Hollocher, T.C. (1984) Heterotrophic nitrification among denitrifiers. *Applied and Environmental Microbiology* **47**(4), 620-623.

Chen, P., Zhou, Q., Paing, J. and Picot, B. (2003) Nutrient removal by an integrated high rate algal pond and macrophyte system in China. *Water Science and Technology* **48** (2) pp 251-257.

Cloete, T.E., Toerien, D.F. and Pieterse, A.J.H. (1984) The bacteriological quality of water and fish of a pond system for the treatment of cattle feedlot effluent. *Agricultural Wastes* **9**(1), 1-15.

Costa, R.H.R., Medri, W. and Perdomo, C.C. (2000) High-rate pond for treatment of piggery wastes. *Water Science and Technology* **42**(10-11), 357-362.

Craggs, R.J., Davies-Colley, R.J., Tanner, C.C. and Sukias, J.P. (2003) Advanced pond system: performance with high rate ponds of different depths and areas. *Water Science and Technology* **48** (2) pp 259–267.

Cromar, N.J., Fallowfield, H.J. and Martin, N.J. (1996) Influence of environmental parameters on biomass production and nutrient removal in a high rate algal pond operated by continuous culture. *Water Science and Technology* **34**(11), 133-140.

Cromar, N.J., Martin, N.J., Christofi, N., Read, P.A. and Fallowfield, H.J. (1992) Determination of nitrogen and phosphorus partitioning within components of the biomass in a high rate algal pond: significance for the coastal environment of the treated effluent discharge. *Water Science and Technology* **25**(12), 207-214.

Curtis, T.P., Mara, D.D. and Silva, S.A. (1992) Influence of pH, oxygen, and humic substances on ability of sunlight to damage fecal coliforms in waste stabilization pond water. *Applied Environmental Microbiology* **58**(4), 1335-1343.

Davies-Colley, R.J., Donnison, A.M. and Speed, D.J. (2000) Towards a mechanistic understanding of pond disinfection. *Water Science and Technology* **42**(10-11), 149-158.

Davies-Colley, R.J., Craggs, R.J. and Nagels, J.W. (2003) Disinfection in a pilot-scale advanced pond system (APS) for domestic sewage treatment in New Zealand. *Water Science and Technology* **48**(2), 81–87.

de la Noüe, J. and De Pauw, N. (1988) The potential of microalgal biotechnology: a review of production and uses of microalgae. *Biotech. Advances* **6**, 725-770.

De Pauw, N. and Van Vaerenbergh, E. (1983). Microalgal waste treatment systems: potential and limits. In *Phytodepuration and the employment of the biomass produced*, (eds Ghetti), vol , pp. 211-287, Centro Ric. Produz. Animali, Reggio Emilia, Italy.

Diab, S., Kochba, M. and Avnimelech, Y. (1993) Nitrification pattern in a fluctuating anaerobic-aerobic pond environment. *Water Research* **27**(9), 1469-1475.

Dodd, J.C. (1986). Elements of pond design and construction. In *CRC Handbook of Microalgal Mass Culture*, (eds Richmond), vol , pp. 265-283, CRC Press Inc., Boca Raton.

Downing, J.B., Bracco, E., Green, F.B., Ku, A.Y., Lundquist, T.J., Zubieta, I.X. and Oswald, W.J. (2002) Low cost reclamation using the advanced integrated wastewater pond systems technology and reverse osmosis. *Water Science and Technology* **45**(1), 117-125.

Duggan, G.L. and Golueke, C.G. (1972) Recycling system for poultry wastes. *Journal Water Pollution Control Federation* **44**, 432-440.

Edwards, P. (1992) *Reuse of Human Wastes in Agriculture, A Technical Review*. International Bank for Reconstruction and Development, Washington, DC.

Edwards, P. and Sinchumpasak, O.A. (1981) The harvest of microalgae from the effluent of a sewage fed high rate stabilization pond by Tilapia Nilotica. Part 1: description of the system and the study of the high rate pond. *Aquaculture* **23**, 1-4.

Eisenberg, D.M., Koopman, B., Benemann, J.R. and Oswald, W.J. (1981) Algal bioflocculation and energy conservation in microalgal sewage ponds. *Biotechnology Bioengineering Symposium* **11**, 429-448.

El Hamouri, B., Khallayoune, K., Bouzoubaa, K., Rhallabi, N. and Chalabi, M. (1994) High-rate algal pond performances in faecal coliforms and helminth egg removals. *Water Research* **28**(1), 171-174.

El Hamouri, B., Jellal, J., Outabiht, H., Nebri, B., Khallayoune, K., Benkerroum, A., Hajli, A. and Firadi, R. (1995) Performance of a high-rate algal pond in the Moroccan climate. *Water Science and Technology* **31**(12), 67-74.

El Ouarghi, H., Boumansour, B.E., Dufayt, O., El Hamouri, B. and Vasel, J.L. (2000) Hydrodynamics and oxygen balance in a high-rate algal pond. *Water Science & Technology* **42**(10-11), 349-356.

El Hamouri, B., Rami, A. and Vasel, J.L. (2003) Comparing first order reaction rate constant and specific removal rate of a high rate algal pond and series of three facultative ponds in Morocco. *Water Science and Technology* **48**(2), 169–276.

El-Gohary, F.A., Abo-Elela, S.I., Shehata, S.A. and El-Kamah, H.M. (1991) Physico-chemical-biological treatment of municipal wastewater. *Water Science and Technology* **24**(7), 285-292.

Esen, I., Puskas, K., Banat, I. and Daher, R.A. (1987) Algal-bacterial ponding systems for municipal wastewater treatment in arid regions. *Water Science and Technology* **19**(12), 341-343.

Esen, I., Puskas, K., Banat, I.M. and Al-Daher, R. (1991) Algae removal by sand filtration and reuse of filter material. *Waste Management* **11**(1-2), 59-65.

Evans, R.A., Fallowfield, H.J. and Cromar, N.J. (2003) Characterisation of oxygen dynamics within a high-rate algal pond system used to treat abattoir wastewater. *Water Science and Technology* **48**(2), 61–68.

Fallowfield, H.J. and Garrett, M.K. (1985) The photosynthetic treatment of pig slurry in temperate climatic conditions: a pilot plant study. *Agricultural Wastes.* **12**, 111-136.

Fallowfield, H.J., Cromar, N.J. and Evison, L.M. (1996) Coliform die-off rate constants in a high rate algal pond and the effect of operational and environmental variables. *Water Science and Technology* **34**(11), 141-147.

Fallowfield, H.J., Martin, N.J. and Cromar, N.J. (1999) Performance of a batch-fed high rate algal pond for animal waste treatment. *European Journal of Phycology* **34**(3), 231-237.

Fallowfield, H.J., Mesple, F., Martin, N.J., Casellas, C. and Bontoux, J. (1992) Validation of computer models for high rate algal pond operation for wastewater treatment using data from Mediterranean and Scottish pilot scale systems: implications for management in coastal regions. *Water Science and Technology* **25**(12), 215-224.

Gaigher, I.G. (1982) The growth and production of Mocambique Tilapia (*Oreochromis mossambicus*) fed on algae in small tanks. *Water South Africa* **8**(3), 142-144.

Garcia, J., Mujeriego, R. and Hernandez-Marine, M. (2000a) High rate algal pond operating strategies for urban wastewater nitrogen removal. *Journal of Applied Phycology* **12**, 331-339.

Garcia, J., Hernandez-Marine, M. and Mujeriego, R. (2000b) Influence of phytoplankton composition on biomass removal from high-rate oxidation lagoons by means of sedimentation and spontaneous flocculation. *Water Environment Research* **72**(2), 230-237.

Garcia, J., Hernandez-Marine, M. and Mujeriego, R. (2002) Analysis of key variables controlling phosphorus removal in high rate oxidation ponds provided with clarifiers. *Water SA* **28**(1), 55-62.

Gerhardt, M.B., Green, F.B., Newman, R.D., Lundquist, T.J. and Tresan, R.B. (1991) Removal of selenium using a novel algal-bacterial process. *Research Journal of the Water Pollution Control Federation* **63**(5), 799-805.

Gloyna, E.F. and Tischler, L.F. (1981) Recommendations for regulatory modifications: the use of waste stabilization pond systems. *Journal of the Water Pollution Control Federation* **53**(11), 1559-1563.

Goldman, J.C. (1979) Outdoor algal mass algal cultures. II: Photosynthetic yield limitations. *Water Research* **13**, 119-136.

Gomez, E., Casellas, C., Picot, B. and Bontoux, J. (1995) Ammonia elimination processes in stabilisation and high-rate algal pond systems. *Water Science and Technology* **31**(12), 303-312.

Green, F.B., Lundquist, T.J. and Oswald, W.J. (1995a) Energetics of advanced integrated wastewater pond systems. *Water Science and Technology* **31**(12), 9-20.

Green, F.B., Bernstone, L., Lundquist, T.J., Muir, J., Tresan, R.B. and Oswald, W.J. (1995b) Methane fermentation, submerged gas collection, and the fate of carbon in advanced integrated wastewater pond systems. *Water Science and Technology* **31**(12), 55-65.

Green, F.B., Lundquist, T.J. and Oswald, W.J. (1996a) *The Capture and Utilization of Methane in Advanced Integrated Wastewater Pond Systems.* Report for the California Energy Commission Energy Technology Advancement Program. ETAP Contract #500-90-026.

Green, F.B., Bernstone, L.S., Lundquist, T.J. and Oswald, W.J. (1996b) Advanced integrated wastewater pond systems for nitrogen removal. *Water Science and Technology* **33**(7), 207-217.

Green, F.B., Lundquist, T.J., Quinn, N.W.T., Zarate, M.A., Zubieta, I.X. and Oswald, W.J. (2003) Selenium and nitrate removal from africultural drainage using AIWPS technology. *Water Science and Technology* **48**(2), 299–305.

Grobbelaar, J.U., Soeder, C.J., Groeneweg, J., Stengel, E. and Hartig, P. (1988) Rates of biogenic oxygen production in mass cultures of microalgae, absorption of atmospheric oxygen and oxygen availability for wastewater treatment. *Water Research* **22**(11), 1459-1464.

Grobbelaar, J.U., Soeder, C.J. and Stengel, E. (1990) Modelling algal productivity in large outdoor cultures and waste treatment systems. *Biomass* **21**, 297-314.

Groeneweg, J., Klein, B., Mohn, F.H., Runkel, K.H. and Stengel, E. (1980). First results of outdoor treatment of pig manure with algal-bacterial systems. In *Algae Biomass*, (eds G. Shelef and C.J. Soeder), vol , pp. 255-264, Elsevier, Nth Holland, Amsterdam.

Groeneweg, J. and Schluter, M. (1981) Mass production of freshwater rotifiers on liquid wastes II. Mass production of *Brachionus rubens* Ehrenberg 1838 in the effluent of high-rate algal ponds used for the treatment of piggery waste. *Aquaculture* **15**(1), 25-33.

Hall, T.W. (1989) *Bioflocculating high rate algal ponds: control and implementation of an innovative wastewater treatment technology.* Ph.D. dissertation, Civil Engineering, University of California, Berkeley, pp. 280.

Jupsin, H., Praet, E. and Vasel, J.-L. (2003) Dynamic mathematical model of high rate algal pond (HRAP). *Water Science and Technology* **48**(2), 197–204.

Kawai, H., Grieco, V.M. and Jureidini, P. (1984) A study of the treatability of pollutants in high rate photosynthetic ponds and the utilization of the proteic potential of algae which proliferate in the ponds. *Environmental Technology Letters* **5**(11), 505-516.

Koopman, B. and Lincoln, E.P. (1983) Autoflotation harvesting of algae from high-rate pond effluents. *Agricultural Wastes* **5**(4), 231-246.

Kroon, B.M.A., Ketelaars, H.A.M., Fallowfield, H.J. and Mur, L.R. (1989) Modelling microalgal productivity in a high rate algal pond based on wavelength dependent optical properties. *Journal of Applied Phycology* **1**(3), 247-256.

Lincoln, E.P. and Hill, D.T. (1980). An integrated microalgae system. In *Algae Biomass: Production and Use*, (eds G. Shelef and C.J. Soeder), vol , pp. 229-244, Elsevier/ Nth Holland Biomedical Press, Amsterdam.

Lincoln, E.P. and Koopman, B. (1986). Bioflocculation of microalgae in mass cultures. In *Algal Biomass Technologies*, (eds Barclay and McIntosh), vol , pp. 207-211, Nova Hedwigia Beih. 83, J. Cramer, Berlin.

Ludwig, H.F. and Oswald, W.J. (1952) The role of algae in sewage oxidation ponds. *Scientific Monthly* **74**(3), 19-25.

Ludwig, H.F., Oswald, W.J., Gotaas, H.B. and Lynch, V. (1950) Sewage works: algae symbiosis in oxidation ponds. I. Growth characteristics of *Euglena gracilis* cultured in sewage. *Sewage and Industrial Wastes* **23**, 1337-1355.

Mahadevaswamy, M. and Venkataraman, L.V. (1986) Bioconversion of poultry droppings for biogas and algal production. *Agricultural Wastes* **18**(2), 93-101.

Marais, G.v.R. (1974) Faecal bacterial kinetics in stabilization ponds. *Journal of the Environmental Engineering Division, Proceedings of the American Society of Civil Engineers.* **100**(EEI), 119-139.

Martin, N.J. and Fallowfield, H.J. (1989) Computer modelling of algal waste treatment systems. *Water Science and Technology* **21**(12), 1657-1660.

McGarry, M.G. (1982) Oxidation ponds and fish culture. *Water Supply and Sanitation in Developing Countries, Ann Arbor Science, Ann Arbor, MI* **1982**, 201-218.

Meiring, P.G.J. and Oellermann, R.A. (1995) Biological removal of algae in an integrated pond system. *Water Science and Technology* **31**(12), 21-31.

Mesple, F., Troussellier, M., Casellas, C. and Bontoux, J. (1995) Difficulties in modelling phosphate evolution in a high-rate algal pond. *Water Science and Technology* **31**(12), 45-54.

Metcalf and Eddy (1991) Wastewater Engineering: Treatment Disposal Reuse. , McGraw-Hill., Inc., Boston, 920.

Metting, B., Rayburn, W.R. and Reynaud, P.A. (1988). Algae and agriculture. In *Algae and Human Affairs*, (eds L.a. Waaland), vol , pp. 335-370, Cambridge University Press, NY, USA.

Middlebrooks, C.H. (1995) Upgrading pond effluents: an overview. *Water Science and Technology* **31**(12), 353-368.

Middlebrooks, E.J., Middlebrooks, C.H. and Reed, S.C. (1981) Energy requirement for small wastewater treatment systems. *Journal of the Water Pollution Control Federation* **53**(7), 1172-1197.

Moraine, R., Shelef, G., Meyden, A. and Levi, A. (1979) Algal single cell protein from wastewater treatment and renovation process. *Biotechnology Bioengineering* **21**, 1191-1207.

Moutin, T., Gal, J.Y., El Halouani, H., Picot, B. and Bontoux, J. (1992) Decrease of phosphate concentration in a high rate pond by precipitation of calcium phosphate: Theoretical and experimental results. *Water Research* **26**(11), 1445-1450.

Nurdogan, Y. and Oswald, W.J. (1995) Enhanced nutrient removal in high-rate ponds. *Water Science and Technology* **31**(12), 33-43.

Nurdogan, Y. and Oswald, W.J. (1996) Tube settling of high-rate pond algae. *Water Science and Technology* **33**(7), 229-241.

Oron, G. and Shelef, G. (1982) Maximizing algal yield in high-rate oxidation ponds. *Journal of the Environmental Engineering Division, Proceedings of the American Society of Civil Engineers* **108**(EE4), 730-738.

Oron, G., Shelef, G., Levi, A., Meydan, A. and Azov, Y. (1979) Algae/bacteria ratio in high-rate ponds used for waste treatment. *Applied Environmental Microbiology* **38**(4), 570-576.

Oron, G., Shelef, G. and Levi, A. (1981a) Algal polymorphism in high rate wastewater treatment ponds. *Hydrobiologia* **77**(2), 167-175.

Oron, G., Shelef, G. and Levi, A. (1981b) Environmental phenotypic variation of *Scenedesmus dimorphus* in high-rate algae ponds and its relationship to wastewater treatment and biomass production. *Biotechnology and Bioengineering* **23**(10), 2185-2198.

Oswald, W.J. (1963) High rate pond in waste disposal. *Developments in Industrial Microbiology* **4**, 112-119.

Oswald, W.J. (1970) Growth characteristics of microalgae cultured in domestic sewage. In *IBP/PP Technical Meeting*, Trebon.

Oswald, W.J. (1978) The engineering aspects of microalgae. In *Handbook of Microbiology*, (eds , vol 2, pp. 519-552, C.R.C. press, Boca Raton, FL, West Palm Beach Florida.

Oswald, W.J. (1980) Algal production: problems, achievements and potential. In *Algae Biomass: Production and Use*, (eds G. Shelef and C.J. Soeder), pp. 1-8, Elsevier biomedical press, Nth Holland, Amsterdam.

Oswald, W.J. (1988a) Micro-algae and waste-water treatment. In *Micro-algal Biotechnology*, (eds M.A. Borowitzka and L.J. Borowitzka), pp. 305-328, Cambridge University Press, Cambridge.

Oswald, W.J. (1988b) Role of Microalgae in Liquid Waste Treatment and Reclamation. *Algae and Human Affairs. Cambridge University Press New York*, 255-281.

Oswald, W.J. (1989) Use of wastewater effluent in agriculture. *Desalination* **72**(1), 67-80.

Oswald, W.J. (1990) Advanced integrated wastewater pond systems. In *Supplying Water and Saving the Environment for Six Billion People: Proceedings of the 1990 ASCE Convention, Env. Eng. Div. New York, USA.*

Oswald, W.J. (1991) Introduction to advanced integrated wastewater ponding systems. *Water Science and Technology* **24**(5), 1-7.

Oswald, W.J. (1995) Ponds in the twenty-first century. *Water Science and Technology* **31**(12), 1-8.

Oswald, W.J. (1996) *A syllabus on Advanced Integrated Wastewater Pond Systems.* American Society of Civil Engineers. Notes from ASCE course on Wastewater treatment with Advanced Integrated Wastewater Pond Systems (AIWPS) and Constructed Wetlands. 323 pp.

Oswald, W.J. and Gotaas, H.B. (1955) Photosynthesis in sewage treatment. In *Prodeedings of the American Society of Civil Engineers.*

Oswald, W.J. and Gotaas, H.B. (1957) Photosynthesis in sewage treatment. *Transcripts of the American Society of Civil Engineers* **122**, 73-105.

Oswald, W.J., Gotaas, H.B., Ludwig, H.F. and Lynch, V. (1953) Algal symbiosis in oxidation ponds III. Photosynthetic oxidation. *Sewage Industrial Wastes* **25**, 690-692.

Oswald, W.J., Gotaas, H.B., Golueke, C.G. and Kellen, W.R. (1957) Algae in waste treatment. *Sewage Industrial Wastes* **29**, 437-457.

Oswald, W.J., Golueke, C.G., Cooper, R.C., Gee, H.K. and Bronson, J.C. (1963) Water reclamation, algal production and methane fermentation in waste ponds. *International Journal of Air and Water Pollution* **7**, 627-648.

Oswald, W.J., Green, F.B. and Lundquist, T.J. (1994) Performance of methane fermentation pits in advanced integrated wastewater pond systems. *Water Science and Technology* **30**(12), 287-295.

Pagand, P., Blancheton, J.P., Lemoalle, J. and Casellas, C. (1998) *High rate algal pond for the treatment of marine effluent from a semi-closed fish rearing system*, Ices, Copenhagen (Denmark).

Pagand, P., Blancheton, J., Lemoalle, J. and Casellas, C. (2000) The use of high rate algal ponds for the treatment of marine effluent from a recirculating fish rearing system. *Aquaculture Research* **31**(10), 729-736.

Paniagua-Michel, J., Farfan, B.C. and Buckle-Ramirez, F. (1987) Culture of marine microalgae on natural biodigested resources. *Aquaculture* **64**, 249-256.

Parhad, N.M. and Rao, N.U. (1974) Effect of pH on survival of *Escherichia coli* . *Journal of the Water Pollution Control Federation* **46**, 980-986.

Phang, S.-M. (1990) Algal production from agro-industrial and agricultural wastes in Malaysia. *Ambio. Stockholm* **19**(8), 415-418.

Picot, B., El Halouani, H., Casellas, C., Moersidik, S. and Bontoux, J. (1991) Nutrient removal by high rate pond system in a Mediterranean climate (France). *Water Science and Technology* **23**(7-9), 1535-1541.

Picot, B., Bahlaoui, A., Moersidik, S., Baleux, B. and Bontoux, J. (1992) Comparison of the purifying efficiency of high rate algal pond with stabilization pond. *Water Science and Technology* **25**(12), 197-206.

Pieterse, A.J.H., LeRoux, J. and Toerien, D.F. (1982) The cultivation of algae using waste water from feedlots. *Water South Africa* **8**(4), 202-207.

Pinheiro, H.M., Reis, M.T. and Novais, J.M. (1987) Study of the performance of a high-rate photosynthetic pond system. *Water Science and Technology* **19**(12), 237-241.

Poelman, E., De Pauw, N. and Jeurissen, B. (1997) Potential of electrolytic flocculation for recovery of micro-algae. *Resources, Conservation and Recycling* **19**(1), 1-10.

Puskas, K. and Esen, I. (1989) Possibilities of utilizing municipal and industrial wastewater effluents to condition soil. *Desalination* **72**(1), 125-132.

Quinn, J. M. and Hickey, W. (1993) Effects of sewage waste stabilisation lagoon effluent on stream invertebrates. *Journal of Aquatic Ecosystem Health* **2**,205-219.

Richmond, A. (1983). Phototrophic microalgae. In *Biomass: microorganisms for special applications; microbial products 1; Energy from renewable resources,*, (eds H. Dellweg), vol 3, pp. 119-143, Verlag Chemie, Weinheim.

Richmond, A. (1986). Cell response to environmental factors. In *CRC Handbook of Microalgal Mass Culture*, (eds A. Richmond), 69-99, CRC Press inc., Boca Raton, Florida.

Richmond, A. and Grobbelaar, J.U. (1986) Factors affecting the output rate of *Spirulina platensis* with reference to mass cultivation. *Biomass* **10**, 253-264.

Rodrigues, A.M. and Santos Oliveira, J.F.S. (1987a) Treatment of wastewaters from the tomato concentrate industry in high rate algal ponds. *Water Science and Technology* **19**(1), 43-49.

Rodrigues, A.M. and Santos Oliveira, J.F.S (1987b) High-rate algal ponds; treatment of wastewaters and protein production: IV. Chemical composition of biomass produced from swine wastes. *Water Science and Technology* **19**(12), 243-248.

Rose, P.D., Maart, B.A., Phillips, T.D., Tucker, S.L., Cowan, A.K. and Rowswell, R.A. (1992) Cross-flow ultrafiltration used in algal high rate oxidation pond treatment of saline organic effluents with the recovery of products of value. *Water Science and Technology* **25**(10), 319-327.

Rose, P.D., Maart, B.A., Dunn, K.M., Rowswell, R.A. and Britz, P. (1995) High rate algal oxidation ponding for the treatment of tannery effluents. *Water Science and Technology* **33**(7), 219-227.

Rose, P.D., Boshoff, G.A., Van Hille, R.P., Wallace, L.C.M., Dunn, K.M. and Duncan, J.R. (1998) An integrated algal sulphate reducing high rate ponding process for the treatment of acid mine drainage wastewaters. *Biodegradation* **9**(3-4), 247-257.

Rovirosa, N., Sanchez, E., Benitez, F., Travieso, L. and Pellon, A. (1995) An integrated system for agricultural wastewater treatment. *Water Science and Technology* **32**(12), 165-171.

Ryther, J.H. (1983) The evolution of integrated aquaculture systems. *Journal of the World Mariculture Society* **14**, 473-484.

Sandbank, E. and Shelef, G. (1987) Harvesting of algae from high-rate ponds by flocculation-flotation. *Water Science and Technology* **19**(12), 257-263.

Schanz, F. and Dubinsky, Z. (1988) Afternoon depression in primary productivity in a high rate oxidation pond (HROP). *Journal of Plankton Research* **10**(3), 373-383.

Schluter, M., Groeneweg, J. and Soeder, C.J. (1987) Impact of rotifer grazing on population dynamics of green microalgae in high-rate ponds. *Water Research* **21**(10), 1293-1297.

Sebastian, S. and Nair, K.V.K. (1984) Total removal of coliforms and *E. coli* from domestic sewage by high-rate pond mass culture of *Scenedesmus obliquus*. *Environmental Pollution* **34**(3), 197-206.

Shelef, G. (1982) High-rate algae ponds for wastewater treatment and protein production. *Water Science and Technology* **14**(1), 439-452.

Shelef, G. (1989). Engineering of microalgae mass culture for treatment of agricultural wastewater, with special emphasis on selenium removal from drainage waters. In *Biotreatment of Agricultural Wastewater. CRC Press, Inc., Boca Raton Florida*, (eds , vol , pp. 143-148

Shelef, G. and Azov, Y. (1987) High-rate oxidation ponds: the Israeli experience. *Water Science and Technology* **19**(12), 249-255.

Shelef, G. and Azov, Y. (2000) Meeting stringent environmental and reuse requirements with an integrated pond system for the twenty-first century. *Water Science and Technology* **42**(10-11), 299-305.

Shelef, G., Azov, Y. and Moraine, R. (1982) Nutrients removal and recovery in a two-stage high-rate algal wastewater treatment system. *Water Science and Technology* **14**(4-5), 87-100.

Shelef, G., Moraine, R., Meydan, A. and Sandback, E. (1977). Combined algae production wastewater treatment and reclamation systems. In *Microbial Energy Conversion*, (eds H.G. Schegel and J. Barnea), vol , pp. 427-442, Pergamon Press, West Germany

Shelef, G., Moraine, R. and Orone, G. (1978a). Photosynthetic biomass production from sewage. In *Microalgae for Food and Feed*, (eds C.J. Soeder and R. Binsack), vol 11, pp. 3-13, Ergebn. Limnol.

Shelef, G., Oron, G. and Moraine, R. (1978b) Economic aspects of microalgae production of sewage. *Arch. Hydrobiol. Beih. Ergebn. Limnol.* **11**, 281-294.

Shelef, G., Azov, Y., Moraine, R. and Orone, G. (1980). Algal mass production as an integral part of a wastewater treatment and reclamation system. In *Algae Biomass*, (eds G. Shelef and C.J. Soeder), vol , pp. 97-113, Elsevier, Nth Holland Biomedical Press, Amsterdam.

Sobsey, M.D. and Cooper, R.C. (1973) Enteric virus survival in algal-bacterial wastewater treatment systems. *Water Research* **7**, 669-685.

Soeder, C.J. (1980) Massive cultivation of microalgae: results and prospects. *Hydrobiologia* **72**(1), 197-209.

Tadesse, I., Isoaho, S.A. and Puhakka, J.A. (2003) Removal of organics and nutrients from tannery effluent by an advanced integrated wastewater pond system. *Water Science and Technology* **48** (2) pp 307-314.

Taiganides, E.P. (1997) *Pig Waste Management and Recycling: The Singapore Experience*, International Development Research Centre, Ottawa, Canada.

Tate, R.L.I. (1980) Variation in heterotrophic and autotrophic nitrifier populations in relation to nitrification in organic soils. *Applied Environmental Microbiology* **40**(1), 75-79.

Terry, K.L. and Raymond, L.P. (1985) System design for the autotrophic production of microalgae. *Enzyme Microb. Technology* **7**(October), 474-487.

van de Graaf, A.A., Mulder, A., de Bruijn, P., Jetten, M.S., Robertson, L.A. and Kuenen, J.G. (1995). Anaerobic oxidation of ammonium is a biologically mediated process. *Appl. Environ. Microbiol.* **61**(4), 1246-1251.

Verstraete, W. and Alexander, M. (1973) Heterotrophic nitrification in samples of natural ecosystems. *Environmental Science and Technology* **7**(1), 39-42.

Von Hippel, D.F. and Oswald, W.J. (1985) *Collection of methane evolved from waste-treatment ponds.* , University of California, Berkeley, USA.

Von Hippel, D.F. and Oswald, W.J. (1989) *Heating facultative ponds to increase rates of gas evolution: a step in the integration of waste-treatment and biomass energy production using microalgae.* Report 89-3, Sanitary Engineering and Environmental Health Research Laboratory, University of California, Berkeley.

WHO. (1987) *Wastewater Stabilization Ponds: principles of planning and practice*, WHO, EMRO technical publication No. 10, Alexandria.

Wood, A., Scheepers, J. and Hills, M. (1989) Combined artificial wetland and high rate algal pond for wastewater treatment and protein production. *Water Science and Technology* **21**(6-7), 659-668.

14

Pond(s) integrated with trickling filter and activated sludge processes

Oleg Shipin and Pieter Meiring

14.1 INTRODUCTION

New stricter discharge regulations for treated wastewater, particularly in terms of P and N, are being introduced in many parts of the world. Though meeting stricter standards is technologically feasible for a modern state-of-the-art activated sludge plant by chemical dosing it is still a weakness of waste stabilisation pond technology. However, it is still paramount to provide low cost technologies, particularly in the developing world that also faces other serious challenges (housing, diseases, *etc*). "Older" low cost technologies, such as waste stabilisation ponds and trickling filter plants incorporating ponds for maturation have served thousands of communities for many decades but are coming to a crossroads, either because of decommissioning or upgrade. Can one ensure a low-cost conversion followed by a low-cost operation and maintenance?

ISBN: 1843390205. Published by IWA Publishing, London, UK.

In the light of current and future predicaments faced by the water authorities, it is difficult to overemphasise the importance of existing treatment assets. Integration of these traditional technologies into innovative systems looms as a way forward. Since the 1970s a number of hybrid systems marrying upfront pond treatment with a polishing unit process, be it a trickling filter or activated sludge, has been introduced.

14.2 ANAEROBIC POND(S)/TRICKLING HYBRID

In contrast to the American experience with horizontal submerged rock filters used downstream of the stabilisation ponds (Chapter 11), only vertical rock filters have been used in Africa (South Africa, Zimbabwe). The first hybrid of this type described in literature (Meiring, 1992) was employed in Phola (former Transvaal, South Africa) in 1991. Conceptually this hybrid was developed as a lower cost alternative to a conventional treatment facility, which usually combines a proper anaerobic digester (plus primary sedimentation) with a trickling filter and is illustrated in Figure 14.1.

For a primary treatment step a deep anaerobic pond is usually recommended, preferably featuring a bottom-fed digestion pit. In order to maintain strictly anaerobic conditions in the pit, the depth should be at least 4m. Highly loaded ponds produce sludge that should be removed on a regular basis (from weekly to yearly intervals, depending on loading conditions). Sludge is removed from the ponds by gravitation or pumping as a 3-5 % slurry and dried on the dedicated drying beds or in a sludge lagoon.

Usually two anaerobic ponds are used in series. Since the bulk of BOD is always removed in the first pond, the second pond fills a useful role of a sludge trap minimising sludge carryover to the trickling filter. These anaerobic ponds are also used to stabilize biofilm humus sloughed off the trickling filter. An important concomitant beneficial effect is that this nitrate-rich stream from the underflow of a humus settling tank to the surface of an anaerobic pond can help allay odours produced by anaerobic digestion. At the same time denitrification of nitrate can occur in the pond enhancing overall N removal in the system.

Figure 14.1 Basic flow diagrams of pond(s) integrated with a trickling filter (TF) or an activated sludge process (ASP). A: Primary (anaerobic) pond; A*: Primary (Anaerobic-Aerobic) pond with a fermentation pit; B_n: Secondary pond(s); C: TF + humus tank; D: Activated sludge process (ASP) or BNR reactor + clarifier; a: algae-rich inter-pond recirculation; b: primary pond effluent directed to secondary ponds; c: primary pond effluent; d: stream rich in readily biodegradable organic matter extracted from the lower strata of fermentation pit to boost denitrifiers or P-removal; e: nitrate-rich humus sludge or waste activated sludge recycle for their disposal; f: algae-rich secondary pond effluent; g: raw sewage as an organic-rich supplement for an anaerobic zone of a Bardenpho reactor (0.3Q). *Streams a and f in the PETRO TF/ASP variants may alternate, depending on necessity of the algae-rich water on the surface of A to allay odours.*

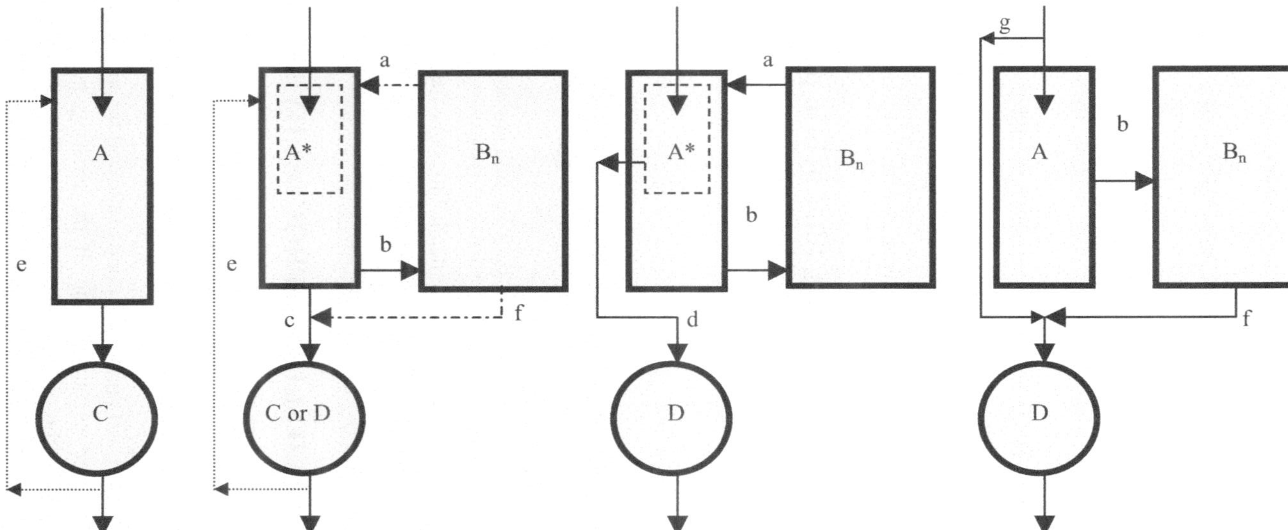

1. Anaerobic pond(s)-trickling filter (TF) hybrid

2. a. PETRO TF or b. ASP variant

3. PETRO BNR

4. Anaerobic pond(s)-ASP

A disadvantage of the hybrid is that potential odours from the facility would not be acceptable in close proximity to human habitation. In order to overcome this problem a substantial stream of final effluent equal up to an Average Dry Weather Flow (ADWF) should be recycled back to help prevent these odours. Anaerobic ponds do not produce microalgae and therefore this recycle system for odour control is inferior to systems featuring secondary algae-rich stabilisation ponds. Furthermore, high sulphide toxicity that can cause scouring of trickling water biofilm is reported to contribute to the unreliability of this hybrid system (Hoffmann, 1997).

The use of the vertical rock TF ensures superior aeration and prevents the sludge build-up in the filter that is more typical of submerged horizontal rock filters which may be functioning as an anaerobic system.

Overall, though the hybrids in South Africa achieve reasonably good BOD and nitrogen removal, the very few existing systems are testament to their limitations.

14.2.1 Case study - Phola ponds/trickling filter hybrid plant (Mpumalanga province, South Africa)

- Two parallel series each consisting of 2 sequential anaerobic ponds followed by 2 rock trickling filters.
- Flow: Average Dry Water Flow 2.8 megalitre/d, Peak Dry Weather Flow 8.4 megalitre/d.
- Raw sewage concentrations: BOD 500 mg/L, SS 410 mg/L, TKN 63 mg/L.
- Final (humus tank) effluent concentrations: BOD 10 mg/L, SS 40 mg/L, TKN 8 mg/L, Nitrate-N 37 mg/L.
- Primary ponds: carbonaceous organic load 0.23 kg BOD/m^3.d, HRT in 2 successive ponds 1.6 days.
- Trickling filters: diameter 28 m, stone media depth 4 m; stone media area to volume ratio 60 m^2/m^3, carbonaceous organic load 3.7 g BOD/m^2.d, nitrogenous load 0.9 g TKN/m^2.d.

14.3 PONDS/ACTIVATED SLUDGE PROCESS HYBRID

An anaerobic pond followed by an activated sludge reactor is a theoretically possible treatment configuration. Although the authors are not aware of any full-scale facilities, there is, however, at least one hybrid consisting of ponds and activated sludge process system.

The Kwe-Kwe system (Zimbabwe) comprises ponds followed by an activated sludge process and was designed as an upgrade of the existing stabilisation ponds in 1979 (Barnard, 1984). Although it appears to be the only hybrid of the type described in the literature, it is possible that similar hybrids have been designed elsewhere. Conceptually this approach is a logical result of an attempt to upgrade existing stabilisation ponds, which can fail to comply with ever-stringent discharge regulations. The activated sludge process was chosen to follow the ponds and provide for biological nutrient removal as well as polishing the pond effluent for improved BOD and suspended solids removal. Because the extensive pond system achieves very high removal of the readily biodegradable organic matter (measured as BOD), the provision of the sufficient BOD for denitrification was a concern. The existing ponds (surface area 4 ha) received average dry weather flow of 4 megalitres/day of mainly municipal sewage mixed with a steel plant effluent containing substantial amount of ferrous sulfate (average influent BOD 400 mg/L, TKN 45 mg/L with Fe at times exceeding 200 mg/L). New effluent quality standards applied at the time were BOD <10 mg/L, SS <25 mg/L; ammonia-N <0.5 mg/L; total N <10 mg/L, P <1.0 mg/L, thus an activated sludge polishing system was an unavoidable choice.

To ensure a consistent stream of readily biodegradable BOD for denitrifying bacteria, a variable proportion of raw sewage (initially 42%, then 50%) was fed directly into the anoxic zone of the activated sludge reactor. The balance was treated at the low cost ponds into which effluent was then fed into the aerobic zone of the reactor for SS (mainly microalgae) and N removal. Besides reducing overall treatment costs, the pond system served as a buffering system allowing for a balanced flow to be discharged into the activated sludge. Efficient removal of the pond microalgae was reported. Adsorption of the algae onto the bacterial floc rendered the activated sludge a pea-soup appearance with little colour remaining in the clarified effluent. Unfortunately it is impossible to ascertain the exact efficiency at this part of the algal removal process, since a parallel chemical precipitation of solids by ferrous sulphate also was used. This configuration led to the well-known phenomenon of concomitant flocculation of microalgal biomass. As a result, most of the incoming Fe was removed with sludge (down to 0.3 mg/L).

The operation of this plant was successful and extensions were designed around the same concept. New ponds (2 ha), incorporating anaerobic ponds that could be desludged on-line, were added. Although discharge standards set a requirement for P to be less than 1 mg/L, no provision was made for biological P removal since it appeared that sufficient concentration of $FeSO_4$ was available for concomitant chemical precipitation of phosphate. The aeration basin (designed as a three stage reactor configuration) consisted of a concrete-lined earth basin with three surface aerators (30 kW each) providing aeration power of 22.5 kW/megalitre wastewater treated. As a result of the incorporation of the ponds into this hybrid system the required standards were achieved at less than 25 % of the cost of a stand-alone biological nutrient removal plant (Barnard and Meiring, 1995). Due to the upfront treatment of sewage in the ponds, even in the case of occasional power failures, the plant could still produce a final effluent of a reasonable quality. As it will become obvious from Section 14.4.2, research has shown that the ability of activated sludge to trap and remove microalgae should not be taken for granted. Firstly, it depends on the 'trophic level' of the influent, i.e. on the quantity of readily biodegradable BOD that it contains. Additionally, another critical factor is the heterotrophic status of the microalgal cells to be removed. In turn, this is a function of the organic loading on the ponds. It has been discovered (Shipin *et al.*, 1999a, b) that the algae from the pond effluent are removed through the process of heterotrophic autoflocculation. It is, therefore, a critical prerequisite that the ponds select for species of heterotrophic rather than phototrophic algae, which pass through without flocculation/removal. A high loading rate has to be imposed on the pond system in order to avoid generation of the phototrophic difficult-to-remove species. Obviously the high loading is conducive to the malfunction of ponds and long-term observations on pond systems under the different loading conditions demonstrate that stability of the pond ecosystem is compromised if recirculation is not a feature of the design (Shelef and Kanarek, 1995). Hence, there are limitations of the Kwe-Kwe hybrid, which does not feature the algal recirculation. Odours can occur under such arrangement. The Kwe-Kwe hybrid system can be seen as an immediate precursor of the PETRO system (Section 14.4), which refined the features of this design and overcame the shortcomings.

14.4 PONDS FOLLOWED BY TRICKLING FILTER/ACTIVATED SLUDGE PROCESS

14.4.1 Introduction to the PETRO concept

The patented PETRO system is an 'appropriate technology' equally applicable in the developed and developing world (Meiring, 1992). It bridges the gap between high and low technologies by combining 'low tech' simplicity with 'high tech' performance. The system has been in use for more than two decades under the demanding conditions of developing communities in South Africa where power failures are not infrequent and the availability of skilled personnel is a problem.

From the perspective of the final effluent quality, ponds have the drawback of producing large quantities of microalgal biomass. Microalgal biomass is difficult to remove from the final effluent at low cost. For this reason secondary polishing processes, such as rock trickling filters (horizontal and vertical), have been employed to deal with the problem. Horizontal submerged (predominantly anaerobic) filters are discussed in Chapter 11. As to the use of vertical rock trickling filters, numerous efforts in many countries have generally failed to remove pond microalgae (Oswald, 1993).

Stabilisation pond(s) treat up to 70% of the incoming carbonaceous organic load. This allows for a considerable decrease in the size of the high cost secondary facility thereby enhancing efficiency of the entire integrated system. Hence the acronym 'Pond Enhanced Treatment and Operation', PETRO. The first installation was designed and constructed in 1978 as an upgrade of an existing pond system for the developing community of Kanyamazane (Mpumalanga Province, South Africa). Conceptually the primary PETRO pond combines all the features of the anaerobic pond system described in Section 14.2, and the incorporation of an algae-rich pond recirculation (Section 14.4.3) brings about important changes making the PETRO primary pond a so-called anaerobic-aerobic pond (Figures 14.1 and 14.2).

Presently there are eighteen plants in South Africa and Australia treating wastewater of municipal and agri-industrial (dairy, abattoir) origin. The majority of these plants were constructed in the 1990s after the long-term operation of the first prototype plants proved the concept's viability. Typical final effluent quality is shown in Table 14.1.

Table 14.1 Typical effluent quality of the activated sludge (AS) and trickling filter (TF) PETRO systems (after humus tank/clarifier)

Parameter	AS (BNR) variant	TF variant
COD/BOD	45/10	45/10
Suspended solids	10	8
Ammonia-N	1	2 (<1)
Nitrate-N	5	15 (5)*
Inorganic P	1	7 **

* The lower value is achieved, if simultaneous nitrification/denitrification is encouraged in the TF.

** Only 20-30% removal of incoming inorganic P concentration is achieved in the ponds.

14.4.2 Biological phenomena underlying PETRO

There are a number of key phenomena related to biological removal of microalgae that underlie the performance of the PETRO system (Figure 14.2). They include:

- Algal heterotrophy;
- Production of exopolysaccharides (extracellular polysaccharides);
- Predation on algae by protozoa/rotifers.

The phenomenon of algal heterotrophy has been known to botanists for several decades (Nielson and Lewin, 1974). A large number of microalgal species are capable of growth in the dark in addition to normal photosynthetic activity. Using heterotrophic metabolism they utilise a variety of dissolved organic compounds such as simple sugars, short-chained volatile fatty acids, alcohols, and so on, substrates which are common products of wastewater degradation. Under stress conditions microalgae are also known to produce secondary metabolites and excrete them into the medium, notably, exopolysaccharides These viscous polymers act as highly efficient flocculants enhancing aggregation of single cells into readily settleable flocs. As has been demonstrated in radioactively labelled tracer studies, the dominant microalgae of PETRO ponds (*Euglena sp., Chlorella sp., Phacus sp., Chlamydomonas sp.*) act in this manner (Shipin *et al.*, 1999a).

Under the conditions characteristic of the PETRO system and as a result of the stress of transfer from a phototrophic (or heterotrophic/phototrophic) mode of metabolism in the ponds to a heterotrophic mode in the PETRO reactor (trickling filter or activated sludge reactor) microalgae radically change their metabolic behaviour. Due to exopolysaccharide production from simple dissolved organic compounds a *heterotrophic autoflocculation* occurs and flocculated algal biomass embedded onto sloughed-off biofilm (humus) or waste activated sludge is thereby removed from the final effluent. Moreover, additional generation of flocculating polymers by microalgae imparts a superior settleability to humus or activated sludge, which manifests itself during the clarification (settling) process.

Properties of exopolysaccarides in general are such that even a small increase in the PETRO reactor leads to significantly improved sludge characteristics (settleability, dewaterability *etc*). A high final effluent clarity is attributed to these microalgal phenomena underlying the process. Algal exopolysaccharides also enhance the activated sludge biomass's capacity to harbour larger microbial populations, since a boost in nitrifying population was seen when increased algal biomass entered an activated sludge reactor from ponds experiencing algae blooms (Shipin *et al.*, 1999b). An important consequence of having exopolysaccharides enter the reactor is a much lower susceptibility of the PETRO activated sludge variants to foaming and bulking which is a common problem of conventional plants. These microbial phenomena were known separately to scientists for decades and are selected by the PETRO design without any need for operator input. This fortunate combination of advantageous mechanisms is achieved by natural selection, which secures a stable and robust operation (Barnard and Meiring, 1995).

14.4.3 Algae-rich inter-pond recirculation

The phenomenon of inter-pond recirculation is as important for the PETRO concept as it is for any conventional pond system. The PETRO system features a number of recirculation streams (see Figures 14.1 and 14.3). The high rate recirculation of water from one of the final secondary ponds to the surface of the first (primary) pond is by far the most important one.

The initial purpose of adding recirculation to a pond system was to suppress offensive odours emanating from the often-overloaded primary (anaerobic) ponds. Recycled water, which is highly oxygenated due to the algal activity in the secondary ponds, oxidises sulphides and organic compounds that can cause unpleasant odours from the anaerobic ponds. This transforms a primary pond into an anaerobic-aerobic pond. Later it transpired that the beneficial effect of the recycle is not limited to combating odours it also affected the characteristics and performance of the pond system as a whole. Long-term full-scale experience and research show that one of the major benefits of recycle is an enhanced biological stability of the entire pond ecosystem. This has been observed on the PETRO plants for decades, and confirmed in pilot and field studies of conventional pond series in Israel (Shelef and Kanarek, 1995) that overall seasonal stability improves.

Overall, a high rate recycle (ratio beyond 1:1) allows for the substantially greater organic loads on the primary (up to 0.4 kg BOD/m^3.d in warm climates) and secondary ponds (up to 250 kg BOD/ha.d) since it spreads the load over much greater volume, provides odour control, brings about greater stability to the pond ecosystem and more consistent quality of the pond effluent.

For PETRO and its downstream microalgal removal, recirculation plays a beneficial role of preconditioning of microalgae generated in the secondary ponds. Passage of microalgae through the anaerobic and highly organically loaded primary pond selects for the microalgae featuring stronger heterotrophic and EPS production traits, i.e. the metabolic characteristics which facilitate microalgal removal in the downstream (post-pond) PETRO trickling filter or activated sludge reactor. Although the presence of the recycle is a strongly recommended feature it is not an obligatory item in the PETRO concept. Recirculation can be omitted in some cases if simplicity is a critical requirement, but it will to an extent compromise high final effluent quality. Recycle is a site-specific feature and can be detrimental if used excessively or not applied properly.

Despite the fact that the concept of recirculation since its inception in the 1950s has demonstrated benefits for a pond-based wastewater treatment system, it still awaits wider application, which it seems to deserve.

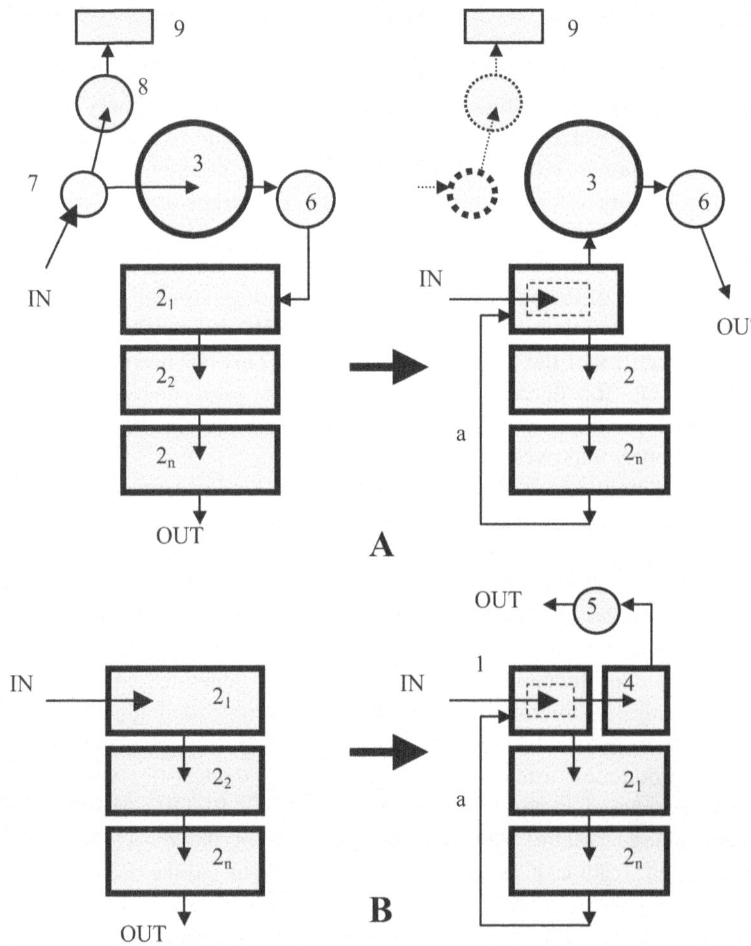

Figure 14.3 Some of the PETRO upgrade options. A. Conventional (vertical) rock trickling filter plant retrofitted into the PETRO TF variant. B. Conventional stabilisation pond system retrofitted into the PETRO Activated Sludge (ASP) variant. 1. Primary (anaerobic-aerobic) pond with a digestion pit; 2_n. Stabilisation ponds or PETRO secondary ponds; 3. TF; 4. AS reactor; 5. Clarifier; 6. Humus tank; 7. Primary settling tank; 8. Anaerobic digester; 9. Sludge drying beds; a. algae-rich inter-pond recirculation

14.4.4 System variants

As mentioned, ponds can be followed in series by a trickling filter or an activated sludge reactor. Nevertheless, the key features of the PETRO concept, which includes the removal of microalgae, reciprocal unit process synergy and lower loading rate on the secondary facility, are common to both variants. The choice of the variant is site-specific and is determined by a number of factors, amongst which the quality of the final effluent required is the most important.

Figure 14.4 One of the first PETRO plants (trickling filter variant). Letlhabile (South Africa). ADWF 4 megalitre/day, built 1981.

The primary pond due to the high rate algal recirculation provided to its surface combines features of a conventional anaerobic and a facultative pond. Oxygen-rich recycle affects changes in pond microbiology leading to an enhanced production of readily biodegradable organic matter in the lower pond strata (digestion pit) during the anaerobic digestion (BOD removal) of sewage. A mild inhibition of strictly anaerobic methanogens by oxygen results in a reasonable accumulation of the products of acidogenesis such as white fatty acids. They are used by microalgae and phosphate accumulating organisms in the downstream PETRO reactor. This process in the pit takes place at the same time with the reduction of overall amount of organic matter in the ponds.

PETRO trickling filter variant

Represented in Fig 14.1, the PETRO trickling filter variant comprises a vertical rock trickling filter preceded with waste stabilisation ponds of which the primary one is a deep pond (with or without a digestion pit). It is the most economical and 'low tech' variant, and as such is the most appropriate for developing communities.

The variant overcomes limitations of the anaerobic ponds/trickling filter hybrid (as discussed in Section 14.2) by the use of the high rate algae-rich recirculation providing odour control. Due to the presence of microalgae the TF variant produces a final effluent of superior quality and with greater reliability than the simpler anaerobic ponds/trickling filter hybrid.

Due to microalgae-induced high pH volatilisation of ammonia up to 10% of the total, ammonia-N is removed from the wastewater. A large majority of the balance is converted to nitrate by the nitrifying bacteria in the trickling filter. If simultaneous nitrification-denitrification is encouraged through the provision of readily biodegradable organic matter, a low nitrate concentration can be achieved in the effluent (Table 14.1). Therefore availability of a certain portion of the readily biodegradable organic matter from the lower strata of the primary pond (pit) is important for the concomitant removal of microalgae and nitrogen.

PETRO activated sludge variant

The PETRO activated sludge variant (as shown in Figure 14.1) comprises primary and secondary ponds followed by an activated sludge reactor. The PETRO activated sludge reactor size and oxygenation requirements are considerably lower since the bulk of the carbonaceous load (up to 80% BOD) is removed by the ponds. The PETRO approach overcomes the limitations of the non-PETRO ponds/activated sludge hybrid (Section 14.3) providing control over algal population and speciation, greater stability for the entire system as well as it solves odour problems and allows construction in proximity to human habitation.

Waste activated sludge can be aerobically stabilised by aerobic composting or in a dedicated aerobic sludge digester (SRT=8 days) or, if no P removal is required, be recycled into the primary pond for anaerobic stabilisation and eventually handled along with the anaerobic sludge. Algae play an important role in the reactor producing additional flocculants that impart greater settleability to the sludge, in turn harbouring larger nitrifier population. Field observations indicate that activated sludge produced by the variant (i.e. in a secondary reactor) settles better (SVI < 100), than the sludge from a conventional (primary) reactor.

PETRO Biological Nutrient Removal (BNR) variant

Stricter discharge standards are being introduced in many countries. Standards featuring total N and P requirements of 5.0 and 1.0 mg/L, respectively, necessitate a technology that can provide efficient nitrogen and phosphorus removal that is integrated with ponds. Biological nutrient removal (N and P) can be achieved by the addition of a modified activated sludge process comprising anaerobic, anoxic and aerobic zones.

Incorporation of the biological excess P removal into the PETRO system is a logical development leading to a relatively 'low tech' pond-based BNR treatment facility. It produces final effluent comparing favourably with the state-of-the-art 'high tech' BNR plants. A basic PETRO BNR flow diagram and performance data are presented in Figure 14.1 and Table 14.1, respectively.

Generation of readily biodegradable organic matter (volatile fatty acids, *etc*) is of critical importance in the concept of biological N and P removal. At the same time supply of readily biodegradable organic matter to the PETRO reactor is an inherent feature of the PETRO concept. The organics generated in the primary pond boost PETRO microflora and ensure removal of microalgae. On the other hand readily biodegradable organics cause the release of phosphate by phosphate accumulating organisms in the BNR reactor. Thus the production of readily biodegradable BOD plays a crucial role both in the PETRO and biological nutrient removal concepts. Despite fairly complete removal of this matter in upstream ponds, if necessary provisions are made, there is still ample amount of organics to remove both microalgae, N and P. The production of readily biodegradable organics (measured as readily biodegradable BOD) in the framework of the concept is an issue closely related to the sludge production in the primary pond. In this light the sludge removal and disposal strategies require attention. The approach to provide readily biodegradable organics for phosphate accumulating organisms is based on the generation of readily biodegradable BOD in the primary pond pit (Shipin *et al.*, 1999a, 2000). Variations in BOD/TKN ratio and the lack of readily biodegradable organics in the BNR reactor feed are major problems in the nutrient removal plants (Henze, 1996). The PETRO ponds attenuate peaks and stabilise these variations while attention to the point of withdrawal of organics from the lower strata of the pit is of paramount concern for the delivery of this matter to the PETRO reactor where microalgae, N and P are removed.

14.4.5 Retrofit guidelines

Retrofitting a wastewater treatment plant is usually undertaken to achieve:

1. a better quality of the final effluent to meet new discharge standards, or
2. an increase of the plant treatment capacity.

Needless to say, a low cost retrofit is always a preference.

Upgrading conventional trickling filter plants

A conventional trickling filter plant typically comprises a primary settling tank, an anaerobic digester, a trickling filter and a humus tank. Sometimes a series of maturation ponds follows to polish the somewhat inferior effluent of the trickling filter. In many countries the final effluent of trickling filter no longer meets the requirements of water authorities.

A cost-efficient method of retrofitting a trickling filter plant is presented in Figure 14.3 - diagram A. A sequence of 'a trickling filter followed by ponds' is reversed and a sequence of 'ponds followed by a trickling filter' is introduced. The primary sedimentation and formal anaerobic digestion stages are replaced by a primary pond with a deep fermentation pit, which gives significant gains in terms of operational costs and simplicity.

The removal of the carbonaceous load now achieved in the ponds significantly decreases the size requirement of the secondary trickling filter, which serves mainly as a polishing facility for the removal of microalgae and nitrogen. Overall, the treatment capacity of the retrofitted plant can potentially be doubled or tripled and the better final effluent produced (Table 14.1).

Upgrading conventional WSP series

The PETRO retrofit (shown in Figure 14.3 - diagram B) of a series of existing waste stabilisation ponds involves a commissioning of a relatively small activated sludge reactor (or trickling filter) downstream of the ponds.

In Melbourne, Australia, an upgrade of an existing stabilisation pond (lagoon) system was undertaken enhancing its capacity from 90 to 195 (and eventually up to more than 500) megalitre/d. This upgrade allows the plant to meet new stricter discharge standards without compromising the international bird sanctuary status of the world's biggest pond-based WWT plant (McLean and Scott, 2002). A cost of up to Aus$750 million was estimated for a conventional treatment. The PETRO upgrade cost was Aus$124 million.

This modern upgrade of traditional ponds highlights another important advantage of any pond-based system, its *eco-friendliness* and demonstrates that the PETRO process is a viable technology for a large city as well as small developing communities.

14.4.6 Design guidelines

Table 14.2 presents design criteria for the different unit processes of the PETRO system: primary pond, secondary pond(s), trickling filter and activated sludge reactor.

Table 14.2 Principal design criteria of the PETRO system

Ranges	
Organic:	1400-100 000 kg BOD/d; 100-11200 kg TKN/d
Hydraulic:	1.0-190 megalitre/d
ADWF : PDWF : PWWF*:	1.0 : 2.0 : 5.0
Unit Process I. PRIMARY POND	
Consists of an anaerobic pond with a digestion pit, featuring a high rate algae-rich recirculation to the surface from the secondary ponds. Pond depth: 2-3 m; Digestion pit depth: 4-6 m; Volumetric ratio, $V_{total\ pond}$: V_{pit} ranging from 2.0 to 4.5.	
Organic loading:	0.1-0.4 kg BOD/m^3.d (fed to the bottom of digestion pit)
HRT $_{prim.pond}$:	4.0-8.0 d (at ADWF)
Unit Process II. SECONDARY POND(S)	
Optimal number: 2-4; Pond depth: 1.2-2.5 m; $HRT_{sec.\ pond(s)}$ 2.0-7.0 d (at ADWF)	
Surface organic loading:	100-300 kg BOD/ha.d (with good influent mixing)
Pond volumetric ratio:	$V_{sec.\ pond(s)}$: $V_{prim.pond}$ ranging from 2.0 to 5.0
Recirculation rate:	b:c ranging from 0 to 2.0 (b. primary pond effluent directed to secondary ponds; c. primary pond effluent directed to the PETRO reactor, Figure 14.1)
Pond surface ratio:	$S_{sec.\ pond(s)}$: $S_{prim.pond}$ ranging from 3.0 to 7.0
Unit Process III. PETRO REACTOR	
Trickling filter variant	
Organic surface loading rates for a 60 m^2 surface/m^3 stone media**:	
Carbonaceous:	2.5 g BOD/m^2 rock surface. d
Nitrogenous:	1.0 g TKN/m^2 rock surface. d
Organic volumetric loading:	150 g BOD/m^3 rock media. d
Humus tank upflow velocity at (at ADWF): 1 m/h	
Faecal coliforms and other pathogens after secondary clarification are the same as for a conventional low rate biofiltration.	
Activated sludge reactor variant	
Specific organic loading**:	
Carbonaceous:	0.22-0.46 kg BOD/m^3.d
Nitrogenous:	0.09 - 1.9 kg TKN/m^3.d
Aeration power:	8.4-14.0 kW/megalitre treated
MLSS:	3500-5000 mg/L
SRT (HRT):	15-35 days (10-15 hours at ADWF)
Faecal coliforms and other pathogens after secondary clarification are the same as for a conventional low rate activated sludge process.	

*ADWF, Average dry weather flow; PDWF, Peak dry weather flow; PWWF, Peak wet weather flow.
** Fed directly from the lower strata of the primary pond (digestion pit).

14.4.7 Capital and operational costs

The PETRO process lends itself well to phased development in cases where financial constraints and issues of capacity building preclude immediate and full implementation of the system.

In retrofit applications existing treatment infrastructure (ponds, trickling filters, activated sludge reactors) can often be incorporated into the upgraded systems thereby yielding significant efficiencies. As a result, a lower cost facility with a higher treatment capacity and superior effluent quality can be constructed. The economics of the PETRO applications are always site-specific, though generalising, it is safe to state that in terms of capital costs and total treatment costs the system can usually offer cost savings of up to 40% in comparison to conventional systems producing effluent of a similar quality.

14.5 SUMMARY AND FUTURE RESEARCH NEEDS

The PETRO process offers a relatively low cost pond-based alternative waste WWT plant which:

- is straight forward to operate and maintain in both developing and developed communities due to its stability and robustness;
- reduces capital, operation and maintenance costs compared to conventional alternatives while easily lending itself to phased development and stage-wise upgrading of the treatment capacity;
- provides a high quality final effluent via efficient removal of organic matter, microalgae, N and (with additional engineering fine tuning) P;
- provides a substantial degree of treatment even during power failures which would cripple any conventional treatment plant;
- minimises water losses through infiltration and evaporation due to intensification of treatment rates and minimisation of the pond surface area.

The concept of the pond-trickling filter/activated sludge hybrid requires further development and refinement of the following aspects:

- combined P removal model for stabilisation ponds and BNR facility;
- further testing on a scope of heavy industrial effluents;
- incorporation of simultaneous nitrification-denitrification (SND) into the trickling filter PETRO variant to provide for the comprehensive N removal in the trickling filter.

REFERENCES

Barnard, J.L. (1984) Design and operation of Bardenpho plants in an African country. *J. Inst. Wat. Pollut. Control,* **83**(4), 443-448.

Barnard, J.L and Meiring, P.G.J. (1995) Algae for advanced wastewater treatment: turning a stumbling block into a stepping stone. *Proc. Ann. Conf. Amer. Wat. Environ. Federation*, Miami, USA.

Hoffmann, J.R. (1997) Application of biofiltration in South Africa. *Technology Transfer Workshop.* Midrand, South Africa.

Meiring, P.G.J. (1992) PETRO process. South African Patent No 92/9644. Australian Patent No. 30074/92.

McLean, B. and Scott P. (2002) Melbourne's Western Treatment Plant. Innovation and cooperation: the keys to upgrade. *Water (Australia),* March.

Meiring P.G.J. (1992) Introducing the PETRO process. *Proceedings 3rd SA Anaerobic Digestion Symposium.*13-16 July 1992. Pietermaritzburg, South Africa.

Neilson, A.H. and Lewin, R.A. (1974) The uptake and utilization of organic carbon by algae:essay in comparative biochemistry. *Phycologia,* **13**(3), 227-264.

Oswald, W.J. (1993) Presentation of AIWPS technology at the SA Water Research Commission. Pretoria, South Africa.

Shelef, G. and Kanarek, A. (1995) Stabilisation ponds with recirculation. *Wat. Sci. Tech.* **31**(12), 389-397.

Shipin, O.V., Meiring, P.G.J. and Rose, P.D. (1999a) Microbial processes underlying the PETRO concept (trickling filter variant). *Wat. Res.* **33**(7), 1645-1651.

Shipin, O.V., Meiring, P.G.J, Phaswana, R. and Kluever, H. G. (1999b) Integrating ponds and activated sludge process in the PETRO concept. *Wat. Res.* **33**(8), 1767-1774.

15

Integrated pond/wetland systems

Chongrak Polprasert, Thammarat Koottatep and Chris Tanner

15.1 INTRODUCTION

Integrated pond/wetland systems have been used for wastewater treatment for decades in both temperate and tropical regions. Recognition of the improved treatment achieved during passage of wastewater discharges through natural wetlands (e.g. Nichols, 1983) has led to development of a range of constructed wetland (CW) approaches. These aim to enhance treatment performance by optimising wetland loading rates, flow distribution, water depths and vegetation characteristics.

Wetlands are passive 'natural' systems that can cost-effectively complement pond treatment processes. Effluents from facultative ponds often contain high algal concentrations that exceed effluent standards for suspended solids and associated BOD, and may detrimentally affect receiving waters and aquatic life. Vegetated wetland systems provide quiescent, shaded conditions conducive to

ISBN: 1843390205. Published by IWA Publishing, London, UK.

sedimentation of algae and other suspended solids. Additionally, these systems provide for further biodegradation of organic matter, assimilation and transformation of nutrients, and reduction of pathogens. CW may also provide ancillary benefits such as wildlife habitat, livestock forage and materials for weaving, and enhance the biodiversity, aesthetics and cultural acceptability of wastewater treatment facilities. Polishing of pond effluent by constructed wetlands, including types, treatment mechanisms, design considerations, vegetation, and reuse is discussed in this chapter.

15.2 CONSTRUCTED WETLANDS

Constructed wetlands (CW) are wastewater treatment system consisting of shallow (usually less than 1 m deep) ponds or channels planted with aquatic plants. The processes by which wastewater is treated include a wide range of interacting biological, physical and chemical mechanisms.

15.2.1 Types of constructed wetlands

Constructed wetlands are often classified into three main types, as detailed in the following sections (US.EPA 2000; IWA 2000), but a range of hybrids and variants (e.g. focussing on floating or submerged aquatic plants) have also been used.

Free water surface (FWS)

Free water surface or surface-flow constructed wetlands closely resemble natural marshes in appearance. They generally contain emergent aquatic plants growing in shallow water (commonly 0.3-0.5 m depth). The plants are rooted in a soil layer (at least 0.2 m deep) on the bottom of the wetland and wastewater flows horizontally through the shoots of the plants. Free-floating and submerged plants may also form part of the vegetation. A typical FWS constructed wetland is shown in Figure 15.1. The components of such a wetland include embankments to hold and channel water, a clay or geotextile liner to retain water, and inlet and outlet structures to distribute and collect wastewater, control water levels and maintain desired hydraulic retention times. Water levels should be able to be adjusted to facilitate planting and plant management, and to accommodate gradual accumulation of solids and plant organic matter in the base of the wetland. Deeper (≥ 1.2 m) open water zones, covering 20-30% of the wetland area may be included to redistribute and reaerate water passing through the wetland, and promote nitrogen removal (Hammer and Knight 1994; US EPA 2000).

Vegetated sub-merged bed (VSB)

Vegetated submerged beds or subsurface-flow wetlands have no standing water (Figure 15.2). Hydraulic problems with early VSB systems employing soil media have led to predominant use of coarser media such as crushed rocks, small stones, gravel or sand. The 0.3 – 0.9 m deep beds are generally lined with low-permeability clay or geotextile and planted with emergent aquatic plants, which grow hydroponically in the saturated media. When properly designed and operated, wastewater remains beneath the surface of the media and flows horizontally in contact with biofilms growing on media, and the roots and rhizomes of the plants. A typical VSB system, like the FWS systems described above, has inlet and outlet structures for distribution of wastewater flows and adjustment of water levels. A range of modified VSB systems have been developed, generally aiming to increase organic matter and ammonia processing rates through physical enhancement of bed aeration. In addition to the vertical-flow systems described below, various approaches to provide fluctuating or tidal water levels have shown promise (e.g. Behrends *et al.,* 2001; Sun et al., 2005; Tanner *et al.* 1999).

Although VSB constructed wetlands are generally more expensive to construct because of the cost of the media, they generally have a smaller 'footprint' than free water surface wetlands and have the advantage that they avoid the potential for direct contact with wastewaters or proliferation of mosquitoes and other potential disease vectors and pests. This is likely to be particularly important in tropical and arid regions.

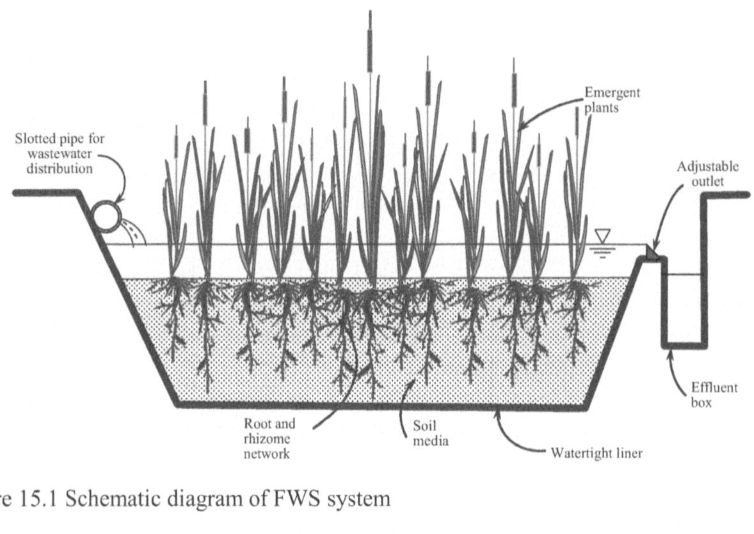

Figure 15.1 Schematic diagram of FWS system

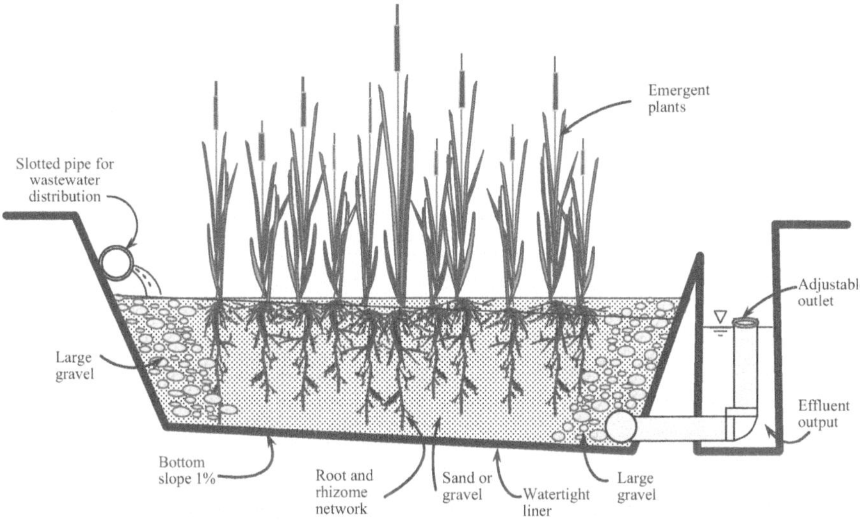

Figure 15.2 Schematic diagram of VSB system

Vertical flow (VF)

Vertical-flow wetlands consist of an intermittently dosed sand and/or gravel-bed planted with aquatic emergent plants (Figure 15.3) through which the wastewater travels in the vertical direction. Typically, pulses of wastewater are uniformly distributed onto the CW surface, and then flow down through the roots and rhizomes of the plants and the media layer.

VF wetlands are essentially simple biological trickling filters. Movement of the plant shoots in the wind helps maintain percolation through the sludge layer that develops on the surface of the bed. Plant transpiration and root-zone oxygen release assists with dewatering and aeration of the upper layers of the bed. The media typically consists of an upper layer of coarse filter sand overlying 0.4 – 1.2 metres of gravel. The sand layer temporarily detains the dosed wastewater on the surface, distributing it across the bed and releasing it gradually into the bed below. The treated effluent, or percolate, is collected at the bottom by an under-drain system such as a perforated-pipe set in coarse gravel or hollow-concrete block network. As the beds drain air is drawn back into them, often assisted by perforated ventilation pipes set into the bed (Figure 15.3). This promotes aerobic decomposition and maintenance of nitrifying bacteria in properly loaded VF beds. Depending on pre-treatment levels and loading rates, a number of VF beds may need to be cycled and periodically rested to maintain permeability.

Design information for VF systems is not further discussed here. Additional information on these systems is available in IWA (2000) and Cooper (2004). VF constructed wetlands also have application for sludge and septage treatment (IWA 2000; Koottatep et al. 2001; Nielsen 2003).

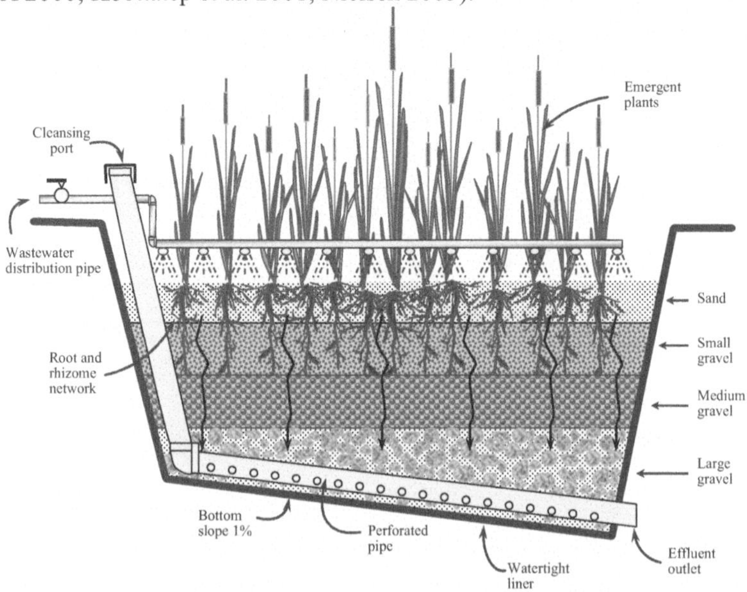

Figure 15.3 Schematic diagram of VF system treating wastewater

15.2.2 Vegetation

Hardy emergent aquatic plants that are able to grow and withstand inundated conditions are most commonly used in CW systems. These emergent plants can grow within a water-table range from 0.5 m below the soil surface to a water depth of up to 1.5 m or more (Brix and Schierup, 1989), but growth rates and shoot densities tend to decline at deeper depths. The most common varieties used in CW units (Figure 15.4) include reeds (*Phragmites* spp.), cattail or reedmace (*Typha* spp.) and bulrush (e.g. *Scirpus or Schoenoplectus* spp.), which have environmental requirements as shown in Table 15.1.

The plants used in constructed wetlands need to be tolerant and adaptable to various kinds of wastewaters, and be able to produce roots and rhizomes in inundated soil conditions. Specific anatomical structures of these plants enhance effective internal ventilation and aeration of the plant roots (Armstrong et al. 1990). As well as enabling growth in flooded conditions, oxygen release into the root-zone can promote aerobic microbial transformations, such as nitrification and methane oxidation.

Although it is often tempting to increase the water depth of FWS and the bed depth of VSB wetlands to provide greater wastewater residence times, plant shoot and root densities generally decline at deeper depths under continuously flooded conditions, and treatment performance is more related to wetland area than volume (Garcia et al. 2004; Kadlec and Knight, 1996)

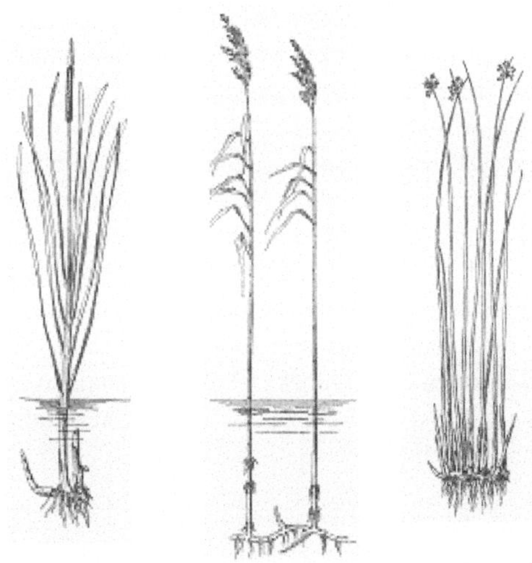

Figure 15.4 Common emergent plants in CW systems

Table 15.1 Emergent vegetation in CW and their environmental requirements

Emergent species	Environmental requirements			
	Water depth range *(m)*	Maximum salinity tolerance *(ppt)*	Optimum pH	Optimum temperature *(°C)*
Reedmace or Cattails (*Typha spp.)*	0.10 – 0.75	30	4 – 10	12 – 24
Bulrushes (*Scirpus* or *Schoenoplectus spp.)*	0.10 – 1.50	20	4 – 9	16 – 27
Reeds (*Phragmites spp)*	0.05 – 0.50	45	2 – 8	10 – 30

Source: Modified from Lim and Polprasert (1996)

15.2.3 Treatment mechanisms

The treatment efficiency of a CW is a function of the complex interplay of physical, chemical and biological processes, and is influenced by wastewater characteristics, hydraulic and organic loading rates, CW configuration, media type, plant species and cover.

Solids and organic matter removal mechanisms

CWs are generally very efficient in suspended solids removal (IWA 2000; Kadlec and Knight, 1996; US.EPA 2000). Low water velocities enhanced by passage through vegetation, and in SVB's the gravel/sand media, promote settling and filtration. Fine particles adsorb to bacterial flocs suspended in the water and biofilms growing on the surfaces of plants, organic detritus and media.

BOD associated with settleable solids is removed by sedimentation. Particulate and soluble organic matter is degraded by both suspended and biofilm-associated bacteria (Polprasert *et al.*, 1998) Decomposition rates are enhanced by atmospheric reaeration, both via diffusion through the water surface and transport through the internal air-spaces of emergent plants and root-zone release (Armstrong et al., 1990). In the less shaded, open-water zones of FWS wetlands oxygen produced by suspended or attached algae and, where present, submerged aquatic plants promote diurnal periods of aerobic conditions.

Nutrient removal mechanisms

Nitrogen (N) is removed in CW systems by four main mechanisms:

(i) Sedimentation of particulate organic nitrogen, which can be mineralised to ammonia by hydrolysis and bacterial degradation, with a proportion retained in recalcitrant organic matter and microbial biomass;

(ii) Plant uptake and either subsequent sequestration in accumulating organic detritus and sediments, or removal via harvesting;

(iii) Nitrification, denitrification and associated microbial processes (such as anaerobic ammonium oxidation or Anammox)) resulting in transformation to nitrogen gases, which are returned to the atmosphere. Nitrification is the process of bacterial oxidation in which NH_4^+ ions are converted into nitrite (NO_2^-) and nitrate (NO_3^-) ions, which can be reduced to gaseous N forms (mainly N_2 and N_2O) via denitrification reactions;

(iv) Ammonia volatilization can become important at elevated pH (e.g. associated with algal photosynthesis) and temperatures.

At normal wastewater loading rates for primary and secondary effluents denitrification is generally the dominant N removal mechanism in CW, with potentially harvestable plant biomass commonly accounting for < 10-20 % of annual wetland N removal (Kadlec and Knight, 1996; Tanner, 2004). As such, harvesting is generally only carried out where the resulting herbage has economic value, or as a means to manage wetland vegetation. CW are predominantly anaerobic environments, so that nitrification (requiring oxygen) is often the rate limiting step in removal of N from ammonium-rich wastewaters. Superior CW nitrogen removal generally occurs when nitrified wastewaters are applied (Bachard and Horne, 2000; van Oostrom and Russell, 1994), when more aerobic open-water zones are incorporated in FWS wetlands (Hammer and Knight 1994), and when VSB wetlands are shallow and well vegetated (Garcia *et al.*, 2004; Tanner, 2001).

Sustainable phosphorus removal is generally relatively limited in CW (US.EPA, 2000), requiring large wetland areas and extended residence times for significant P reduction at concentrations characteristic of pond-treated sewage wastewaters (3-7 g TP m^{-3}). Short-term phosphorus (P) removal mechanisms include sedimentation, plant and microbial uptake, chemical adsorption and precipitation reactions in the presence of aluminium (Al^{3+}), calcium (Ca^{2+}), iron (Fe^{3+}), magnesium (Mg^{2+}), and manganese (Mn^{2+}) ions. Ongoing plant uptake and available sorption sites in wetland soils or media are generally limited, so in the longer-term sediment accretion (via particulate sedimentation, chemical precipitation and accumulation of refractory plant detritus) is the dominant sustainable P removal mechanism in CW (IWA, 2000; Kadlec and Knight, 1996; Reed *et al.*, 1995; US.EPA, 2000). Special P-retaining media with enhanced sorption potential and/or high levels of reactive ions may be able to provide improved longer-term P retention (e.g. Brix *et al.*, 2001; Sakadevan and Bavor 1998). However, such media will eventually become saturated, and eventually require removal and replacement to maintain ongoing P removal.

Pathogen removal mechanisms

Numerous studies have shown that the removal of a wide range of pathogens in CW systems is a function of retention time and temperature (IWA, 2000; Kadlec and Knight, 1996). Tanner and Sukias, (2003), found faecal indicator bacteria levels were commonly reduced by around one log unit when sewage pond effluents were applied to a range of constructed wetlands in New Zealand. Kadlec (2005) found superior removal of faecal coliform bacteria in surface-flow constructed wetlands compared to pond systems, over the inlet concentration range of $10^4 - 10^6$ (100 mls)$^{-1}$. However, background levels of faecal coliforms still remain in CW effluents, presumably contributed by waterfowl and animals in the wetland.

The mechanisms for pathogen removal in CW systems are similar to those in ponds (see Chapter 6) and other natural treatment technologies, including:

(i) Physical processes, such as flocculation and aggregate formation followed by sedimentation, filtration and adsorption;

(ii) Die-off and inactivation as a result of prolonged exposure to hostile environmental conditions (including solar radiation) and;

(iii) Grazing and predation by zooplankton and protozoa.

The relative importance of sunlight-mediated disinfection processes in CW is likely to be much less than in pond systems (see Chapter 6) because exposure to sunlight is markedly reduced due to shading by plants in FWS wetlands and by both the plants and media in VSB wetlands. Other removal mechanisms particularly enhanced settling and retention, and grazing and predation (Decamp and Warren, 1998; Stott *et al.*, 2004) are believed to be important disinfection processes in CW systems.

15.3 APPLICATION OF POND AND CW SYSTEMS

Algal suspension in the pond water increases suspended solids (SS) content of the pond effluent. Polishing of the pond effluent, SS in particular, can be achieved by using FWS, VSB or VF systems. The combination of ponds and CW systems can be applied for treatment of a variety of wastewaters.

15.3.1 Domestic wastewater treatment

Integrated ponds and CW systems have been used for decades to treat domestic wastewaters in various locations around the world. In particular, CW can reduce algal SS in pond effluents, and provide additional removal of BOD, nutrients and faecal microbes. Enhanced effluent quality is reported when ponds and FWS wetlands are employed in series for domestic wastewater treatment at numerous sites in the USA and New Zealand (Kadlec 2003a, Surampalli *et al.* 2002; Tanner and Sukias, 2003). SS and BOD are commonly reduced by 60-90%, ammonia and TN by 40-60%, and faecal indicator bacteria by an order of magnitude or more in appropriately loaded wetlands. Kadlec (2005), comparing the relative performance of ponds and wetlands, reported that both showed similar BOD removal efficiencies, except at higher loading where ponds excelled. Wetlands produced distinctly lower SS concentrations than ponds due to suppression of algal growth and enhanced settling, whilst well-vegetated wetlands generally provided superior ammonia removal.

Other examples where ponds and wetlands have been used together for domestic waste treatment include rural areas of Spain where a series of FWS and

VSB wetlands were used to polish the effluent of waste stabilization ponds; the Asian Institute of Technology campus in Bangkok, Thailand, where anaerobic and facultative ponds are supplemented by FWS wetlands (Polprasert and Koottatep 2004), and in the final stages of advanced pond systems (El Hafiane and El Hamouri, 2004; Tanner et al. 2005).

15.3.2 Agricultural/industrial wastewater treatment

Several researchers reported that CW systems could supplement pond treatment of wastewaters from various agricultural and industrial sources, including pig/swine and dairy farms (Sezerino *et al.*, 2003; Stone *et al.* 2002; Tanner *et al.*, 2003), meat processing facilities (van Oostrom and Russell, 1994), and petrochemicals (Knight *et al.* 1999). Polprasert *et al.* (1996) found the performance of CW beds treating an industrial wastewater to depend on the concentrations of toxic heavy metals, most of which accumulated in the roots rather than in the stems and leaves of the emergent plants. Subject to operating conditions such as retention time, loading rates and concentrations of toxic compounds, the CW systems can be very efficient in removing the remaining toxic compounds or heavy metals in the effluent from ponds treating industrial wastewaters.

15.4 DESIGN CONSIDERATIONS

This section describes environmental requirements and design criteria for constructed wetlands when they are used as a polishing step for pond systems. The design of pond treatment systems are summarised in Chapters 8, 9, 10 and 11.

For CW systems, the emergent plant species, as described in Section 15.2.2, are known to be tolerant of low oxygen levels, high nutrient concentrations, a wide pH range of 2 – 8, salinity up to 40 g/L, and temperatures of 12 – 23°C (Angelakis, 2001).

15.4.1 Process design criteria

There are a number of guidelines for the process design of CW systems available in the literature (IWA, 2000, US.EPA 2000; Kadlec and Knight, 1996; Polprasert, 1996). The general characteristics of CW systems, including organic and N loading rates, retention times, hydraulic loading rates, water depth and bed depth, are shown in Table 15.2. In the past hydraulic and contaminant loading rates and hydraulic retention times have often been used as a basis for various 'rule of thumb' and 'cookbook' approaches to treatment wetland design. For example, much of the European literature on constructed wetlands has focussed on small-scale treatment of primary-settled domestic sewage and expresses loading rates in terms of m^2 of wetland area required per person equivalent. Obviously, such design criteria are highly dependant on the characteristics, strength, and preceding treatment of the wastewater being treated, and assumptions regarding the desired quality of the discharge. They are therefore difficult to translate and apply to other situations. A review by Rousseau *et al.* (2004) illustrates some of the challenges and problems with application of such 'rules of thumb' as well as some more detailed models for VSB design.

The US.EPA (2000) guidelines provide pragmatic guidelines for FWS treatment of waste stabilisation pond and primary-treated effluents. They do not recommend use of VSB constructed wetlands for algae-rich pond effluents. The guidelines are based only on data for North American systems operating mainly in continental temperate and arid climatic conditions. For FWS wetlands they propose a 3-zone design, incorporating fully vegetated inflow and outflow zones (depth ≤ 0.75 m) each with hydraulic residence times of at least 2 days and an intermediate open-water zone (depth ≥ 1.2 m) with a maximum hydraulic residence time of 2-3 days supporting submerged plants. Recommended maximum loadings over the entire system are 45 kg BOD ha^{-1} d^{-1} and 30 kg SS ha^{-1} d^{-1} to achieve a 20:20 effluent standard and 60 kg BOD ha^{-1} d^{-1} and 50 kg SS ha^{-1} d^{-1} to achieve a 30:30 effluent standard. Based on limited data an effluent TKN (organic and ammonia N) of < 10 g m^{-3} is predicted for this type of system at TKN loadings of < 5 kg ha^{-1} d^{-1}.

Table 15.2 Typical range of loading rates and basic design characteristics for free water and vegetated submerged bed CW systems

Parameters	Types	
	FWS	VSB
Organic loading rates, kg BOD ha^{-1} day^{-1}	5 – 110	10 – 200
N loading rate, kg N ha^{-1} day^{-1}	0.5 – 60	2 – 80
HRT, days	3 – 10	2 – 7
Hydraulic loading rate, cm d^{-1}	2.5 – 10	2.5 – 20
Water depth range, cm	20 – 50	2 – 10 (below surface)
Aspect ratio (length/width)	4:1-6:1	2:1
Bed depth*, cm	-	30 – 90

* Depending on depth to which plant roots can grow for the particular wastewater and bed characteristics.

Source: Adapted from Polprasert (1996), Kadlec and Knight (1996), US.EPA 2000)

The most commonly used design model currently used for constructed wetlands is the areal *k-C** kinetic model (Kadlec and Knight, 1996; IWA, 2000), which accounts for temperature-based, first order reduction to background levels (C^*). C^* accounts for substances generated in the wetland by biological activity, sediment release etc., as well as the non-degradable fraction of the contaminant. Knowing the average wastewater flow (Q, $m^3\ d^{-1}$), the appropriate area based first order removal rate constant (k, $m\ yr^{-1}$) and the average influent concentration (C_i), the wetland area (A, m^2) required to meet the target outflow concentration (C_o) can then be calculated (assuming neutral water balance and plug flow) using the following equation:

$$A = -\frac{Q}{k}\ln\left(\frac{C_o - C^*}{C_i - C^*}\right) \tag{15.1}$$

Somewhat surprisingly in comparison to other wastewater treatment systems, many constructed wetlands pollutant removal rates aren't particularly sensitive to temperature (IWA, 2000; Kadlec and Reddy, 2001). Temperature effects on k are most important for removal of the various forms of nitrogen, and are generally summarised using the following modified Arrhenius equation in which k_t is the rate constant at temperature T (°C) and k_{20} is the rate constant at 20 °C:

$$k_t = k_{20}\theta^{(T-20)} \tag{15.2}$$

The k-C^* model gives a good first approximation for likely seasonal and annual treatment performance. It, however, has limitations related to hydraulic assumptions, variability of rate constants and the effects of pollutant speciation (Kadlec, 1999, 2003b). Kadlec (2003b) has argued for the advantages of a 'relaxed' tanks-in-series model, where the 'number of tanks' accounts for the inherent variability of both the detention time distribution and the k values. This has been termed the P-k-C* model and is destined to replace the k-C* model in the forthcoming 2nd Edition of Kadlec and Knight, (1996; in preparation, R.H. Kadlec pers. com., January 2005).

Characteristic k-C^* rate constants, background concentrations and temperature factors for FWS and VSB systems are shown in Table 15.3 It should be noted that these are the central tendencies derived from a large dataset, and may vary substantially for individual wetlands, depending on wastewater characteristics, specific wetland design and other variables. For pollutants which undergo sequential reactions (notably nitrogen) it is important that the k-C* is applied to each step of the sequence, and production rates included in the calculations (e.g. in the case of N, ammonia N produced by mineralisation of Organic N and nitrate produced by nitrification of ammoniacal N). It is always advisable to check predictions against measured results for systems that are similar to the model you are proposing. Regression equations and graphs summarising the performance of large datasets are given in Kadlec and Knight (1996), IWA (2000) and Rousseau *et al.* (2004).

Vertical flow CW systems have not been commonly used to treat pond effluents, being mostly applied for small-scale treatment of primary settled sewage and septic tank effluents. However, their enhanced aeration and nitrification capabilities are likely to be useful for treatment of anaerobic pond effluents and higher strength agricultural and industrial wastes (e.g. Sezerino *et al.* 2003). Recommended design criteria are outlined in IWA (2000) and Cooper (2004). Koottatep *et al.* (2001) found nitrification reactions could take place in vertical flow CW beds treating septage, as was evidenced by the significant increase of percolate NO_3^- concentrations. Reduction in NH_4^+ content can be obtained through nitrification if the pond effluent is fed into vertical flow CW system. Denitrification reactions in which the NO_3^- compounds are converted to N_2 can occur if the percolate is retained in the CW beds for 5 – 7 days.

Table 15.3 Typical k-C* rate constants (*k*), background concentrations (*C**) and temperature factors (θ) reported for FWS and VSB constructed wetlands by Kadlec and Knight (1996), unless otherwise noted

	FWS			VSB		
	k_{20} (m yr^{-1})	θ	C^*	k_{20} (m yr^{-1})	θ	C^*
BOD	36.5†	0.9-1.015††	$3.5+0.053C_i$	22-66†	1-1.05†	$3.5+0.53C_i$
TSS	1000	1.00	$5.1+0.16C_i$	1000	1.00	$7.8+0.63C_i$
Organic N	17	1.01-1.07††	1.5	35	1.05	1.5
NH_4-N	18	1.04-1.16††	0	34	1.05	0
NO_x-N	35	1.04-1.11††	0	50	1.05	0
TN	22	1.08-1.09††	1.5	27	1.05	1.5
TP	12	1.00	0.02	12	1.00	0.02
Faecal coliforms	75	1.00	300	95 (50-300†)	1.00	10

C_i = influent concentration
† from IWA (2000)
†† from Kadlec and Reddy (2001)

15.4.2 Design example – free water surface wetland

Design a FWS-CW system in a tropical region (mean cool-season air temperature of 25 °C) to supplement waste stabilization pond treatment for a sewage discharge to a moderate-sized river with effluent characteristics and treatment targets as shown in Table 15.4.

Table 15.4 Hypothetical pond effluent contaminant concentrations, FWS CW treatment targets and estimates of the wetland area required to achieve them

Parameter	Pond effluent	Final discharge mean annual target concentration	Estimated FWS area required* (ha)
Flow (m^3 d^{-1})	200	-	-
BOD_5 (g m^{-3})	80	30	0.24
TSS (g m^{-3})	100	30	< 0.1
Total N (g m^{-3})	40	15	0.23**
NH_3-N (g m^{-3})	30	10	0.28
NO_3-N (g m^{-3})	2.0	2.0	-
Total P (g m^{-3})	10	8	0.14
Faecal coliforms (MPN (100 mls)$^{-1}$)	10,000	1,000	0.26

* calculated from equations 15.1 and 15.2 using typical parameter values from Table 15.3

** it is preferable to calculate sequential nitrogen reaction sequence as proposed in Kadlec and Knight (1996) if sufficient information is available

Solution

Use first-order design model to calculate wetland areas required to meet the required effluent concentrations (C_o) using the first order areal k-C* design model. For simplicity we will assume that rainfall additions and evapotranspiration losses are roughly equal.

1. Estimate the likely wetland effluent background concentration C* from Table 15.3 or using local knowledge for wetland systems treating similar wastewaters; e.g. calculate C* for BOD using equation given in Table 15.3:

$$C^* = 3.5 + (0.053\ C_i) = 3.5 + (0.053 \text{ x } 80) = 7.74 \text{ g m}^{-3}$$

2. Calculate the temperature-corrected first-order rate constants, k_t using equation 15.2; e.g. for ammonium-N, where $k_{20} = 18$ and $\theta = 1.10$:

$$k_t = 18 \times 1.10^{(25-20)} = 29$$

3. Calculate FWS wetland area (A) required to meet each effluent contaminant concentration (C_o) using equation 15.1 (ensure equivalent units are used for all parameters). For example, for BOD: assuming flow (Q) = 200 m^3 d^{-1} = 73,000 m^3 yr^{-1}, influent concentration (C_i) = 80 g m^{-3}, C^* = 7.74 g m^{-3} and first-order rate constant (k_t) = 36.5 m yr^{-1} (assumes negligible temperature effect, $\theta = 1$):

$$A = -\frac{73{,}000}{36.5}\ln\left(\frac{30-7.74}{80-7.74}\right) \qquad = 2355\text{m}^2 \approx 0.24 \text{ ha}$$

4. Compare wetland area estimates for all target contaminants. Calculated wetland areas required to meet effluent targets for all contaminants are shown in Table 15.4. These show a range of areas up to 0.28 ha, with ammonium and faecal coliforms requiring the largest areas. Note that these calculations are based on annual average performance. If effluent standards need to be stringently met on a short-term basis then additional wetland area will be required to account for the normal variability in treatment performance. Expected treatment performance variability for CWs is discussed in Kadlec and Knight (1996), and IWA (2000).
5. Assume, based on above calculations that a wetland area of 0.28-0.30 ha will be required. Check that basic design features fit with guidelines shown in Table 15.2. For example:
    a. The theoretical hydraulic residence time (HRT) = wetland area (A, m^2) x water depth (0.3-0.4 m to promote good plant growth) x porosity (assume 0.9 to account for volume taken up by plants and detritus) ÷ flow (Q, m^3 d^{-1}) = 4.4–4.7 d., which equates to hydraulic loading rates (HLR or q) of 6.7–7.7 cm d^{-1}. These are both within, but at the lower (HRT) or upper (HLR) end, respectively, of the range typical for FWS systems.
    b. The calculated BOD and TN mass loading rates ($=(Q \times C_i) \div A$) for the proposed wetland are 53-61 kg BOD ha^{-1} d^{-1} and 27-31 kg TN ha^{-1} d^{-1}, which are both near the middle of the range given in Table 15.2.
6. Check performance expectations against available local data and regression equations for general constructed wetland performance in Kadlec and Knight (1996) and IWA (2000).
7. Develop basic design concept, focussing on optimising treatment for key contaminants of concern. Split flow equally between 2 or more CW beds to allow operational flexibility. Disperse influent evenly across the width of each bed and maintain aspect ratios (length to width) between 4 and 6:1 to promote plug-flow. Line base of beds with 200 mm of compacted local clay subsoil and cover with 400 mm layer of topsoil to provide rooting medium with good anchorage. Select hardy tall-growing local varieties of cattail, reed or bulrush plants to establish in the CWs. Provide short residence-time, deep-water zones in wetland to enhance aeration and enhance ammoniacal N removal, whilst limiting algal regrowth.

15.4.3 Resource recycling and reuse

Integrated pond/wetland systems can be applied not only for wastewater treatment but also for recycling and reuse of resources in terms of harvested plant biomass. There are techniques available for processing them into the reusable forms. These include mulching and composting to produce soil additives; hydrolysis and fermentation to form ethanol; pulping to yield fiber; and silaging to produce livestock fodders (Lim and Polprasert, 1996).

15.5 SUMMARY AND FUTURE RESEARCH NEEDS

Constructed wetlands are able to simply and cost-effectively complement pond treatment systems. In particular, they can readily provide additional removal of suspended solids, BOD, nitrogen and faecal pathogens. The optimum design considerations and operating conditions for CW systems treating pond effluent to reliably meet reuse and discharge standards is an on-going research challenge. Improved nitrogen and pathogen removal requires further information on the most favourable mix of wetland and open-water zones for FWS systems, and the best practical strategies to increase physical aeration of VSB systems using intermittent vertical or tidal flow regimes. Given the variability in reported performance of different wetland systems, development of generic modular designs for common applications may simplify application and enhance reliability.

REFERENCES

Angelakis, A.N. (2001) Management of wastewater by natural treatment systems with emphasis on land-based systems. *Decentralized Sanitation and Reuse,* (eds) Lens, P., Zeeman, G. and Lettinga, G., IWA Publishing, London, UK.

Armstrong, W., Armstrong, J. and Beckett, P.M. (1990) Measurement and modelling of oxygen release from roots of *Phragmites australis*. In: *Constructed wetlands in water pollution control*, (eds P.F. Cooper and B.C. Findlater) Pergamon Press, Oxford, pp. 41-52.

Bachard, P.A.M. and Horne, A.J. (2000) Denitrification in constructed free-water surface wetlands: II. Effects of vegetation and temperature. *Ecological Engineering* 14, 17-32

Brix, H. and Schierup, H. H. (1989) Sewage treatment in constructed wetlands - Danish experiences. *Water Science and Technology* **21**, 1665-1668.

Brix, H., Arias, C.A., del Bubba, M. (2001) Media selection for sustainable phosphorus removal in subsurface flow constructed wetlands. *Water Science and Technology* **44**(11/12), 47-54.

Behrends L., Houke, L., Bailey E., Jansen, P. and Brown, D. (2001) Reciprocating constructed wetlands for treating industrial, municipal and agricultural wastewater. *Water Science and Technology* **44**(11/12), 399-405.

Cooper, P.F. (2004) The performance of vertical flow constructed wetland systems with special reference to the significance of oxygen transfer and hydraulic loading rates. *Proceedings of the 9th IWA International Conference on Wetland Systems for Water Pollution Control.* Avignon, France, 26-30 September, 2004, **1**, 153-160

Decamp, O. and Warren, A. (1998) Bacterivory in ciliates isolated from constructed wetlands (reedbeds) used for wastewater treatment. *Water Research* **32**, 1989-1996.

El Hafiane, F. and El Hamouri, B. (2004) Subsurface-horizontal flow wetland for polishing high rate ponds effluent. *Proceedings of the 6th International Conference on Waste Stabilisation Ponds and the 9th IWA International Conference on Wetland Systems for Water Pollution Control.* Avignon, France, 26 Sept -1 October, 2004. Joint volume , pp. 141-147.

Garcia, J., Morato, J., Bayona, J.M. and Aguirre, P. (2004) Performance of horizontal subsurface flow constructed wetlands with different depth. *Proceedings of the 9th IWA International Conference on Wetland Systems for Water Pollution Control.* Avignon, France, 26-30 September, 2004, **1**, 269-276.

Hammer, D.A. and Knight, R.L. (1994) Designing constructed wetlands for nitrogen removal. *Water Science and Technology*, **29**(4), 15-27.

IWA (2000) *Constructed Wetlands for Pollution Control: Processes, Design and Operation.* International Water Association Scientific and Technical report No. 8. IWA Publishing, London, UK.

Kadlec, R.H., (1999) The inadequacy of first order treatment wetland models. *Ecological Engineering* **15**, 105-120.

Kadlec, R.H. (2003a) Pond and wetland treatment. *Water Science and Technology* **48**(5), 1-8

Kadlec, R.H. (2003b) Effects of pollutant speciation in treatment wetland design. *Ecological Engineering* **20**, 1-16.

Kadlec, R.H., 2005. Wetland to pond treatment gradients. *Water Science and Technology* **51**(9), 291-298.

Kadlec, R.H., and Knight, R.L. (1996). *Treatment Wetlands.* Lewis Publishers, Boca Raton, FL.

Kadlec, R.H. and Reddy, K.R., 2001. Temperature effects in treatment wetlands. *Water Environment Research* **73**, 543-557.

Knight, R.L., Kadlec, R.H. and Ohlendorf, H.M., 1999. The use of treatment wetlands for petroleum industry effluents. *Environmental Science and Technology* **33**, 973-980.

Koottatep, T., Kim Oanh, N.T., Polprasert, C., Heinss, U., Montangero A. and Strauss, M. (2001) Potential of Vertical-flow Constructed Wetlands for Septage Treatment in Tropical Regions. In *Advances in Water and Wastewater Treatment Technology – Molecular technology, nutrient removal, sludge reduction, and environmental health,* (ed. T. Matsuo), Elsevier Science, Amsterdam.

Lim, P.E. and Polprasert, C. (1996) Constructed wetland for wastewater treatment and resource recovery. *Environment Systems Reviews*, No. 41, Asian Institute of Technology, Bangkok, Thailand.

Nichols, D.S., 1983. Capacity of natural wetlands to remove nutrients from wastewater. *Journal of the Water Pollution Control Federation* **55**, 495-505

Nielsen, S. (2003) Sludge drying reed beds. *Water Science and Technology* **48**(5), 101-109.

Polprasert, C. and Koottatep, T. (2004) Integrated pond and constructed wetland systems for sustainable wastewater management. *Proceedings of the 6th International Conference on Waste Stabilisation Ponds and the 9th IWA International Conference on Wetland Systems for Water Pollution Control.* Avignon, France, 26 Sept -1 October, 2004. Joint volume , pp. 25-33.

Polprasert, C., Dan, N.P. and Thayalakumaran, N. (1996) Application of constructed wetlands to teat some toxic wastewater under tropical conditions. *Water Science and Technology* **34**(11), 165-171.

Polprasert, C., Khatiwada, N.R. and Bhurtel, J. (1998) A Design Model for BOD Removal in Constructed Wetlands Based on Biofilm Activity. *Journal of Environmental Engineering.* ASCE **124**(9), 838-843.

Reed, S. C. Middleborooks, E.J., and Crites, R. W. (1995) *Natural Systems for Waste Management and Treatment,* 2nd Edition, McGraw-Hill, Inc., New York.

Rousseau, D.P.L., Vanrolleghem, P.A., De Pauw, N. (2004) Model-based design of horizontal subsurface flow constructed treatment wetlands: a review. *Water Research* **38,** 1484-1493.

Sakadevan K. and Bavor, H.J. (1998) Phosphate adsorption characteristics of soils, slags and zeolite to be used as substrates in constructed wetlands. *Water Research* **32**, 393-399.

Sezerino, P.H., Spiller, V.R., Santos, M.A., Kayser, K., Kunst, S. Philippi, L.S. and Soares H.M. (2003) Nutrient removal of piggery effluent using vertical constructed wetlands in South Brazil. *Water Science and Technology* **48**(2), 129-136.

Stone, K.C., Hunt, P.G., Szogi, A.A., Humenik, F.J., Rice, J.M. (2002) Constructed wetland design and performance for swine lagoon wastewater treatment. *Transactions of the ASAE* 45, 723-730.

Stott, R., May, E., Matsushita, E., Warren, A. (2004) Protozoan predation as a mechanism for the removal of cryptosporidium oocysts from wastewaters in constructed wetlands. *Water Science and Technology* **44**(11/12), 191-198.

Sun, G., Zhao, Y. and Allen, S. (2005) Enhanced removal of organic matter and ammoniacal-nitrogen in a column experiment of tidal flow constructed wetland system. *Journal of Biotechnology* **115**, 189-197.

Surampalli R.Y., Banerji, S.K., McCallister, D., and Tyagi, R.D. (2002) Upgrading Waste Stabilization Ponds for Nutrient Removal. *Proc. of the 5th International IWA Specialist Group Conference on Waste Stabilization Ponds: Pond Technology for the New Millennium*, Auckland, New Zealand. 2-5 April.

Tanner, C.C. (2001). Plants as ecosystem engineers in subsurface-flow constructed wetlands. *Water Science and Technology* **44**(11/12), 9-17.

Tanner, C.C. (2004). Nitrogen removal processes in constructed wetlands. In: *Wetland ecosystems in Asia.* (ed M.H.Wong), pp. 331-346, Elsevier Science, The Netherlands.

Tanner, C.C. and Sukias, J.P.S. (2003) Linking pond and wetland treatment: performance of domestic and farm systems in New Zealand. *Water Science and Technology* **48**(2), 331-340.

Tanner, C.C.; Craggs, R.J.; Sukias, J.P.S. and Park, J. (2005) Comparison of maturation ponds and constructed wetlands as the final stage of an Advanced Pond System. *Water Science and Technology* **51**(12), 307-314.

Tanner, C.C.; D'Eugenio, J.; McBride, G.B.; Sukias, J.P.S. and Thompson, K. (1999) Effect of water level fluctuation on nitrogen removal from constructed wetland mesocosms. *Ecological Engineering* **12**, 499-520.

US.EPA (2000) *EPA Manual: Constructed wetlands treatment of municipal wastewater.* Report No. EPA/625/R-99/010. Office of Research and Development, Cincinnati, Ohio, USA.

van Oostrom, A.J. and Russell, J.M. (1994) Denitrification in constructed wetlands receiving high concentrations of nitrate. *Water Science and Technology* **29**(4), 7-14.

16

Integrated pond/aquaculture systems

Chongrak Polprasert and Thammarat Koottatep

16.1 AQUACULTURE PONDS

Although the algal cells photosynthetically produced during pond treatment contain protein, their small sizes (generally less than 10 μm) makes harvesting for reuse economically unviable. One of the drawbacks of utilizing wastewater-grown algae as animal feed (with the exception of *Spirulina*) is the low digestibility of the algal cell walls. Thus, the culture of phytoplankton-feeding (herbivorous) fish that feed on the algae is attractive for the production of fish protein biomass, which is then easily harvestable for subsequent utilization as an animal feed or for human consumption.

16.1.1 Types of aquaculture ponds

Depending on fish farming techniques, wastewater-fed aquaculture ponds can be divided into two types:

ISBN: 1843390205. Published by IWA Publishing, London, UK.

- ***Monoculture ponds*** are those in which only one fish species is reared in the pond, whereby supplementary feed can be applied to increase fish yield. Management, processing and marketing may be simpler for such single species production systems.
- ***Polyculture ponds*** are those in which more than one fish species are raised in the same pond. In the ecosystem of polyculture pond, different locations/niches are filled by different fish species (Table 16.1). By combining different species in the same pond, the total fish production can be raised to a higher level than that achieved with only a single species.

Table 16.1 Common fish species used in polyculture ponds

Location	Common name	Scientific name	Food habits
Surface-water	Grass carp (Chinese carp)	*Ctenopharyngodon idella*	Macrophytes, vascular plants and phytoplankton
	Mrigal (Indian carp)	*Cirrhina mrigala*	
	Silver carp (Chinese carp)	*Hypophthalmichthys molitrix*	
Mid-water	Catla (Indian carp)	*Catla catla*	Phytoplankton and zooplankton
	Rohu (Indian carp)	*Labeo rohita*	
	Bighead carp (Chinese carp)	*Arigtichthys nobilis*	
Bottom	Mud carp (Chinese carp)	*Cirrhina molitorella*	Sediment, detritus, faeces of upper level animals
	Calbasu (Indian carp)	*Labeo calbasu*	
	Common carp (Chinese carp)	*Cyprinus carpio*	

Source: Adapted from Korn (1996)

16.1.2 Treatment mechanisms

A wastewater-fed pond for aquaculture is functioning well when a symbiotic balance is established between the algae, bacteria and fish. The treatment mechanisms in pond/aquaculture systems are very similar to those described in earlier chapters for facultative ponds. In essence, the algae produce oxygen that promotes decomposition by aerobic and facultative bacteria, and the algal cells that grow on the released nutrients. The algae and zooplankton, which graze on them, serve as feed for the fish (Figure 16.1). Three zones exist in the ponds:

- In the aerobic zone the oxygen is supplied partly by natural surface re-aeration and from algal photosynthesis, which is then used by the bacteria in the aerobic decomposition of the organic matter. The nutrients and carbon-dioxide (CO_2) released in this decomposition are, in turn, used by the algae for growth;
- The facultative (intermediate) zone is partly aerobic and partly anaerobic, often varying during the course of the day. Waste breakdown and stabilization is carried out by facultative bacteria;

- In the anaerobic bottom zone, anaerobic bacteria decompose accumulated solids.
-

Fish normally reside in the aerobic and facultative zones where oxygen and algae (food) are present.

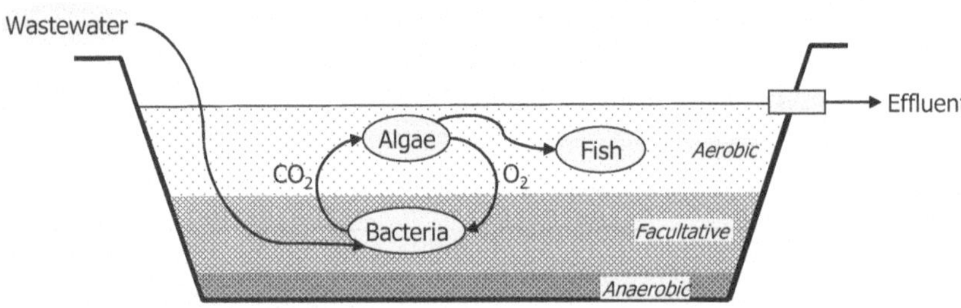

Figure 16.1 Schematic diagram of treatment mechanisms in aquaculture ponds

16.1.3 Fish species

Fish can be classified into three groups on the basis of food source, as shown in Table 16.2.

Most fish are adaptable in their feeding habits and utilize the most readily available foods (Lagler et al., 1962). Relatively few kinds are strict herbivores or carnivores, and none feed solely on one food source. Food sources also may change with seasonal variation and can depend on the stage of the fish's development.

Based on their feeding characteristics, Polprasert (1996) suggested that herbivorous or omnivorous species such as tilapia, Chinese carp, and Indian carp are suitable to be cultivated in wastewater-fed ponds. These species are widely used because they feed on waste-grown phytoplankton, are tolerant to low dissolved oxygen (DO) level, diseases and other adverse environmental conditions.

Table 16.2 Examples of common types of fish and their food source

Types	Common name	Scientific name	Food habits
Herbivorous	Grass carp Silver carp	*Ctenopharyngodon idella* *Hypophthalmichthys molitirix*	Algae and higher plants
Carnivorous	Snakehead	*Ophicephalus striatus*	Insects, zooplankton, bacteria, trash fish.
Omnivorous	Tilapia*	*Oreochromis nicloticus* *Oreochromis aurea* *Oreochromis mossambicus* *Oreochromis hornorum*	Animals and plants

* http://aquanic.org/publicat/state/il-in/as-494.htm

16.1.4 Fish cultivation/stocking

To cultivate fish in wastewater-fed ponds, fish sex is an important consideration for optimizing yield. Owing to genetic reasons, male tilapias generally grow faster than their female counterparts (Guerrero, 1982). Monosex schools of male tilapias possess faster growth and eliminate unnecessary reproduction. Monosex tilapias can be obtained by: (i) manual separation of the sexes, which can be difficult when small fish fry is used; (ii) production of monosex broods thorough hybridization; or (iii) sex reversal with the use of sex steroids (such as androgen, methyltestosterone, and ethyltestosterone) in the diet for fish fry during the period of gonadal differentiation (the reproduction of male or female organ). The effectiveness of these steroids was reported to be 93 – 100% (Nacario, 1987).

The tilapia species is believed to be most suitable for waste-fed ponds due to its hardiness. Other herbivorous fish, such as silver carp and bighead carp, are also promising species. Polyculture is recommended where benthic feeding fish such as mud carp are stocked together with omnivorous fish such as common carp.

Stocking density is an important factor in fish culture. As shown in Figure 16.2, high stocking density gives high yield per pond volume, but with a small mean size of fish. To produce 'trash fish' for animal feed, a stocking density of 10 or more fish/m^2 is recommended and the period of fish culture may be as short as 3 months. Ponds producing marketable-size fish, for possible use as human food, should be stocked at a lower density such as 1 or 2 fish/m^2 or even less, and the rearing period may be up to 1 year, or longer. However, several commercial fish farms are reportedly able to stock fish at densities greater than 50 fish/m^2, provided that the pond water quality and fish feed are maintained at the levels favorable for fish growth.

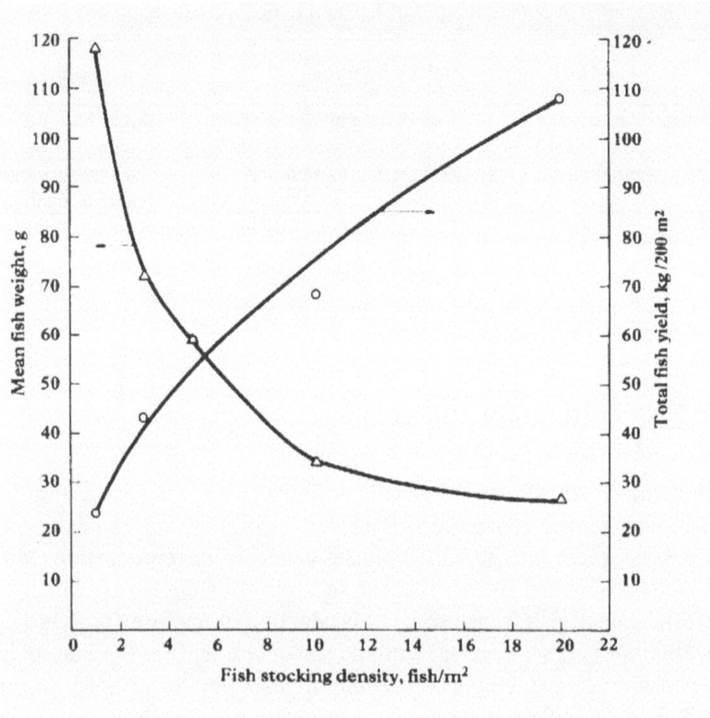

Figure 16.2 Effects of fish stocking density on mean fish weight and total fish yield after 6 months of pond operation (Polprasert 1996)

16.2 APPLICATIONS OF PONDS AND AQUACULTURE SYSTEMS

Because pond effluent normally contains high concentrations of suspended algal cells, 'feeding' this effluent to aquaculture ponds for grazing by herbivorous fish can reduce the suspended solids and nutrient content in the final effluent whilst producing valuable fish protein biomass for use as animal feeds or human foods. To safeguard public health, it has generally been considered that pond effluents applied to fishponds should meet the WHO microbiological guidelines (Table 16.3). The level of pre-treatment required to achieve such criteria, with concomitant loss of organic matter and nutrients, has been considered by some to be overly restrictive (Mara *et al.* 1993; Edwards, 2001), significantly reducing the potential fish yield. Edwards (2001) instead proposes treatment in primary ponds with a minimum of 8-10 days residence time to reduce trematode worms, suspension of wastewater loading for two weeks prior to fish harvest and a short period of gut depuration before consumption. Appropriate hygiene practices should be followed during harvesting, depuration, transport, handling, storage, processing and marketing to protect the health of aquacultural workers and consumers (see microbiological guidelines for edible fish and crustacean summarised in Strauss, 1996).

Particular care should also be taken where industrial discharges, sludges and septage may contribute toxic loads of heavy metals and/or organic compounds, causing chronic and long-term health effects (Edwards, 2001; Strauss 2000). Potential organic contaminants of concern include volatile organic compounds (VOCs), organic solvents, polychlorinated aromatic hydrocarbons (PAHs), and pesticides.

Table 16.3 Guidelines for microbiological quality for aquaculture reuse (WHO, 1989)

Reuse process	Viable trematode eggs [a] (average no./L or kg)	Fecal coliforms (geometric mean no./ 100 mL or 100g) [b]
Fish culture	0	$< 10^4$
Aquatic macrophyte culture	0	$< 10^4$

[a] *Clonorchis, Fasciolopsis,* and *Schistosoma.* Consideration should be given to this guideline only in endemic area.

[b] This guideline assumes that there is one $\log_{10}$ unit reduction in faecal coliforms occurring in the pond, so that in-pond concentrations are < 1,000 per 100 mL. If consideration of pond temperature and residence time indicates that a higher reduction can be achieved, the guideline may be relaxed accordingly.

The Calcutta, India, wastewater fisheries are reported to be the largest wastewater-fed aquaculture systems in the world (Edwards, 1985 and 1990). The 5.7-ha fishponds (0.7 m deep) receive 740 m^3/day of raw domestic wastewater. Algae is photosynthetically produced which serve as food for Indian carps stocked in these ponds. Estimated yields of fish from these ponds vary from 0.6 ton/ha.year to 4–9 ton/ha.year (Edwards, 1990). The wastewater-fed aquaculture in Fonyod city, Hungary, is also able to produce 5–7 ton/ha.year of silver carps and tilapia, with its wastewater being purified to a level comparable to those of conventional secondary processes as shown in Table 16.4 (Olah, 1990).

Table 16.4 Treatment performance of integrated pond/aquaculture fed with domestic wastewater

Parameter	Concentration, mg/L		Removal (%)
	Sewage	Effluent	
Total solids	500 – 1,000	150 – 200	70 – 80
BOD	110 – 120	9 – 10	90 – 92
Total N	50 – 55	2.0 – 3.0	95 – 96
Total P	10 – 12	0.7-1.0	92 – 93

Source: Adapted from Olah (1990)

Aquaculture ponds can also be integrated with constructed wetland systems for wastewater treatment and recycling with economic benefits achievable through production of fish, ducks and commercial plants. In Tianjin city, China, fish ponds with an area of four hectares can yield 25,000 kg of fish as well as 750 kg of ducks, while 30,000 kg of dried reeds could be harvested annually in the CW units that have a total area of about 10,000 m^2 (Xianfa and Chuncai, 1995).

16.3 DESIGN CONSIDERATIONS

This section describes environmental requirements and design criteria of aquaculture systems when they are used as a polishing step of the pond systems. The design considerations for pond treatment systems are given in Chapters 8, 9, 10 and 11.

Critical factors for aquaculture ponds include dissolved oxygen and ammonia, which correlate to the organic loading rates and stocking density. To avoid fish suffocation, the dissolved oxygen in wastewater-fed aquaculture ponds should not drop below 1 – 2 mg/L (dawn is a critical period in this regard). If this is not obtained naturally, then mechanical aeration should be provided. To avoid toxic effects on the fish, the ammonia concentration in aquaculture ponds should not exceed 1.0 mg/L, by which the substantial amount of ammonia be removed in the former pond system as well as maintaining a neutral pH in the pond water by herbivorous grazers. Within pH range of 7 – 8 and proper control of organic loading, the toxic effect of H_2S should be minimal. The presence of other toxic materials in the pond water requires proper pre-treatment to remove these toxic compounds and avoid their bioaccumulation and biomagnification in the food chain.

Typical design criteria for wastewater-fed aquaculture ponds are given in Table 16.5.

Table 16.5 Design criteria for wastewater-fed aquaculture ponds

Parameters	Range
Organic loading rates	
- kg BOD/ha.day	25 – 75
- kg COD/ha.day	50 – 150
Minimum DO at dawn, mg/L	1 – 2
NH_3 concentration of pond water, mg/L	< 0.02
pH of pond water	6.5 – 9.0 (optimum 7.0 – 8.0)
Fish stocking density, fish/m^2	0.5 – 50
H_2S concentration, mg/L	Nil
Culture period, month	3 – 12
Fish yield, kg/ha.year	1,000 – 10,000
Pond width, m	< 30
Pond depth, m	1 – 1.5
Area for each pond*, m^2	400 – 4,000

* To enable effective fish harvesting

Source: Adapted from Polprasert (1996)

To ensure ongoing reliability it is important that these systems are properly operated and maintained. The basic operation and maintenance required for these systems includes:

- Water level maintenance;
- Avoidance of short-circuiting flow;
- Embankment and dike maintenance;
- Control of nuisance pests and insects;
- Removal of undesirable vegetation and accumulated sludge; and
- Periodical harvesting of the emergent plants or fishes.

16.3.1 Design example – aquaculture ponds

Design aquaculture ponds fed with pond effluent having characteristics as follows:

- Quantity 120 m^3/d
- COD 100 mg/L
- Organic N 110 mg/L
- NH_3-N 60 mg/L
- Total P 10 mg/L

Solution

From data presented in Table 16.5, use organic loading = 50 kg COD ha^{-1} d^{-1}.

$$\text{Area requirement} = \frac{120 \times 100}{50 \times 1{,}000} = 0.24\,\text{ha} = 2{,}400\,\text{m}^2$$

Select a fish pond size of 20 m (width) x 20 m (length) x 1.5 m (depth); water depth = 1.0 m

$$\text{Number of ponds} = \frac{2{,}400}{20 \times 20} = 6$$

Each pond is separated from the others to facilitate harvesting and pond drying operations. Select tilapia (*Oreochromis niloticus*) with a stocking density of 5 fish/m^2. The products will be sold as trash fish to produce animal feed. This will reduce possible health hazard from direct consumption of these fish.

Number of fingerling (0.5 – 5 g) /fish used = 20 x 20 x 5 = 2,000 fish/pond

Based on the data shown in Figure 16.2, a fishpond at stocking density of 5 fish/m^2 can yield total fish of 50 kg/200 m^2 after 6 month of operation.

$$\text{Total fish yield} = \frac{2{,}400 \times 50}{200} = 600 \text{ kg or } 0.6\,\text{ton}$$

These wastewater-fed fishponds with total area of 2,400 m^2 can produce tilapia at the yield of 0.6 x 2 = 1.2 tons/year. At the relatively low organic loading rate applied in these ponds, DO concentration at dawn should be adequate for the tilapia growth. However, to avoid H_2S and NH_3 toxicity, pH of the pond water should be monitored and maintained at the neutral range.

16.3.2 Resource recycling and reuse

Integrated pond/aquaculture systems can be applied not only for wastewater treatment but also for recycling and reuse of resources in terms of waste-grown fish. To directly use waste-grown fish for human consumption, they obviously must be safe from pathogens and helminthes, which should be achieved by undertaking the following measures:

- control microbiological quality of pond effluent to meet the WHO standards for feeding to fish ponds (see Table 16.3);
- undertake 'depuration' by putting waste-grown fishes in clean water for 1 – 2 weeks.

In order to avoid public health risks, the waste-grown fish can instead be used as feed for carnivorous fish or shrimps. The high protein content of about 80 percent in fishmeal is also reportedly suitable for feeding to chicken and pigs (Polprasert, 1996).

16.4 SUMMARY AND FUTURE RESEARCH NEEDS

As described in Section 16.2, the fish yield from aquaculture systems may be contaminated with pathogens. Future research should focus on improving pathogen removal in these integrated systems prior to reuse or discharge of the treated effluent. Public health concerns should not only be limited to pathogenic organisms, but should also consider toxic compounds that may contaminate the harvested products. Defining the optimum design considerations and operating conditions that are able to meet reuse standards is an ongoing research challenge.

ACKNOWLEDGEMENTS

The authors are grateful to Mr. Abu Syed Md. Kamal for his valuable assistance in data and information collection.

REFERENCES

Edwards, P. (1985) Aquaculture: A component of low-cost sanitation technology. *World Bank Technical Paper No. 36 Integrated Resource Recovery Series*. The World Bank, Washington D. C.

Edwards, P. (1990) *Reuse of human waste in aquaculture: A technical review*. World Bank, Washington D.C.

Edwards, P. (2001) Public health issue of wastewater-fed aquaculture. *Urban Agriculture Magazine*, **3**, 20-22 (available at: http://www.ruaf.org/1-3/20-22.pdf)

Guerrero, R.D. III. (1982) Control of tilapia production. In *The Biology and Culture of Tilapias* (eds. R.S. V. Pullin and R.H. Lowe-McConnell), 309-15, International Center for Living Aquatic Resources Management, Manila.

Korn, M. (1996) The pond dyke concept: Sustainable agriculture and nutrient recycling in China. *AMBIO - A journal of the Human Environment* **25**(1).

Lagler, K.F., Bardach, J.E. and Miller, R.R. (1962) *Ichthyology: The study of fishes*, Wiley, New York.

Mara, D.D., Edwards, P., Clark, D., Mills, S.W. (1993) A rational approach to the design of wastewater-fed fishponds. *Water Research* **27,** 1797-1799.

Nacario, E.N. (1987) Sex reversal of Nile tilapia in cages in ponds. AIT master's thesis No. AE-87-35, Asian Institute of Technology, Bangkok.

Olah, J. (1990) Wastewater-fed fish culture in Hungary. In: *Wastewater-fed aquaculture*, (eds) Edwards, P. and Pullin, R.S.V., Asian Institute of Technology, Bangkok.

Polprasert, C. (1996) *Organic Waste Recycling – Technology and Management*, 2nd edition, John Wiley *and* Sons.

Strauss, M. (1996) Health (Pathogen) considerations regarding the use of human waste in aquaculture. In *Recycling the Resource–Ecological Engineering for Wastewater Treatment*, eds Stadenmann, J., Schonborn, A.. and Etnier, C. Environmental Research Forum, Proceedings of the 2nd International Conference on Ecological Engineering for Wastewater Treatment, Wadenswill, Switzerland, 18-20 September. Transtec Publications, Zurich (available at http://www.sandec.ch/).

Strauss, M. (2000) Human waste (excreta and wastewater) reuse. In *ETC/SIDA Bibliography on Urban Agriculture* (available at http://www.sandec.ch/).

World Health Organizations (WHO), (1989) *Guidelines for the use of wastewater in the aquaculture.* Report of a WHO Scientific Group, Technical Series No. 778, WHO, Geneva, Switzerland.

Xianfa, L., and Chuncai, J. (1995) Constructed wetlands systems for water pollution control in north China. *Water Science and Technology* **32**(3), 349-356.

17

Wastewater reservoirs

Marcelo Juanicó

17.1 INTRODUCTION

17.1.1 What are wastewater reservoirs?

Wastewater reservoirs have been operating in Israel for 30 years for the storage and treatment of wastewater effluents during the rainy winter months, in order to re-use them for agricultural irrigation during the dry summer months. There are more than 200 of these reservoirs operating in the country and they have proved to be a practical, reliable and cost effective wastewater treatment and storage technology.

While Israel is today the leading country in the research and use of this technology, other countries already used wastewater reservoirs before their wide scale introduction in Israel in the seventies. Many countries continue, or are starting, to use wastewater reservoirs nowadays. Juanico and Friedler (1999) have published a review on the use of these reservoirs outside Israel.

The biological processes in wastewater reservoirs resemble that of waste stabilisation ponds, but the increased depth (average between 6m and 8m, maximum up to 20m), greater volumes (up to several million cubic meters) and the non-steady-state hydraulic regime (Figures. 17.1 and 17.2) introduce several differences in both the hydrology and the biological community.

These reservoirs perform simultaneously as storage volumes for better flow management, and as deep stabilisation ponds. They have a high treatment capacity for both high-rate decaying pollutants such as pathogens, and low-rate decaying pollutants such as refractory compounds. Reservoirs are a key component in almost any wastewater reuse project.

Juanico and Dor (1999) have edited a major review on wastewater reservoirs. For further detail on the microbiology of these systems refer to Chapter 2, and to Chapter 6 for discussion on pathogen removal.

17.1.2 Water demand for irrigation and the hydrological cycle

The Mediterranean climate has a rainy winter and a dry summer. Thus, seasonal storage of wastewater in reservoirs is imperative if wastewater is to be reused in irrigation. The seasonal operation of the irrigation determines the annual cycle of the reservoirs: filling during the winter, emptying during the summer (Figure 17.1).

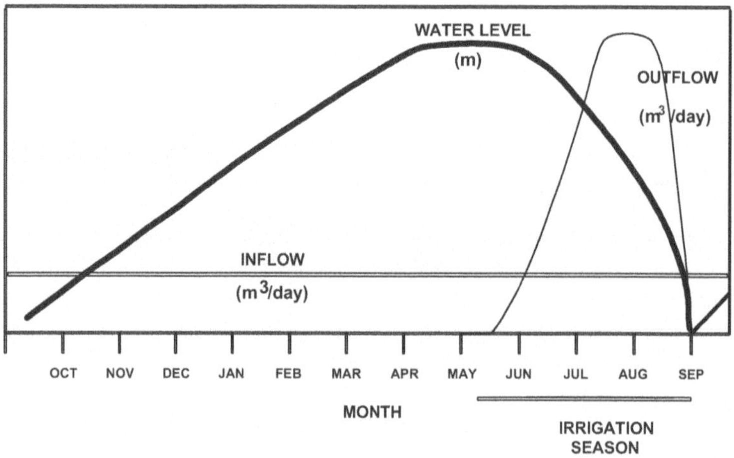

Figure 17.1 Typical annual cycle of a continuous-flow wastewater reservoir in Israel (rainy winter – dry summer)

17.1.3 Basic concepts in designing wastewater reservoirs

There are two main differences between the design of wastewater reservoirs and the design of classic wastewater treatment units such as waste stabilisation ponds or activated sludge, as discussed in the following sections:

Non-steady-state reactors

Most biological treatment units are steady-state reactors with a fixed volume: continuous flow sewage enters and leaves the reactor every day. There are some variations in the quantity and quality of sewage that enters the reactor from day to day, but the overall performance of the reactor can be analysed by assuming steady-state conditions. The assumption of steady-state conditions allows the use of simple analytical solutions for kinetic rates in the process design of the reactor. On the contrary, wastewater reservoirs share with sequential batch reactors (e.g., SBR activated sludge) the category of non-steady-state reactors whose process design requires special equations that cannot be solved by analytical procedures. Modelling and numerical solutions are required (Juanico and Friedler, 1994).

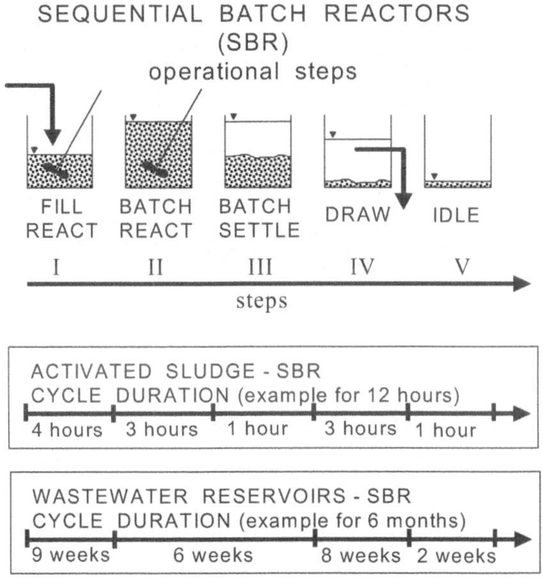

Figure 17.2 Operational cycle of sequential batch activated sludge and sequential batch wastewater reservoirs

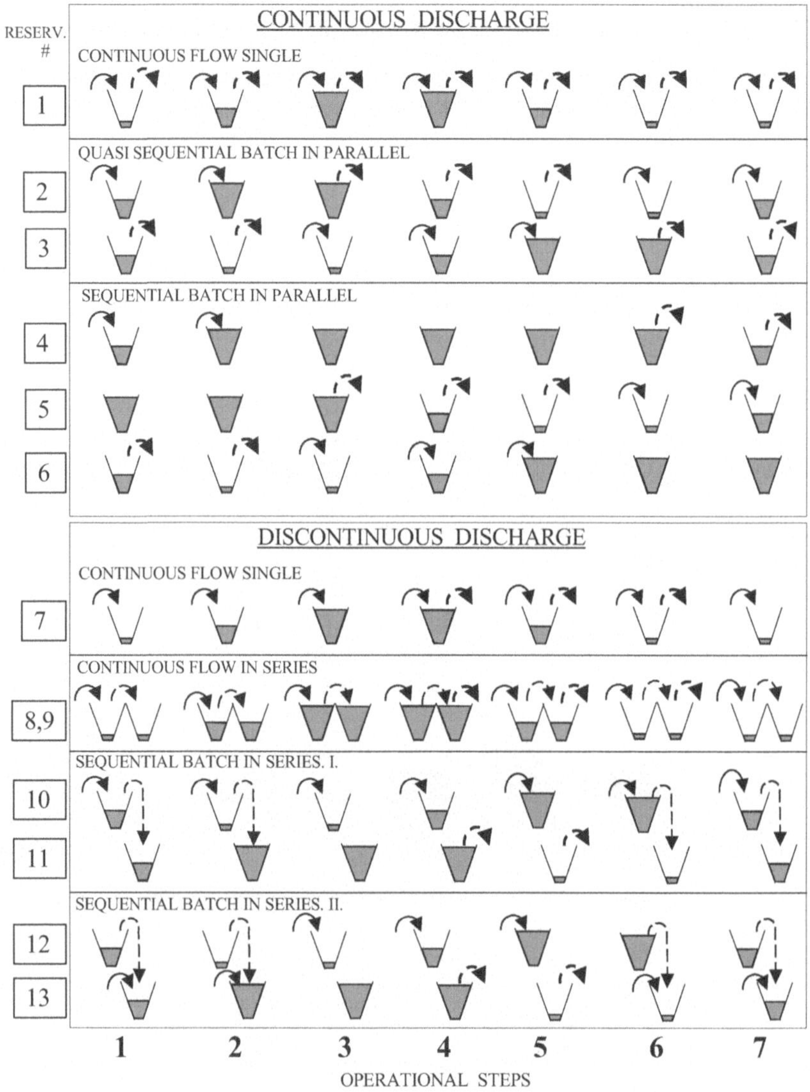

Figure 17.3 Some (7) basic operational regimes of wastewater reservoirs. Inflow and outflow rates (plain and dotted arrows) are not necessarily equal

Together with waste stabilisation ponds and constructed wetlands, wastewater reservoirs may be referred to as 'natural' wastewater treatment systems. All these systems are large, but wastewater reservoirs are especially large with volumes from 50,000 m^3 to 12 million m^3 (typical values in Israel range between half to two million m^3). As a consequence of their large size, reservoirs behave not only as chemical reactors but also as limnological units like lakes, water supply reservoirs and other large water bodies. Some of the limnological elements which affect the behaviour of the reservoirs are evaporation, solar radiation, stratification (Juanico 1994c), winds, waves, currents, tides, settling of particles and living organisms (Avnimelech and Wodka, 1988; Avnimelech, 1989), succession of plankton communities (Eren, 1978), and so on. Water quality management in 'limnological units' is based on different concepts and tools than the ones used for the control of chemical reactors; the designer must take this point into consideration.

17.2 OPERATIONAL REGIMES AND WATER DEMAND

Figure 17.3 describes some basic operational regimes combining different input and discharge possibilities. Continuous-flow (or continuous input) regimes (reservoirs # 1,7,8,9) receive wastewater all the year round.

Sequential batch regimes are operated in such a way that only one of the multiple reservoirs receives wastewater while any reservoir which is discharging effluent stops receiving wastewater before its outlet is opened. In the quasi-sequential batch regime (reservoirs # 2,3) the input of wastewater to the reservoir is stopped simultaneously with the opening of the outlet. In the sequential batch in series regimes, the input may always enter the same reservoir of the series (reservoir # 10) or alternatively to both of them (reservoirs # 12,13).

Different discharge curves result from different situations (Figure 17.4). For example, horticulture irrigation in arid zones may have an almost homogeneous demand for water all the year round. In semi-arid regions (such as those of Mediterranean climate) golf courses (shallow roots) may demand irrigation all the year round but much more intensively during the summer, while some fruit trees (deeper roots) may demand irrigation only during spring-summer-fall but not in winter. Under the same conditions, quick growing crops such as cotton may need large amounts of water concentrated during a short irrigation season of 3-4 months. Wastewater storage reservoirs constructed to avoid the discharge of effluents to the sea, lakes or rivers during some periods of the year, may also have a short discharge season such as that of cotton irrigation. The operational regime and active volume of wastewater reservoirs is a function of the requirements of the water discharge curves.

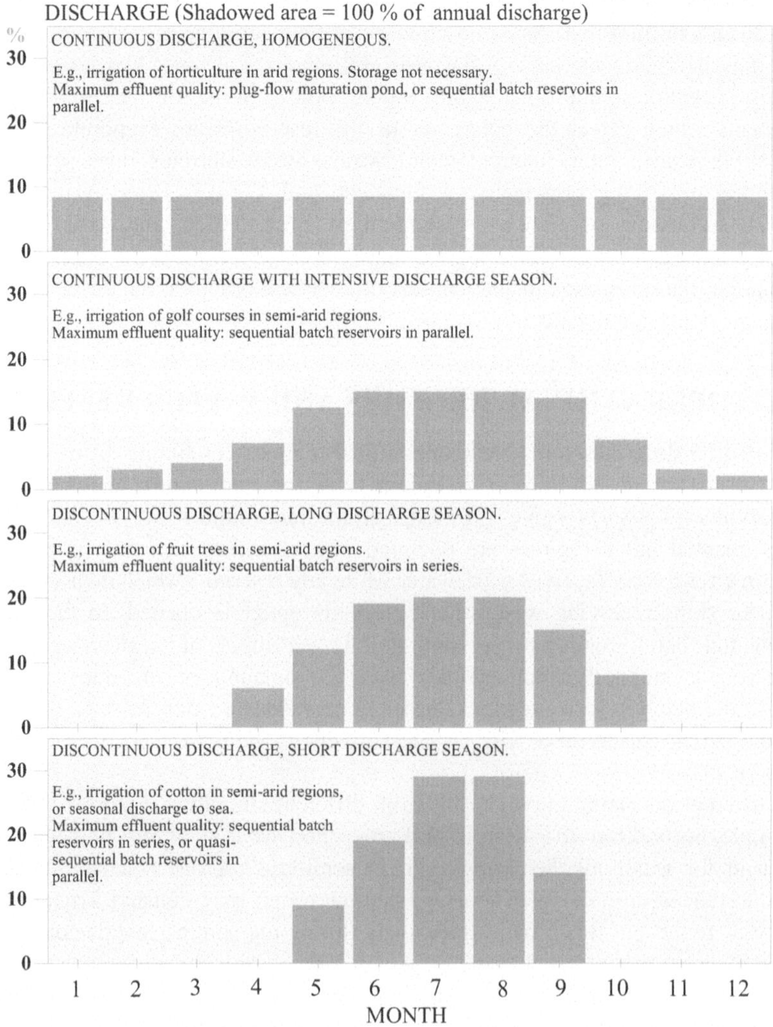

Figure 17.4 Typical effluent discharge curves from wastewater reservoirs and the operational regime that will render maximum effluent quality for each discharge curve

17.3 THE 'OLD' CONTINUOUS-FLOW SINGLE RESERVOIR

Most of the reservoirs constructed in Israel during the seventies (Pano, 1975; Berend and Pano, 1976) and eighties followed the operational regime # 1 (see Figure 17.3). A single reservoir was constructed, wastewater entered the reservoir all the year round, and wastewater was released from the reservoir only during the cotton irrigation season (generally May to September).

The criteria to design these reservoirs are empirical and were developed by both researchers and farmers, by trial and error and common sense.

17.3.1 Volume and depth

The 'active' volume of the reservoir is computed by making a balance between the gains (inflow and rain) and the losses (evaporation and seepage) during the non-irrigation season. A 'dead' volume of about 1 m depth must be added to account for the effluents which remain in the reservoir at the end of the irrigation season due to the impossibility of pumping them out without dragging the bottom sediments. A freeboard (or 'dry' volume) is also needed.

Some reservoirs may receive floodwaters as well as wastewater. If the determination of the active volume is based on the average flow of these water sources, it will happen that in rainy years the reservoir will be full by the end of the winter, before the beginning of the irrigation season. As a consequence, the operator will be compelled to release some effluents to the closest water body. If the goal of the project is that all the wastewater goes to irrigation with no release to water bodies, then the calculation of the active volume must be based on maximum flows and not on the average ones.

Deep reservoirs with a small area/volume relationship are recommended in semi-arid regions where evaporation losses may account for more than 15% of inputs to the reservoir, and as a consequence, increase the salinity of the remaining water (Meron and Eren, 1985). Reservoirs with an active depth of 7 to 15 m are normally used in Israel. Deeper reservoirs up to 20 m active depth are being designed nowadays. Dissolved oxygen in reservoirs comes from photosynthetic activity of algae and from diffusion of atmospheric oxygen. Thus, deeper reservoirs with a small area/volume ratio have a poorer supply of oxygen than shallow ones. Deep reservoirs require a lower surface organic loading or may require aerators to maintain aerobic conditions.

17.3.2 Outlet and inlet location

In most reservoirs the outlet is suspended from a raft, with its opening about 1 m below the water surface. This ensures the effluent is discharged from the oxygen rich epilimnion, which has better quality than the anaerobic hypolimnion layer.

The outlet must not be located at the leeward side of the reservoir where the wind-induced currents concentrate suspended solids and resuspended pollutants (Figure 17.5).

The inlet should be located as far as possible from the outlet in continuous-flow reservoirs. However, in a sequential batch reservoir a single device may act as both the inlet and outlet because inflow is stopped before outflow starts.

Both field data and simulation modelling show that continuous-flow reservoirs with the inlet located at the bottom of the reservoir perform better because this location improves the oxygen balance of the reservoir and avoids hydraulic short-circuiting between inlet and outlet.

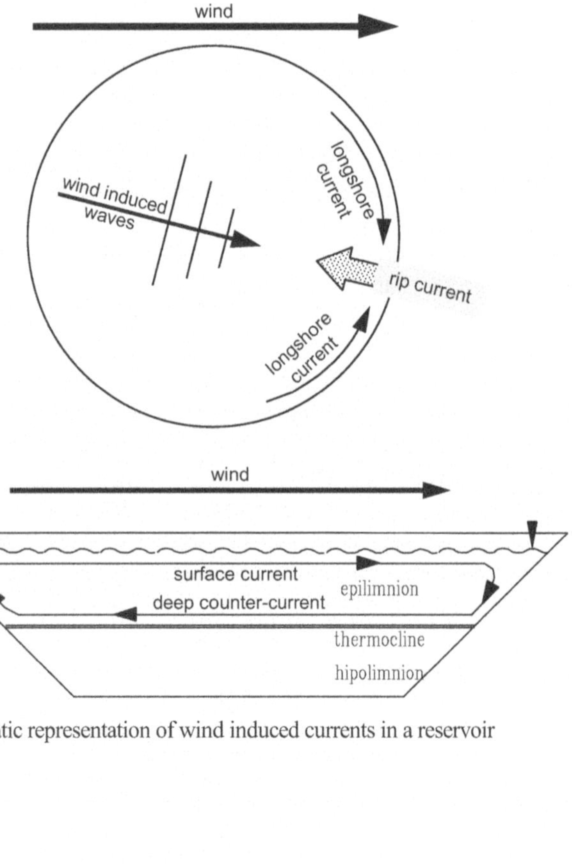

Figure 17.5 Schematic representation of wind induced currents in a reservoir

In well-designed continuous-flow reservoirs the inlet-outlet axis is perpendicular to the direction of dominant winds in order to avoid inlet-outlet hydraulic short-circuiting due to wind induced surface currents and deep counter-currents (Figure 17.5).

17.3.3 The hydraulics of the continuous-flow reservoirs

Waste stabilisation ponds are designed as steady-state flow reactors, with almost constant inflow, outflow, volume and depth. Physico-chemical and biological processes and the resultant pond performance follow an annual cycle related to temperature and/or solar radiation (in tropical areas the performance of the pond is homogeneous during the year). Deep stabilisation ponds also belong to this category (Wachs and Berend, 1968) because they are steady-state reactors in spite of their depth.

On the contrary, continuous-flow stabilisation reservoirs have an annual empty-full-empty cycle (Figure 17.1), the residence time, volumetric hydraulic loading and surface organic loading change during the year, as a function of changes in inflow-outflow rates, water level, area and volume of the reservoir.

The hydraulic mixing pattern of three stabilisation reservoirs has been studied by using a tracer pulse-injection technique while the reservoirs were maintained in steady-state flow: Argaman *et al.*, (1988) used Rhodamine B to study the Northern and Southern Kishon reservoirs (6 million m^3 each), and Moreno *et al.*, (1988) used tritiated water in a 50,000 m^3 reservoir. In both cases it was concluded that reservoirs were perfectly mixed or close to that.

The tools for the analysis of the hydraulic age distribution of effluents within non-steady-state reactors with perfect mixing were developed by Juanico and Friedler (1994) and are now reviewed in the sections below.

The mean residence time (MRT)

MRT of effluents within a steady-state flow reactor (such as a stabilisation pond) is computed as follows under the assumption of perfect mixing:

$$\mathrm{MRT(in days)} = \overline{V} / \overline{Q}$$

where: $\overline{V}$ is mean volume (m^3) and $\overline{Q}$ is mean flow (m^3/day).

This equation is not applicable to a non steady-state flow reactor. Thus, another equation was developed to calculate MRT in stabilisation reservoirs. At the end of a given day d, when there is no outflow from the reservoir, under the assumption of perfect mixing:

$$\mathrm{MRT_d} = \frac{\left[\left(\mathrm{MRT_{d-1}} + 1\right) * \mathrm{VOL_{d-1}}\right] + \left(0.5 * \mathrm{IN_d}\right)}{\mathrm{VOL_{d-1}} + \mathrm{IN_d}}$$

where:

d = end of a given day d
MRT = mean residence time of water within the reservoir (days)
VOL = volume of water in reservoir (m^3)
IN = inflow (m^3) that enters the reservoir during day d.

The MRT of the previous day (d-1) is needed in the equation, so the calculation must be started when the reservoir is empty and $MRT_d = 0$ at the end of the irrigation season of the previous year, and computed forward on a daily basis. As the system is not steady state, daily values of volume and inflow are needed. When outflow occurs, this equation introduces a small error in the calculation of MRT due to the one-day time-step used while changes in inflow, outflow and volume are continuous. The difference between the values of MRT obtained when using a step of one-day or a step of one hour is less than 4 % over a one-year simulation.

The percentage of fresh effluents (PFE)

The PFE refers to the small fraction of recent inflow to the reservoir with a relatively short residence time, expressed as a percentage of the whole volume of the reservoir contents. The percentage of these 'fresh' effluents having less than 30 days within the reservoirs is called PFE30. PFE of effluents having less than 5 days within the reservoir is called PFE5, and so on. The analysis of several reservoirs has indicated that PFE is the parameter which best represents the hydraulic operation of the reservoir, and the main factor affecting the reservoir's behaviour and water quality (see the section on 'performance' below). The concentration of organic matter in the effluents from the reservoir is closely related to PFE30 and PFE10. For example, the BOD in the effluent from a reservoir is affected by the quantity and quality of wastewater introduced in the reservoir during the last 30 or 10 days. On the contrary, the concentration of faecal coliforms in the effluent from a reservoir is more related to PFE5 and PFE1 (the quantity and quality of wastewater introduced to the reservoir during the last 5 or 1 days). The reason for the difference is that the rate of faecal coliforms die-off is much higher than the rate of organic matter degradation. This relationship between hydraulics, reaction rate and the resultant efficiency is discussed further in Chapter 10. Juanico and Friedler (1994) provide a computer algorithm to calculate PFE.

MRT and PFE in continuous-flow reservoirs

Figure 17.6 presents typical values of MRT and PFE30 in continuous-flow reservoirs operated in Central-Northern Israel.

MRT varies from 30-40 days in early winter to 130-160 days in middle summer. PFE values at the end of the irrigation season are four or more times higher than at the beginning. The four reservoirs present very similar MRT and PFE30 values in spite of big differences in volume. These two parameters depend on the relationships between volume, inflow and outflow of the reservoir and not on the absolute values themselves.

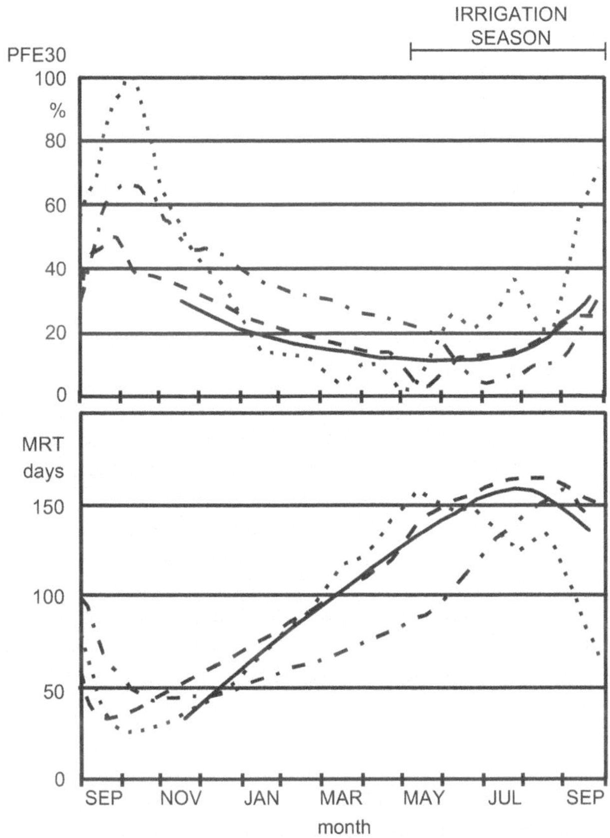

Figure 17.6 The percentage of fresh effluents (PFE30) and mean hydraulic retention time (MRT) during the hydrological year, in four wastewater reservoirs of different sizes (50,000m^3 to 6 million m^3)

17.3.4 Performance of continuous-flow reservoirs

In reservoirs with an irrigation season of 4-5 months, the MRT is 80-130 days; the annual mean BOD removal percentage is 70-85%, COD removal is 50-85%, TSS removal is 40-80% and detergents removal is 50-60%. Removal of N and P in two stabilisation reservoirs in series with a joint annual MRT of 180 days and some batch periods (Kishon North and Kishon South) is 70% and 60% respectively (Table 17.1).

Sedimentation is a major process for the removal of different pollutants. The sludge layer in the bottom of wastewater reservoirs in Israel is typically 20-40 cm high even in those reservoirs receiving huge organic loadings. Organic matter accumulates in the bottom until the degradation rate equals the sedimentation rate, reaching a steady-state after 2-3 years (Avnimelech, 1989). Other pollutants such as heavy metals accumulate indefinitely in the bottom of the reservoir (Juanico *et al.*, 1995).

The main limitation of continuous-flow reservoirs is the decline in effluent quality during the irrigation season (Figure 17.7). The quality of the effluent is optimal at the beginning of the irrigation season when the reservoir contents have a long residence time and the percentage of fresh effluents within the reservoir is minimal. Then the water quality begins to deteriorate as water level drops while new influent continues to enter the reservoir. By the end of the irrigation season when the reservoir is almost empty and 30-50 % of the effluents have 30 days or less within it the effluent quality is at its worse. Figure 17.7 shows that, at this time, the concentration of COD within the reservoir may be even higher than that in the influent entering the reservoir. This is due to the feedback of organics, such as volatile fatty acids, to the water column from the anaerobic decay of the bottom sediments. Other parameters including pathogens behave in the same way compared with COD (Juanico and Shelef 1991, 1994; Liran *et al.*, 1994; Juanico, 1995; Bar-Or and Keshet, 1996; Nature Reserves Agency, 1997). The differences in the quality of the outflow between the beginning and the end of the irrigation season are so conspicuous, that it is impossible to assign the quality of the outflow from a particular reservoir to any standard of water quality for irrigation. Under these conditions, the continuous-flow reservoirs can be considered as storage units only, because their treatment capacity is good in average but so irregular in time.

The maximum removal percentages, which are obtained at the beginning of the irrigation season, reach one order of magnitude for BOD, COD, TSS and detergents, and 3-4 orders of magnitude for coliforms. These maximum removal percentages are much higher than the mean ones.

Table 17.1 Removal efficiencies of continuous-flow and batch wastewater reservoirs. All results are from real scale reservoirs except those quoted as "experimental"

Parameter	Continuous-flow	Batch 30-50 days	Sources
BOD	70 %	90 %	Juanico&Shelef (1991) Soler *et al.* (1991)
COD	50 %	80 - 90 %	Juanico&Shelef (1994)
MBAS (detergents)	50 %	90 %	Juanico&Shelef (1991) Juanico&Shelef (1994)
Nitrogen		70 % – 80 % (1)	Juanico (1999)
Phosphorus	< 30 %	60 % – 85 % (1)	Avnimelech (1999) Bahri *et al.* (2000)
		10 - 30 %	Sala *et al.* (1994) Araujo *et al.* (2000) (experimental)
Faecal coliforms	90 - 99 %	99.99 % - total	Kott *et al.* (1978) Felgner & Sandring (1983)(experimental) Juanico&Shelef (1991) Juanico&Shelef (1994) Liran *et al.* (1994) Indelicato *et al.* (1996) Athayde *et al.* (2000) (experimental)
Streptococcus and Clostridium		total	Berná *et al.* (1986)
Giardia and Cryptosporidium		99.99 %	Nasser *et al.* (2000)
Polivirus I - Chat		total	Funderburg *et al.* (1978) (experimental)
Nematode eggs		total	Kouraa *et al.* (2002) Barbagallo *et al.* (2002)
Heavy metals	down to background concentration in unpolluted waters	down to background concentration in unpolluted waters	Juanico *et al.* (1995)
Organic micropollutants : -- phthalates -- alkyl phenols -- alkyl benzenes -- hydrocarbons	60 – 75 % (2)		Muszkat (1999)

(1) Data by Juanicó and Avnimelech are from two deep reservoirs in series, operated as continuous-flow reactors but with some short periods of batch operation. Data by Bahri are from shallow reservoirs.

(2) Soils irrigated with effluents from reservoirs did no present accumulation of studied organic micropollutants. Those irrigated with effluents from activated sludge plants presented build-up of some organic micropollutants.

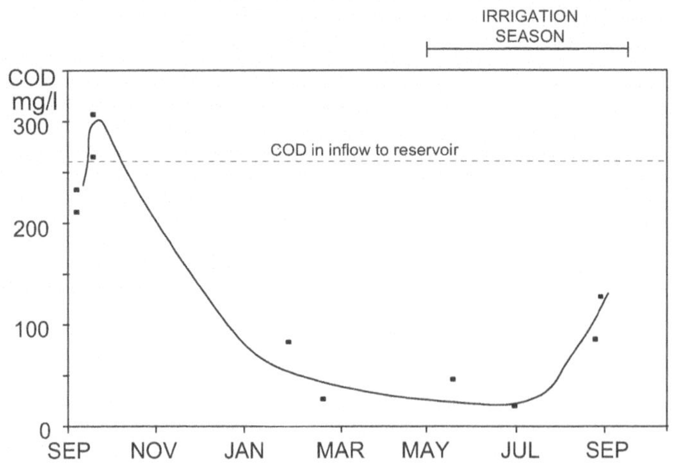

Figure 17.7 The concentration of COD in the epilimnion of the Getaot reservoir during the year

The removal of coliforms

The annual mean removal of coliforms in these reservoirs is only one-two orders of magnitude, a low value for treatment units with so long mean residence time (Dor *et al.*, 1987; Perle, 1988; Liran *et al.*, 1994; Goldberg, 1994; and others).

Fattal *et al.* (1993), Liran *et al.* (1994) and Goldberg (1994) performed independent studies on the parameters affecting the die-off of faecal coliforms and E. coli in wastewater reservoirs, finding that die-off is closely related to photosynthetic activity. Die-off rate is higher in summer than in winter, due to the higher photosynthesis that is a function of solar radiation and water temperatures. These findings agree with those of Curtis *et al.* (1992), who identified the production of singlet oxygen associated with high pH values as main parameters in the die-off of pathogens in stabilisation ponds (see Chapter 3).

Liran *et al.* (1994), found that the removal of faecal coliforms in the reservoir significantly changes throughout the year following the changes of the operational parameters of the reservoir (input/output rates and volume).

There are also sharp differences in coliform removal efficiency between the epilimnion and the hypolimnion layers. Coliform removal is high in the epilimnion of the reservoir, which is characterized by high pH values generated by the photosynthetic consumption of CO_2. When the reservoir is operated as a continuous-flow reactor, the removal of coliforms in the epilimnion is related mainly to the percentage of fresh effluents with five days or less within the reservoir (PFE1 and PFE5). Removal of coliforms is low in the hypolimnion with low pH values and no photosynthesis. When the reservoir is operated as a continuous-flow reactor, the removal of coliforms in the hypolimnion is related mainly to PFE20-PFE30.

The low coliform removal/mean residence time performance ratio of stabilisation reservoirs operated as continuous-flow reactors is due to:

- Their relatively large hypolimnion with no algae activity and low pH values where coliform removal is minimal (this effect can be reduced by mixing the reservoir).
- The drastic increase in the percentage of fresh effluents within the reservoir (PFE) by the end of the irrigation season (the operation of the reservoirs as sequential batch reactors solves this problem).

17.4 THE 'NEW' BATCH RESERVOIRS

17.4.1 A single batch reservoir

Batch reservoirs are not considered merely storage units but an integral part of the sewage treatment system (Juanico, 1994d; Friedler and Juanico, 1996) and, as shown in Table 17.1, can produce better quality effluent than continuous flow systems.

In the batch operational mode, the inflow to the reservoir is stopped before the reservoir starts to release effluents (Figure 17.2). The extent of the filling and batch storage period determines the final quality of the effluents. Batch periods between 30 and 50 days are generally used.

The rate of BOD degradation or faecal coliforms die-off in batch reservoirs is the same as in continuous-flow ones, but the overall removal obtained is much higher because there are no new (fresh) effluents with high concentration of pollutants entering the reservoir (PFE is zero). Several real-scale experiments performed by independent research teams in Israel and abroad indicate that the removal of pollutants with low degradation rates such as BOD, COD and detergents in batch reservoirs is up to one order of magnitude, and the removal of pollutants with high degradation (die-off) rate such as faecal coliforms is up to five orders of magnitude or until pathogens are no longer detectable (Table 17.1). The operation of reservoirs in batch mode also potentially enables the removal of phosphorous by flocculant addition and settling.

17.4.2 Several reservoirs working in sequential batch

When the input to the reservoir is closed, effluents must be stored in another reservoir. This requires more than one reservoir in each system. Optimal number of SBR (sequential batch reservoirs) is 3-4, but 5-6 may be used in some cases. At any given time effluents are taken from only one reservoir while the others are in filling, batch or idle stages. The fill-batch-emptying-idle cycles of the reservoirs are shorter during the high water demand periods and some reservoirs may be emptied twice during the irrigation season.

SBRs require more storage capacity than continuous-flow reservoirs due to the period of no inflow, which must be compensated by supplementary storage capacity in another reservoir. Continuous-flow reservoirs perform the poorest as treatment units, but are relatively smaller in relation to the flow. In addition, their operation is simple and flexible. SBRs render effluents of very good quality, but are larger and their operation is more complicated. An optimisation of the operation of these reservoirs shows that SBR in series perform better when the effluents are released over a period of a few months (Figure 17.3 and 17.4) while SBR in parallel perform better when the discharge season is longer or ongoing throughout the year.

17.5 ORGANIC LOADING

17.5.1 Calculating the surface organic loading

Surface organic loading (expressed as kg BOD/ha/day) is a parameter normally used for the design of stabilisation ponds, which have a constant water surface area. In stabilisation reservoirs the water surface area is not constant but varies with water level during the year. The reduction in water surface area during the irrigation season (a function of water level and the profile of the reservoir) may double the surface organic loading even in situations where the inflow amount and BOD concentration are constant during the whole year, thus, the calculation of surface organic loading in stabilisation reservoirs should be made separately for each day of the year. The annual mean of these daily organic loading values can be then computed for the whole hydrological year.

Another difficulty arising from the non steady-state condition of these systems is that surface organic loading values do not have the same meaning when the reservoir is full and when the reservoir is almost empty and therefore similar surface loadings may correlate to very different volumetric loadings.

17.5.2 Maximum surface organic loading

The relationship between the organic loading of the reservoir and its oxygen regime is more complicated than in conventional waste stabilisation ponds, due to two main factors:

(1) The cumulative effect of the wastewater entering a reactor of changing volume. The impact of the organic load during winter months (low temperatures and solar radiation; a reservoir of increasing volume) compared to summer months (high temperatures and solar radiation; reservoir of decreasing volume), is obviously quite different.

(2) The deeper the reservoir, the lower the surface organic loading that it can support without developing anaerobic conditions. Dissolved oxygen in reservoirs is provided by photosynthetic activity of algae and diffusion of atmospheric oxygen. Thus, deeper reservoirs with a small area/volume ratio have less oxygen input than shallow water bodies. Additionally, deeper reservoirs have a larger anaerobic hypolimnion with its associated high oxygen demand. Deeper reservoirs, therefore, require lower surface organic loading or the use of aerators and/or mixers to maintain aerobic conditions.

Another factor affecting the allowable loading to the reservoir is the quality of the input wastewater. Wastewater with BOD=100 mg/l entering the reservoir from a stabilisation pond imposes less oxygen demand than wastewater with BOD=100 mg/l entering from an activated sludge or aerated lagoons reactor. This is due to the oxygen produced by the algae present in the incoming pond wastewater. This is especially valid in autumn when the reservoir is almost empty.

Thus, the proper determination of the maximum surface organic loading requires simulation of the behaviour of the whole reservoir (Friedler, 1999). However, a 'rule of thumb' may be useful as a first approximation. Engineers in Israel have used an average surface organic loading of 50 kg BOD/ha/day as the maximum allowable loading. A survey by Dor and Raber (1990) on twelve real-scale reservoirs has confirmed that most reservoirs which receive this loading have good performance. Figure 17.8 shows that reservoirs receiving lower surface organic loadings (annual mean up to 30 Kg BOD/ha/day) are fully aerobic or facultative which will prevent any odour problems. However, in the Eliahu reservoir (a 9 m deep reservoir receiving 40 kg BOD/ha/day) dissolved oxygen at surface may drop to zero at night with sporadic emissions of offensive odours.

Reservoirs receiving organic loadings similar to those of facultative stabilisation ponds (150 kg BOD/ha/day) or more are anaerobic most of the time with strong emissions of offensive odours. These findings point out that the value of 50 kg BOD/ha/day may be too high in some cases. A value between 30 and 40 kg BOD/ha/day would be a safer upper limit when the non-emission of offensive odours is imperative. Athayde *et al.* (2000), have proposed a much higher organic loading (about 650 kg BOD/ha/day) for northeast Brazil, but their results are based on experiments run in a tank 15 m diameter and can not be extrapolated to full scale reservoirs.

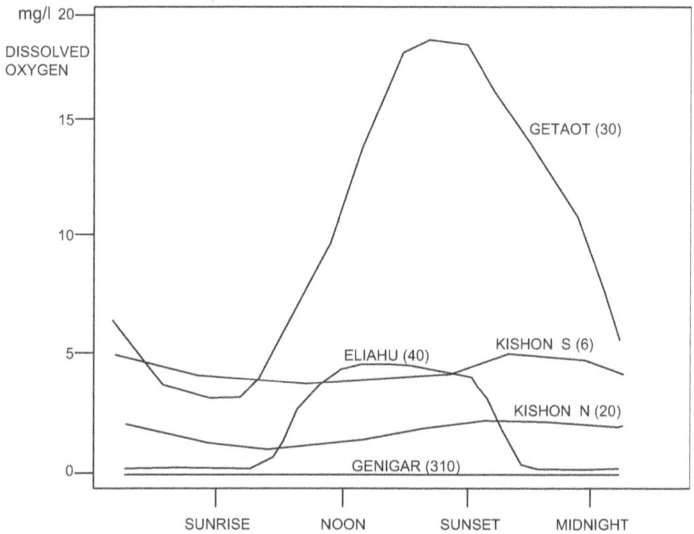

Figure 17.8 The daily cycle of dissolved oxygen in five reservoirs in Israel at 20 cm below the surface, in July (middle summer). The values in parentheses are the annual mean organic surface loading of the reservoirs in kg BOD/ha/day (from Juanico and Shelef, 1991, 1994)

17.5.3 Increasing the surface organic loading

It is possible to increase the surface organic loading on wastewater reservoirs beyond the limits recommended above, by adding mixers and aerators to the reservoirs. Unfortunately, there is no 'rule of thumb' to calculate the relationship between maximum organic loading and the required mixing and aeration power; simulation by modelling in a case-by-case basis is still required.

17.5.4 Shocks of high organic loading

Wastewater reservoirs can receive sporadic shocks of high organic loading without any conspicuous change to their oxygen balance. This is due to the buffering effect the large reservoir volume has. Essentially the reservoir acts like a very large equalisation tank. This capacity is important in the design of a sewage treatment and reuse system, because intensive sewage treatment systems such as activated sludge and others, are known to suffer from periodic process failure with resultant release of low quality effluents. Reservoirs receiving effluent from such a system will 'absorb' these failures without problem. Most sewage treatment and reuse systems in Israel are made of a series of intensive of semi-intensive treatment units such as activated sludge or anaerobic+aerated lagoons, followed by a series of extensive units such as wastewater reservoirs (Juanico, 1994a, 1994b; Aharoni and Kanarek, 1994).

17.6 THE TOOLS FOR DESIGN

17.6.1 Three complementary models

The unsteady-state condition of wastewater reservoirs does not enable the use of simple kinetic models with analytical solutions for process design. Instead, simulation by means of mathematical modelling is required.

Three main models are necessary for the optimal design and operation of wastewater reservoirs:

1. *Hydraulic age distribution*. This model (Juanico and Friedler, 1994) is mainly used for the design of continuous-flow reservoirs. It describes the changes in the hydraulic age distribution under different operational regimes. It allows the quick elimination of design and operational alternatives, which lead to high PFE values in the effluents discharged by the reservoirs.
2. *Ecological simulation*. This model (Friedler, 1999) simulates the behaviour of the reservoir and the quality of the released effluents. It requires time and effort to run and interpret the results, but it allows fine-tuning of different design and operational alternatives and accurately predicts effluent quality.
3. *Multiple-reservoirs operation*. This model (not yet published) simulates the simultaneous operation of several reservoirs in sequential batch mode, and matches the discharge from the reservoirs with the water demand curve. It allows comparison between parallel and in-series alternatives, determines the optimal number of reservoirs, and selects the optimal sequence of fill-batch-emptying-idle cycles.

17.6.2 Prior and post-treatment of effluents

Wastewater reservoirs can be integrated with several different intensive, semi-intensive or extensive sewage treatment units for prior and post-treatment. Figure 17.9 highlights a few of the numerous possibilities.

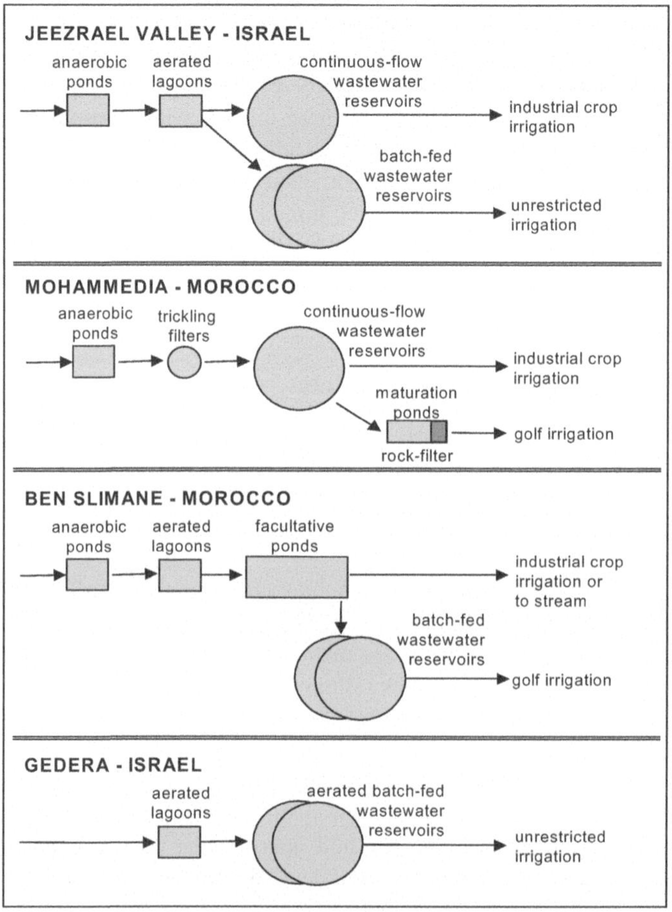

Figure 17.9 Some examples of wastewater reservoirs integrated with semi-intensive sewage treatment units and waste stabilisation ponds

17.7 SUMMARY AND FUTURE RESEARCH NEEDS

Wastewater reservoirs perform simultaneously as storage volumes for better flow management, and as deep stabilisation ponds with a non-steady-state hydraulic regime. They are necessary in wastewater reuse projects, and can also be used advantageously in many other situations when a controlled release of the treated effluents is needed (seasonal discharge to rivers or coastal areas, etc.).

Wastewater reservoirs can be designed as continuous-flow reactors (optimised for water quantity) or as batch reactors (optimised for water quality). There are numerous alternative operational regimes matching different quality requirements and water release schedules.

The removal of pollutants with low degradation rates such as BOD, COD and detergents is 50-70% in continuous-flow reservoirs, and about 90% in batch. The removal of pollutants with high degradation (die-off) rates such as faecal coliforms and other pathogens is about 1-2 orders of magnitude in continuous-flow reservoirs, and up to five orders of magnitude (or until pathogens are no longer detectable) in batch systems. Removal rates for heavy metals and organic micropollutants are also high.

Because wastewater reservoirs are non-steady state reactors, the classic tools developed for steady-state reactors cannot be used to design them. Modelling and numerical solutions are used instead. The reservoirs are big water bodies, and thus limnological concepts and tools also require consideration in their design. Wastewater reservoirs can be combined with different intensive, semi-intensive or extensive sewage treatment units for prior and/or post treatment.

The most urgent needs for future research and development in wastewater reservoirs are:

(1) Removal of nematode eggs. A few available field studies and all theoretical considerations point out that it is as good as removal of other pathogens, but more detailed studies are necessary to give further confidence in this respect.

(2) Nutrient removal. It has been long claimed that the nutrients in the treated wastewater are fertilizers for agriculture and thus must not be removed. This assumption has proved to be false. N and P concentration in treated wastewater can be too high in many cases, leading to problems with crop development and pollution of aquifers and water bodies. The processes involved in the removal of nutrients in the reservoirs are not yet clearly understood. Nitrification/denitrification may be a major path for nitrogen removal but the nitrification step is irregular (Azov and Tregubova, 1995; Abeliovich, 1999). The introduction of floating substrates to allow attachment of nitrifying bacteria and encourage nitrification has been proposed but not studied yet.

(3) Standards for control and monitoring. Some of the available data on wastewater reservoirs are flawed due to sampling mistakes. Some data analyses are also wrong due to misunderstanding of reservoir behaviour. Juanico (1999) proposed a few recommendations to avoid these problems, but a standard procedure to sample and analyze data is needed.

(4) Clogging capacity of effluents. Suspended solids in reservoir effluents (particles, algae, zooplankton) have a high clogging potential of filters and drips, and several recommendations to cope with this problem have been

proposed (Teltsch *et al.*, 1991; Teltsch, 1999; Adin, 1999) but more work on algae and zooplankton control is still needed.

(5) Destratification. Stratification breakdown has been applied for decades as a tool to improve water quality in lakes and water supply reservoirs. All available data and modelling simulation proves that it should also be effective in wastewater reservoirs, but research on this issue has not yet been performed.

REFERENCES

Abeliovich, A. (1999) Nitrogen and nitrification. In *Reservoirs for wastewater storage and reuse* (eds. Juanico and Dor,) pp. 159-172, Springer-Verlag, Germany.

Adin, A. (1999) Particle characterization and filtration. In *Reservoirs for wastewater storage and reuse* (eds. Juanico and Dor,.) pp. 247-262, Springer-Verlag, Germany.

Aharoni, A. and Kanarek, A. (1994) The wastewater reclamation system of Natania. Performance of the Southern Reservoir during the 1993 year. *Water and Irrigation* **338,** 42-45 (in Hebrew).

Araujo, A., de Oliveira, R., Mara, D., Pearson, H. and Silva, S. (2000) Sulphur and phosphorus transformations in wastewater storage and treatment reservoirs in northeast Brazil. *Water Sci. Technol* **42** (10-11), 203-210.

Argaman, Y., Redlich, E., Juanico, M. and Rom D. (1988) Distribution of Residence Time in the Reservoirs. In *The Kishon Complex Monitoring Program*, Fifth Annual Rep., Technion, Haifa, pp. 111-141 (in Hebrew).

Athayde, G., Mara, D., Pearson, H. and Silva, S. (2000) Faecal coliform die-off in wastewater storage and treatment reservoirs. *Water Science and Technology* **42**(10-11), 139-147.

Avnimelech, Y. and Wodka, M. (1988) Accumulation of nutrients in the sediments of Maaleh Hakishon reclaimed effluents reservoir. *Water Research* **22**(11), 1437-1442.

Avnimelech, Y. (1989). Modelling the accumulation of organic matter in the sediments of a newly constructed reservoir. *Water Research* **23**(10), 1327-1329.

Avnimelech, Y. (1999) Sediment-water interrelationships. In *Reservoirs for wastewater storage and reuse* (eds. Juanico and Dor,) pp. 145-152, Springer-Verlag, Germany.

Azov, Y. and Tregubova, T. (1995) Nitrification processes in Stabilization Reservoirs. *Water Sci. Technol.* **31**(12), 313-319.

Bahri, A., Basset, C. and Jrad-Fantar, A. (2000) Agronomic and health aspects of storage ponds located on a golf course irrigated with reclaimed wastewater in Tunisia. *Water Science and Technology* **42**(10-11), 399-406.

Barbagallo, S., Cirelli, G., Consoli, S. and Somma F. (*2002*) Wastewater quality improvement through storage: a case study in Sicily. Pre-prints IWA 3rd World Water Congress, Melbourne, Australia.

Bar-Or, Y. and Keshet, N. (1996) Water quality in wastewater reservoirs. *Water & Irrigation.* **358**, 29-37 (in Hebrew).

Berend, J. and Pano A. (1976) *Winter Storage of Wastewater Effluents.* Tahal Report, Tel Aviv, 50 pp. (in Hebrew).

Berná, L., Torrella, F., Soler, A., Saez, J., Llorens, M. and Martinez, I. (1986) Study of the biological and physico-chemical selfdepuration of wastewater in deep lagoons. *Anales de Biología* **10** (Biología General 2), 49-59 (in Spanish).

Curtis, T., Mara, D., and Silva, S. (1992) Influence of pH, oxygen, and humic substances on ability of sunlight to damage coliforms in waste stabilization pond water. *Applied Envir. Microbiol.* **58**(4), 1335-1343.

Dor, I. and Raber, M. (1990) Deep Wastewater Reservoirs in Israel: Empirical Data for Monitoring and Control. *Water Research* **24**(9), 1077-1084.

Dor, I., Schechter H. and Bromley, H. (1987) Limnology of a hypertrophic reservoir storing wastewater effluent for agriculture at Kibbutz Na'am, Israel. *Hydrobiologia* **150**, 225-241.

Eren, J. (1978) Succession of phyto - and zooplankton in a wastewater storage reservoir. *Verh. Internat. Verein. Limnol.* **20**, 1926 - 1929.

Fattal, B., Puyeski, Y., Eitan, G. and Dor, I. (1993) Removal of indicator microorganisms in wastewater reservoirs in relation to physico-chemical variables. *Water Science and Technology* **27**(7-8), 321-329.

Felgner, G. and Sandring, G. (1983) Wastewater storage - a way to ensure wastewater treatment and utilization over the whole year. *Wasserwirtsch.-Wassertech. WWT* **33**(9), 321-323 (in German).

Friedler, E. (1999) Modelling. In *Reservoirs for wastewater storage and reuse* (eds. Juanico and Dor,) pp. 105-144, Springer-Verlag, Germany.

Friedler, E. and Juanico, M. (1996) Treatment and storage of wastewater for agricultural irrigation. *Water Irrig. Review* **16**(4), 26-30.

Funderburg, S., Moore, B., Sorber, C. and Sagik, B. (1978) Survival of polivirus in model wastewater holding ponds. *Progress Water Technol.* **10**(5-6), 619-629.

Goldberg, T. (1994) Model for the forecasting of die-off rate of E. coli in the wastewater reservoir of the kibbutz Naan. M.Sc. Thesis, The Hebrew University of Jerusalem, 152 pp. (in Hebrew).

Indelicato, S., Barbagallo, S., Cirelli, G. and Zimbone, Z. (1996) Reuse of municipal wastewater for irrigation in Italy. *Procc. 7th. Internat. Conf. Water and Irrigat.*, Tel Aviv, pp. 210-221.

Juanico, M. (1994a) Alternative schemes for municipal sewage treatment and disposal in industrialized countries: Israel as a case study. *Ecol. Eng.* **2**, 101-118.

Juanico, M. (1994b) The role of stabilization reservoirs on sewage treatment systems. *Water and Irrigation* **335**, 19-21 (in Hebrew).

Juanico, M. (1994c) Limnology of a warm hypertrophic wastewater reservoir in Israel. I. The physical environment. *Int. Revue ges. Hydrobiol.* **79**(3), 423-436.

Juanico, M., (1994d) The role of stabilization reservoirs on sewage treatment systems. *Water and Irrigation* **335**, 19-21 (in Hebrew).

Juanico, M. (1995) Limnology of a warm hypertrophic wastewater reservoir in Israel. II. Changes in Water Quality. *Int. Revue ges. Hydrobiol.* **80**(3), 415-428.

Juanico, M. (1999) Process design and operation. In *Reservoirs for wastewater storage and reuse* (eds. Juanico and Dor,) pp. 61-84, Springer-Verlag, Germany.

Juanico, M. and I. Dor (Eds.) (1999) *Reservoirs for wastewater storage and reuse.* Springer-Verlag, Environmental Science Series, Germany, 394 pp.

Juanico, M. and Friedler, E. (1994) Hydraulic age distribution in perfectly mixed non-steady-state reactors. *ASCE J. Environ. Eng.* **120**(6), 1427-1445.

Juanico, M. and Friedler, E. (1999) Experiences outside Israel. In *Reservoirs for wastewater storage and reuse* (eds. Juanico and Dor,) pp. 283-302, Springer-Verlag, Germany.

Juanico M., Ravid R., Azov Y. and Teltsch B. (1995) Removal of Trace Metals from Wastewater during Long-Term Storage in Seasonal Reservoirs. *Wat., Air and Soil Pollut.* **82**, 617-633.

Juanico, M. and Shelef, G. (1991) The Performance of Stabilization Reservoirs as a Function of Design and Operation Parameters. *Water Science and Technology* **23**(7-9), 1509-1516.

Juanico, M. and Shelef, G. (1994) Design, Operation and Performance of Stabilization Reservoirs for Wastewater Irrigation in Israel. *Water Research* **28**(1), 175-186.

Kott, Y., Ben-Ari, H. and Betzer, N. (1978) Lagooned, secondary effluents as water source for extended agricultural purposes. *Wat. Res.* **12**, 1101-1106.

Kouraa, A., Fethi F., Fahde A., Lahlou A. and Ouazzani N. (2002) Reuse of urban wastewater treated by a combined stabilization pond system in Benslimane (Morocco). *Urban Water* **4**, 373-378.

Levenspiel, O. (1972) *Chemical Reaction Engineering*. John Wiley & Sons, 2nd Edition, 578 pp.

Liran, A., Juanico, M. and Shelef, G. (1994). Bacteria Removal in a Stabilization Reservoir for Wastewater Irrigation in Israel. *Water Research* **28**(6), 1305-1314.

Meron, A. and Eren, J. (1985) Effect of Salinity on Agricultural Reclamation. *Proc. Water Reuse Symp.* **III** (1), pp. 543-553.

Moreno, M., Medina, M., Moreno, J., Soler, A. and Saez, J. (1988) Modeling the performance of deep waste stabilization ponds. *Water Resour. Bull.* **24**(2), 377-387.

Muszkat, L. (1999) Degradation of Organosynthetic Pollutants. In *Reservoirs for wastewater storage and reuse* (eds. Juanico and Dor,) pp. 205-218, Springer-Verlag, Germany.

Nasser, A., Greenfeld, S., Molgen, S. and Huberman, Z. (2000) Removal of Giardia & Cryptosporidium from wastewater in activated sludge and stabilization reservoirs. *Water Technologies* **48,** 14-18 (in Hebrew).

Nature Reserves Agency (1997) *Monitoring of Streams, Reservoirs and Wastewater Irrigation*, Report on 1996 Activities, Jerusalem, 90 pp. (in Hebrew).

Pano, A. (1975) *Storage of wastewater in Sarid and Mizra reservoirs.* Tahal Report, Tel Aviv, 26 pp.(in Hebrew).

Perle, M. (1988) The fate of indicator and pathogenic microorganisms in a wastewater renovation system. Master Thesis, Technion, Haifa, 130 pp. (in Hebrew).

Sala, L., Garcia, J., Mujeriego, R. and Hernandez, M. (1994) Phytoplankton studies in hypertrophic lakes used for irrigation. *Verh. Internat. Verein. Limnol.* **25**, 1983-1988.

Soler, A., Saez, J., Llorens, M., Martinez, I., Torrella F. and Berna, L. (1991) Changes in Physico-Chemical Parameters and Photosynthetic Microorganisms in a Deep Wastewater Self-Depuration Lagoon. *Wat. Res.* **25**(6), 689-695.

Teltsch B., Juanico M., Azov Y., Ben Harim I. and Shelef G. (1991) The clogging capacity of reclaimed wastewater: a new quality criterion for drip irrigation. *Wat. Sci. Tech.*, 24(9):123-132.

Teltsch B. (1999) The clogging capacity of effluents. In *Reservoirs for wastewater storage and reuse* (eds. Juanico and Dor,) pp. 233-246, Springer-Verlag, Germany.

Wachs, A. and Berend, A. (1968) Extra Deep Ponds. In *Advanc. Water Quality Improv.*, E. Glyna and W.Eckenfelder (Eds), Univ. Texas Press, pp.450-456.

18

Cold and continental climate ponds

Sonia Heaven and Charles Banks

18.1 INTRODUCTION

18.1.1 History and development

Investigations into the use of waste stabilisation ponds (WSPs) for wastewater treatment in continental climate conditions were carried out at Lyublinsk fields in Moscow from 1913 onwards (Vinberg *et al.*, 1966), but the results were not put into widespread application. In North America, the introduction of treatment ponds began in the early 1940s. Originally these were simple storage lagoons that held wastewater until self-purification made it fit to discharge into the natural environment. It was soon realised, however, that biological purification processes were occurring in these ponds even in the winter months, and that the processes accelerated through the spring and into the summer period. This realisation and the empirical development of designs to improve treatment efficiency led to a dramatic growth in the use of WSPs in the 1950s-60s. By 1990 approximately 80% of municipally owned treatment systems in Alberta

ISBN: 1843390205. Published by IWA Publishing, London, UK.

and Saskatchewan (504 in number) were engineer-designed waste stabilisation pond systems (Prince *et al.*, 1994). In Canada and Alaska there are more than 1000 WSP systems for domestic/industrial wastewater treatment, representing at least 50% of the total treatment capacity in this region and in many cases outperforming conventional plant. The majority (70%) of these systems serve populations of less than 1000. They have also been used for large communities such as Winnipeg, Canada, although this is less common partly because of the large land areas required. Use of WSP systems in extreme climates elsewhere is less widespread, but examples exist in all the main low temperature regions (Pearson *et al.*, 1987, Juanico *et al.*, 2000, Heaven *et al.*, 2003).

18.1.2 Pond types and operating mode

The pond types used in extreme climates are essentially the same as in temperate or tropical areas: anaerobic, facultative and maturation/storage ponds. The main difference is in the discharge mode. Most warm climate ponds operate as continuous discharge systems, where treated effluent discharges into a watercourse at a rate dependent on the inflow. In cold and extreme climates intermittent discharge systems are more common, in which the wastewater is retained for long periods and is only released once or twice a year, usually in spring and/or autumn. The long retention time is based on the fact that in winter the degree of treatment and the capacity of the receiving watercourse are sharply reduced, and ice cover may make discharge impossible. A third category of total containment ponds exists, in areas where evaporation is greater than precipitation: these do not discharge to a water body at all. Suitable conditions for these ponds often arise in continental climates with their warm summers and cold winters, especially in arid regions: but evaporation of the water in this case may represent loss of a valuable resource. Containment or evaporation ponds are relatively rare in North America and northern Europe but are common in the former Soviet Union, especially for industrial enterprises.

A typical extreme climate engineered WSP system consists of all three types of pond: depending on the flow and loading, there may be multiple examples of each type, working in series or in parallel. Figure 18.1 shows the sequence of events in a one-year retention WSP system including anaerobic, facultative and maturation/storage ponds. There is a single batch discharge in the autumn, although continuous discharge through the summer is also possible if the effluent quality is satisfactory. The diagram assumes that at the end of the summer period the soluble BOD in the maturation/storage pond is at a very low level (a reasonable assumption considering the extended storage under oxygen-enriched conditions at warm temperatures). When the temperature falls in autumn, algae in both the facultative and maturation/storage ponds begin to settle, leaving a water body that is free of suspended solids (Prince *et al.*, 1995b). At this point the water in the maturation/storage pond is decanted leaving the algal/bacterial sludge and a minimum volume of bottom waters, to maximise storage capacity. In some cases part of the water in the facultative pond is also decanted, but this is not common.

During the winter wastewater continues to flow through the anaerobic cells, where solids settle out, into the facultative pond, where ice cover exists and water temperatures are around 0-4°C. The treated water from the facultative pond is diluted and washed out into the maturation/storage pond.

By spring the contents of the facultative pond are more or less equal in quality to untreated settled sewage, given the minimal rates of biological decomposition under ice cover. The water in the maturation/storage pond is similar to partially diluted sewage. Hence in spring when the ice melts, the organic load in the system is at its maximum and dissolved oxygen levels are at a minimum. It is at this point that the system can become odorous.

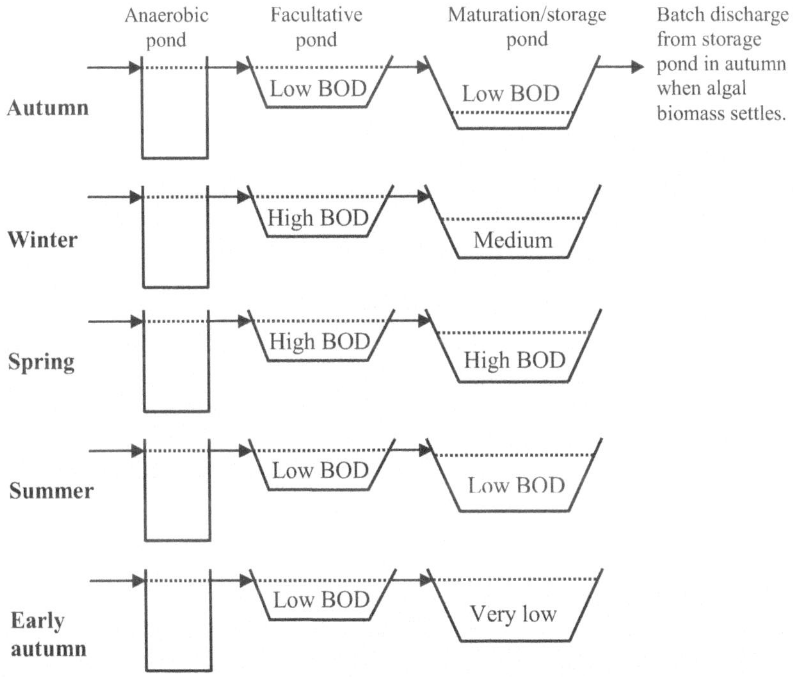

Figure 18.1 Schematic of typical operational pattern for an annually discharged WSP

During the spring 'warm up' period the BOD in both the facultative and maturation/storage ponds will start to be reduced as a result of aerobic heterotrophic microbial utilisation of the dissolved organic matter. Oxygen for this will be supplied by the population explosion of algae that grow photosynthetically utilising macronutrients and inorganic carbon sources. During this time the facultative pond will continue to receive settled wastewater whilst the maturation/storage pond will receive a reduced organic load due to its influent having been 'pre-treated' in the facultative pond. In this spring period, both the facultative and maturation/storage ponds will be operating in 'facultative mode', with an excess of soluble organic carbon and nutrients. As spring turns to summer the depth of water in the maturation/storage pond will increase beyond the optimum for a facultative pond. By this time, however, the organic load on it will be very low, and it will start to function as a maturation pond reducing soluble BOD to low levels. Depending upon temperature, initial organic load and other factors both the facultative and maturation/storage ponds should reach a steady state by early summer and continue to operate like conventional WSPs through the summer period. By the end of the summer the soluble BOD in the maturation/storage pond will be very low (e.g. 5-15 mg l^{-1}) allowing a high quality autumn discharge, while that in the facultative pond will also be substantially reduced (e.g. to around 30 mg l^{-1}).

Figure 18.2 qualitatively depicts the fate of key parameters in the facultative pond during the 12-month period. These are descriptively divided into three phases: 'accumulation' where BOD load is added but not reduced; 'non steady state' where BOD load is progressively reduced as temperature and microbial numbers (algal and bacterial) increase; and 'steady state' where BOD is consistently low, microbial numbers are in equilibrium, and summer temperatures are more or less constant.

The critical phase is during the 'non steady state' period as the initial organic load is high, the algal oxygen production potential is at its lowest, and the potential for bacterial growth (aerobically or anaerobically) is at its maximum. If the initial load is such that the rate of oxygen depletion is greater than the rate of replenishment then the system will be predominantly anaerobic, leading to lower rates of reaction and potentially to odours. If the initial load and the continuing input are low then the rate of oxygen replenishment may exceed the oxygen demand and the system will become oxygen saturated. In terms of treatment this latter state does no harm, but it does represent process inefficiency, as the system is larger than it needs to be. A well-designed pond system should therefore aim to balance the oxygen input in relation to the oxygen demand exerted by the system. In other words, exactly the same design basis should be adopted for WSPs as for conventional mechanical and forced aeration aerobic biological treatment plant.

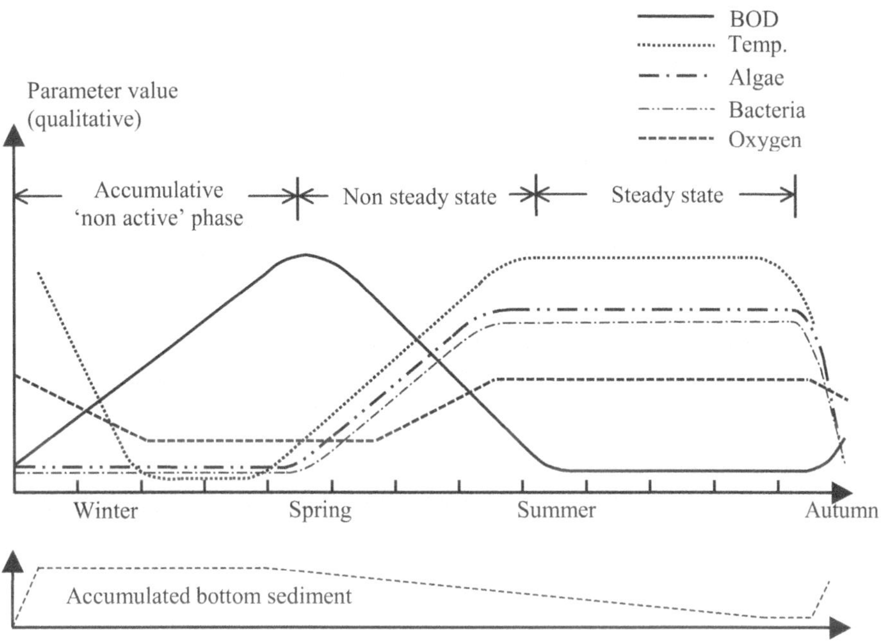

Figure 18.2 Seasonal fluctuations in key parameters in a cold climate facultative pond working in batch mode with a once-yearly discharge

18.2 PROCESS DESIGN

A range of design guides and standards are available for extreme climate WSPs. A key document is the US Environmental Protection Agency's design manual for municipal wastewater stabilization ponds (USEPA, 1983). In Canada, design standards are not consolidated in a single document, but central and provincial authorities have published a wealth of guidance material (Environment-Canada, 1985, 1987, Heinke and Smith, 1988, Heinke *et al.*, 1988, NovaTec, 1996, Saskatchewan, 1996, Newfoundland, 2002). Design and construction in the former Soviet Union is regulated by documents known as the Construction Norms and Regulations (SNiPs). WSP design is covered in SNiP 2.04.03-85 on Water Drainage: External Networks and Structures (SNiP, 1996). Few recent guidelines are available from northern Europe, reflecting the decreasing number of unmodified WSP systems operating in the region (Ødegaard *et al.*, 1987).

18.2.1 Design methods and equations

Various analytical and empirical design methods have been suggested for cold climate facultative ponds. The US EPA manual gives sample calculations at low temperatures based on five design methods: Areal Loading Rate, Gloyna, Marais-Shaw, Plug Flow, and Wehner-Wilhelm (USEPA, 1983). Additional methods have been proposed by Thirumurthi and others (Thirumurthi, 1974, Smith and Finch, 1983, Environment-Canada, 1987). Many of these are not specifically applicable to intermittent discharge ponds or low temperatures, however. Smith and Finch (1983) discuss design and modelling approaches, and conclude that no one method is adequate and simpler methods based on retention time are likely to perform as well as others. Middlebrooks (1987) compared various rational and empirical methods with actual performance data including data from cold or continental climate areas; and concluded that surface loading rates produced the best results, with plug flow the best of the rational methods. Arceivala, 1998, offers a rational method of calculating algal oxygen production per hectare (and hence permissible BOD loading) based on the sky clearance factor and visible radiation at different latitudes, with values up to 50° N. A number of attempts have been made to model processes within the pond system, as a basis for rational design (Moreno-Grau *et al.*, 1996, Kayombo *et al.*, 2000, Giraldo and Garzon, 2002, Banks *et al.*, 2003).

In practice, however, the majority of currently used design methods are based on plug flow with a first order reaction rate, or on a combination of surface loading rate with hydraulic retention time (HRT) or depth. The US EPA design methods for areal loading and plug flow are summarised in Table 18.1. Suggested values for loading rates, pond depths and retention times are available from a wide range of sources. There seems to be a consensus that loading rates of 11-22kg BOD ha^{-1} day^{-1} are suitable for intermittent discharge ponds in cold regions, and that loading on the first pond should not exceed 40 kg BOD ha^{-1} day^{-1} to prevent odours (Dawson and Grainge, 1969, USEPA, 1983, NovaTec, 1996). Recommended depths vary depending on severity of the climate: typical working depths of 1-1.5m for facultative and maturation ponds in temperate climates may be increased to 2-3 m allow for ice cover and sludge accumulation (see Section 18.3 below).

The plug flow reaction rate constant varies depending on the BOD loading and the temperature. Values and adjustments widely used in North America are shown in Table 18.1. Reaction rate constants and adjustments used in the former Soviet Union differ from those in the west, particularly at low temperatures (Alferova *et al.*, 1973, Samokhin, 1981). SNiP 2.04.03-85 (SNiP, 1996) requires design to be based on plug flow, and states that WSPs can be used for full or tertiary treatment, but there appear to be inconsistencies: in particular, the very limited depths and BOD loadings specified suggest it is mainly intended to apply to high-rate ponds or to tertiary treatment (Heaven *et al.*, 2003).

Table 18.1 Areal Loading Rate and Plug Flow design methods (based on USEPA, 1983)

Areal Loading Rate method for average winter air temperature < 0 °C

BOD_5 loading on whole system limited to 11-22 kg ha^{-1} day^{-1}

BOD_5 loading on first pond limited to 40 kg ha^{-1} day^{-1}

HRT = 120-180 days, depending on period of ice cover and discharge conditions (in practice once-per-year discharge systems are often recommended)

Plug Flow method

$$\frac{C_e}{C_o} = e^{-k_p t} \quad \text{(equation 18.1)}$$

where
C_e = effluent BOD_5, mg l^{-1} and C_o = influent BOD_5, mg l^{-1}
k_p = plug flow 1st-order reaction rate, day^{-1}
t = HRT, days

BOD_5 loading on first pond limited to 40 kg ha^{-1} day^{-1}

Temperature adjustment $k_{pT} = k_{p20}(1.09)^{T-20}$ (equation 18.2)

where
T = minimum operating water temperature °C

k_p varies with BOD loading rate as shown:

BOD_5 kg ha^{-1} day^{-1}	22	45	67	90	112
K_{p20} day^{-1}	0.045	0.071	0.083	0.096	0.129

Design of anaerobic ponds is usually based on hydraulic retention time and depth, as in warm climates, although surface or volumetric loading rates are sometimes quoted (Gray, 1999). Dawson and Grainge (1969) suggested depths of around 3-7.5m for short-retention anaerobic ponds in northern latitudes, to conserve heat and allow for sludge accumulation. More recently, Environment Canada recommended depths of 3-5m and a minimum HRT of 2-5 days (NovaTec, 1996). Abdrashitova *et al.* (2001) suggest the anaerobic pond should be designed to balance the rates of accumulation and digestion of organic solids. In order to reduce the frequency of de-sludging, in future this may lead to a rational basis for the sizing of anaerobic ponds, based on factors such as the temperature and duration of the warmer months.

Design of the maturation/storage pond for an intermittent discharge system is also based on HRT, determined by climatic conditions (e.g. the period of ice cover) and the required frequency of discharge (see Section 18.4.2 below). The most common values are 6 and 12 months, sometimes with a safety margin of 2-3 months to allow for conditions such as volume of flow, BOD concentration, or break-up of ice cover in the receiving watercourse. The maximum recommended working depth of maturation/storage ponds is around 2.5m (Heinke *et al.*, 1991, Prince *et al.*, 1994). This is greater than for facultative ponds, since storage rather than treatment is the main purpose (Prince *et al.*, 1995b), and since for much of the year this type of pond contains treated water with a low BOD.

18.3 SPECIAL ASPECTS OF CONSTRUCTION

18.3.1 Configuration and orientation

Prince *et al.*, 1994, 1995a, b, carried out a major review of the effect of system configuration on performance of cold climate ponds, based on data from 190 WSPs in Alberta, Canada. It was concluded that the most robust system consisted of four anaerobic ponds, one facultative pond and one maturation/storage pond with 12 months' storage capacity. This system is often described as 4S, 1T, 1L, using a notation where S refers to sedimentation, T to treatment and L to storage by lagooning, and is widely considered as a standard design in North America. Heinke *et al.*, 1991, give design criteria and a flowchart for selection of constructed ponds and systems making use of natural lakes. With respect to the dimensions of individual ponds, there is a sharp division between practice in the west and in the former Soviet Union (Heaven *et al.*, 2003). SNiP 2.04.03-85 requires a length:width ratio of 20:1 or more for naturally aerated ponds. Much smaller ratios are accepted in North America, with 3:1 considered as adequate to promote good plug flow (Environment-Canada, 1985, NovaTec, 1996).

Ponds in extreme climates need additional volume to accommodate ice formation. Design values for ice thickness can be calculated from reference sources (Smith, 1996), or obtained from typical values for the region (Environment-Canada, 1985, SNiP, 1996). Ice depths of over 2 m can occur in ponds in northern Canada; thicknesses in WSPs are usually less than in natural water bodies, however, because of the warm influent (Hanæus, 1991b).

Like WSPs in warm and temperate climates, it is recommended that cold climate ponds be sited downwind of inhabited areas; in cold regions the risk of odour release is highest on thawing, so the direction of the prevailing wind in spring should be considered. Recommended widths of buffer zones range from 30 to 300 m for different purposes and in different jurisdictions (NovaTec, 1996, SNiP, 1996).

Winter wind direction and the location of physical barriers such as fencing around the site are also important factors with respect to snow drifting. A layer of snow improves insulation, reduces ice thickness and may therefore lead to an earlier spring warm-up; but cuts down light, reduces disinfection and prevents photosynthetic aeration, which can occur even under ice and may contribute to odour reduction (Environment-Canada, 1987, Heinke *et al.*, 1991). One investigation found 30% of light penetrated a layer of ice 350mm thick, whereas only 0.3% passed through 100mm of snow (Environment-Canada, 1985). Different approaches to snow accumulation may be appropriate in different locations. In WSPs located in the far north, for example, the amount of light available in winter is limited and there may be greater benefits from promoting snow cover. In more southerly continental climates, the temperature may briefly rise above freezing in the middle of the day, allowing the surface to thaw and re-freeze and produce very transparent ice. In arid regions such as central Asia, there is relatively little snow, and clear skies and bright sunshine are common in winter, accompanied by periodic strong winds. Under these conditions light penetration can be significant and it may be better to design layout to minimise cover by drifting snow. Much more work is needed, however, to establish the potential influence of layout and operating practice on snow cover and pond performance.

18.3.2 Earthworks and lining

Extensive work has been carried out to establish guidelines for construction in permafrost regions (Environment-Canada, 1985, Smith, 1996). WSPs are particularly sensitive, as storage of large volumes of relatively warm wastewater can cause thawing after construction, leading to catastrophic embankment or liner failures. Damage to liners can also be caused by 'ice rafting', where wind-blown ice rides up the side of the pond, and side slopes of 4:1 are recommended to prevent this (Heinke *et al.*, 1991). Merrill and Stephl (1996) describe a case of liner failure in southern Alaska, to which ice rafting may have contributed. Embankment design may need to provide truck access not only for normal maintenance and de-sludging, but also for discharge of trucked wastes such as septic tank sludges, which are often added to WSPs in remote northern areas with scattered populations.

18.3.3 Hydraulics and pipework

The Cold Regions Utilities Monograph (1996) suggests design details for a wide range of pipework and valving systems appropriate to cold climate ponds, including control structures and pond connections. Overflow pipes are particularly problematic, because of their high level: ice may form in or around the pipe and then lift it as the pond level rises. It is therefore good practice to build the pipe level with the berm (NovaTec, 1996). Heinke *et al.* (1991) recommend large submerged inlet pipes to avoid inlet freezing. In cold conditions a submerged inlet may also promote mixing and prevent short-circuiting by formation of a layer of warmer influent above the pond contents.

Ideally pipework and valving arrangements should allow series or parallel operation of ponds, with the option of isolating any section (USEPA, 1983, SNiP, 1996). This is particularly useful in cold climate ponds because of the potential for using empty cells to dewater sludges (see Section 18.4.1.below).

In sharply continental climates, the spring warm-up can occur very rapidly. This may impose significant additional hydraulic and/or BOD loads, due to the volume of melt water and the effect of the spring turnover. A common solution in Scandinavia is to install adjustable outlet weirs, so that the water volume remains constant (Hanæus, 1991b). The use of baffles within a pond can help to prevent stratification and reduce short-circuiting, but design of membrane baffle systems is difficult due to ice loading.

Heinke *et al.* (1991) recommend that the maturation/storage pond outlet pipe is 0.5 m above the liner to provide a buffer layer of water that will prevent immediate freezing of the pond bottom after an autumn discharge.

18.4 OPERATION OF EXTREME CLIMATE PONDS

18.4.1 Sludge accumulation and disposal

Rates of sludge accumulation in all pond types are greater in cold climates and under winter conditions, due to the reduced rate of solids digestion at low temperatures. Typically sludge accumulates during the winter period, and may then decrease during the summer (see Figure 18.2). Accumulation rates vary depending on a number of factors, including wastewater temperature, degree of insolation, organic loading, and concentration of inorganic solids in the influent.

Middlebrooks *et al.* (1982) quote accumulation rates for facultative ponds in Canada and Alaska ranging from 0.25 - 0.4 m^3 day^{-1} per 1000 population, with higher values at more northerly sites. Fieldwork on facultative ponds in Utah by Schneiter *et al.* (1983) indicated values of 0.7-0.8 cm $year^{-1}$ and 1.1-1.7 kg total solids m^{-2} $year^{-1}$. For anaerobic ponds, the range of volumetric accumulation rates quoted by (Middlebrooks *et al.*, 1982) is similar to facultative ponds, but there is no clear geographical pattern.

A study of the anaerobic ponds in Whitehorse, Yukon, conducted after 7 years of operation found a considerably higher sludge accumulation value of 0.65 m^3 day^{-1} per 1000 population (Allan and Jeffreys, 1987). A summary of results from the 1985 Winnipeg workshop on cold climate WSPs shows a range of 0.17-0.5m^3

day^{-1} per 1000 population (Environment-Canada, 1987). The above figures give a rather wide range, and for the purposes of estimating sludge accumulation it may be simpler to use a value of 35g day^{-1} settleable solids per person, and a solids concentration of 5-9% after settlement and consolidation. Assuming a long-term solids degradation rate of 40% (and no wash-out of settled sludge), this gives an accumulation rate of 0.23-0.42m^3 day^{-1} per 1000 population.

Freezing is well known as an effective means of dewatering sludges. Much of the work carried out has concerned sludges from aerated ponds, chemically-conditioned sludge from precipitation ponds, and septic sludges (e.g. (Hanæus, 1991b, Desjardins and Briere, 1996, Hedstrom and Hanæus, 1999). It can be assumed that the properties of ordinary WSP sludges will not differ significantly from these (Schneiter *et al.*, 1984). Extreme climate options for WSP sludges include the use of special sludge freezing beds (Martel, 1993, Hellstrom, 1997). Alternatively, since winter in cold regions is a season of low flows, one option is partially or completely to empty the first pond in autumn and to stockpile the sludge to allow natural freezing and drainage (Ødegaard *et al.*, 1987). A solids content of 50% is reported to have been obtained using this method (Hanæus, 1987); and see Section 18.7.2 below).

18.4.2 Pond discharge

Both spring and autumn discharges are common practice in intermittent discharge WSP systems. Spring discharge may have the advantage of coinciding with peak flows and maximum dilution in the receiving watercourse, and of allowing a reduction in storage volume if discharges can be made twice a year. Care must be taken, however, to avoid discharging during a spring turnover when the pollutant load will be high. Autumn discharges, on the other hand, can be made when the algal suspended solids are at a minimum and effluent quality is at its peak. In this case, rather than being a problem the discharge can potentially make a useful contribution to maintaining water volume and quality, especially in arid or sharply continental climates where river flows during this period are low. There is a growing consensus, however, that under most conditions an operating regime consisting of a full 12 months of storage and an autumn discharge provides the best results. Extensive surveys carried out by Prince *et al.* (1994, 1995a, b) confirmed that with a single autumn discharge the 4S, 1T, 1L system can out-perform most conventional wastewater treatment plants over a wide range of parameters.

Control of the discharge operation is the most critical factor in performance of extreme climate intermittent-discharge ponds. The US EPA gives detailed guidelines on the requirements for achieving good results (USEPA, 1983). The operator should be provided with discharge instructions and a typical schedule.

Pond and receiving water quality must be carefully assessed before and during the discharge. The normal procedure is to isolate the pond to be discharged and to measure a range of parameters including BOD, suspended solids (SS), volatile suspended solids (VSS), pH and dissolved oxygen (DO). Colour, turbidity, and any unusual factors are also noted. Provided that the effluent meets regulatory standards discharge can begin, and can continue as long as the weather is favourable, DO is near or above saturation, and turbidity satisfactory. During the discharge period samples should be taken three times daily in the receiving water near the outlet, and analysed for DO and SS. A typical operating pattern where there are multiple ponds is to draw down the last two ponds to a depth of 0.45-0.60 m. Once one is empty, discharge is interrupted while flow is diverted into the drawn-down pond and the remaining pond is rested before emptying. Similar guidelines have been developed by a number of Canadian provinces (Heinke and Smith, 1988, Saskatchewan, 1996).

18.4.3 Monitoring and maintenance

The majority of monitoring and maintenance tasks for extreme climate WSPs are similar to those for tropical and temperate ponds. Apart from the desludging and discharge operations described above, the main difference is the risk of odour release in spring, at or around the time of ice break-up. Oleszkiewicz and Sparling (1987) looked at factors affecting odour release, and concluded that surface-loading rate was the main controllable variable. Experiments with over-winter low-intensity aeration of highly loaded ponds indicated that, while the concentration of sulphides in the aerated ponds was reduced, perceived odour was not significantly affected. A natural odour protection mechanism occurs when, as the pond thaws, a layer of meltwater with very low BOD forms above the ice providing a barrier to release of odours. Possible control mechanisms for investigation therefore include supply of clean oxygen-rich water to form a surface layer, addition of chemicals such as sodium or potassium nitrate, or high-rate aeration, though the latter options are likely to be expensive.

Winter monitoring data should include the thickness of ice and the percentage of the surface area frozen. The Winnipeg workshop on cold climate ponds (Environment-Canada, 1987) recommended monitoring DO levels below the ice: although the results cannot be used immediately for odour control, they add to the database of available information. In general very little information is available on the physical, chemical and microbiological processes occurring in cold climate ponds during the winter period, and adequate monitoring is essential to establish better understanding and control. Other areas requiring monitoring are ground temperatures and embankment faces for signs of icing that may indicate seepage or melting of a permafrost ice lens.

18.5 POND MICROBIOLOGY AND PATHOGEN REMOVAL

18.5.1 Pond biology and microbial activity in cold WSPs

There is little evidence to suggest that microorganisms found in cold climate ponds are significantly different from those in other regions, and the common genera of algae observed during the summer period are those typically encountered in freshwater habitats worldwide. Henry and Prasad, 1986, suggested that relative proportions of psychrophilic and mesophilic bacteria may change in the winter and summer months. Detailed studies on the macrobiology of ponds are rare, but any species differences that occur may be expected to reflect the natural distribution of populations of aquatic invertebrates such as fly larvae and worms.

Biologically the most striking feature of an extreme climate pond is the rapid growth and equally rapid death of the population, brought about by transitions in temperature, light intensity and day length. It is clear that formation of ice over the pond surface during winter does not kill all the microorganisms in the pond, and once the ice melts and temperatures increase a rapid increase in numbers of both bacteria and algae follows. This is not only a case of survival at low temperatures: there is also continuing microbial activity within both the ice and the unfrozen water mass. Mackenthun and McNabb (1961) found sub-ice photosynthetic aeration in the Junction City WSP system, with substantial quantities of DO confined to a narrow stratum just below the ice. Concentration increased during daylight hours when the percentage of incident light reached 4-7%. Studies on the ice-covered Siberian Lake Baikal show that convective mixing can occur if the near-ice layer of water warms to the point of maximal density. The effect may be to suspend non-motile phytoplankton in the upper water column, providing cells with enough light for growth during ice-covered periods (Kelley, 1997).

Although there is ample evidence for continuing microbial activity in both ice slush and free water at or near freezing point (e.g. (Felip *et al.*, 1995, McKnight *et al.*, 2000, Phillips and Fawley, 2000), rates of reaction are sub-optimal and only a minimal amount of treatment through biological mechanisms occurs during the winter months. During the spring period as water temperatures increase there is typically an explosion of both photoautotrophic and heterotrophic activity within the pond, followed by a lag period which may be as long as 2 to 3 months before steady state conditions are achieved with cyclic oscillations or succession of secondary feeders (Banks *et al.*, 2002).

18.5.2 Survival of pathogenic micro-organisms

WSPs are quoted as being highly effective in the removal of pathogenic organisms including bacteria, viruses and helminth eggs. The reasons for this are discussed in Chapter 6 and are attributed to a combination of factors such as long retention periods, fluctuations in pH and dissolved oxygen concentration, exposure to UV radiation, and the antagonistic effect produced by certain algal species. In a long retention cold climate WSP these mechanisms are most likely to be effective during the warmer months of the year, when there is no ice cover and microbial population dynamics are similar to those in a conventional WSP. In fact, the total pathogen load discharged is likely to be lower than that from a continuous discharge pond as the likelihood of by-pass through hydraulic short-circuiting is eliminated by the mode of operation. In a once-per-year discharge pond the pathogens enter the environment over a relatively short period, however, and with twice-yearly discharge the springtime pathogen load is likely to be high as some of the destructive mechanisms mentioned above do not function when the pond is ice-covered.

The effect of cold on pathogen survival has been extensively researched due to its importance in food preservation. Extension of the results to WSPs must be treated with caution, because of the very different environment: likewise studies in natural surface and ground waters may not be applicable to wastewater. It is clear, however, that bacterial pathogens such as *Salmonella*, *E.Coli* and *Streptococcus* have adaptation mechanisms to withstand cold shock and sub-optimal temperatures, including freezing, which allow not only survival but also growth (Thieringer *et al.*, 1998, Phadtare *et al.*, 1999, Wouters *et al.*, 1999, Horton *et al.*, 2000). There is also evidence that bacteria such as *Salmonella*, *Shigella* and *E.Coli* that can carry a resistance (R) factor as an extra plasmid in their cells may live longer in wastewater treatment systems (Environment-Canada, 1985).

Viruses can survive for much longer periods and at lower temperatures than bacterial pathogens, and are regularly found in WSP effluents (Environment-Canada, 1985). Viruses can persist and remain infectious at temperatures near freezing for several months: the die-off of Polio 1 and Hepatitis A (HAV) viruses was found to be lower in wastewater at 10°C than at temperatures of 20 and 30°C, and much lower than that of *E.Coli* (Nasser *et al.*, 1993, Nasser and Oman, 1999). It has been suggested that *E Coli* is not a reliable indicator of the presence and persistence of enteric viruses, particularly with decreasing temperature. Castillo and Trumper (1991) showed that the die off rates of coliphages in WSPs did not reflect seasonal climate changes as much as *E.Coli*, and also found a threefold difference in die-off rates between summer and winter for both faecal coliforms (FC) and *Salmonella*. Work by Torrella *et al.*,

2003, confirmed the long-term low-temperature survival potential of bacterial and viral indicator organisms isolated from wastewater, and the importance of wastewater components in enhancing this survival potential.

Data on pathogen survival over the winter period in cold climate WSPs is limited. Vinberg *et al.* (1966) state that the percentage removal of *E.Coli* from ponds in Minsk was 99.99% in summer, 99.9% in autumn and 55% in winter. Prince *et al.*, 1994, looked at faecal coliform data from Alberta for different pond types and discharge periods. The results showed FC numbers in a spring discharge are an order of magnitude greater than for effluents discharged in the autumn. Soniassy and Lemon (1986) looked at total coliform (TC), FC and faecal streptococci (FS) numbers in raw sewage and treated effluent for a WSP at Yellowknife, Canada. Total coliform counts reached a peak in March with numbers around 1×10^6 counts per 100 ml, dropping to 20 counts per 100 ml in August, followed by a very rapid rise between September and December. Similar trends were found with both FC and FS, and the ratio between counts of these two groups remained around 4, indicating similar die-off rates throughout the year. Mackenthun and McNabb (1961) reported Junction City WSP coliform counts of 4 per 100 ml in June to 4.9×10^4 per 100 ml in February.

It is clear that from a public health perspective, in the majority of cases discharge of effluent from a cold climate WSP should not be undertaken in spring. The practice is sometimes justified, however, by the argument that the turbidity of the receiving water at this time of the year is too high, due to the thawing of snow and ice, for it to be considered for drinking water.

18.6 MODIFICATIONS AND TRENDS IN DESIGN OF EXTREME CLIMATE PONDS

As with warm climate WSPs, a number of modifications and additions can be applied to extreme climate ponds. The two most common are artificial aeration and chemical precipitation. Aerated ponds perform satisfactorily in cold climate regions, although with reduced efficiency in the cold season: the main difficulties are with power supplies in remote areas and operation of mechanical plant at low temperatures. Surface aerators are usually unsuitable because of icing (Environment-Canada, 1987). In Scandinavia precipitation ponds using alum, ferric salts or lime have almost entirely replaced conventional WSPs due to the need to protect inland waters from nutrient enrichment (Ødegaard *et al.*, 1987, Hanæus, 1991b). Precipitation ponds are also suitable for small settlements that experience major fluctuations in population during cold periods, such as winter skiing and recreational areas. Hanæus (1991b) suggests that the future for pond systems may lie in systems combining conventional, aerated and precipitation ponds. Browne and Jenssen (2001) report on a system in Norway which uses vertical and horizontal flow constructed wetlands, flowform cascades, an advanced facultative pond and three facultative ponds, and achieves year-round treatment to a high standard including nutrient removal. Macrophyte ponds have been used for some time in central Asia and new work is in progress (Yunusov, 1983, Sarsenbaev and Atabaeva, 2000), but the severe winters in the region present difficulties with the survival of common temperate-climate macrophyte species.

The idea of introducing additional or special microorganisms has been raised at various periods. It was particularly popular in the former Soviet Union (VNII-SIS, 1987, Zhirkov *et al.*, 1987), but no large-scale applications seem to have been reported. More recently references have appeared to the possibility of adding special enzymes to the ponds at Whitehorse, Canada (TaigaNet, 2002).

18.7 CASE STUDIES

The following section contains two case studies, one of a basic layout successfully replaced by a more modern design, the other of a conventional pond system replaced by chemical precipitation technology.

18.7.1 Whitehorse, Yukon - continuing with WSPs

The city of Whitehorse is located in the Yukon Territory of Canada at latitude 60° 43' N. Mean temperatures range from -18.7°C in January to 14°C in July. Per capita wastewater flows are high due to the practice of continuous bleeding from water supply pipes to prevent freezing: this practice is gradually being phased out, but the wastewater remains dilute and cool (Butt and Enns, 2001).

The city's original wastewater treatment plant began operation in 1979. It was designed for a population of 29,700, and at the time of construction served about 10,000 people. The plant consisted of four so-called anaerobic ponds, each with an operating volume of 57,100m^3 and a depth of 6.1 m intended to promote heat conservation and provide sludge storage capacity. The side slopes were 1:4, with the inlet and outlet 60m apart and 4m deep (see Figure 18.3). Design removal values were 60% for BOD_5 and 50% for suspended solids. The ponds were operated in two parallel pairs, giving a retention time of about 20 days. Effluent was discharged continuously into the Yukon River. Discharge conditions based on Yukon Water Board standards allowed an effluent BOD_5 of 45mg l^{-1}, suspended solids of 60mg l^{-1} and a faecal coliform (FC) count up to 200,000 per 100ml. The ponds were monitored over a 2-year period from first filling, to provide data on performance and start-up (Bethell, 1981).

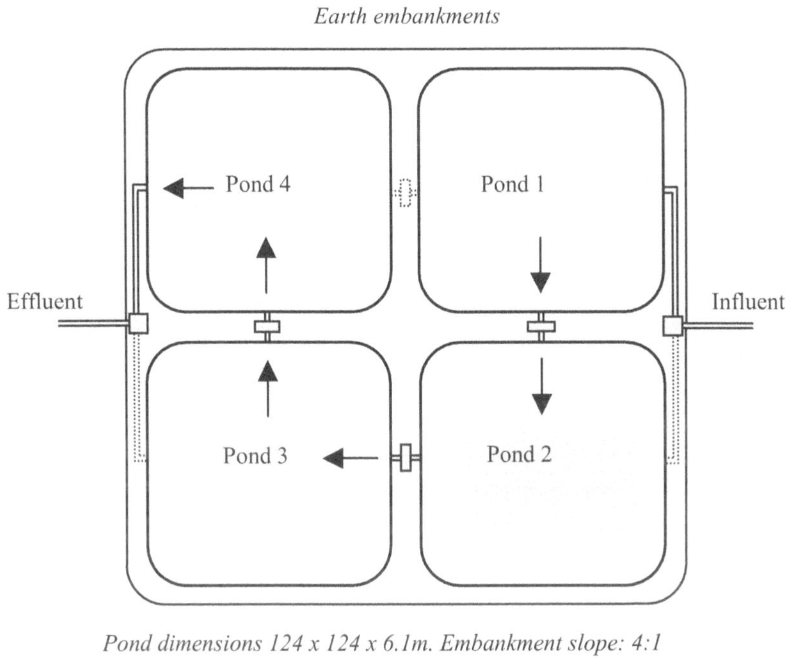

Figure 18.3 Layout of the original pond system at Whitehorse (based on Alan and Jeffreys, 1987 and Whitley and Thirumurthi, 1992)

By 1985 the ponds were serving a population of 14,800 with a flow of about 12,500m^3 day^{-1}. In July 1985 studies revealed short-circuiting and accumulation of sludge (Allan and Jeffreys, 1987). The first two ponds were desludged, and weirs and valving arrangements were modified to change the flow pattern and allow series operation. This led to improvements, but follow-up investigations in 1986 still found hydraulic problems with only 60% utilisation of pond volume. Installation of baffles was recommended, but was not implemented. Meanwhile there were growing concerns about environmental performance and impact, in particular about possible toxicity of the effluent to fish due to ammonia concentrations. In 1989 new standards were introduced requiring a 96-hour LC_{50} bioassay and an FC count of <100,000 per 100ml.

Research carried out in 1989-90 found an average effluent BOD_5 of 46mg l^{-1} and suspended solids of 15mg l^{-1} (Whitley and Thirumurthi, 1992). Removal efficiencies of 52% for BOD5 and 82% for suspended solids compared well with typical figures for this type of plant, in spite of a high system loading of 193kg BOD_5 ha^{-1} day^{-1}. The ponds were found to be not entirely anaerobic: oxygen was present in pond 4 to the full depth of 6 m from May to July, and pond 3 was partially aerobic in the same period. Measurements of photosynthetically active radiation (PAR) showed penetration to 1-1.5m in ponds 3 and 4. Chlorophyll concentration in the effluent was zero between December and March and low in October – November and April, but reached a maximum of 190μg l^{-1} between May –and September.

The need to conform to tighter discharge standards and the city's continuing growth led to a series of studies in the early 1990s aimed at identifying ways to meet projected needs to the year 2012. The review was based on a design population of 33,800 and flows of 19,300m^3 day^{-1} (NovaTec, 1992). Options considered included mechanical plant, WSPs, constructed wetlands, infiltration basins, river outfalls and even snowmaking. The recommended solution was a new WSP system consisting of anaerobic, facultative and maturation/storage ponds with once-yearly discharge into the Yukon River.

The new system began operation in 1996, serving a population of 18,464. It consists of two anaerobic ponds 6m deep, followed by four facultative ponds and a long-detention maturation/storage pond. The facultative ponds range from 125-350m wide and from 600-1000m long (see Figure 18.4). The length-to-width ratio of 7:1 was chosen to minimise short-circuiting. The facultative ponds are 2.5m deep and have a surface area of 42ha. Wastewater then flows into the maturation/storage pond with a mean depth of about 4m and a maximum depth of 17m. The total surface area of the maturation/storage pond is about 177ha, larger than necessary for 12 months' storage as it occupies a natural depression. Seepage losses are low as the pond lies above an extensive silt deposit. Annual discharge takes place over about 60 days, beginning in August or September and finishing by the end of October. Discharge limits for the new WSP system are specified in the Water Use License issued by the Yukon Territory Water Board (see Table 18.2).

Table 18.2 Discharge limits for the new Whitehorse WSP system

Parameter	Limit
BOD_5, mg l^{-1}	45
Suspended Solids, mg l^{-1}	60
Oils and Grease, mg l^{-1}	5
pH	6-9
Faecal Coliforms MPN 100 ml^{-1}	2000
96-hour static LC_{50} %	100

* based on effluent grab sample

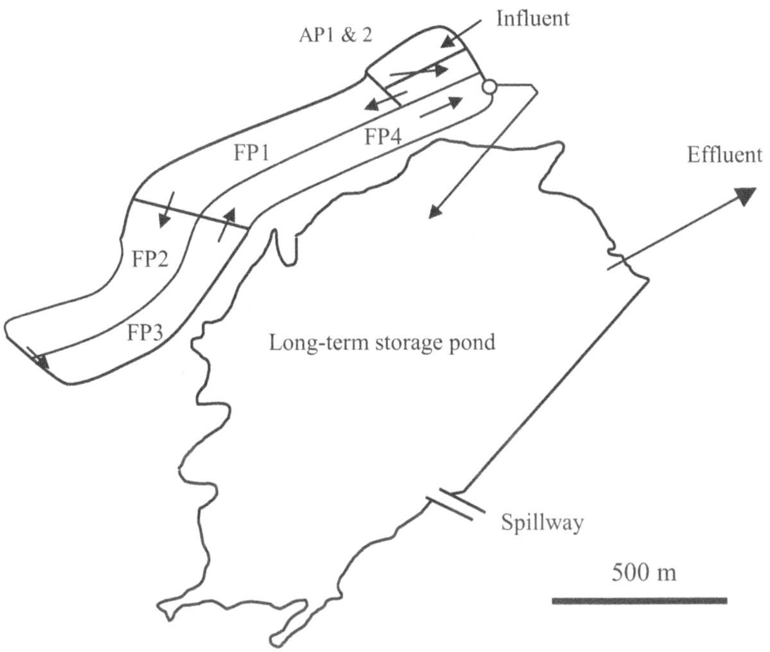

Figure18.4 City of Whitehorse new WSP system (based on Butt and Enns, 2001)

A study of the new pond system in 1999 and 2000 looked at the pollutant removal performance of the WSP and found effluent concentrations of less than 10mg l^{-1} were consistently achieved for BOD5 and suspended solids, and NH_3-N levels were below 5mg l^{-1}. The study also found that primary mechanisms for pollutant removal were uptake into algal biomass and volatilisation of NH_3-N stemming from photosynthesis-induced high pH (Butt and Enns, 2001).

The Whitehorse ponds, both old and new, have been the focus of considerable study and extensive public debate. The choice of a new WSP system to replace the old one in preference to other competing technologies is a clear indication of the satisfactory performance and the high degree of local acceptance achieved in this case.

18.7.2 Stugun, Sweden - from WSPs to precipitation ponds

The town of Stugun is situated in the central inland region of Sweden, approximately 450km northwest of Stockholm. Air temperature ranges from 18°C in July to -15°C in January. Stugun has a year-round population of about 1000, which does not vary greatly as it is not in a major tourist region. The main factors affecting wastewater flows are therefore seasonal, particularly snowmelt which occurs in late April.

Wastewater is treated in a series of three ponds that were originally designed as WSPs, but later modified to operate as chemical precipitation ponds. Pond layout and dimensions are shown in Figure 18.5. The total surface area of the system is 0.93ha and the mean retention time is approximately 30 days (Ødegaard *et al.*, 1987). In the period while they operated as conventional WSPs the ponds received both wastewater, and about 500m^3 of sludge from individual septic tanks discharged into the first pond in summer (Hanæus, 1987). Influent COD varied from 264mg l^{-1} in summer to 752mg l^{-1} in winter, with an annual mean of 451mg l^{-1}. Pond water temperatures were found to vary from 17°C in June-August to 1.5°C in December-March (Hanæus, 1991b).

In the 1960s and 1970s there was increasing concern throughout Scandinavia about the effect of nutrients on inland waters (Ulmgren, 1973; Viitasaari, 1973; Balmer and Vik, 1978). The Stugun wastewater treatment plant discharges into the River Indalsälven, which is used for fishing and recreation. In 1980 a trial was carried out to investigate the possibility of using alum dosing for chemical precipitation, to reduce effluent organic and nutrient concentrations. Results for influent and effluent quality with and without alum dosing are shown in Table 18.3. As a result of the above trials, an alum-dosing rate of 150gm^{-3} was recommended.

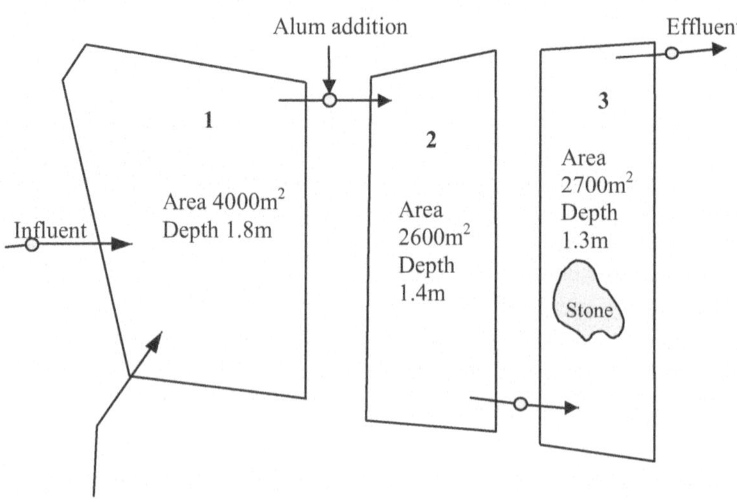

Figure 18.5 Layout of pond system at Stugun (based on Hanaeus, 1991)

Trials were also carried out in summer to investigate the hydraulic performance of the ponds. The wastewater flow in this period averaged 400m^3 day^{-1}. A preliminary survey showed that approximately 15% of the volume of the first pond was occupied by sludge from biological treatment. Tracer tests were carried out with dye released from the inlets and from the corner opposite, and detected at the inlet to pond 2. These findings indicated a significant degree of short-circuiting: the theoretical detention time was 15 days, but the peak passed in 12-24 hours. A test of the whole system was carried out by adding tracer dye at the inlet and monitoring the outlet: the peak passed in 5-10 days, in comparison with a design detention time of 33 days.

Table 18.3 Results of alum dosing trials at Stugun (Hanæus, 1987)

Date	07.01.80 - 27.04.80		28.04.80 - 13.05.80	
Measurement	Mean	Median	Mean	Median
Flow m^3/day	260	260	325	295
Alum dose g/m^3	0	0	114	100
Influent temperature °C	6.0	6.0	8.5	7.0
Effluent temperature °C	1.5	1.5	6.0	3.0
Influent COD mg/l	866	590	652	280
Effluent COD mg/l	171	170	109	110
COD reduction %	80	71	83	61
Influent total phosphorus mg/l	9.6	9.6	7.1	6.3
Effluent total phosphorus mg/l	6.0	6.5	1.1	1.0
Total phosphorus reduction %	38	32	85	84

Sludge from biological treatment in the first pond was removed by isolating the pond and emptying it (Hanæus, 1987, 1991a). Earth-moving equipment was then used to move the sludge into one quarter of the pond. This area was subsequently separated from the rest of the pond by the construction of an earth dyke. The sludge was left to freeze and dewater during the winter. Through this process a dry solids content of 50% was achieved.

The Stugun ponds now operate as a chemical precipitation system: details of the modifications and subsequent performance can be found in Hanæus (1987). There are approximately 20 plants of this type serving communities of 200-2000 population equivalent in Sweden and Finland, and many more working as one element in a combined system (TemaNord, 1995). Hanæus (1991b) suggests that operation of chemical precipitation ponds could be optimised, for example by having a dose-free period in summer to utilise the biological potential and reduce sludge production. While traditional WSP systems are now rare in Scandinavia, the basic principles may therefore have a continuing role to play.

18.8 FUTURE DIRECTIONS

Despite their widespread use, many uncertainties remain concerning appropriate design procedures for cold and extreme climate ponds. Although empirical methods have developed over the years, reliable relationships still need to be established between temperature, loading rates and pond depths. Because of the variability induced by both geographical and local factors, there is no certainty that generally applicable empirical equations can be obtained. One profitable approach might be to obtain a more fundamental understanding of the kinetics of both algal and bacterial growth at low temperatures and under non steady state conditions, from which kinetic models can be derived to simulate critical phases in the annual cycle within both facultative and maturation/storage ponds. Advances in our understanding of complex systems frequently occur through the interaction of empirical and analytical approaches. A great deal of practical experience has been gathered in recent years, and it may now be time for new developments in the field of modelling and analysis to make their contribution to the process and engineering design of cold climate ponds.

Anaerobic ponds play an important role in the overall pond system, but again there is little information on rates of degradation or stabilisation of WSP sludges at low temperatures. For example, it is uncertain whether sludges in a cold-climate pond contribute toward a net loss of organic carbon from the system through methane production, or whether they serve purely to solubilise settleable organics and pass the carbon load through to other parts of the system. It is quite possible that they act as a slow-release carbon sink, helping to balance the load to the aerobic parts of the system. If this is so, more work is needed on optimisation of the design to fulfil this purpose.

Springtime odour is probably one of the biggest problems faced by operators of cold climate pond systems. Several interesting observations suggest that melting of a low-BOD, well-oxygenated layer on top of the ice can produce an effective 'odour buffer'. It is possible that an oxygenated layer created by algal photosynthesis below the ice cover could also act in this way. More research is needed in this area as it could help to resolve the question of whether to design ponds to maximise or minimise snow cover and ice formation. Trapping the odour is one solution but preventing odour formation in the first instance is equally important. Again this requires a better understanding and control of the anaerobic reactions within the accumulating sludge layers of both facultative and anaerobic ponds. Further work is also needed on the possible advantages and disadvantages of supplementing natural aeration with mechanical aeration to meet the demand of the springtime oxygen deficit.

There is clear evidence that over-winter storage of wastewaters can result in a high concentration of pathogens, especially viruses, within the pond system.

This raises concern over the potential for reuse of this water for irrigation, especially in arid regions with sharply continental climates where this application appears very promising. Work is needed to determine die-off rates during the non steady state springtime acclimatisation period, to identify alternative indicator organisms, and to provide guidance on the minimum holding period before continuous summer-time discharge to irrigation systems can be permitted.

In recent years the Scandinavian countries have moved away from biological pond treatment in favour of chemical treatment, but there appears to be scope for both types of systems to work in a complementary manner to maintain or increase performance while reducing chemical usage and overall sludge production. Pond systems working in conjunction with wetlands provide further potential for improving final effluent water quality, even in cold climates. The design of hybrid systems, their mode of operation and optimum configuration is a promising area that is receiving growing attention.

WSPs have served well communities in extreme climate regions for more than half a century by providing a low-cost, low-maintenance and reliable means of treating wastewater. The case histories presented above are good examples of the manner in which these systems have developed, not always working adequately at the first attempt! They do work and indeed work extremely well; but to obtain the best results and full benefits, empirical approaches and trial-and-error need to be replaced by a robust design methodology based on an improved fundamental understanding of the system and its biology.

REFERENCES

Abdrashitova, S. A., Banks, C.J., Pak, L.N. and Koloskov, G.B. 2001. An evaluation of waste stabilisation ponds for extreme continental climates: the design of laboratory and pilot scale trials. *Isvestia* (Biological and Medical series) **6,** 20-29.

Alferova, L. A., Skirdov, L.V., Ponomarev, V.G., Hudenko, B.M., Gladkov, V.A. and Rogovskaya, Z. I. 1973. Sewage treatment in the northern areas of the USSR, pp 64-74 *in* E. Davis, editor. *International Symposium on Wastewater Treatment in Cold Climates.* Environment Canada, Saskatoon.

Allan, R. B., and Jeffreys, Y. 1987. Performance Evaluation of the City of Whitehorse Sewage Treatment Lagoons. *Regional Program Report* No. 87-17, Environment Canada.

Arceivala, S. J. 1998. Wastewater treatment for pollution control. Tata McGraw-Hill, New Delhi.

Balmer, P. and Vik, B. 1978. Domestic wastewater treatment with oxidation ponds in combination with chemical precipitation. *Progress in Water Technology* **16**(5-6), 867-880.

Banks, C. J. Koloskov, G. B., Lock, A.C. and. Heaven, S. 2003. A computer simulation of the oxygen balance in a cold climate winter storage WSP during the critical spring warm-up period. *Water Science and Technology* **48**(2), 189-196.

Banks, C. J., Pak, L.N. and Rspaev, M.K. 2002. Springtime acclimatisation of a winter ice-covered waste stabilisation pond: operational data from 4 experimental units, pp 671-678 *in* 5th International IWA Conference on Waste Stabilisation Ponds, Auckland, New Zealand.

Bethell, G. 1981. Start-up characteristics and performance evaluation of an anaerobic sewage lagoon, north of 60, White Horse YT. Regional Program Report No. 81-9, Environment Canada.

Browne, W. and Jenssen P.D. 2001. Exceeding tertiary standards with a pond/reedbed system in Norway. *in* Ecological Engineering for Landscape Services and Products, New Zealand.

Butt, C. and Enns V. 2001. Treatment performance of a sub-arctic sewage lagoon: Whitehorse, Yukon. Regional Program Report No. 01-13, Environment Canada.

Castillo, G.C. and Trumper, B.A. 1991. Coliphages and other microbial indicators in stabilization ponds. *Environmental Toxicity and water quality* **6**(2), 197-207.

Dawson, R.N., and Grainge, J.W. 1969. Proposed design criteria for wastewater lagoons in arctic or sub-arctic regions. *Journal of the Water Pollution Control Federation* **41,** 237-246.

Desjardins, M.A., and Briere, F.G. 1996. Conditioning and hydration of facultative aerated lagoon sludge using natural freeze-thaw processes - Experimental results. *Canadian Journal of Civil Engineering* **23**(2), 323-339.

Environment-Canada. 1985. Sewage lagoons in cold climates. *Report No. EPS 4/NR/1,* Environment Canada.

Environment-Canada. 1987. Cold Climate Sewage Lagoons. *Report No. EPS 3/NR/1,* Environment Canada.

Felip, M., Sattler, B., Psenner, R, and Catalan, J. 1995. Highly-Active Microbial Communities in the Ice and Snow Cover of High-Mountain Lakes. *Applied and Environmental Microbiology* **61**(6), 2394-2401.

Giraldo, E. and Garzon, A. 2002. Compartmental model for organic matter digestion in facultative ponds. *Water Science and Technology* **45**(1), 25-32.

Gray, N.F. 1999. Water Technology: an introduction for scientists and engineers. Arnold, London.

Hanæus, J. 1987. Swedish field experiences with chemical precipitation in stabilization ponds. *Canadian Journal of Civil Engineering* **14**(1), 33-40.

Hanæus, J. 1991a. Sludge accumulation in ponds for wastewater treatment using alum precipitation. *Vatten* **47,** 181-188.

Hanæus, J. 1991b. Wastewater treatment by chemical precipitation in ponds. PhD. thesis, Lulea University of Technology, Lulea.

Heaven, S., Lock, A. C., Pak, L. N. and Rspaev, M.K. 2003. Waste Stabilisation ponds in extreme continental climates: a comparison of design methods from the USA, Canada, northern Europe and the former Soviet Union. *Water Science and Technology* **48**(2), 25-33.

Hedstrom, A. and Hanæus, J. 1999. Natural freezing, drying, and composting for treatment of septic sludge. *Journal of Cold Regions Engineering* **13**(4), 167-179.

Heinke, G.W. and Smith, D.W. 1988. Guidelines for the Planning, Design, Operation and Maintenance of Wastewater Lagoon Systems in the Northwest Territories: Vol 2 Operations and Maintenance. Department of Municipal and Community Affairs, Government of the Northwest Territories, Yellowknife.

Heinke, G.W., Smith, D.W. and Finch, G. R. 1988. Guidelines for the Planning, Design, Operation and Maintenance of Wastewater Lagoon Systems in the Northwest Territories:

Vol 1 Planning and Design. Department of Municipal and Community Affairs, Government of the Northwest Territories, Yellowknife.

Heinke, G.W., Smith, D.W. and Finch, G.R. 1991. Guidelines for the planning and design of wastewater lagoon systems in cold climates. *Canadian Journal of Civil Engineering* **18,** 556-567.

Hellstrom, D. 1997. Natural sludge dewatering. 2. Thawing-drying process in full-scale sludge freezing ditches. *Journal of Cold Regions Engineering* **11**(1), 15-29.

Henry, J.G. and Prasad, D. 1986. Microbial Aspects of the Inuvik Sewage Lagoon. *Water Science and Technology* **18**(2), 117-128.

Horton, A.J., Hak, K.M..,Steffan, R.J., Foster, J.W. and Bej. A.K. 2000. Adaptive response to cold temperatures and characterization of cspA in Salmonella typhimurium LT2. *Antonie van Leeuwenhoek International Journal of General and Molecular Microbiology* **77**(1), 13-20.

Juanico, M., Weinberg, H and Soto, N. 2000. Process design of waste stabilization ponds at high altitude in Bolivia. *Water Science and Technology* **42**(10-11), 307-313.

Kayombo, S., Mbwette, T.S.A., Mayo, A.W., Katima, J.H.Y and Jorgensen, S.E.. 2000. Modelling diurnal variation of dissolved oxygen in waste stabilization ponds. *Ecological Modelling* **127**(1), 21-31.

Kelley, D.E. 1997. Convection in ice covered lakes: effects on algal resuspension. *Journal of Planktonic Research* **19**(12), 1859-1880.

Mackenthun, K.M. and McNabb, C.D. 1961. Stabilization pond studies in Wisconsin. *Journal of the Water Pollution Control Federation* **33**(12), 1234-1250.

Martel, C.J. 1993. Fundamentals of sludge dewatering in freezing beds. *Water Science and Technology* **28**(1), 29-35.

McKnight, D.M., Howes, B.L., Taylor, C.D. and Goehringer, D.D. 2000. Phytoplankton dynamics in a stable stratified Antarctic lake during winter darkness. *Journal of Phycology* **36**(5), 852-861.

Merrill, K.S. and Stephl, M. 1996. Case history of a lined wastewater treatment lagoon failure, pp 518-532 *in* R.F. Carlson, editor. 8th International Conference on Cold Regions Engineering. ASCE, Fairbanks, Alaska.

Middlebrooks, E. J. 1987. Design equations for BOD removal in facultative ponds. *Water Science and Technology* **19**(12), 187-193.

Middlebrooks, E.J., Middlebrooks, C., Reynolds, J.H., Watters, G.Z., Reed, S.C. and George, D.B. 1982. Wastewater stabilization lagoon design, performance and upgrading. Macmillan, New York.

Moreno-Grau, S., Garcia-Sanchez, A., Moreno-Clavel, J., Serrano-Aniorte, J. and Moreno-Grau, M.D. 1996. A mathematical model for waste water stabilization ponds with macrophytes and microphytes. *Ecological Modelling* **91,** 77-103.

Nasser, A.M. and Oman, S.D. 1999. Quantitative assessment of the inactivation of pathogenic and indicator viruses in natural water sources. *Water Research* **33**(7), 1748-1752.

Nasser, A.M., Tchorch Y. and Fattal, B. 1993. Comparative survival of Escherchia-Coli, F+bacteriophages, HAV and Poliovirus-1 in wastewater and groundwater. *Water Science and Technology* **27**(3-4), 401-407.

Newfoundland. 2002. Guidelines for the design, construction and operation of water and sewerage systems. Government of Newfoundland and Labrador.

NovaTec. 1992. Sewage Treatment Feasibility Study: Selection of Final Treatment Option. Novatec, Vancouver.

NovaTec. 1996. Sewage lagoon design using wetlands and other upgrading technologies to achieve non-acutely toxic effluent. Report No. DOE FRAP 1994-34, Environment Canada, Vancouver.

Ødegaard, H., Balmer, P. and Hanæus, J. 1987. Chemical precipitation in highly loaded stabilisation ponds in cold climates: Scandinavian experiences. *Water Science and Technology* **19**(12), 71-77.

Oleszkiewicz, J.A. and Sparling, A.B. 1987. Wastewater lagoons in a cold climate. *Water Science and Technology* **19**(12), 47-53.

Pearson, H.W., Mara, D.D., Thompson, W. and Maber, S.P. 1987. Studies on high altitude waste stabilization ponds in Peru. *Water Science and Technology* **19**(12), 349-353.

Phadtare, S., Alsina, J. and Inouye, M. 1999. Cold shock response and cold shock proteins. *Current Opinion in Microbiology* **2**(2), 175-180.

Phillips, K.A. and Fawley, M.W. 2000. Diversity of coccoid algae in shallow lakes during winter. *Phycologia* **39**(6), 498-506.

Prince, D.S., Smith, D.W. and Stanley, S. J. 1994. Evaluation of lagoon treatment in Alberta. Environmental Engineering Technical Report 94-1, Department of Civil Engineering, University of Alberta, Edmonton.

Prince, D.S., Smith, D.W. and Stanley, S.J. 1995a. Intermittent discharge lagoons for use in cold regions. *Journal of Cold Regions Engineering* **9**(4), 184-194.

Prince, D.S., SmithD.W. and Stanley, S.J. 1995b. Performance of lagoons experiencing seasonal ice cover. *Water Environment Research* **67**(3), 318-326.

Samokhin, V.N., editor. 1981. Wastewater systems for inhabited places and industrial enterprises (in Russian), 2nd edition. Stroyizdat, Moscow.

Sarsenbaev, B.A. and Atabaeva, S.D. 2000. Biotechnology of phyto-treatment of the wastewater of the City of Almaty, pp 43-45 *in* International Ecological Forum on Problems of Sustainable Development of the Ili-Balkhash Basin, Almaty.

Saskatchewan. 1996. Lagoon Operation and Maintenance. EP 115A, Saskatchewan Environment and Resource Management.

Schneiter, R.W., Middlebrooks, E.J. and Sletten, R.S. 1983. Cold region wastewater lagoon sludge accumulation. *Water Research* **17**(9), 1201-1206.

Schneiter, R.W., Middlebrooks, E.J. and Sletten, R. S. 1984. Wastewater lagoon sludge characteristics. *Water Research* **18**(7), 861-864.

Smith, D.W., editor. 1996. Cold Regions Utilities Monograph, 3rd edition. ASCE, New York.

Smith, D.W. and Finch, G.R. 1983. A critical evaluation of the operation and performance of lagoons in cold climates. University of Alberta, Edmonton.

SNiP. 1996. SNiP 2.04.03-85: Water Drainage: External Networks and Structures. Construction Norms and Regulations. (In Russian). *in*. MinStroi, Moscow.

Soniassy, R.N. and Lemon, R. 1986. Lagoon treatment of municipal sewage effluent in a sub-arctic region of Canada (Yellowknife, NWT). *Water Science and Technology* **18**(2), 129-139.

TaigaNet. 2002. Lagoon works wonders. *in*. Your Yukon (http://www.taiga.net/yourYukon/col79.html).

TemaNord. 1995. Small Wastewater Treatment Plants. TemaNord 1995:650, Nordic Council of Ministers, Copenhagen.

Thieringer, H.A., Jones, P.G. and Inouye, M. 1998. Cold shock and adaption. *BioEssays* **20**(1), 49-57.

Thirumurthi, D. 1974. Design criteria for waste stabilization ponds. *Journal of the Water Pollution Control Federation* **46**(9), 2094-2106.

Torrella, F., Lopez, J.P. and Banks, C.J. 2003. Survival of indicators of bacterial and viral contamination in wastewater subjected to low temperatures and freezing: application to cold climate waste stabilisation ponds. *Water Science and Technology* **48**(2), 105-112.

Ulmgren, L. 1973. Swedish experiences in sewage treatment, pp 45-63 *in* E. Davis, editor. International Symposium on Wastewater Treatment in Cold Climates. Environment Canada, Saskatoon.

USEPA. 1983. Design manual: Municipal wastewater stabilization ponds. Report No. 625/1-83-016, US Environmental Protection Agency, Cincinnati, Ohio.

Viitasaari, M. 1973. Sewage treatment methods in Finland, pp 29-44 *in* E. Davis, editor. International symposium on wastewater treatment in cold climates. Environment Canada, Saskatoon.

Vinberg, G.G., Ostapenya, P.V., Sivko, T.N. and Levina, R.I. 1966. Biological ponds in the practice of wastewater treatment (in Russian). Belarus, Minsk.

VNII-SIS. 1987. Recommendations for construction of biological oxidation contact stabilisation (BOCS) ponds in the USSR for small populated places (in Russian). Report No. 121-12/701-14, All-Union Scientific-Production Institute for Agricultural Use of Wastewater, Ministry of Melioration and Water Management USSR, Moscow.

Whitley, G. and Thirumurthi, D. 1992. Field monitoring and performance evaluation of the Whitehorse sewage lagoon. *Canadian Journal of Civil Engineering* **19**(5), 751-759.

Wouters, J.A., Rombouts, F.M., de Vos, W.M., Kuipers, O.P. and Abee, T. 1999. Coldshock proteins and low temperature response of streptococcus thermophilus CNRZ302. *Applied and Environmental Microbiology* **65**(10), 4436-4442.

Yunusov, I.I. 1983. Flora and plants of biological ponds and wastewater evaporation fields in Uzbekistan (in Russian). Fan, Tashkent.

Zhirkov, E.I., Tereshina, A.N., Dolivo-Dobrovolskiy, L.B., Baltrukonis, Y.A. and Lyusenko, V.P. 1987. Recommendations for treatment and disinfection of wastewater from populated places and from poultry enterprises in biological ponds (in Russian). All-Union Scientific-Production Institute for Agricultural Use of Wastewater, Ministry of Melioration and Water Management USSR, Moscow.

19

Ponds for livestock wastes

James Sukias and Chris Tanner

19.1 INTRODUCTION

Ponds are a proven practical, low cost, low maintenance treatment option to store and/or treat agricultural wastes in many areas of the world. Livestock wastes requiring treatment are generated primarily where animals are housed, or held for periods in paved yards and buildings. Common applications of ponds include: piggeries (hog or swine farms), cattle feed lots, dairy barns and milking parlours, and poultry farms. Cleaning of faecal matter, urine and excess feed deposited in barns and yards is frequently achieved by flushing or use of high-pressure hoses, generating large volumes of high strength wastewater. Where appropriate these wastes are often applied back to agricultural land. While pond storage/treatment before land application will generally result in some nutrient losses, it has the following practical advantages:

(1) Provides flexibility and efficiency, as land application can be carried out intermittently, when the water and nutrients can be best utilized by crops

ISBN: 1843390205. Published by IWA Publishing, London, UK.

or pasture, and when the risk of soil damage (e.g. due to pugging or waterlogging) and surface run-off is minimized (i.e. during dry periods).

(2) Nutrient reduction during pond treatment can reduce the land area required for land application (when limited by nitrogen loading) and minimise the potential for "crop-burn" from applying excessive amounts of ammoniacal nitrogen in fresh manure (e.g. from poultry).

(3) Coarse solids settle-out and are retained in the pond, reducing the potential for nozzle blockage and for pump wear. Biosolids and associated nutrients can be land-applied at optimum times of the year.

(4) Treatment of pathogenic micro-organisms can reduce health risks to agricultural workers and consumers of irrigated crops.

(5) The potential for odour emission during and after land application can be reduced.

In contrast to the advantages of treating wastes in ponds before application, the potential adverse effects of applying untreated manures include: damage to pasture, crops and soil structure; reduced feed palatability; spread of plant and animal diseases and parasites; encouragement of weed species (Ap Dewi, 1994).

The configuration of ponds used for livestock wastes depends upon their intended purpose. Where the waste is to be irrigated onto a crop or pasture, the pond may be a single storage pond (ASAE, 1998) where settling and some anaerobic digestion will occur. However, a higher level of treatment may be necessary when the quantity and strength of waste generated exceeds the capacity of the available land to sustainably absorb and process them, or where the effluent is to be discharged to surface waters. These pond treatment systems may include two or more stages, and be of different depths and design in order to maximise their treatment capacity. In the following sections we will focus on ponds specifically designed for wastewater treatment. Depending on the level of treatment and local regulations (and/or codes of practice), the pond effluent could be applied to land, recycled for yard washing or discharged to waterways.

19.2 CHARACTERISTICS OF LIVESTOCK WASTES AND WASTEWATERS

19.2.1 Oxygen demand

Livestock wastewaters tend to be more concentrated than domestic sewage wastewaters, and exhibit a range of different contaminant and physico-chemical characteristics that can influence their treatment in pond systems.

BOD_5 (biochemical oxygen demand measured over 5 days) is the most commonly used measure of oxygen demand for domestic wastewaters. The

BOD_5 in domestic wastewaters commonly comprises around 68% of the BODu (BOD, ultimate) (Marais, 1970) and 60% of the COD (Gray, 1987). In contrast, for raw livestock wastes, the BODu is often 3-5 times higher than the BOD_5, and the COD can be 7 times higher (Table 19.1). This is because livestock wastewaters contain a higher proportion of more slowly degradable organic constituents and so they exert a more sustained oxygen demand. Allowances must be made for this, both in design of treatment systems, and in assessment of effluent pollution potential of livestock wastewaters.

Table 19.1 Oxygen demand of raw livestock wastes in kg per 1000 kg live animal mass per day, and as influent wastewater to an anaerobic pond (g m^{-3})

	Raw waste characteristics						
	Average animal weight	BOD_5		COD		NOD	
	(kg)	(kg)	(g m^{-3})	(kg)	(g m^{-3})	(kg)	(g m^{-3})
Dairy, (harvested ration)	600	1.6	17700	11	122000	1.95	21600
Dairy, (pasture fed)	500	2.0	20000	8.6	86000	2.1	21000
Beef	450	1.6	25400	7.8	12600	1.47	23300
Pig	50	3.1	27300	8.4	74000	2.25	20000
Poultry	1.5	3.3	43600	11	145300	3.63	48000

Values taken from ASAE (2000) and Vanderholm *et al.*, (1984). NOD based on Total Kjeldahl Nitrogen values. Poultry values for laying hens.

Table 19.2 Common oxygen demand characteristics of pond-treated effluents

	Anaerobic pond effluents (g m^{-3})			Anaerobic/Facultative pond effluent (g m^{-3})		
	BOD_5	COD	NOD	BOD_5	COD	NOD
Dairy Cow	200-1200	600-2100	300-3000	65–240	400-1400	300–2136
Beef	200-2500	1400	780-1150			
Pig	230-3600	1000-3600	600-4800	100–1700	380–3600	600–2550
Poultry[1]	600-3800	3000-10000	3500-6000	500-1500	1000-2000	3500–4300

[1] Using under-cage flush system, which may be combined with egg-wash water

The organic nitrogen components of the wastes exert an additional nitrogenous oxygen demand (NOD) when released to surface waters (Cooper,

1986), over that measured by COD or BOD_5 tests[1]. This NOD occurs when ammonium is biologically oxidised to nitrate (nitrification), consuming ~4.3 g of oxygen per g of NH_4^+-N (Henze *et al.*, 1995). The additional NOD of livestock wastes is frequently similar to the carbonaceous BOD_5 (Table 19.1) and may become 5-10 times higher than the BOD during primary and secondary treatment (Table 19.2). Mineralisation of organic nitrogen during storage and treatment can also release additional quantities of ammoniacal-N, providing further substrate for nitrification (and thus NOD).

There are many texts that have comprehensively characterised the constituents of livestock wastes (ASAE, 2000; Loehr, 1984; Overcash *et al.*, 1983; USDA/SCS, 1992; Vanderholm *et al.*, 1984). Table 19.1 summaries the oxygen demand of raw faecal matter, while Table 19.2 gives characteristic values for these wastes after treatment in anaerobic and facultative ponds. The treatment levels found in different studies are influenced by different WSP design approaches and local environmental influences, as well as the way the animals are kept, feeds they are given, and the methods and volumes of water used to clean stalls and yards.

19.2.2 Solids

The quantity and characteristics of solids in livestock faecal wastes is of interest for a range of reasons. Recalcitrant solids can accumulate in ponds, reducing their storage and treatment capacity. Where wastes are to be irrigated, solids such as grain husks and animal hair can cause blockages of spray-nozzles or drippers. Excessive application of solids to pasture and crops can cause smothering and damage to plants and soils. In ponds, the volatile (organic) components (VSS) are commonly used as the basis for sizing of anaerobic ponds (e.g. ASAE, 2000). Lastly, where treated wastes are discharged to receiving waters, solids can be unsightly, cause conspicuous changes in water colour and clarity, limit light penetration, and cause smothering of benthic aquatic life. Table 19.3 presents a range of characteristic solids values for typical raw livestock wastes. Table 19.4 summarises influent and effluent solids concentrations for various livestock anaerobic and anaerobic/facultative pond treatment systems.

---

[1] If nitrification is not specifically inhibited in the BOD_5 test then it may include some proportion of NOD exertion.

Table 19.3 Solid content of raw livestock wastes per 1000 kg live animal mass per day

	Raw waste characteristics		
	TS	VS	Percentage VSS % of TS
Dairy, (harvested ration)	12	10	83
Dairy, (pasture fed)	8.8	6.4	73
Beef	8.5	7.2	85
Pig	11	8.5	77
Poultry	16	12	75

TS= Total Solids; VS = Volatile Solids; VSS = Volatile Suspended Solids

Table 19.4 Suspended solids concentrations in pond influent/effluents

	Influent wastewater $g\ m^{-3}$	Anaerobic pond effluent $g\ m^{-3}$	Facultative pond effluent[1] $g\ m^{-3}$
Dairy Cow	133000	1100-47000	150–300
Beef	135000	500-1300	300
Pig	97000	460-9200	100–450
Poultry	211000	2000-8000	–

[1] After pre-treatment in an anaerobic pond

19.2.3 Nutrients

The nutrient content of a livestock waste is of interest because of its potential value as a fertiliser for crop and pasture irrigation. In a receiving water, the "fertiliser" component of effluent can, however, contribute to eutrophication, causing excessive algal and plant growth. Table 19.5 presents typical values for total nitrogen (N), phosphorus (P) and potassium (K) in raw livestock wastes. Values for other macro- and micro-nutrients can be found in ASAE (2000).

Table 19.5 Nutrient content of raw livestock wastes in kg per 1000 kg live animal mass per day, and as influent wastewater to a treatment system ($g\ m^{-3}$)

	Raw waste characteristics					
	TN		TP		TK	
	(kg)	(g m-3)	(kg)	(g m-3)	(kg)	(g m-3)
Dairy, (harvested ration)	0.45	5000	0.094	1000	0.29	3200
Dairy, (pasture fed)	0.48	4800	0.050	500	0.32	3200
Beef	0.34	5400	0.092	1450	0.21	3300
Pig	0.52	4600	0.18	1600	0.29	2550
Poultry	0.84	11100	0.30	4000	0.30	4000

Table 19.6 Nutrient concentrations in pond effluents

	Anaerobic pond effluent ($g\ m^{-3}$)				Anaerobic/facultative pond effluent ($g\ m^{-3}$)			
	Total N	NH_4^+	Total P	Total K	Total N	NH_4^+	Total P	Total K
Dairy cow	75–1750	50–720	25–200	500–860	69–480	40–100	57–240	360
Beef	180–530	180–265	150–480	350–900	–	–	–	–
Pig	300–3000	200–2000	75–300	480–587	300–480	220–300	75–240	380–480
Poultry	800–3600	330–1500	200–360	480–1170	200–900	200–900	80–360	280–500

Table 19.6 shows typical concentrations of N, P and K in effluent from various agricultural anaerobic and facultative treatment ponds.

During storage and/or treatment (Table 19.6) the relative N content of livestock wastewaters declines, due primarily to ammonia volatilisation and sedimentation of solids.

During irrigation of wastes additional ammonia can volatilise, reducing the inorganic nitrogen by 25–50% from the above figures. With direct incorporation of the effluent into the soil, this loss of available nitrogen can reduce to between 3–25% (Sutton *et al.*, 2001; Tyson, 1996). Once applied to the soil, nitrogen can be lost via denitrification, and when applied at excessive loadings by leaching. Loss of N exacerbates the already low N:P ratios of livestock wastes, which can result in P levels above that which the crop is able to utilize unless additional N is applied or a nitrogen-fixing, leguminous crop grown. P build-up can thus occur in soils irrigated with manure, eventually causing excessive P run-off and leaching into receiving waters. In addition, there is also a potential problem of salt accumulation in these soils, which can reduce cropping rates and affect animal health (e.g. K build-up, Wang *et al.*, 2004). Detailed guidance on irrigation of agricultural wastes can be found in Ayers and Westcot (1985) and USDA/SCS (1992).

Table 19.6 demonstrates the high concentrations of ammoniacal-N frequently associated with livestock faecal wastewaters. Un-ionised ammonia, which is prevalent at elevated pH (>8), is toxic to fish, stream invertebrates and algae (USEPA, 1998). Discharge of poorly treated faecal wastes containing high concentrations of ammonia can thus cause fish mortalities and severely impact on stream ecosystems.

19.2.4 Pathogens

Livestock waste has the potential to transmit pathogenic bacteria, intestinal worm parasites and protozoan parasites to humans, as well as to other livestock and wild animal populations. Transmission may occur via occupational exposure (e.g. farm workers, Cliver and Moe, 2004). More indirectly problems may arise from irrigation of wastewaters by aerosols or via seepage or effluent discharges from ponds that contaminate waterways that are used for water supply, recreation or as an aquatic food source (e.g. shellfish). Pathogens of major concern are zoonotic organisms such as *Cryptosoridium*, *Giardia*, *Campylobacter*, *Salmonella*, *Leptospira*, enteropathogenic *Eschericihia coli* and parasitic worms (helminths), where livestock are important reservoirs of infective organisms. Natural disinfection processes that operate in WSP systems are discussed in Chapter 6.

Despite improvements in the ability to detect these pathogens directly, we are still largely reliant upon the use of indicator organisms (e.g. faecal coliforms, faecal enterococci and *E. coli*) to assess the presence and potential disease risk from faecal contamination of drinking and recreational waters. Domesticated animals such as cows, pigs and poultry each excrete differing proportions of the above indicator organisms, and thus determining the potential disease risk from the presence of the indicator is not a straightforward procedure. Despite the common presumption that human illness risks for livestock will be lower than for sewage wastewaters, recent studies have reported no substantial differences between indicator associated illness risks from bathing at beaches contaminated with human or animal waste (Cheung *et al.*, 1990; McBride *et al.*, 1998).

Current WHO (1989) guidelines for the reduction of health risks associated with wastewater reuse in agriculture are ≤1000 faecal coliforms $(100\text{mls})^{-1}$ and ≤ 1 nematode egg L^{-1}. Recent recommendations (Blumenthal *et al.*, 2000) propose that the nematode guideline should be reduced to ≤ 0.1 egg L^{-1} for unrestricted irrigation, and that these guidelines should also apply to so-called "restricted" irrigation (of crops that are processed before human consumption or not eaten by humans) where occupational exposure occurs (e.g. where workers, particularly children, are engaged in flood or furrow irrigation). Other regions of the world have developed their own, generally more stringent, standards for wastewater reuse, for example NZLTC (2000) and State of California (2001).

19.2.5 Other contaminants

In addition to the potentially polluting biotic factors in animal wastes, various industries supply additives to the animal directly or in feed to enhance productivity. When these additives are excreted they may adversely affect the pond operation and can accumulate to environmentally unacceptable concentrations. For instance pig feed can include copper and zinc additions (Gilley *et al.*, 2000) and antibiotics (Hermanson and Kalita, 1994). Cattle and other agricultural animals are also routinely treated with agrochemicals to remove helminths, mites and insects. Other abiotic chemicals can enter the waste while cleaning housing (e.g. pesticides used in the poultry industry) or milking plant equipment (dairy plant cleaners, sanitizers, acids and alkalis). Livestock wastes may also contain biologically significant quantities of natural steroidal estrogen hormones that can cause reproductive abnormalities in aquatic organisms (Hanselman *et al.*, 2003). As practices vary geographically and between industries, agricultural waste managers and regulatory personnel need to use knowledge of the specific practices at each site to ensure that these substances do not interfere with the ponds (and other treatment systems), or cause harm if discharged to the environment.

19.2.6 Physico-chemical factors affecting pond treatment of livestock wastewaters

In addition to the high total oxygen demand and concentration of ammonium and other nutrients already noted, livestock wastes frequently contain high amounts of dissolved organic matter or gilvin (dissolved yellow matter) (Davies-Colley *et al.*, 2005) which imparts a strong yellow-brown colour to the wastewater and is highly light absorbing. Plant fibres and seed husks from the animal feed also contribute to high levels of light attenuation in these ponds. Because they are of a refractory nature these compounds influence the performance of algal dominated ponds and can persist through several ponds in series.

High light attenuation substantially reduces the euphotic zone where algae can sustain photosynthesis, (typically $\approx$1% light penetration depth). Algae within ponds also absorb light and, assuming nutrients are not limiting, will grow until light is limited. Where the euphotic depth is very shallow, restricted algal growth and photosynthetic oxygen production may, in turn, limit BOD removal and nutrient assimilation within the pond.

A reduced euphotic zone also limits microbial pathogen removal in two ways. Firstly, it reduces the penetration of UV light, which inactivates pathogens by disrupting their nuclei and cellular repair mechanisms (Davies-Colley *et al.*, 1999). Secondly, high pH's (associated with algal absorption of CO_2), which in the presence of photo-oxidisers and UV light enhances bacterial die off, is also restricted to a narrower zone of activity.

High light absorption near the surface restricts the depth of heating from the sun. Therefore, any thermocline that develops may tend to be closer to the surface than in ponds with greater light penetration. This could potentially result in a shallower epilimnion (warmer upper layer), and a reduction of the aerobic treatment provided by this oxygenated zone.

19.3 LIVESTOCK POND DESIGN AND OPERATION

19.3.1 Anaerobic ponds

Anaerobic ponds are commonly used for initial treatment of high strength agricultural wastes. Solids are removed by settling with subsequent anaerobic digestion in the settled sludge layer (Mangino *et al.*, 2002). In order to maximise treatment, these ponds are deep (minimum depth 2–3m, but frequently >5m) with as small a surface area as practical to minimise wind mixing and oxygen input which can be inhibitory to the anaerobic digestion process. Because livestock wastes have high concentrations of slowly degradable solids, recommended anaerobic pond loading rates differ significantly from those used for domestic wastewaters, generally being around 10-fold lower when compared on the basis of BOD_5.

Anaerobic lagoon design for livestock wastes ideally needs to take into account the following factors:

(1) Treatment requirements;
(2) Odour control;
(3) Sludge accumulation rates;
(4) Provision of adequate storage volume where land-applied.

Design criteria for anaerobic pond treatment of livestock wastewaters have been developed largely on the basis of empirical experience dealing with these factors. Required pond sizes are determined by the digestible organic component of the wastes, either measured as VS (volatile solids), COD or BOD_5 (ASAE, 2001; DEC, 1996) and requirements for sludge, wastewater and storm run-off (commonly based on the theoretical 25 yr, 24 hour storm) storage. A minimum retention time of 50 days is recommended by ASAE (2001).

Local climatic conditions must be taken into account when sizing ponds. The rate at which wastes are digested is temperature dependent, slowing down drastically below 10°C and practically ceasing below 4°C. Slower digestion in colder areas results in faster accumulation of solids, requiring larger ponds or more frequent desludging. The ASAE (1999) and USDA (1992) guidelines recommend similar maximum loading rates which vary between 48–96g VS m^{-3} d^{-1} over continental USA, and rise to 192g VS m^{-3} d^{-1} in tropical zones (Table 19.7).

Table 19.7 Summary of maximum recommended anaerobic pond loading rates for livestock wastes

Location	Latitude	Approx. mean annual air temperature (°C)	Climate[1]	VS Loading ($g\ m^{-3}\ d^{-1}$)	BOD Loading ($g\ m^{-3}\ d^{-1}$)
United States[2]					
Northern USA	47° (50°+)[a] N	5-10	Te/Co	48	8[b]
Mid-latitude USA	37° (42°)[a] N	12	Te/Co	70	11[b]
Southern USA	32° (37°)[a] N	17	Wt/Co	84	13[b]
Southern Texas / Florida	28° N	23	Ar/Co, St/Oc	96	15[b]
Hawaii/Puerto Rico	18-22° N	23-27	Tr/Oc	192	31[b]
New Zealand[3]	(Specifically for dairy parlour wastewaters)				
Southern South Island	44-47°S	7.5-10	Te/Oc	125[b]	20
Central	39-44°S	10-12.5	Te/Oc	150[b]	24
Northern North Island	34-39°S	12.5-15.5	Wt/Oc	175[b]	28

[1] Basic climate types: Te = temperate, Wt = warm temperate, Ar = arid, St =sub-tropical, Tr = tropical; Climatic influences: Co = continental; Oc =oceanic

[2] ASAE (2001), extrapolated to tropical areas based on USDA/SCS (1992).

[3] DEC (1996)

[a] Proposed latitudinal ranges vary across the continent. Values are estimates for mid-continental and eastern states, with approximations for western coastal states given in parenthesis.

[b] Calculated based on VS:BOD ratios for raw wastes ASAE (2001)

Guidelines developed in New Zealand specifically for dairy parlour wastes, but based on BOD loading, are also shown for comparison in Table 19.7 using raw waste VS:BOD_5 ratios for dairy cows (ASAE, 2000) to enable comparison. The New Zealand guideline loadings appear to be much higher than those proposed for regions in the USA with similar mean annual temperatures. This may reflect the reduced seasonal temperature variation of New Zealand's oceanic climate. There are shorter periods when anaerobic digestion is arrested by low temperatures and, possibly, an acceptance of shorter desludging intervals (maximum recommended interval = 4 years; DEC, 1996).

Solids separation by screening prior to treatment in anaerobic ponds is sometimes practised where wastes have a high percentage of slowly digestible (e.g. grain husks) or inorganic components. Where such solids enter a treatment pond they can rapidly accumulate reducing the retention time and efficiency of the system. Settling tanks and simple plate-screens can remove up to 50% of incoming solids and associated BOD (Tyson, 1996). Separated solids can be directly land applied, composted or dried.

The production of sulphides and mercaptans from sulphur-containing amino acids in these ponds can result in odour production (Stucky, 1990). The high protein diet of factory-farmed animals may exacerbate this problem. Anaerobic pond odour problems appear to be most commonly associated with piggery wastes (Galvin *et al.*, 2002). They are much less common for ponds treating the more digested wastes of ruminants (e.g. beef and dairy cows). However, care should be taken to site ponds away from areas where they can cause a nuisance.

Buffer zones of at least 400m (and up to 760m) from residences are recommended by ASAE (2001), while New Zealand guidelines for farm dairy ponds (DEC, 1996) propose a minimum distance of 300 m from public areas and at least 45m from the outside edge of dairy parlours and associated yards. Pond desludging can result in odour release, as surface crusts, scum layers or overlying aerobic layers are disrupted, and settled wastes are mixed and exposed to the air. Desludging, therefore, needs to be undertaken with particular care.

Seepage from anaerobic ponds has the potential to contaminate groundwater and affect the safety of wells and other water supplies. Anaerobic ponds should normally be constructed in low permeability soils, away from areas with high water tables. Compaction of soils, or lining with clays or synthetic membranes may be required to provide a suitable seal. ASAE suggests a minimum distance of 90 m from other groundwater abstractions.

Stormwater or surface runoff entering the ponds reduces waste retention times. Ponds and inflow devices should be designed and positioned to prevent this extraneous water from entering the system.

19.3.2 Facultative ponds

Facultative ponds contain an anaerobic zone overlain by an aerobic layer. The combination of these conditions within a single pond, in theory, can result in a synergy between aerobic and anaerobic digestion processes. In reality, the various microbiological processes frequently do not operate at high efficiencies due to fluctuating environmental conditions. In addition, the high oxygen demand of livestock wastes, even after pre-treatment in anaerobic ponds, makes it difficult to consistently maintain surface aerobic conditions. In order to maximise oxygen input, facultative ponds have a larger surface area and are shallower (<1.5m deep) than anaerobic ponds. Oxygen enters via diffusion through the water/air interface and from algal photosynthesis.

The design criteria for these ponds are based on areal organic loading. USDA (1992) recommends loading rates of 22–67 kg BOD_5 ha^{-1} d^{-1} for various climatic regions in the United States (Table 19.8), whilst a single rate of 84.3 kg BOD_5 ha^{-1} d^{-1} has been recommended (DEC, 1996) throughout New Zealand for dairy parlour wastewaters.

In addition to the high oxygen demands, animal wastewaters can strongly absorb light, reducing its availability for algal growth and photosynthetic oxygen production. The algae found in these ponds may thus be found only in a very shallow zone close to the surface. Euphotic depths of only 10–15cm can cause instability in the algal population, resulting in highly variable performance (Hickey *et al.*, 1989a; Sukias *et al.*, 2001).

Table 19.8 Summary of maximum recommended facultative pond loading rates for livestock wastes

Location	Latitude	Approx. mean annual air temperature (°C)	Climate[1]	BOD Loading ($kg\ ha^{-1}\ d^{-1}$)
United States[2]				
Northern USA	47° (50°+)[a] N	5-10	Te/Co	28
Mid-latitude USA	37° (42°)[a] N	12	Te/Co	45
Southern USA	32° (37°)[a] N	17	Wt/Co	56
Southern Texas/Florida	28° N	23	Ar/Co, St/Oc	67
Hawaii/Puerto Rico	18-22° N	23-27	Tr/Oc	67
New Zealand[3]	(Specifically for dairy parlour wastewaters)			
Whole country	39-47°S	7.5-15.5	Wt-Te/Oc	84

[1] Basic climate types: Te = temperate, Wt = warm temperate, Ar = arid, St =sub-tropical, Tr = tropical; Climatic influences: Co = continental; Oc =oceanic

[2] ASAE (2001), extrapolated to tropical areas based on USDA/SCS (1992).

[3] DEC (1996)

[a] Proposed latitudinal ranges vary across the continent. Values are estimates for mid-continental and eastern states, with approximations for western coastal states given in parenthesis.

[b] Based on VS:BOD ratios for raw wastes ASAE (2001)

19.3.3 Mechanically aerated ponds

Mechanical aeration, either using floating aerators or submerged diffusion systems (spargers), can be used to increase the supply of oxygen available for degradation of organic matter by facultative heterotrophic bacteria. Aerated ponds can be constructed both deeper and with much smaller surface areas than conventional facultative ponds since algal oxygenation is less important. Once the oxygen demand required for organic matter degradation is satisfied, microbial nitrifiers become established utilising the available oxygen to convert ammonium to nitrate. This process proceeds best at oxygen concentrations of about 2g m^{-3}, although low levels of nitrification have been reported at concentrations below 1g m^{-3}. Microbial oxidation of sulphides also occurs where free-oxygen is available, reducing the potential for odour release.

"High performance" or "complete-mix" aerated ponds include a high degree of mixing combined with aeration, so that a high proportion of solids are kept in suspension. These ponds, which are essentially a low-rate activated sludge system, have hydraulic residence times (HRT) of 1–10 days, a depth of 3–5m (Loehr, 1984), and are more common in intensive livestock sites. They generally require a post-aeration settling basin or clarifier to remove solids from the effluent. These systems are capable of high levels of treatment with very low odour risk, although they have high energy and management requirements.

"Partial-mix" aerated ponds are more common for livestock waste treatment. Aeration is at a much lower level, and may be achieved by addition of an aerator to an existing facultative pond with an HRT of 20-40 days. Aeration may be used to provide enhanced BOD_5 and ammonium reduction, or may be installed for odour reduction (Schulz and Barnes, 1990). Basic design recommendations for BOD reduction are outlined in USDA/SCS (1992) and Reed (1985). High levels of nitrification of ammonium in the effluent are possible (see Section 19.4.2 for a case study), substantially reducing the potential toxicity of discharges to surface waters. Subsequent denitrification using constructed wetlands or wood-chip filters can potentially be used to reduce nitrogen concentrations to low levels.

Intermittent aeration has also been used to generate sequential aerobic and anoxic conditions conducive to nitrification/denitrification within a single pond (Araki *et al.*, 1990; Koottatep *et al.*, 1993), and this technique may also prove suitable for livestock applications. Addition of biofilm attachment surfaces has been shown to enhance nitrification in aerated (and facultative) ponds by promoting development of stable populations of slow-growing nitrifiers (Shin and Polprasert, 1988) (see Section 19.4.2).

19.3.4 Maturation ponds

As proposed in Chapter 9 for ponds treating domestic wastewater, a series of shallow (1–1.5 m) short residence-time (~3–4 d.) maturation ponds can be employed after facultative ponds to enhance disinfection and further reduction of SS, BOD and nutrients. Their use, however, appears to be relatively rare for livestock wastes. In New Zealand they have been proposed (DEC, 1996) as a practical way of dealing with increasing farm dairy waste loads to two-pond systems (e.g. due to gradually increasing herd size) and improving overall pond performance where final discharge is to surface waters. Limited monitoring data for such a system showed enhanced treatment performance compared to two-pond systems (Selvarajah, 1996). Maturation ponds are also a key component of Advanced Integrated Wastewater Pond Systems (see Sections 19.3.5, 19.4.3 and

Chapter 13). Basic design procedures should follow the guidelines outlined in Chapters 9 and 13, as appropriate.

19.3.5 High rate pond systems

High rate ponds (HRP's, or high rate algal ponds), initially developed by Oswald and colleagues (Oswald and Gotaas, 1955; Oswald *et al.*, 1957) comprise a shallow pond (0.2–0.8 m; residence time 2-8 days) mixed by a paddle wheel that promotes the development of an intensive algal culture (see Chapter 13).

HRP's have been used to effectively treat piggery wastes, cattle feedlot wastes (Fallowfield *et al.*, 1992), poultry (Oswald and Green, 1992) and dairy parlour wastes (Craggs *et al.*, 2003, and see Section 19.4.3).

Nutrients and CO_2 are assimilated by algal growth, resulting in supersaturation of oxygen during the daytime and elevation of pH to 9 and above. Organic matter is decomposed by bacteria, ammonia nitrogen volatilised and phosphorus precipitated with calcium (provided alkalinity is sufficient). High exposure to light, combined with high pH and dissolved oxygen can also promote rapid inactivation of faecal micro-organisms (Davies-Colley *et al.*, 2003).

HRP's are generally combined with a preceding anaerobic treatment system, and a subsequent algal settling basin, in which the majority of nutrients are removed as biosolids (50–70% of TN and TP). Algal settling basins may be followed by maturation ponds, and/or constructed wetlands, and in combination, are generally referred to as Advanced Integrated Ponds. Such systems offer significant potential for resource recovery. Methane capture from the anaerobic pond can provide an energy source for use on the farm. Periodic removal, and application to the land, of settled solids allows nutrients to be recycled and provides beneficial organic soil amendments (see Chapter 13). Alternatively the algal biomass can be reused as a high protein (~50%) livestock feed (Duggan *et al.*, 1972).

Advanced integrated pond systems have been used to treat a variety of livestock wastes. Fallowfield *et al.* (1992) describe a treatment system for pig slurry, using an aerobic digester system followed by two HRP's. BOD in the final effluent averaged between 10–80 g m^{-3}. On a dairy farm, where milking parlour wastes were treated in an anaerobic and facultative pond, Craggs *et al.* (2003) replaced the existing facultative pond with an HRP, algal settling ponds and maturation ponds to achieve a higher quality effluent (see Section 19.4.3 below for more details on this system).

19.4 FARM DAIRY CASE STUDY – NEW ZEALAND

19.4.1 Anaerobic and facultative pond systems

In New Zealand, dairy cows are largely pastured with supplementary feed produced "off-farm" providing a relatively small component of the animal diet. As a result, the cows spend the majority of their time on the pasture, and little time in housed areas. The exception is during milking, when the cows enter the milking parlour (or "milking-barn" or "dairy-shed"), generally twice a day, during a 9-month milking season. The wastes generated in the milking parlour constitute about 10% of the daily waste production of urine and faecal matter, and these have commonly been treated in two-pond (anaerobic and facultative pond) systems. The anaerobic ponds achieve 85% removal of BOD_5 (Ministry of Agriculture and Fisheries: Policy, 1994).

Despite good removal in the preceding anaerobic pond, incoming wastewater to the facultative pond is still very high in total oxygen demand due to the high COD:BOD_5 ratio of dairy wastes (12.1, c.f. 1.5-1.8 in domestic sewage, Bucksteeg, 1987; Llorens *et al.*, 1992). Thus BOD_5 removal in the facultative pond (generally 40–50%) is much lower. However the two ponds in combination generally achieve close to 95% removal of BOD_5 and SS.

In addition, the two ponds in combination remove around 70% of total N and P (based on data summarised in Longhurst *et al.*, 2000), much of which is eventually recycled to land during desludging operations.

Evaluation of dairy facultative ponds revealed euphotic depths of only 10–15cm, indicating considerably higher light attenuation than that found in equivalently loaded sewage facultative ponds. This was due to median beam attenuation coefficients and turbidity's 5–7 times higher than in sewage ponds (Sukias *et al.*, 2001). Despite this, depth integrated measurements of algal abundance, as indicated by Chl *a*, had a median of 1041 mg m^{-3}, similar to that reported for New Zealand sewage ponds (Davies-Colley *et al.*, 1995; Hickey *et al.*, 1989b). The algae were able to receive sufficient light by concentrating in a very restricted zone near the surface during the day.

Faecal coliform concentrations correlated positively with variables associated with light attenuation, such as turbidity, and correlated negatively with 1% light penetration depth, suggesting much of the measured faecal coliform removal is associated with UV light exposure (Davies-Colley *et al.*, 1997).

Figure 19.1 Standard combination of anaerobic (left) and facultative (right) ponds used widely for treatment of farm dairy waste in New Zealand since the 1970's. The treatment performance of these systems, particularly in terms of ammoniacal N, nutrient and faecal microbial contaminants, is increasingly being challenged as the environmental standards required for surface water discharges become more stringent.

19.4.2 Mechanical aeration of facultative ponds

Mechanical aeration of a facultative pond offers the potential to reduce effluent oxygen demand and promote nitrification of ammonium. Sukias *et al* (2000) split an existing large facultative pond (treating waste from a 350 cow dairy farm) with an impermeable barrier, and equally divided the inflow between the two halves. One side was continuously aerated with a 1.5 kW horizontal aspirator-type aerator, while the other side was not aerated. Within 6 weeks, ammoniacal-N in the aerated side had reduced by 99% via nitrification. BOD_5 in the aerated side was also half of that in the non-aerated side (see Table 19.9).

In an attempt to reduce aeration costs and promote sequential denitrification processes, trials were conducted with night-only aeration, allowing daytime algal photosynthesis to replace mechanical aeration during the day. However under this regime the hypolimnetic zone, (where most of the nitrifying bacteria probably resided at the interface with the pond base), became fully anoxic in the daytime, and nitrification rates were reduced. Ammoniacal-N removal, however, was still relatively high and ranged from 84–90% (over two successive dairy seasons, Table 19.9).

Table 19.9 Summary of BOD and ammoniacal-N removal with various aeration regimes (from Sukias *et al.*, 2000)

	Influent BOD_5	Effluent BOD_5		Influent Ammoniacal-N	Effluent Ammoniacal-N	
	($g m^{-3}$)	($g m^{-3}$)	% removal	($g m^{-3}$)	($g m^{-3}$)	% removal
On-aerated	130	43	67[a]	100	40	60
Continuously aerated	130	22	83	100	<1.0	99
Night aerated[b]	84–130	35–49	53–59	95-134	10–17	84–90
Night aerated (with biofilm supports)	93	37	60	116	8	93

[a] This well-operated pond achieved much higher removal than usual for facultative ponds.

[b] Operated over two successive seasons.

Addition of "biofilm attachment surfaces" in the ponds, such as geotextile material suspended on frames, in combination with night-only aeration enabled the slow growing nitrifying bacteria to remain permanently in the aerobic surface waters. This improved ammoniacal-N removal to 93%. In all of the aeration regimes tested, denitrification was not significant. Overall, continuous aeration provided the highest level of ammoniacal-N removal for these dairy ponds.

19.4.3 Advanced integrated ponds

Advanced integrated ponds treating dairy wastes provided considerably improved treatment compared with standard ponds, see Table 19.10 (Craggs *et al.*, 2003) with 65% lower BOD_5, 71% lower SS, 54% lower TKN, 74% lower NH_4, 24% lower TP, and a two fold reduction in *E. coli*. Secondly, variability in effluent quality was considerably reduced, as shown by comparison of differences between median and 95 percentile concentrations. Advanced pond systems at full scale (Figure 19.2) have shown similar levels of improvement (Craggs *et al.*, 2004).

Figure 19.2 HRP in a full-scale Advanced Integrated Pond system in New Zealand. Note slow-moving paddlewheel (left background) and baffles to direct flow

Table 19.10 Effluent characteristics of advanced pond and conventional pond systems (Craggs *et al*, 2003)

	Advanced pond system		Two-pond system	
Effluent characteristic	Median	95 percentile	Median	95 percentile
Temperature (°C)	18.0	26.3	18.3	26
pH	8.7	9.8	8.1	8.6
Dissolved oxygen (% saturation)	54	132	36	106
Conductivity (μS cm^{-1})	1000	1171	1142	1669
Alkalinity (g $CaCO_3$ m^{-3})	360	551	537	884
BOD_5 (g m^{-3})	34	76.8	108	250
Filter BOD_5 (g m^{-3})	8.0	28.2	20.0	149.1
Suspended solids (g m^{-3})	64	340	220	523
Volatile suspended solids (g m^{-3})	55	319	190	480
Chlorophyll *a* (mg m^{-3})	142	2924	1375	10296
Total Kjeldahl nitrogen (g m^{-3})	25.3	80.1	55.0	88.1
Ammoniacal-nitrogen (g m^{-3})	7.5	28.8	28.9	51.3
Oxidised nitrogen (g m^{-3})	0.9	0.9	0.9	1.1
Total phosphorus (g m^{-3})	15.2	31.9	20.0	73.9
Dissolved reactive phosphorus (g m^{-3})	12.8	22.9	17.1	24.4
E. coli (MPN $100ml^{-1}$)	$1.46x10^2$	$5.75x10^3$	$1.62x10^4$	$1.09x10^5$

19.4.4 Supplementary wetland treatment

Constructed wetlands have been trialled in North America for livestock waste treatment after settling basins and anaerobic ponds (Knight *et al.*, 2000), and in New Zealand as a means of upgrading farm dairy effluent quality from two-pond systems (Tanner and Sukias, 2003; Tanner *et al.*, 1998). Surface-flow wetlands are generally recommended for treatment of livestock wastewaters, because they are cheaper to construct than subsurface-flow systems, and are less susceptible to clogging through accumulation of refractory herbage solids (Tanner *et al.*, 1998).

Guidelines have been developed to achieve 3 different levels of supplementary wetland treatment under New Zealand conditions (Tanner and Kloosterman, 1997). Addition of a small surface-flow wetland (415 m^2 for the milking parlour wastes of a 200 cow herd) can provide basic buffering of effluent flow and water quality from a two-pond system, and remove an additional 30-70% of BOD and suspended solids, and 70-85% additional removal of faecal coliforms, but contribute little to reduction of nutrients. Larger surface-flow wetlands followed by a horizontal subsurface-flow gravel-bed wetland (500 m^2 and 215 m^2, respectively, for the milking parlour wastes of a 200 cow herd) will provide an additional 60-75% removal of BOD and suspended solids, 35-50% removal of ammoniacal and total nitrogen, and 85-95% removal of faecal coliforms compared with a two-pond system alone. Considerably higher levels of treatment, particularly ammoniacal and total nitrogen removal (65-80% greater than in two-pond systems alone), are possible using a combination of surface followed by subsurface flow constructed wetlands after mechanically-aerated facultative farm dairy ponds that provide pre-nitrification (Tanner and Kloosterman, 1997).

Refer to Chapter 15 for further information on integrated pond and wetland systems.

19.5 PIGGERY CASE STUDY

19.5.1 Anaerobic and facultative ponds

Anaerobic and facultative waste stabilisation ponds in series have commonly been used to treat piggery wastewater in New Zealand, because of their relatively low cost and maintenance requirements and their ability to significantly reduce BOD and SS levels (generally >90% removal). Performance of a three-stage pond system has been studied in detail at Massey University (Warburton, 1983). The system treated wastes from 1,786 pigs, (loading of 0.14 kg BOD per pig d^{-1} and 25 L pig d^{-1}) in an anaerobic lagoon (4.5 m deep, 10,000

m^3) followed by two 1 m deep facultative ponds of 8,400 and 3,240 m^2 respectively. The measured annual average flows into and out of the combined pond system were 47 and ~20 m^3 d^{-1}, with minimal outflow measured from the facultative ponds during summer. Oxygen demand and solids concentrations were reduced significantly (Table 19.11), with greatest efficiency in the anaerobic pond. Average final effluent BOD_5 and SS for the WSP system were below 100 g m^{-3}, providing 98-99% removal overall. Mean N and P loads were reduced by 86 and 60 % respectively.

Final discharge quality for three other NZ piggery WSP systems (at Paerata, Ramarama and Hautapu) is shown in Table 19.12. Median ammoniacal N levels exceeded 200 g m^{-3} at two of the sites, suggesting aeration would have been an appropriate pre-treatment system. Also faecal coliform bacterial concentrations in the final effluent remained relatively high. At Paerata a solids-separator was operated prior to the wastewater entering the ponds, while at Ramarama the wastewater was reused as flushing water, resulting in higher salinity (electrical conductivity 360 mS m^{-1} compared to 180 mS m^{-1} at Paerata).

Table 19.11 Mean annual flow-proportional concentrations of contaminants and % mass removal (in brackets) during WSP treatment (Warburton, 1983)

	Raw	Anaerobic		Facultative 1		Facultative 2		Total system
	(g m^{-3})	(g m^{-3})	%	(g m^{-3})	%	(g m^{-3})	%	%
COD	11,210	1,565	(85)	654	(49)	398	(53)	96
BOD	5,500	439	(91)	187	(48)	72	(70)	99
SS	4,364	256	(94)	173	(18)	85	(44)	98
TS	7,940	1,690	(76)	1,170	(16)	840	(47)	95
VS	5,876	43	(65)	38	(4)	36	(62)	89
TN	174	80	(49)	30	(54)	23	(40)	86
TP	211	151	(21)	88	(29)	81	(28)	60

Table 19.12 Mean piggery effluent characteristics (g m^{-3}) for 4-stage WSP treating piggery wastewaters, North Island, New Zealand (Tanner and Sukias, 2003)

Site	$CBOD_5$	SS	TP	TN	NH_3-N	NO_x-N	FC
	(g m^{-3})	(g m^{-3})	(g m^{-3})	(g m^{-3})	(g m^{-3})	(g m^{-3})	cfu (100mls)$^-$
Paerata	185	309	42.4	81	43.6	0.39	2.9 x 10^3
Ramarama	70	1204	22.6	273	252	<1	1.0 x 10^4
Hautapu	98	155	39	204	220	0.07	3.9 x 10^4

Wastewater strength and composition are known to be influenced by pig diet (Clanton *et al.*, 1991), waste removal and separation procedures (Hill and Bolte, 1986; Payne, 1986), the effectiveness of aerobic pond treatment systems (Oleszkiewicz, 1986) and recycling of pond wastewater, and these factors are likely to have contributed to the differing wastewater composition at the three sites

All of the piggeries facultative pond waters were conspicuously red brown in colour, consistent with the presence of autotrophic sulphur bacteria, which tend to be common in piggery WSP's. These bacteria use the sulphur in H_2S (rather than the oxygen in H_2O) as an electron donor and hence remove hydrogen sulphide, reducing the potential for odours. These bacteria have poor settling characteristics, reducing the potential for subsequent removal in constructed wetlands (Tanner and Sukias, 2003). At wetland retention times of 11-15 days median additional mass removals ranged between 40-56% SS, 49-57%, $CBOD_5$, 19-45% ammoniacal N, and 16-37% TN for the three systems. Median faecal coliforms increased during passage through one of the wetland systems and were reduced by 50-60% at the others. Overall these studies concluded that higher levels of pretreatment (e.g. aeration) would be required before wetland treatment to substantially reduce ammoniacal and total N levels, and enable them to be safely discharged to surface waters.

19.6 SUMMARY AND FUTURE RESEARCH NEEDS

Ponds are a widely used and effective means of treating livestock wastes. Livestock wastes are of significantly higher strength and differing characteristics to domestic wastes, requiring pond loading criteria specific to each wastewater type. Use of anaerobic ponds to treat livestock wastes is more prevalent than for domestic sewage, and their final effluents are often land applied as a fertilizer. Addition of facultative, maturation, mechanically aerated and advanced integrated ponds, and complementary systems, such as constructed wetlands, provide additional treatment and greater levels of environmental protection. This becomes important where the potential for land application is limited or effluents are discharged to surface waters.

Specific microbiological risks associated with the use and discharge of livestock wastes are poorly understood, and epidemiological studies are needed to understand the health implications of exposure to these wastes by humans, livestock and wild animals.

Because of increasingly stringent environmental requirements, particularly in developed regions of the world, further research is needed to develop and field test pond systems capable of advanced levels of treatment (including nutrient removal and disinfection) and resource recovery. The increasing trend towards large concentrated livestock operations in many regions provides new challenges for pond treatment, whilst offering considerable potential for innovative approaches for water reuse, recovery of energy, nutrients, and organic compounds, and reduction in greenhouse gas emissions.

REFERENCES

Ap Dewi, I., 1994. The use of animal waste as a crop fertilizer. In: I.F.M.M. I. Ap Dewi, and H.M.Omed (Editor), *Pollution in livestock production systems*. CAB International, Wallingford, Oxon, UK, pp. 309-331.

Araki, H., Koga, K., Inome, K., Kusuda, T. and Awaya, Y., 1990. Intermittent aeration for nitrogen removal in small oxidation ditches. *Water Science and Technology* **22**(3/4), 131–138.

ASAE, 1998. *Manure storages ASAE EP393.3 DEC98*, St. Joseph, MI.

ASAE, 1999. *Design of anaerobic lagoons for animal waste management EP403.3 JUL99*, ASAE, St Joseph, MI.

ASAE, 2000. *Manure production and characteristics; ASAE D384.1 DEC99*, St. Joseph, MI.

ASAE, 2001. *Design of anaerobic lagoons for animal waste management*, ANSI/ASAE EP403.3 JUL99, St. Joseph, MI., pp. 686-690.

Ayers, R.S. and Westcot, D.W., 1985. *Water Quality for Agriculture*. FAO Irrigation and Drainage Paper 29, Revision 1, Food and Agriculture Organisation of the United Nations, Rome, Italy.

Blumenthal, U.J., Mara, D.D., Peasey, A., Ruiz-Palacios, G. and Stott, R., 2000. Guidelines for the microbiological quality of treated wastewater used in agriculture: recommendations for revising WHO guidelines. *Bulletin of the World Health Organisation* **78**(9), 1104-1116.

Bucksteeg, K., 1987. German experiences with sewage treatment ponds. *Water Science and Technology* **19**(12), 17–23.

Cheung, W.H.S., Chang, K.C.K., Hung, R.P.S. and Kleevens, J.W.L., 1990. Health effects of beach water pollution in Hong Kong. *Epidemiology and Infection* **105**, 139-162.

Clanton, C.J., Nichols, D.A., Moser, R.L. and Ames, D.R., 1991. Swine manure characterization as affected by environmental temperature, dietary level intake, and dietary fat addition. *Transactions of the ASAE* **34**(5), 2164–2170.

Cliver, D.O. and Moe, C.L., 2004. Prospects of waterborne viral zoonoses. In: J.A. Cotruvo *et al.* (Editors), *Waterborne zoonoses: identification, causes and control*. Emerging issues in water and infectious diseases series. IWA Publishing, Colchester, pp. 242-254.

Cooper, A.B., 1986. Developing management guidelines for river nitrogenous oxygen demand. *Journal of the Water Pollution Control Federation* **58**, 845–852.

Craggs, R.J., Sukias, J.P., Tanner, C.C. and Davies-Colley, R.J., 2004. Advanced pond system for dairy-farm effluent treatment. *New Zealand Journal of Agricultural Research* **47**, 449–460.

Craggs, R.J., Tanner, C.C., Sukias, J.P.S. and Davies-Colley, R.J., 2003. Dairy farm wastewater treatment by an advanced pond system. *Water Science and Technology*, 48(2), 291-298.

Davies-Colley, R.J., Craggs, R.J. and Nagels, J., 2003. Disinfection in a pilot-scale "advanced" pond system (APS) for domestic treatment in New Zealand. *Water Science and Technology* **48**(2), 81-87.

Davies-Colley, R.J., Craggs, R.J., Park, J. and Nagels, J.W., 2005. Optical characteristics of waste stabilization ponds - recommendations for monitoring. *Water Science and Technology* **51**(12), 153-161.

Davies-Colley, R.J., Donnison, A.M., Speed, D.J., Ross, C.M. and Nagels, J.W., 1999. Inactivation of faecal indicator micro-organisms in waste stabilisation ponds: interactions of environmental factors with sunlight. *Water Research* **33**(5), 1220–1230.

Davies-Colley, R.J., Hickey, C.W. and Quinn, J.M., 1995. Organic matter, nutrients, and optical characteristics of sewage lagoon effluents. *New Zealand Journal of Marine and Freshwater Research* **29**, 235–250.

Davies-Colley, R.J., Speed, D.J. and Donnison, A.M., 1997. Sunlight wavelengths inactivating faecal microbial indicators in waste stabilisation ponds. *Water Science and Technology* **35**(11–12), 219–225.

DEC, 1996. *Dairying and the environment– managing farm dairy effluent*, Dairying and the Environment Committee of the New Zealand Dairy Board, Palmerston North, NZ.

Duggan, G.L., Golueke, C.G. and Oswald, W.J., 1972. Recycling system for poultry wastes. *Journal of the Water Pollution Control Federation* **44**, 432–440.

Fallowfield, H.J., Svoboda, I.F. and Martin, N.J., 1992. Aerobic and photosynthetic treatment of animal slurries. In: G.M.G. J.C. Fry, R. A. Herbert, C. W. Jones and I. A. Watson-Craik (Editor), *Microbial Control of Pollution*. Society for General Microbiology Symposium. Cambridge University Press, Cambridge, pp. 171–197.

Galvin, G., Lowe, S., Atzeni, M. and Casey, K., 2002. *The effect of loading rate on odour emissions from anaerobic effluent ponds in South-East Queensland*, 4th Queensland environmental conference "Environmental solutions - meeting the challenge in 2002". Environmental Engineering Society, Brisbane, Australia.

Gilley, J.E., Spare, D.P., Koelsch, R.K., Schulte, D.D., Miller, P.S. and Parkhurst, A.M., 2000. *Copper and zinc in swine diets affect phototrophic anaerobic lagoons*. In: J.A. Moore (Editor), 8th International Symposium on Animal, Agricultural and Food Processing Wastes. American Society of Agricultural Engineers, Des Moines, Iowa, USA, pp. 656–663.

Gray, N.F., 1987. *Biology of wastewater treatment*. Oxford University Press, Oxford.

Hanselman, T.A., Graetz, D.A. and Wilkie, A.C., 2003. Manure-borne estrogens as potential environmental contaminants: a review. *Environmental Science and Technology* **37**, 5471-5478.

Henze, M., Harremoës, P., Jansen, J.l.C. and Arvin, E. (Editors), 1995. *Wastewater treatment: Biological and chemical processes*. Environmental Engineering. Springer-Verlag, Berlin, pp 383.

Hermanson, R.E. and Kalita, P.K., 1994. *Animal manure data sheet*. Clean Water for Washington, Cooperative Extension, Washington State University.

Hickey, C.W., Quinn, J.M. and Davies-Colley, R.J., 1989a. Effluent characteristics of dairy shed oxidation ponds and their potential impacts on rivers. *New Zealand Journal of Marine and Freshwater Research* **23**, 569–584.

Hickey, C.W., Quinn, J.M. and Davies-Colley, R.J., 1989b. Effluent characteristics of domestic sewage oxidation ponds and their potential impacts on rivers. *New Zealand Journal of Marine and Freshwater Research* **23**, 585–600.

Hill, D.T. and Bolte, J.P., 1986. Characteristics of whole and scraped swine waste as substrates for continuously expanding anaerobic digestion systems. *Agricultural Wastes* **16**, 147–156.

Knight, R.L., Payne, V.W.E.J., Borer, R.E., Clarke, R.A. and Pries, J.H., 2000. Constructed wetlands for livestock wastewater management. *Ecological Engineering*, 15, 41-55.

Koottatep, S., Leesanga, C. and Araki, H., 1993. Intermittent aeration for nitrogen removal in small aerated lagoon. *Water Science and Technology*, 28(10), 335–341.

Llorens, M., Sáez, J. and Soler, A., 1992. Influence of thermal stratification on the behaviour of a deep wastewater stabilization pond. *Water Research* **26**(5), 569–577.

Loehr, R.C., 1984. *Pollution control for agriculture*. Academic Press Inc., Orlando.

Longhurst, R.D., Roberts, A.H.C. and O'Conner, M.B., 2000. Farm dairy effluent: A review of published data on chemical and physical characteristics in New Zealand. *New Zealand Journal of Agricultural Research* **43**, 7–14.

Mangino, J., Bartram, D. and Brazy, A., 2002. *Development of a methane conversion factor to estimate emissions from animal waste lagoons*, 11th International emission inventory conference - "Emission inventories - partnering for the future". EPA, Atlanta, GA.

Marais, G.v.R., 1970. *Dynamic behavior of oxidation ponds*, 2nd International Symposium of Waste Treatment Lagoons. Missouri Basin Engineering Health Council & Federal Water Quality Administration, Kansas City, Missouri, pp. 15–46.

McBride, G.B., Salmond, C.E., Bandaranayake, D.R., Turner, S.J., Lewis, G.D. and Till, D.G., 1998. Health effects of marine bathing in New Zealand. *International Journal for Environmental Health Research* **8**(3), 173–189.

Ministry of Agriculture and Fisheries: Policy, 1994. *Dairy shed wastewater treatment ponds*. ISBN 0478-07381-X, ISSN 1171-4662, Ministry of Agriculture and Fisheries, Wellington, NZ.

NZLTC, 2000. Public health guidelines for safe use of sewage effluent and sewage sludge on land. In: L.J. Whiltehouse, H. Wang and M.D. Tomer (Editors), *Guidelines for utilisation of sewage effluent on land, Part 2: Issues for design and management*. New Zealand Land Treatment Collective, Rotorua, New Zealand, pp. 172-180.

Oleszkiewicz, J.A., 1986. Aerated lagoon treatment of piggery wastes: kinetics of carbon removal. *Agricultural Wastes* **16**, 121–134.

Oswald, W.J. and Gotaas, H.G., 1955. Photosynthesis in sewage treatment. *Proceedings of the American Society of Civil Engineers* **81**(686), 1–34.

Oswald, W.J., Gotaas, H.G., Golueke, C.G. and Kellen, W.R., 1957. Algae in waste treatment. *Sewage and industrial wastes* **29**, 437–457.

Oswald, W.J. and Green, F.B., 1992. *Advanced integrated ponds systems for dairies and feedlots: a new waste treatment and reclaimation technology for the farm*, Food and Agriculture Organization of the United Nations, Santiago, Chile.

Overcash, M.R., Humenik, F.J. and Miner, J.R., 1983. *Livestock waste management*. CRC Press Inc, Boca Raton, FL.

Payne, R.W., 1986. Characteristics of faeces from institutional and commercial piggeries. *Agricultural Wastes* **16**, 1–11.

Reed, S.C., 1985. Nitrogen removal in wastewater stabilization ponds. *Journal of the Water Pollution Control Federation* **57**(1), 39-45.

Schulz, T.J. and Barnes, D., 1990. The stratified facultative lagoon for the treatment and storage of high strength agricultural wastewater. *Water Science and Technology* **22**(9), 43–50.

Selvarajah, N., 1996. *Dairy farm effluent treatment pond performance in the Waikato region: A preliminary review of the regional survey*. In: I.G. Mason (Editor), Tertiary treatment options for dairyshed and piggery wastewaters. Department of Agricultural Engineering, Massey University, Palmerston North, NZ, pp. 18a–18h.

Shin, H.K. and Polprasert, C., 1988. Ammonia nitrogen removal in attached-growth ponds. *Journal of Environmental Engineering* **14**(4), 846–863.

State of California, 2001. California health laws related to recycled water, *The Purple Book, Excerpts from the Health and Safety Code, Water Code and Titles 22 and 17 of the California Code of Regulations, State of California*, Sacramento, CA.

Stucky, G.E., 1990. Waste Treatment Lagoons, *Agricultural waste management field manual.* United States Soil Conservation Service, pp. 12.25–12.48.

Sukias, J.P.S., Craggs, R.J., Tanner, C.C., Davies-Colley, R.J. and Nagels, J.W., 2000. *Continuous and night-only aeration of farm dairy lagoons to promote nitrification.* In: J.A. Moore (Editor), 8th International Symposium on Animal, Agricultural and Food Processing Wastes. American Society of Agricultural Engineers, Des Moines, Iowa, USA, pp. 142–150.

Sukias, J.P.S., Tanner, C.C., Davies-Colley, R.J., Nagels, J.W. and Wolters, R., 2001. Algal abundance, organic matter, and physico-chemical characteristics of dairy farm facultative ponds: implications for treatment performance. *New Zealand journal of agricultural research* **44**, 279–296.

Sutton, A.L., Jones, D.D., Joern, B.C. and Huber, D.M., 2001. *Animal manure as a plant nutrient resource.* Cooperative Extension Service, Purdue University, West Lafayette, Indiana.

Tanner, C.C. and Kloosterman, V.C., 1997. *Guidelines for constructed wetland treatment of farm dairy wastewater in New Zealand.* NIWA Science and Technology Series No. 48, National Institute of Water and Atmospheric Research, Hamilton, NZ.

Tanner, C.C. and Sukias, J.P.S., 2003. Linking pond and wetland treatment: Performance of domestic and farm systems in New Zealand. *Water Science and Technology* **48**(2), 331–339.

Tanner, C.C., Sukias, J.P.S. and Upsdell, M.P., 1998. Relationships between loading rates and pollutant removal during maturation of gravel-bed constructed wetlands. *Journal of Environmental Quality* **27**(2), 448–458.

Tyson, T.W., 1996. *Using irrigation to renovate livestock lagoons.* Alabama Cooperative Extension System, Agricultural Engineering, Auburn University, Alabama.

USDA/SCS, 1992. *Agricultural waste management field handbook.*

USEPA, 1998. *1998 Update of ambient water quality criteria for ammonia*, Criteria and Standards Division, U S Environmental Protection Agency, Washington D. C.

Vanderholm, D.H., Dakers, A.J., Drysdale, A.B., Giffney, A.R., Painter, D.J., Smith, K.A. and Warburton, D.J., 1984. *Agricultural waste manual.* New Zealand Agricultural Engineering Institute, Lincoln College, Canterbury, NZ.

Wang, H., Magesan, G.N. and Bolan, N., 2004. An overview of the environmental effects of land application of farm effluents. *New Zealand Journal of Agricultural Research*, 47, 389-403.

Warburton, D.J., 1983. *Lagoon performance treating livestock waste.* In: P.N. McFarlane (Editor), 15th New Zealand Biotechnology Conference. Massey University, Palmerston North, New Zealand, pp. 89–122.

WHO, 1989. *Health guidelines for the use of wastewater in agriculture and aquaculture.* Report series 778, World Health Organisation, Geneva, Switzerland.

20

Stormwater management ponds

Jiri Marsalek, Ben Urbonas and Ian Lawrence

20.1 INTRODUCTION

Stormwater management ponds (SMPs) represent a special class of ponds, which are commonly used for storage and treatment of stormwater. SMPs differ from traditional wastewater ponds by the type of medium treated (stormwater as opposed to sewage), pollutants of concern (typically solids and adsorbed chemicals as opposed to biodegradable organic waste and pathogens), short hydraulic residence times (for some parts of the stored stormflow, less than 2 days), emphasis on creation of recreational and habitat amenities, and other related aspects.

SMPs function by providing storage for stormwater drainage, therefore buffering flows and reducing runoff peaks, and enhancing stormwater quality by various treatment processes, among which settling is the most important. Furthermore, they offer additional benefits in the form of aesthetic/recreational amenities, groundwater recharge, habitat creation and protection of downstream

ISBN: 1843390205. Published by IWA Publishing, London, UK.

receiving streams against pollution and erosion. Numerous types of SMPs are described in the literature (Schueler, 1987), but for brevity, only the most common types are discussed in this chapter. In general, SMPs are classified with respect to the nature of storage, as stormwater detention or retention ponds. Stormwater detention ponds store stormwater only temporarily (for 6 to 48 hours), and following the storm, drain almost completely, except for a small permanent storage pool (also called micro-pool) by the outlet. On the other hand the stormwater retention ponds maintain fairly large permanent stormwater storage. Stormwater ponds can be located on-stream or off-stream; with each type having some advantages and disadvantages. On-stream stormwater ponds are built along the existing urban streams, by enlarging the stream channel. Besides the stormflow, on-stream stormwater ponds receive a continuous baseflow generated in the stream watershed, which may extend beyond the urban area. Off-stream stormwater ponds are excavated in new developments, can be designed for extended detention times, and typically receive only stormflow and no significant baseflow. SMPs can also incorporate elements of wetland biological processes (see Chapter 15 for further detail). In cases where the catchment discharge is attenuated and low in suspended solids, the stormwater ponds will function mostly as 'wetlands'. Some specific features of stormwater detention and retention ponds are discussed in Section 20.4.2 - 20.4.4.

The use of SMPs is common practice in the USA, Canada, Australia and Sweden, where tens of thousands of stormwater ponds have been built during the last 30 years. Their design has evolved mostly through empiricism, practical experience and ongoing research. Consequently, there is a great wealth of information on planning, designing and operating stormwater ponds, the most pertinent of which is summarised in this chapter. It must be emphasised that the brief expose presented herein is by no means complete and that additional information can be found in numerous references on this subject.

20.2 STORMWATER POND PROCESSES

In this section, processes that contribute to effective treatment in stormwater ponds are described. Understanding these processes provides the scientific basis underpinning stormwater pond design.

20.2.1 Hydraulics

Stormwater pond influent should be spread laterally by the pond inlet to dissipate the inflow momentum and to spread incoming particulate matter over the wider pond area. Outside of this inlet zone, conditions favouring quiescent settling should be encouraged to enhance settling and prevent hydraulic short-circuiting. In practical terms, this is usually achieved by empirical design, based on simple rules of thumb (pond length/width > 3) and some basic understanding of hydraulics. Refer to Chapter 10 for further details on hydraulic design.

Flow distribution in stormwater ponds is affected by a number of phenomena, including:

(a) inflow discharge, velocity and density;
(b) wind; and
(c) densimetric stratification of the stormwater pond.

High inflow discharges create elevated velocities within the stormwater pond which tend to disturb sediments and exacerbate short-circuiting problems. This situation may be aggravated by relative density differences between the inflow compared to the stormwater pond. Such density differences may be caused by differences in water temperature or chloride concentrations, especially in cold climates, where road salts are used in road maintenance. Thus, a chloride laden or cooler inflow may enter the stormwater pond as a sinking jet, or a lower density (warmer) inflow may enter as a buoyant jet. Both cases imply reduced mixing of the inflow with stormwater pond water and short-circuiting of a large part of the pond's volume. Regardless of the inflow, water bodies like stormwater ponds can become density stratified in their own right. Densimetric stratification of stormwater pond water impedes vertical mixing and makes them less efficient at settling out solids. This is a critical issue for stormwater ponds in inland areas in Australia, where there is low prevailing wind and extreme summer solar radiation conditions.

Effects of wind shear stress on the flow field in water bodies have been studied extensively for large lakes. Sustained wind across the stormwater pond surface will generate a flow field, with a significant surface velocity in the wind direction (up to 5% of the wind speed), sinking flow at the down-wind end, and return flow in the lower layer. This flow field affects stormwater pond performance in several ways, including

(a) short-circuiting of the surface layer;
(b) disturbance of settling; and
(c) effective turnover of stormwater pond water, with oxygenated surface water brought to the pond bottom and vice-versa.

While practical stormwater pond design will continue to rely on experience when choosing pond shape and orientation, recent publications on this subject describe the use of computational fluid dynamics (CFD) modelling for the assessment of flow fields in research studies of stormwater ponds (Shaw *et al.*, 1997). CFD modelling is undoubtedly a powerful tool for this type of analysis, but limitations remain with respect to the lack of verification of model results by field data (Marsalek *et al.*, 2000; Persson, 1999).

Finally, it should be noted that with respect to stormwater quality enhancement by storage, no stormwater pond is 100% efficient in utilisation of the "treatment" volume. Studies of cooling water ponds (Thackston *et al.*, 1987) showed that even the favourably shaped ponds were at best about 70% effective, with the remaining pond area occupied by relatively ineffective "dead" or recirculation zones. Field studies of poorly performing stormwater ponds indicate similar concerns for stormwater ponds. In practical terms, it means that the distribution of hydraulic residence times with respect to individual water particles is highly non-uniform with some fraction of the total inflow receiving less treatment than expected. Some allowance for these considerations is included in selection of the water quality capture volume (WQCV), as discussed in Section 20.4.1. In existing stormwater ponds with poor flow distribution, some improvement can be achieved by retrofitting flow baffles, which lengthen the flow path and contribute to better settling (Matthews *et al.*, 1997).

20.2.2 Water quality processes

Water quality processes in stormwater ponds and wetlands depend on the characteristics of the drainage area, with respect to flow rates and suspended solids concentrations, and the type of stormwater pond or wetland. Figure 20.1 summarises the different discharge patterns and pollutant compositions for each of the catchment conditions, the subsequent pollutant transport, transformation and interception process conditions, and the treatment measures appropriate to each set of conditions.

In the case of the 'impervious area' runoff, which may be high in trace metals, a zone of macrophytes, selected to generate a high biomass of plant stems or cellulose material, may be incorporated into the stormwater pond to adsorb metals or organic pesticides.

Often the available land area for siting the stormwater ponds is constrained. Occasionally, availability of land allows adoption of a large wetland system where a stormwater detention pond would normally be employed.

In practice, elements of each of the pollutant capture measures will be present in the final treatment measure whether it be a stormwater pond or constructed stormwater wetland. Stormwater detention ponds incorporate macrophytes to:

- enhance calming of flows through stormwater ponds;
- provide some biofilm adsorption capacity;
- oxygenate sediments in the primary deposition zone; and
- maintain benthic microbial ecology of the sedimentation zones.

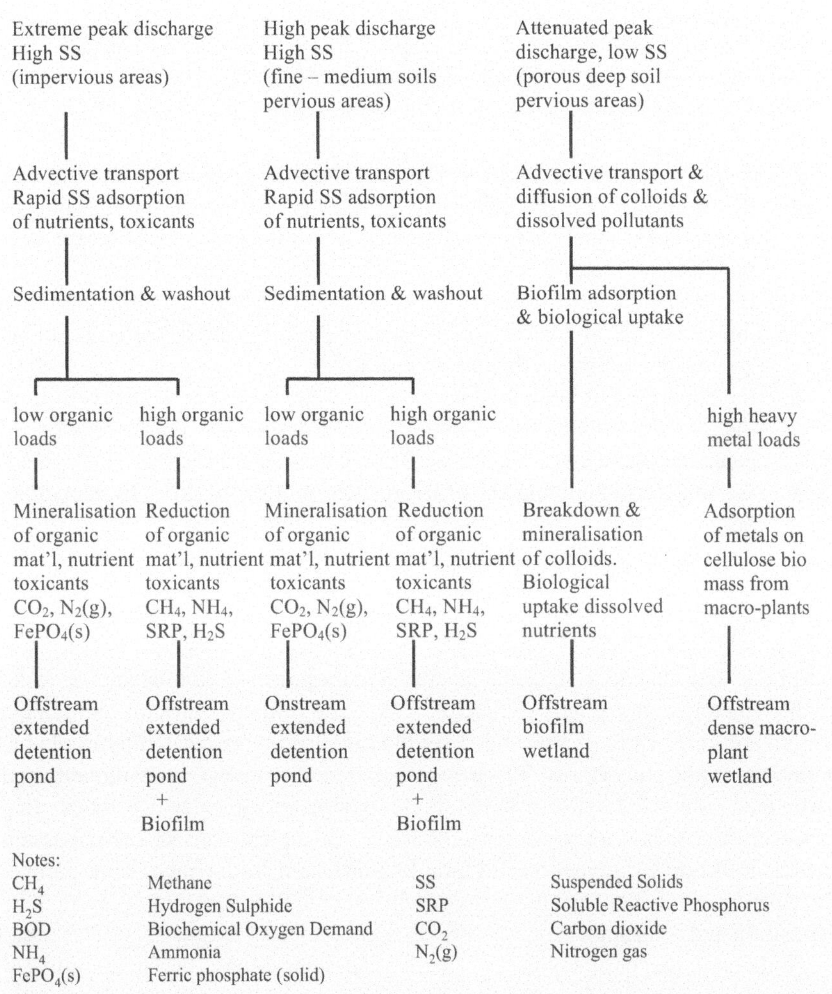

Figure 20.1 Decision tree guiding the selection of stormwater pond/wetland type

Often, a two-stage system will be employed, comprising a sedimentation basin followed by a shallow macrophyte wetland basin to provide further treatment via dissolved and fine colloidal material and denitrification. This approach requires the incorporation of extended detention into the sedimentation basin in order to limit detachment of biofilm within the second (wetland) basin.

The adoption of an extended detention feature in stormwater ponds and wetlands significantly enhances their treatment performance as well as providing attenuation of flows and enhancement of biodiversity. The drying of the extended detention zones after storm events enhances the mineralisation and fixation of a range of pollutants.

The sustainability of the stormwater pond and wetland processes depends on:

- the limitation of sediment load discharged to stormwater ponds or wetlands by incorporation of a sediment trap upstream of stormwater ponds or wetlands;
- the management of flow velocities within stormwater detention ponds to minimise the potential for re-suspension of settled material, or in stormwater wetlands, the loss of biofilm;
- the control of organic loading on stormwater ponds or wetlands to limit the potential for redox conditions and associated remobilization of settled pollutants;
- the long-term containment of intercepted pollutants either by recycling, burial in inert forms in the sediments (lithosphere), or return to the atmosphere as N_2(gas) and CO_2.

20.2.3 Sediment settling

Settling is undoubtedly one of the most important processes for enhancing water quality in stormwater ponds (Schueler, 1987) and removes not only particulate (commonly referred to as suspended solids) but also the associated hydrophobic pollutants (total phosphorus, TP, heavy metals, trace organic contaminants and particulate associated hydrocarbons). Field experience indicates that the settling in stormwater ponds, or similar facilities, is a complex process encompassing particle transport by advection and turbulent diffusion, with concomitant particle flocculation, and floc break-up by turbulence and the resulting particulate deposition and scouring.

To bypass the complexities of theoretical approaches, the treatability of stormwater by settling is commonly estimated empirically by testing stormwater samples in settling columns. Laboratory settleability tests indicate removal of particulate in the range from 0 to 90%, total phosphorus 50%, lead 65 to 85%, zinc 30 to 45% and copper about 40% for settling times ranging from 24 to 40 hours (Randall *et al.*, 1982). However, settling columns do not reproduce disturbance of quiescent settling by velocity fields and turbulence generated in stormwater ponds by flow circulation (driven by the incoming flow) or wind driven waves and currents. Consequently, settling columns may overestimate the settling efficiencies found in field conditions.

Besides flow conditions, settling efficiency in stormwater ponds also depends on physico-chemical characteristics of the local runoff and its particulate matter. Urban stormwater conveys larger quantities of solids, including sand, silt, clay and dissolved solids, which enter ponds in three forms, i.e., as the bed load or as the suspended or dissolved loads. The bed load consists mostly of sand ($D > 62$ μm) and the suspended load consists mostly of silt ($4\ \mu m < D < 62\ \mu m$) and clay ($D < 4\ \mu m$). Sand entering ponds settles quickly by the inlet (Marsalek *et al.*, 1997), the settling of finer particulate is more complicated.

The use of 'percent solids removal' as a measure of settling efficiency is ambiguous and inadequate without specifying the nature of solids. Sandy particles can be removed with almost 100% efficiency by simple designs with short hydraulic residence times in the order of minutes. On the other hand, fine clays or larger flocs of low density may require days to settle, or do not settle at all. Thus, stormwater pond effluents convey some residual (non-treatable) particulate concentrations (20-40 $mg.L^{-1}$), which are attributable to very fine particulate and low-density flocs, kept in suspension by naturally occurring turbulence. Another shortcoming of the percent removal measure is the fact that for high incoming concentrations, even substantial removals (60-70%) may leave residual concentrations which are still too high for the protection of sensitive receiving waters (Strecker *et al.*, 2004).

20.2.4 Ecology

An understanding of constructed stormwater pond and wetland ecology is important, both from the viewpoint of maintaining good treatment, as well as meeting community values with respect to supporting a diversity of plants and animals and landscape visual appeal. As noted previously, the biota play an important role in the transformation and interception of pollutants.

In the case of the stormwater ponds, the dominant ecosystem may be described as secondary benthic (heterotrophic bacteria) production, in association with limited primary production (algae, macrophytes), grazing and predation. For stormwater wetlands, the dominant ecosystem is "primary (benthic biofilm, macrophytes and epiphytic algae) production, in association with grazing and predation". These ecosystems are illustrated in Figure 20.2.

The high frequency of stormwater discharges (causing washout of larger biota) and the pulsed nature of external loads driving the stormwater pond or wetland processes (creating a lag in biota population growth) result in a limited role for animals in ecological processes as compared to natural ponds and wetlands. It is physical adsorption, sedimentation and the benthic microbes, plants and biofilm that are the key elements with respect to achieving treatment effectiveness.

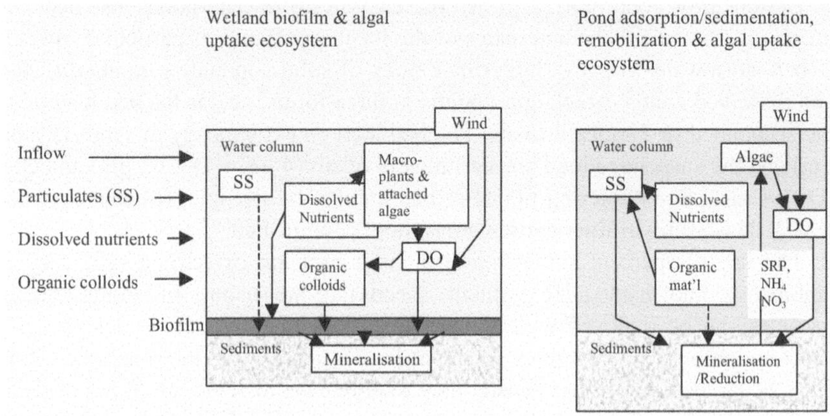

Figure 20.2 Ecosystem categories

Macrophytes play an important role by:

- sustaining attached bio-film systems;
- contributing to oxidation of sediments;
- providing substrate for epiphytes;
- generating the moderately labile carbon promoting denitrification or the active carbon cellulose biomass adsorbing heavy metals and organics.

Algae also play an important role in moderating O_2, CO_2, and pH levels, and in direct uptake of dissolved nutrients in waters, even though this role is not fully understood. It was noted that algal communities can serve as biological indicators of SMP performance and that the pond hydraulic residence time can be used to inhibit undesirable nuisance algae (Olding, 2000).

The two stormwater wetland categories most closely related to stormwater treatment systems are:

- marshes - wetlands dominated by soft-stemmed submerged, emergent and floating herbaceous plants, grasses and sedges;
- ephemeral or occasionally flooded zones - aquatic and terrestrial plants typical of most soil habitats.

While having less diversity of animals than natural ponds and wetlands, constructed stormwater ponds and wetlands nevertheless sustain a wide range of animals, including:

- hydra and sponges, rotifers (wheel animalcules) and mussels grazing the small organic particles;
- worms and snails grazing larger organic material in and on the sediments;

- water fleas and copepods grazing on the algae and biofilm;
- shrimps, scuds and yabbies grazing the larger organic detritus;
- insect nymph forms of Caddisfly, Mayflies, Dragonflies and Damselflies, together with water bugs and beetles predating the smaller macro-invertebrates;
- larger animals, including frogs, toads, salamanders, lizards, turtles, water rats, rodents (lemmings, mice and voles).

With respect to fish and bird communities, stormwater pond populations are often similar to those found in constructed wetlands. Fish commonly established in constructed wetlands comprise bass, carp, eels, goldfish, minnow, mud-minnows, mosquito fish, perch and shiner. Stormwater wetlands may represent significant habitats for water birds, with some 30% of total species associated with wetlands. Constructed urban stormwater wetlands provide a significant refuge during periods of extended drought, diminishing the available area of natural wetland habitats. The major wetland groups of birds include geese, swans, duck, teal, mallard, grebe; cormorants, pelican, egret, heron, bittern, spoonbill, ibis, stork, brolga, crane; moorhen, swamphen, coot, rail, bustard.

20.3 PERFORMANCE OF STORMWATER MANAGEMENT PONDS

The volume of reliable and comprehensive data on performance of stormwater management ponds has been steadily growing during the last 15 years. The collection of such data is undoubtedly complicated by the intermittent nature of stormwater discharges and seasonal variation of stormwater pond treatment processes. Proper monitoring requires continuous gauging of pond inflows and outflows, and water quality sampling during both storm events and intervening low flows over a 6 to 24 month period. Detailed guidance for monitoring stormwater pond performance (as well as the performance of other stormwater management measures) can be found elsewhere (GeoSyntec Consultants, UDFCD and UWRRC, 2002; Urbonas, 1995).

Using data reported in the literature from investigations at 51 separate stormwater detention facilities across four countries, Duncan (1997) reported treatment efficiencies for stormwater wet detention ponds, extended detention stormwater ponds and stormwater wetlands (Table 20.1). The use of the percent removal for assessing stormwater pond performance has some limitations. This parameter depends on the magnitude of the influent concentrations, and therefore, for a highly polluted inflow, even high removals (90%) may not provide an adequate protection for receiving waters and vice versa. Consequently, other performance measures, for example 'event mean concentrations', are also sometimes used. At present, perhaps the best source of stormwater pond performance data is the International Stormwater Best Management Practices (BMP) Database, which was established by the American Society of Civil Engineers (ASCE, 2005; Clary *et al.*, 2001) as an ongoing project (Strecker *et al.*, 2004).

Table 20.1 Summary of stormwater pond and wetland pollutant removal performance and inflow event mean concentrations (Duncan, 1997)

Treatment device	Range of reported pollutant concentration reductions (%)						No. of reference sites
	SS	TP	TN	Pb	Zn	BOD	
Retention stormwater ponds	39–98	0–80	30–85	9–95	24–72	0–69	14
Extended detention stormwater ponds	50–75		10–35	70–90	24–62		9
Stormwater Wetlands	40–98	-33–97	-9–43	6–94	-29–97	18–34	28
Range of inflow (mg/L)	100 – 450	0.15 – 0.9	1.3 – 5.1	0.03 – 0.7	0.1 – 0.8	5 – 30	
Data sets (No.)	(247)	(206)	(139)	(169)	(144)	(114)	

Notes: (a) Range of pollutant removal performance based on published data from 51 separate locations across four countries.
(b) Range of inflow 'event mean concentrations' based on published data for > 67% urbanised catchments from 362 studies across North America, UK, Western Europe and Australia.

Long-term performance of comprehensively monitored and well-designed stormwater ponds and wetlands, is summarised in Table 20.2. The average annual pollutant removal level is estimated by applying the 'storm volumes/loads and removal rates' regression curve to long-term time series of discharges for the sites. Another approach to the assessment of long-term performance of stormwater ponds was presented by Driscoll (1989).

Table 20.2 Estimates of long-term pollutant removals for comprehensively monitored stormwater ponds

Pollutant	Average annual removal range (%)
Suspended solids (SS)	65 – 85
Total Phosphorus (TP)	60 – 75
Total Nitrogen (TN)	40 – 63 (b)
Heavy Metals (Pb, Zn)	70 – 90
Biological Oxygen Demand (BOD)	40 – 70

Notes: (a) Sources: Lawrence *et al.*, (1996); Maxted and Shaver (1997); Nietch *et al.*, (2002); Pettersson *et al.*, (1999); Urbonas and Stahre (1993).
(b) Improved performance can be secured through the incorporation of extended detention capacity into ponds and wetlands.
(c) Higher levels of TN removal are possible where second stage denitrification zones are incorporated into ponds.

The wide range of reported treatment efficiencies reflects a range of factors, including:

- limitations in selection and design of stormwater ponds;
- the size of the stormwater pond relative to storm volume/loads;
- variations in rainfall depth, intensity and entrained pollutant loads;
- situations of sewer overflows (treatment plant bypass, sewer pump station failure, sewer surcharge, sewer blockage) to stormwater; and
- uncertainties in the reported data.

20.3.1 The fate of the pollutants

Recent research has started to focus on the life cycle of captured pollutants. This work has considered the potential for transformation of pollutants and their subsequent remobilisation and environmental impact. From a sustainability perspective, the stormwater management goal is to minimise impacts on receiving waters by either limiting the discharge of pollutants at their source, or by interception and removal of pollutants from drainage.

Trapped pollutants that are not recycled need to be ultimately returned to the lithosphere in an inert form capable of burial, or be incorporated into the atmosphere as N_2 or CO_2. The creation of severe reducing conditions in sedimentation zones of stormwater ponds, due to elevated area loading of sedimented organic material, may lead to transformation of metals, sulphates, nutrients and organic pesticides, leading to their remobilization in highly bio-available or toxic forms.

From an ecological perspective, the consideration of pollutant species is very important. For example, particulate (adsorbed) P is unavailable directly to algae, whereas soluble reactive phosphorus forms are highly bio-available. Total N as nitrate may be beneficial in the case of freshwater in view of its redox buffering role and promotion of de-nitrification processes, whereas N as ammonia may promote nuisance algal species (blue-greens) and toxicity.

Biological monitoring indicates that the comprehensive application of stormwater management ponds and related devices to urban waterways in locations such as Canberra, Australia, has substantially restored the water quality and ecology of urban waterways and downstream receiving waters to pre-development levels (Lawrence, 1999).

20.4 DESIGN OF STORMWATER DETENTION AND RETENTION PONDS

SMP design is typically based on criteria to mitigate the impacts of urban development on receiving streams by runoff flow and quality control. The most important of these criteria are the volume of stormwater to be stored, duration of such storage, maximum post-development discharge for specific return periods T (e.g. T=3 months to 10 years) and conveyance of a critical rare storm (T=100 years). With respect to the design storage volume, the well-documented and accepted (particularly in the U.S.A.) concept of the 'water quality capture volume' (WQCV) developed by Guo and Urbonas (1996) is adopted here. Other approaches are based on the water quality treatment volume (Claytor, 1995) or capture and treatment of runoff from specific "design" storms with depths of rainfall varying from 13 to 25 mm, with further refinements for such area-specific conditions as climate, the level of protection (for specific classes of receiving waters) and the type of a stormwater pond considered (MOEE, 2003).

20.4.1 Volume and depth

Urbonas *et al.*, (1989) analysed the sizing of stormwater storage for capture and treatment of stormwater runoff, using the Denver, Colorado rainfall records, and noted the law of "diminishing returns" – past a certain storage size, the relative benefits, defined as percent runoff capture per storage size increment, gradually decreased. After extending this analysis to other U.S. locations with different meteorological patterns, Guo and Urbonas (1996) introduced the so-called water quality capture volume (WQCV) concept. This concept was then adopted by Water Environment Federation (WEF) and American Society of Civil Engineers (ASCE) in their joint "*Manual of Practice for Stormwater Quality Management*" (WEF, 1998) for sizing stormwater management pond storage. The manual recommends that stormwater quality detention be sized to capture, as a minimum, a volume somewhere between the runoff volume from the "mean" and the "maximised" storm events specific to the design site. Follow-up investigations revealed that, regardless of the location in the United States, the "mean" runoff event volume captures in total and treats 65 to 75 percent of all rainfall runoff events and the "maximised" event volume captures and treats 85 to 93 percent of rainstorm runoff events. Thus, adoption of either criterion will result in capture of the majority of all runoff events that occur after urbanisation. This is important, as these events have been found to destabilise the geomorphology of smaller receiving streams and alter their aquatic habitat.

The water quality capture volume (WQCV) can be calculated from Equation 20.1 as a function of the mean storm runoff corrected for local climate (Guo and Urbonas, 1996):

$$P_0 = (a \cdot C) \cdot P_6 \qquad (20.1)$$

where P_0 = maximised or mean WQCV determined in watershed mm;
P_6 = mean storm precipitation depth, in mm, found from actual rainfall records, using 6-hour separation between storms (i.e. a storm is separated from the preceding and following storms by at least 6 hours of no rain) and filtering out all storms smaller than 2.54 mm in depth;
a = dimensionless constant taken from Table 20.3 either for the maximised storage capture volume, or the storage volume based on mean storm;
C = dimensionless volumetric catchment runoff coefficient.

Table 20.3 Values of coefficient a in Equation 20.1 (after Guo and Urbonas, 1996; and WEF, 1998)

		Drain/Emptying time of capture Volume			
		6-hrs	12-hrs	24-hrs	48-hrs
Maximised storage capture volume	a =	0.8	1.109	1.299	1.545
Storage capture volume based on **Mean** storm	a =	0.53	0.71	0.86	1.01

For the most frequent (minor) storm events, C can be estimated from a regression equation ($r^2 = 0.72$), which was developed in the Nationwide Urban Runoff Program (U.S. EPA, 1983) from data collected at 60 sites:

$$C = 0.858\, i^3 - 0.78\, i^2 + 0.774\, i + 0.04 \tag{20.2}$$

where C = volumetric runoff coefficient; and
i = catchment imperviousness ratio (i.e. total impervious area divided by the total catchment area).

Thus, depending on local requirements on stormwater quality enhancement, the designer chooses an appropriate value of 'a' from Table 20.3, and substitutes it in Equation 20.1 to calculate design WQCV. For retention stormwater ponds, WEF (1998) recommends the size of the storage capture volume to be between the "mean" and the "maximised" runoff event (WEF, 1998) using 6 to 12 hour emptying time. For extended stormwater detention basins, the WQCV should be based on a 24 to 48 hour drain time, and further increased by 20% to account for loss of storage volume due to the accumulation of sediment deposits. Longer 'drain periods' are recommended in areas with larger mean storm depths (i.e. ≥ 20 mm).

Numerical values specified in Table 20.3 produce reasonable estimates of WQCV. However, more accurate results can be obtained using local rainfall data and continuous runoff simulation to establish a runoff event data series, and deriving maximised and mean P_0 values taking into account the appropriate drain/emptying times.

In other cases, the drainage authority may specify the stormwater volume to be stored, V_s [m^3], e.g. as $V_s = 0.0254 \cdot C \cdot A$ (Claytor, 1995), where C is as defined earlier, and A is the drainage contributing area in m^2. This volume is detained in the stormwater pond for 24 hours, i.e., it is drained from the brimful stormwater pond in 24 hours. Documented Australian practice ranges from regional storm intensity and duration probability curve based sizing to continuous simulation of inflows, sedimentation and sediment remobilisation processes (Wong *et al.*, 1998; Lawrence and Breen, 1999; Lawrence, 2001). Longer detention times allow better treatment, but also increase the risk of another storm occurring before the pond storage has been emptied.

Following the determination of the pond volume and duration of detention, the stormwater pond layout and hydraulic structures are designed. Such structures include a sediment forebay (easily accessible for maintenance), inlet (spreading the influent), outlet and outfall (protected by riprap), and an emergency spillway usually designed for a 100-year or larger flood.

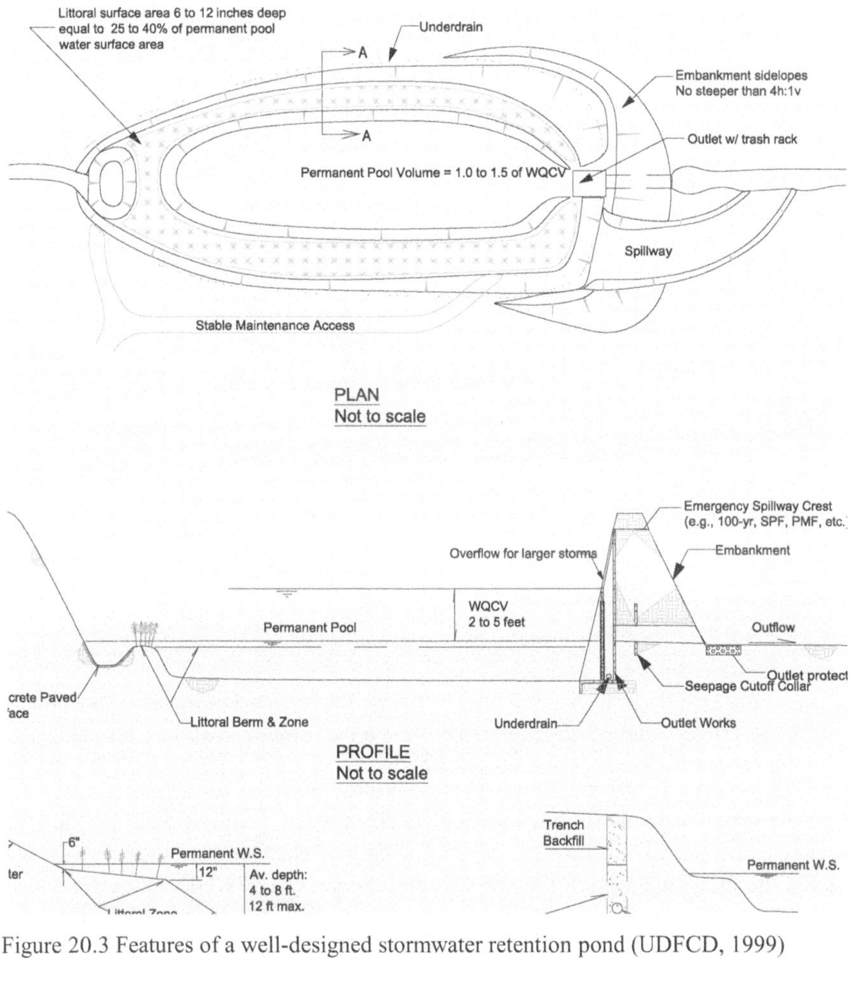

Figure 20.3 Features of a well-designed stormwater retention pond (UDFCD, 1999)

20.4.2 Stormwater retention ponds

Stormwater retention ponds, which are also sometimes called stormwater wet ponds or stormwater wet detention basins, are small constructed reservoirs with a permanent pool of water and often with emergent wetland vegetation around the perimeter. The major features of a stormwater retention pond are illustrated in Figure 20.3. Stormwater ponds with a shallow bench of wetland vegetation around the edges, also called a littoral zone, provide enhanced aquatic habitat and may aid treatment. A larger surcharge storage volume overlying the permanent pool provides WQCV. This extra storage helps improve the hydraulic and treatment efficiency of the stormwater pond and helps to mitigate the effects of urbanisation (such as rapid runoff). If so desired, extra storage can also be provided to assist flood attenuation for larger storms. Figure 20.4 is a photograph of a three-cell stormwater retention pond with shallow wetland benches separating the cells, located in Superior, Colorado, USA. Two of the cells have their inlet pipes and outlet located in the central cell. All cells provide the WQCV as a surcharge above the permanent pool.

Figure 20.4 Multi-cell *Stormwater Retention Pond* for treating stormwater quality and mitigating effects of urban runoff (photo used with the permission of the Urban Drainage and Flood Control District, Denver, Colorado, USA)

The water in the permanent pool of a stormwater retention pond is replaced partly or fully with stormwater runoff during wet weather periods. In the space above the permanent pool, the WQCV storage is created (UDFCD, 1999) which serves to fully capture the more frequently occurring storms. Stored stormwater is then released over periods ranging from 6 to 12 hours, or even longer (up to 24 hours), depending on the required level of mitigation of downstream impacts. In most cases, even the shorter detention times are generally sufficient to permit fine sediments in the influent to settle below the outlet level, thus allowing the cleaner water to decant while retaining the more contaminated water in the permanent pool. During dry weather, complex physical and bio-chemical processes take place in the stormwater pond, which facilitate the removal of fine particulate and some of the dissolved nutrients and metals brought in by the storm runoff (see Sections 20.2.2 and 20.2.3 for specific discussion on these processes in SMPs and Chapters 2 and 3 for a more general review of the physical, chemical and biological environment that exists in treatment ponds).

Retention stormwater ponds can be used to improve the quality of urban runoff from various sources including roads, parking lots, residential and commercial areas and industrial sites. Furthermore, stormwater retention ponds work well in conjunction with other stormwater best management practices (BMPs) such as upstream on-site source controls (e.g. "Soakaways") or as pre-treatment devices for downstream filter basins or wetlands.

Stormwater retention ponds can provide cost-effective runoff control, especially for tributary catchments exceeding 50 hectares. Their primary advantages include moderate to high removal rates for many of the pollutants found in urban stormwater runoff (Section 20.3), wildlife habitat opportunities, recreation, aesthetics and open space amenities and the possibility to be designed in combination with larger flood control facilities serving to reduce runoff hydrograph peak flows (commonly referred to as peak-shaving). Disadvantages may include safety concerns; difficult removal of pollutants accumulated on the pond bottom; public concerns about mosquito breeding; thermal pollution; attraction of waterfowl that add to the nutrient and pathogen loads in the pond and its effluent; and, occasional problems with floating litter and scum, algal blooms and odours. The issue of potential mosquitos breeding in stormwater ponds received much attention in recent years, particularly with the threat of West Nile Virus. Such risks can be minimized by proper design and operation of stormwater detention and retention, including avoidance of shallow wet depressions, nutrient controls, implementation of a mosquito control program, and regular inspections (University of Florida, 1998; Department of Environmental Protection, Montgomery County, MD, 2005).

20.4.3 Extended detention stormwater basins (EDSBs)

Extended stormwater detention basins (ponds) are sedimentation impoundments that are designed to empty most of their stored water in 24 to 48 hours after the cessation of stormwater runoff. To facilitate good pollutant removal, they use very small outlets to extend the emptying time for the most frequent runoff events. A well-designed stormwater basin will have a small storage pool (also called a "micro-pool") at its outlet to improve its pollutant removal capabilities. To a large extent, EDSBs function similarly as the retention stormwater ponds, but provide somewhat lower peak removals of pollutants (see Table 20.1). EDSBs may be used preferentially in drainage catchments where it would be difficult to maintain permanent water storage (because of the climate and a small contributing area), or where there is a strong need for dual use of land to be occupied by the stormwater pond/basin.

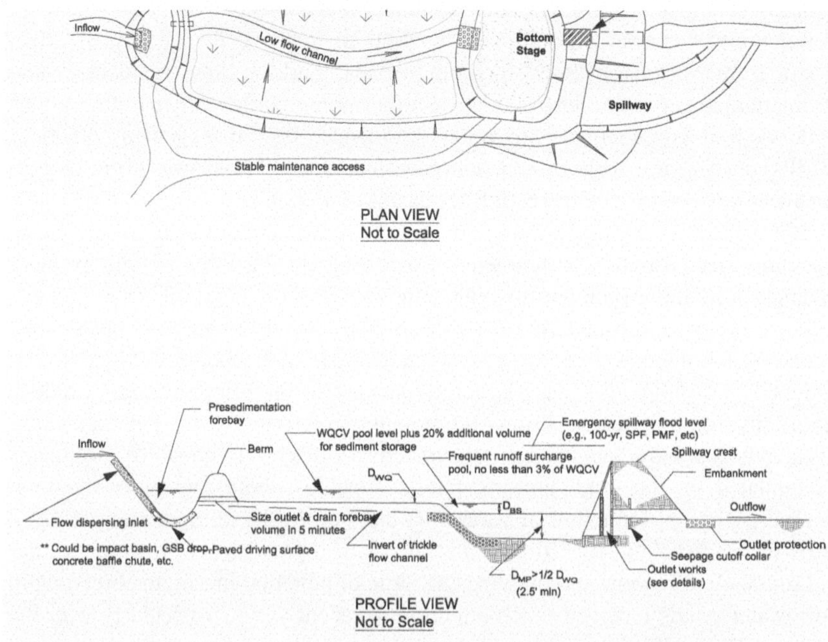

Figure 20.5 Features of a well designed extended detention stormwater basin (UDFCD, 1999)

The major features of an extended detention stormwater basin are illustrated in Figure 20.5. EDSBs are particularly effective in mitigating the effects of hydrologic changes caused by urbanisation in the removal of solids and, if so desired, can provide flood control for larger storms. Figure 20.6 is a photograph of an extended detention stormwater basin with a small outlet pool (micro-pool) and a forebay in Grant Ranch development in Denver, Colorado, USA.

The emptying time of 24 to 48 hours is generally sufficient to allow much of the fine sediment in the influent to settle to the stormwater basin bottom. Applications of extended detention stormwater basins are similar to those of retention stormwater ponds – they can be used to improve the quality of urban runoff from roads, parking lots, residential and commercial areas and industrial sites; and perform well in conjunction with upstream on-site source controls (e.g. soakaways, rain gardens, etc.) and with downstream post-treatment facilities such as filter basins and wetland channels.

Figure 20.6 *Extended Detention Stormwater Basin* 20 hours after 25 mm rain (photo used with the permission of the Urban Drainage and Flood Control District, Denver, Colorado, USA)

Compared to stormwater retention ponds, extended detention stormwater basins can be also designed to provide supplementary benefits in addition to stormwater management, such as recreation and open spaces. They are very effective in removing particulate matter and the associated heavy metals and other pollutants. Australian research indicates that the drying of settled material post-event transforms phosphorus into forms less susceptible to subsequent remobilisation.

20.4.4 Design considerations for stormwater ponds and basins

Land requirements

The land required for extended detention stormwater basins and stormwater retention ponds ranges from 0.5 to 2.0 percent of the tributary urban watershed area, depending on the degree of imperviousness and local rainfall patterns. In most jurisdictions, this land fits under the open space allocation requirements to be met by the developer. Stormwater retention ponds are particularly suitable for areas with high groundwater and dry weather flows, but stormwater detention basins are favoured in drier areas.

Water balance

Stormwater retention ponds need sufficient continuous baseflow or groundwater inflow to prevent the permanent pool from drying up, which would eliminate important pollutant removal feature described for stormwater retention ponds in Section 20.2.2 and in the sub-section immediately below. At sites where this may be a problem, a complete water budget for each month of the year should be evaluated to ensure that the inflows would exceed evapotranspiration and seepage losses.

Pollutant removal

Stormwater retention ponds achieve moderate to high removals of particulate solids and adsorbed chemicals through sedimentation during and after the runoff event. When it rains, all, or at least a portion of the water in the permanent pool is displaced. During the period between storms, the settling process continues in the permanent pool aiding removal of fine solids and biological chemical processes in the pool remove some of the soluble nutrients. A large storm event can resuspend sediment from parts of the pond bottom, though proper hydraulic design and consolidation of sediments does provide some protection against this. Soluble chemical compounds can remobilise when unfavourable chemical conditions exist in the pool, such as low dissolved oxygen and pH.

Extended detention stormwater basins also achieve moderate to high removals of solids and adsorbed chemicals (trace metals, oil and grease, some nutrients), and low to moderate removals of soluble nutrients. The removal of nutrients can be improved when a small shallow pool (a micro-pool) is included as part of the stormwater basin's bottom. Factors that influence the degree of pollutant removal are the emptying time provided by the outlet and the influent particle sizes.

Aesthetics and multiple uses

Stormwater retention ponds provide more opportunities for excellent aesthetic amenities. Both retention and extended detention stormwater ponds have been successfully integrated into the local landscape to provide benefits other than just stormwater management. In stormwater retention ponds most of the sediments removed are retained within the permanent pool zone while in stormwater detention basins they are deposited on the bottom and become visible to the public. In extended detention stormwater basins, the outlet pool offers some aquatic habitat, habitat for waterfowl and for terrestrial species favouring a water environment. However, the latter can pollute the stormwater pond with their droppings.

Permanent pool for stormwater retention ponds

When nutrient removal is not the primary concern, the permanent pool volume should equal 1.0 to 1.5 times the WQCV. Stormwater ponds with smaller volumes can become stagnant. It is more difficult to give definite guidance for sizing the pool for nutrient removal. A protocol based on the lake eutrophication model was suggested by Walker (1987) and is based on the average residence time the water spends in the pond, which is estimated by Equation 20.3.

$$T = 52 \cdot \frac{V_S}{(V_R \cdot n)} \tag{20.3}$$

where T = average hydraulic retention time, (weeks);
V_s = permanent pool storage volume, (m^3);
V_R = runoff volume generated by a mean storm, (m^3);
n = number of runoff producing storms per year at the site.

From field studies, Hartigan (1989) recommended T = 2 to 3 weeks. Stormwater ponds with retention time greater than three weeks are at risk of thermal stratification and development of anaerobic water near the bottom, creating a condition for nutrient remobilisation and export. Thermal stratification was also observed in Australia, where extended dry periods and high summer solar radiation result in temperature stratification of even shallow ponds. Anaerobic conditions may also occur in cold climates during winter months when stormwater ponds become ice covered (Marsalek *et al.*, 2003).

Stormwater retention pond depth zones

The permanent pool should have two depth zones: (1) a littoral zone 0.15 to 0.45 m deep that covers about 23 to 30 percent of the permanent pool surface area, all of it vegetated with emergent aquatic plants, and (2) a deeper zone with a 1.2 to 2.5 m average depth to promote sedimentation, store sediments and provide nutrient uptake by phytoplankton. It is also recommended that the maximum depth of the pond does not exceed 3.6 metres to avoid thermal stratification and anoxia (UDFCD, 1999).

Outlet works

SMPs require multiple function outlets, which provide the required outflow controls. Typically, a low flow perforated plate (or riser) is used for WQCV control, a higher stage orifice or weir serves for peak flow control (usually for 2-10 year return periods), and finally, an emergency spillway is used to pass a 100-year or larger flood. All these devices may be housed in a single outflow structure. Low-flow WQCV control outlets have small openings, are very susceptible to clogging by floating and neutrally buoyant trash, and need to be designed well. The urban drainage and flood control district in Denver, Colorado, USA has developed very practical AutoCAD* details (UDFCD, 1999) that address this problem. Their website (www.udfcd.org) supplies free downloads. The intermediate flow control devices are usually designed as orifices that are placed at appropriate elevations. Finally, the emergency spillway is usually designed as a vegetated side spillway of a width sufficient to convey the 100-year or larger flood, with an appropriate freeboard to the top of the embankment.

Side slopes

Side slopes should be gently sloping to limit rill erosion, assure safety to the public and facilitate maintenance. Such slopes should be no steeper than 4:1 (above the mean water level), and preferably even milder. The littoral zone in a pond should be relatively flat (e.g. 40:1) and the side slope below the littoral zone should be no steeper than 3:1 for safety reasons. Australian practice is to adopt side slopes of not less than 8:1, to minimise the risk of forming isolated depressions around the edge following stormwater pond drawdown - a condition conducive to the breeding of mosquitoes.

Dam embankment

The embankment should be designed to avoid failure during very large storms (e.g. 100-year and larger). Embankment slopes are recommended to be no steeper than 3:1, preferably flatter, planted with dense turf-forming grasses. Embankment soils should be compacted to 95 percent density using the ASTM D 698-70 (modified proctor) standard or equivalent.

Inlet

Preference is given to a single inlet point to optimise separation between the inlet/outlet locations, but multiple inlets are sometimes unavoidable. A submerged pipe inlet, with a headwall, hardened surface below the inlet and possibly an upward flow deflector is preferred to pilot open channels (MOEE, 2003).

Forebay design

It is a common practice to incorporate sediment forebays into stormwater pond design. Such structures are the most cost-effective means of intercepting and removing some 80% of the sediment loading to the stormwater pond and protect the macro-plants from burial by excessive deposits of sediment. Forebays are coarse sediment traps, with a paved bottom, provided at all inlets to the pond. The forebay volume V_f can be estimated by one of the following two methods: (a) V_f = 0.05 WQCV (UDFCD, 1999) (see Section 20.4.1 for calculation of WQCV), or (b) allowing storage of 2.5 to 6 mm per impervious hectare of the catchment (Claytor, 1995). According to the Australian practice, a sediment forebay upstream of the stormwater pond should be designed to remove 80% of the sediment loading. For ease of maintenance, forebays should be easily accessible, have a hard bottom, and be easily drained.

Underdrains adjacent to stormwater ponds

To facilitate restorative maintenance when the pond has to be drained to remove deposited sediment, it is recommended that underdrain trenches be located near the edge of the permanent pool.

20.5 MAINTENANCE OF STORMWATER PONDS AND BASINS

All stormwater management devices, including stormwater management ponds, require regular maintenance to sustain their design performance. Typical maintenance activities for stormwater ponds include regular inspections of the whole facility and individual structures; grass cutting; weed control; replanting of upland vegetation, shoreline fringe and flood fringe vegetation, and aquatic vegetation; trash removal; outlet valve adjustment; and removal of accumulated sediments. Among these activities, sediment removal is probably the most important as it maintains the design storage volume. The size of the tributary catchment and the erosion within it will determine the frequency of sediment removal needed for retention ponds and detention basins. Field observations indicate that sediments accumulate in stormwater ponds at widely varying rates from a few mm per year to over 20 mm/year (Yousef *et al.*, 1994; Marsalek *et al*, 1997). Sediment removal is triggered by various criteria, the State of Florida recommends sediment removal when the pond storage volume is reduced by 10%.

Under normal conditions, sediments need to be removed every 5 to 20 years (UDFCD, 1999) depending on the nature of the activities in the catchment, but more frequent removals may be required if effective erosion control is not achieved in the catchment. To facilitate pond maintenance, allowance for access by heavy maintenance equipment must be considered in the design.

Studies of sediments accumulated in SMPs indicate the greatly varying nature of such materials (Marsalek *et al.*, 1997). Coarse sediments which settle in the sediment forebay, or by the pond inlet, are relatively lightly polluted, have low water content and generally could be reused as an ‘intermediate’ landfill cover or possibly for winter road maintenance (sanding). The disposal of fine-grained materials is more challenging, as they can be more severely polluted and unsuitable for landfill disposal because of their high water content. Under such circumstances, more costly disposal in special facilities may be required.

20.6 SUMMARY

Stormwater management ponds have been widely accepted as cost-effective measures serving to mitigate negative impacts of urbanization on water resources. Since the introduction of stormwater ponds into urban drainage practice about 40 years ago, tens of thousands of stormwater ponds have been built worldwide, and their design has evolved from simple impoundments to complex facilities controlling stormwater flows and quality, reducing impacts on downstream receiving waters, and providing valuable recreational and ecological amenities. Much of the existing knowledge concerning stormwater ponds has been obtained empirically, by building and operating such facilities, learning from this experience, and making design changes as required. This empirical process is enhanced by collection and sharing of field data. There are many examples of well performing stormwater ponds, generally in jurisdictions with well-established progressive stormwater management programs, but there are also examples of stormwater pond failures, either because of poor design or poor operation and maintenance. Currently, there is an opportunity to analyse the existing stormwater pond data, derive general findings and bring the stormwater pond design to a more scientific foundation. This is particularly true for water quality processes and stormwater pond ecology.

Emerging challenges include application of sustainability criteria/assessments to stormwater ponds; advancing the understanding of the role of algae in SMP performance, particularly with respect to dissolved chemicals; mitigation of pollution stresses/loadings on stormwater ponds by improved upstream source controls; improved understanding of secondary stormwater pond impacts on downstream waters (particularly the delayed impacts) with respect to outflow regime, thermal enhancement, and chemical/sediment releases during large storms; managing new chemicals of concern in stormwater (including endocrine disrupters); controlling infectious disease vectors; and, coping with climate change.

REFERENCES

American Society of Civil Engineers (ASCE)(2005). International Stormwater Best Management Practices (BMP) Database. http://www.bmpdatabase.org (visited on Feb. 25, 2005).

Clary, J., Urbonas, B., Jones, J., Strecker, E. and O'Brien, J. (2001) Developing and Evaluating a Stormwater BMP Effectiveness Database. In *Proceedings of NOVATECH 2001, Lyon, France*, June 25-27, pp. 161-168, GRAIE, Lyon.

Claytor, R.A. (1995) *Stormwater Management Pond Design Example*. The Center for Watershed Protection, Silver Springs, MD, USA.

Department of Environmental Protection, Montgomery County (2005). West Nile Virus and Stormwater Management.

Driscoll, E.D. (1989) Long Term Performance of Water Quality Ponds. In *Design of Urban Runoff Quality Controls* (eds. L.A. Roesner, B. Urbonas and M.B. Sonnen), pp. 145-163, ASCE, New York, NY.

Duncan, H. (1997) *Urban Stormwater Treatment by Storage: A Statistical Overview*. Cooperative Research Centre for Catchment Hydrology, Australia.

GeoSyntec Consultants, Urban Drainage and Flood Control District (UDFCD) and Urban Water Resources Research Council (UWWRRC) of ASCE (2002). Urban Stormwater BMP Performance Monitoring. Report EPA-821-B-02-001, U.S. EPA, Washington, D.C.

Guo, C.Y. and Urbonas, B.R. (1996) Maximised Detention Volume Determined by Runoff Capture Ratio. *J. Water Resources Planning and Management Division, ASCE* **122**(1), 24–32.

Hartigan, J.P. (1989) Basis for Design of Wet Detention Basin BMPs. In *Design of Urban Runoff Quality Controls* (eds. L.A. Roesner, B. Urbonas and M.B. Sonnen), pp. 122-143, ASCE, New York, NY.

Lawrence, A.I., Marsalek, J., Ellis, J.B. and Urbonas, B. (1996) Stormwater Detention and BMPs. *Journal of Hydraulic Research* **34**(6), 799-814.

Lawrence, I. and Breen, P. (1998) *Design Guidelines: Stormwater pollution control ponds and wetlands*. Cooperative Research Centre for Freshwater Ecology, Australia.

Lawrence, I. (2001) *Integrated urban land and water management: Planning and Design Guidelines*. Technical Report 1/2001, Cooperative Research Centre for Freshwater Ecology, Australia.

Lawrence, A.I. (1999) Canberra stormwater management strategy. In *Proceedings of the International Conference on Diffuse Pollution*, IAWQ, Perth, Australia.

Lawrence, A.I. and Breen, P.F. (1999) Application of pond and wetland design guidelines. In *Proceedings 8th International Urban Storm Drainage Conference* (eds. I.B. Joliffe and J. Ball), Sydney, Australia, Aug. 30–Sept. 3, 1999, pp. 1066-1072.

Marsalek, J., Watt, W.E., Anderson, B.C. and Jaskot, C. (1997) Physical and Chemical Characteristics of Sediments from a Stormwater Management Pond. *Water Qual. Res. J. of Canada* **32**(1), 89-100.

Marsalek, P.M., Watt, W.E., Marsalek, J. and Anderson, B.C. (2000) Winter Flow Dynamics of an On-Stream Stormwater Management Pond. *Water Qual. Res. J. of Canada* **35**(3), 505-523.

Marsalek, P.M., Watt, W.E., Marsalek, J. and Anderson, B.C. (2003) Winter Operation of an On-Stream Stormwater Management Pond. *Water Sci. Tech.* **48**(9), 133-143.

Matthews, R.R., Watt, W.E., Marsalek, J., Crowder, A.A. and Anderson, B.C. (1997) Extending Retention times in a Stormwater Pond with Retrofitted Baffles. *Water Qual. Res. J. Canada* **32**(1), 73-87.

Maxted, J.R and Shaver, E. (1997) The use of retention basins to mitigate stormwater impacts on aquatic life. In *Effects of Watershed Development and Management on Aquatic Ecosystems* (ed. L.A. Roesner), pp. 494-512, ASCE, New York, NY.

Ministry of Environment and Energy (MOEE) (2003). *Stormwater Management Planning and Design Manual.* MOEE, Toronto, Ontario.

Nietch, C.T., Borst, M. and O'Shea, M.L. (2002) Stormwater Treatment Ponds vs. Constructed Wetlands. In *Linking Stormwater BMP Design and Performance to Receiving Water Impact Mitigation* (ed. B.R. Urbonas), Proceedings of an Engineering Foundation Conference, Snowmass, CO, USA, Aug. 19-24, 2001, ASCE/UEF/EWRI, pp.524-528.

Olding, D.D. (2000) Algal communities as a biological indicator of stormwater management pond performance and function. *Water Qual. Res. J. Canada* **35**(3), 489-503.

Persson, J. (1999) Hydraulic Efficiency in Pond Design. Ph.D. Thesis, Dept. of Hydraulics, Chalmers University of Technology, Goteborg, Sweden.

Pettersson, T., German, J., and Svensson, G. (1999) Pollutant removal efficiency in two stormwater ponds in Sweden. In *Proceedings 8th International Urban Storm Drainage Conference* (eds. I.B. Joliffe and J. Ball), Sydney, Australia, Aug. 30–Sept. 3, 1999, pp. 866-873.

Randall, C.W., Ellis, K., Grizzard, T.J., and Knocke, W.R. (1982) Urban Runoff Pollutant removal by Sedimentation. In *Stormwater Detention Facilities* (ed. W. deGroot), ASCE, New York, NY, 205-219.

Schueler, T.R. (1987) *Controlling Urban Runoff: A Practical Manual for Planning and Designing Urban BMPs.* Washington Metropolitan Water Resources Planning Board, Washington, DC.

Shaw, J.K.E., Watt, W.E., Marsalek, J., Anderson, B.C., and Crowder, A.A. (1997) Flow Pattern Characterization in an Urban Stormwater Detention Pond and Implications for Water Quality. *Water Qual. Res. J. Canada* **32**(1), 53-71.

Strecker, E., Quigley, M., Urboinas, B., Jones, J., Clary, J. and O'Brien, J. (2004). Urban Stormwater BMP Performance: Recent Findings for the International Stormwater BMP Database Project. In *Proceedings of NOVATECH 2004, Lyon, France*, June 2004, pp. 465-472, GRAIE, Lyon.

Thackston, E.L., Shields, F.D., and Schroeder, P.R. (1987) Residence Time Distributions of Shallow Basins. *J. of Env. Engng, ASCE* **113**, 1319-1332.

University of Florida (1998). Mosquitoes and Stormwater Management. IFAS Publication DH 421, Institute of Food and Agricultural Science, University of Florida, Gainesville, FL.

U.S. Environmental Protection Agency (EPA). (1983) *Results of the Nationwide Urban Runoff Program Volume I - Final Report.* Water Planning Division, U.S.EPA, Washington, DC.

UDFCD (1999) *Urban Storm Drainage Criteria Manual Vol. 3.* Urban Drainage and Flood Control District, Denver, Colorado.

Urbonas, B. (1995) Recommended Parameters to report with BMP Monitoring Data. *J. Water Resources Planning and Management, ASCE* **121**(1), 22-34.

Urbonas, B.R and Stahre, P. (1993) *Stormwater: Best Management Practices and Detention for Drainage Water Quality and CSO Management.* 2nd Edition, Prentice Hall.

Urbonas, B.R., Guo, C.Y., and Tucker, L.S. (1989) Optimization of Stormwater Quality Capture Volume. In *Urban Stormwater Quality Enhancement – Source Controls, Retrofitting and Combined Sewer Technology.* American Society of Civil Engineers, Reston, VA, USA, pp. 94-110.

Walker, W.W. (1987) Phosphorus Removal by Urban Runoff Detention Basins. In *Lake and Reservoir Management: Volume III*, pp 314-326. North American Lake Management Society, Washington DC, USA.

WEF (1998) Chapter 5 – Selection and Design of Passive Treatment Controls," *Urban Runoff Quality Management.* WEF Manual of Practice No. 23, ASCE Manual and Report on Engineering Practice No. 87, Water Environment Federation, Alexandria, VA, USA.

Wong, T.H.F., Breen, P.F., Somes, N.L.G. and Lloyd, S.D. (1998) *Managing Urban Stormwater using constructed wetlands.* Industry Report 98/7. Cooperative Research Centre for Catchment Hydrology, Australia.

Yousef, Y.A., Hvitved-Jacobsen, T., Sloat, J. and Lindeman, W. (1994) Sediment Accumulation in Detention or Retention Ponds. *Sci. Tot. Env.*, **146/7**, 451-456.

Index

Note: page numbers in *italics* refer to Figures, those in **bold** denote Tables.

IWA Publishing's authorised EU representative for General Product Safety Regulations is Diane D'Arras, 15 rue Duret, 75116 Paris, France, e-mail: safety@iwap.co.uk.

Printed and bound by CPI Group (UK) Ltd, Croydon, CR0 4YY
16/06/2026
02140384-0001